MALE-MEDIATED DEVELOPMENTAL TOXICITY

REPRODUCTIVE BIOLOGY

Series Editor: Sheldon J. Segal

The Population Council
New York, New York

Current Volumes in this Series

AIDS AND WOMEN'S REPRODUCTIVE HEALTH
Edited by Lincoln C. Chen, Jaime Sepulveda Amor, and Sheldon J. Segal

AUTOCRINE AND PARACRINE MECHANISMS IN REPRODUCTIVE ENDOCRINOLOGY
Edited by Lewis C. Krey, Bela J. Gulyas, and John A. McCracken

CONTRACEPTIVE STEROIDS: Pharmacology and Safety
Edited by A. T. Gregoire and Richard P. Blye

DEMOGRAPHIC AND PROGRAMMATIC CONSEQUENCES OF CONTRACEPTIVE INNOVATIONS
Edited by Sheldon J. Segal, Amy O. Tsui, and Susan M. Rogers

ENDOCRINE AND BIOCHEMICAL DEVELOPMENT OF THE FETUS AND NEONATE
Edited by José M. Cuezva, Ana M. Pascual-Leone, and Mulchand S. Patel

ENDOCRINOLOGY OF EMBRYO-ENDOMETRIUM INTERACTIONS
Edited by Stanley R. Glasser, Joy Mulholland, and Alexandre Psychoyos

MALE-MEDIATED DEVELOPMENTAL TOXICITY
Edited by Andrew F. Olshan and Donald R. Mattison

REPRODUCTIVE TRACT INFECTIONS: Global Impact and Priorities for Women's Reproductive Health
Edited by Adrienne Germain, King K. Holmes, Peter Piot, and Judith Wasserheit

STEROID CONTRACEPTIVES AND WOMEN'S RESPONSE: Regional Variability in Side-Effects and Pharmacokinetics
Edited by Rachel Snow and Peter Hall

UTERINE AND EMBRYONIC FACTORS IN EARLY PREGNANCY
Edited by Jerome F. Strauss III and C. Richard Lyttle

A Continuation Order Plan is available for this series. A continuation order will bring delivery of each new volume immediately upon publication. Volumes are billed only upon actual shipment. For further information please contact the publisher.

MALE-MEDIATED DEVELOPMENTAL TOXICITY

Edited by

Andrew F. Olshan
University of North Carolina
Chapel Hill, North Carolina

and

Donald R. Mattison
University of Pittsburgh
Pittsburgh, Pennsylvania

SPRINGER SCIENCE+BUSINESS MEDIA, LLC

Library of Congress Cataloging-in-Publication Data

On file

Proceedings of an International Conference on Male-Mediated Developmental Toxicity, held September 16–19, 1992, in Pittsburgh, Pennsylvania

ISBN 978-1-4613-5764-3 ISBN 978-1-4615-1877-8 (eBook)
DOI 10.1007/978-1-4615-1877-8

Originally published by Plenum Press, New York in 1994
Softcover reprint of the hardcover 1st edition 1994

PREFACE

The cause of many of the adverse reproductive outcomes and developmental diseases among offspring is not well understood. Most of the epidemiologic and experimental animal research has focused on the relationship between maternal exposures including medications, tobacco smoke, alcohol, infections, and occupation and the occurrence of spontaneous abortion, low birth weight, and birth defects. The potential role of paternal exposures has not been investigated as extensively despite long-standing animal research that demonstrates the induction of mutations in the male germ cell after exposure to certain agents and subsequent reproductive failure or early pregnancy loss. Given this relative lack of interest, acquisition of epidemiologic data and the development of a definitive model or mechanism for potential male-mediated effects has been hindered. However, recent laboratory and epidemiologic investigations have suggested that paternal exposures may be more important than previously suspected.

This topic has been termed by some as "male-mediated developmental toxicity." This is meant to refer to the effects of exposures and other factors relating to the male parent that result in toxicity to the conceptus and abnormal development. The developmental endpoints of interest can include fetal loss, congenital abnormalities, growth retardation, cancer, and neurobehavioral effects. These effects may operate through a variety of mechanisms including gene mutation, chromosomal aberrations, seminal fluid transfer of toxicants and epigenetic events. Since the focus is specifically on the conceptus, semen quality parameters such as count, motility, and morphology are not of direct interest as endpoints, although they are of importance as potential intermediate markers of effect.

This book presents material based upon an international conference on male-mediated developmental toxicity held in Pittsburgh, Pennsylvania, September 16-19, 1992. This was the first conference that focused on this topic. The purpose of this conference was to provide a summary of the current research literature, to describe potential mechanisms, to highlight new data from both laboratory and epidemiologic studies, and to point out limitations of previous studies and gaps in knowledge. In addition, risk assessment, policy issues, and physician and patient education were covered. This was a multidisciplinary conference with speakers and attendees from a variety of disciplines including epidemiologists, toxicologists, geneticists, reproductive biologists and others.

The organization of this book generally follows that of the meeting with major sessions including: 1) Concepts and Mechanisms; 2) Laboratory Evidence; 3) Epidemiologic Findings; and 4) Risk Assessment and Policy Issues. The section on laboratory evidence (Chapters 2-9) provides an overview of the current methods using animal test systems to detect male germ cell damage. In addition, patterns that have emerged from these studies with respect to agent specificity, timing of exposure, and specificity of the mutational effect are presented. New and promising molecular approaches such as the use of transgenic mice that may have application to the study of male germ damage are discussed in Chapter 8.

Laboratory research on specific agents and endpoints is described in Chapters 10-13. The agents include radiation, cyclophosphamide, ethylnitrosourea, and metals. Congenital anomalies, tumors, and neurobehavioral outcomes are the developmental endpoints discussed. Also in this section are two chapters that consider novel pathways or mechanisms that have generated much interest, such as imprinting (Chapter 8) and post-testicular effects (Chapter 9).

An important aspect of the conference was the presentation of the human epidemiologic data so as to allow for a broad comparison with the available animal data and to engage discussion with laboratory workers. The epidemiology chapters (14-21) have been organized by endpoint including fetal loss, birth weight and prematurity, birth defects, and childhood cancer. In addition, two uniquely exposed cohorts were considered: offspring of men treated for cancer (Chapter 18) and atomic bomb survivors (Chapter 19). This section also includes reviews on antioxidants in relation to birth defects and cancer (Chapter 21) and a view of biologic factors and the interpretation of epidemiologic evidence (Chapter 20).

Chapters 22-25 provide an overview of risk assessment, policy and physician and patient education issues. To facilitate discussion of the current evidence and identify future research agendas within a multidisciplinary context, seven "breakout" groups were organized during the meeting. These groups were laboratory research methods, mechanisms, markers and endpoints, epidemiologic approaches, physician and patient education, risk assessment and risk management, and multidisciplinary approaches. The breakout group chairs were asked to prepare a summary of their group deliberations and these have been included in this book as Chapters 33-39. Finally, in addition to the main sessions, poster sessions were held. The authors of the posters were invited to submit a short paper describing their work and those submitted constitute Chapters 26-32.

Much uncertainty remains in the understanding of the potential contribution of paternal genetic and environmental factors to adverse reproductive outcome. Further research is needed to clarify biological pathways and mechanisms as well as an assessment of the public health and clinical significance of male-mediated developmental toxicity. We hope that this book will serve as a starting point by providing a broad view and current synthesis from the perspectives of both laboratory researchers and epidemiologists.

ACKNOWLEDGEMENTS

We gratefully acknowledge financial support of the conference from the following agencies and organizations: Burroughs Wellcome Co.; The Environmental Protection Agency (2D2772NAEX); International Life Sciences Institute (ILSI); Magee Women's Hospital; National Cancer Institute (1 R13 CA58438-01); and Department of Health & Human Services, the Agency for Toxic Substances & Disease Registry (R13/ATR398310-01).

Members of the organizing committee included Robert L. Brent, Robert W. Miller, John J. Mulvihill, and Andrew J. Wyrobek.

We are all indebted to the following individuals for all their activities and support of the conference, preparation of grant applications, organization of schedules and preparation of manuscripts:

Peggy Allport for her unusually acute ability to allow science and public health issues to be debated and discussed without concern for schedules or time.

M. Joyce Smith for her remarkable talents for smoothing ruffled feathers and scientific egos.

Mary C. Chabala and her colleagues in the Department of Conference Management at the University of Pittsburgh Medical Center for their organizational and negotiation skills.

Bernadette M. Miller for transferring the many different languages of science into a coherent volume.

CONTENTS

CONCEPTS AND MECHANISMS

LABORATORY EVIDENCE

LABORATORY EVIDENCE FOR CONGENITAL ANOMALIES, CANCER AND NEUROBEHAVIORAL OUTCOMES

EPIDEMIOLOGIC FINDINGS

RISK ASSESSMENT AND POLICY ISSUES

ABSTRACTS

BREAKOUT GROUPS

METHODS AND CONCEPTS IN DETECTING ABNORMAL REPRODUCTIVE OUTCOMES OF PATERNAL ORIGIN*

Andrew J. Wyrobek

Biodosimetry Group L-452
Biology and Biotechnology Research Program
Lawrence Livermore National Laboratory
University of California
7000 East Avenue, PO Box 808
Livermore, CA 94550

INTRODUCTION

The rapid global expansion in population indicates that human reproduction functions very well. In the course of human reproduction, however, there are many infertile couples, fetal losses, malformed embryos and babies, genetically defective children, and other kinds of abnormal reproductive outcomes (U.S. Congress, 1986; Committee on Life Sciences, 1989). In the U.S.A. it is estimated that more than two million couples who wish to have children are infertile. About 1.9 million conceptions are lost before the 20th week of gestation, and ~30,000 are dead at birth (assuming 4.2 million births each year in the U.S., as for 1990 [U.S. Bureau of the Census, 1992]). About 7% of live births have low birth weight, and 3 to 7% have some birth defect. The causes of most of the ~250,000 defective babies are unknown. Chromosomal abnormalities explain some but not all spontaneous abortions, neonatal deaths, and birth defects. The frequency of chromosomal abnormalities among newborns is about 0.6%, including aneuploidies and structural aberrations. Chromosomal abnormalities in offspring are thought to be primarily *de novo* events that arose in the germ cells of one of the parents or early after conception. In addition, about 1% of newborns carry a gene for autosomal mutation for a genetic

* This work was performed by the Lawrence Livermore National Laboratory under the auspices of the U.S. Department of Energy under contract W-7405-Eng-48, with funding from the National Institute of Environmental Health Sciences (Y01-ES-10203-00) and State of California Tobacco Related Disease Research Program (3RT-0223). This paper was presented at the "International Workshop on In Vitro Methods in Reproductive Toxicology," held in Ottawa, Canada May 19–20, 1992, organized by Dr. D. Villeneuve of Health and Welfare Canada. It is reprinted from Reproductive Toxicology, vol. 7, pp-3-16, 1993 with permission from Pergamon Press Ltd.

disease with only ~20% of these arising anew early in development or in the reproductive cells of one of the parents. The social and medical costs of these abnormal reproductive outcomes are formidable, and there is a critical need to identify the underlying risk factors and to identify the responsible parent.

There is growing worldwide concern over the potential hazards to public health and human reproduction created by industrial chemical release (both old and new chemicals) and lifestyle factors such as smoking. Research with laboratory animals is the basis for assessing the human risk of exposure to toxic agents (Heywood and James, 1985). There are uncertainties, however, in the use of laboratory animals for identifying and characterizing male reproductive toxicants. Animal tests of reproductive function evaluate diverse reproductive outcomes, but they are generally insensitive to specific and subtle changes, and the response seen with the animal may not indicate what is seen in humans. There remains a critical need to improve the risk-assessment procedure for male reproductive toxicity.

The cause of certain abnormal reproductive outcomes have been traced to the father. This paper views the role of the father in reproduction in a multigenerational context, which considers genetic susceptibility and exposure histories of the father, the mother, as well as their offspring. It summarizes the status of the techniques (that is, biomarkers) used to detect early signs of male reproductive dysfunction. It also addresses the following issues in risk assessment: (1) need for efficient biomarkers of exposures and early effects in exposed human males; (2) need for "bridging" biomarkers of exposure and effects between humans and animal models to facilitate the use of animal data in risk assessment and to identify the animal genotypes that best model specific human responses; and (3) role of *in vitro* systems for investigating molecular and cellular mechanisms of male reproductive toxicity.

MULTIGENERATIONAL VIEW OF HUMAN REPRODUCTION

Reproduction can be viewed as the cycling of the germ line from generation to generation. In considering a developing child's health and subsequent reproductive fitness as an adult, we take into account the health, genetic susceptibility, and exposure history of the maternal and paternal germ cells and reproductive tracts. The relevant time period during which toxicity could occur can be thought to begin with conception of both of the child's parents and end when the child becomes an adult, ready to begin the cycle anew (Figure 1). This is a duration of 36 years, probably more. Important aspects of this process are:

1. Development during the respective grandmother's pregnancy of the mother's and father's precursor germ cells, reproductive tracts, and endocrine systems.
2. The exposure histories of the maternal and paternal germ cells during development, childhood, adolescence, and adulthood;
3. Effective gametogenesis in the adult father and mother producing intact gametes suitable for fertilization;
4. Effective mating and sperm transport through the female reproductive tract and fertilization of the egg;
5. Implantation of the egg and uterine environment that supports normal development and birth of the offspring;
6. A postnatal environment that supports a healthy childhood, adolescence and adulthood, and fertility as an adult.

Defined in this manner, one cycle of human reproduction involves aspects of the lives of five individuals from three generations: maternal and paternal grandmothers, mother, father, and the offspring (Figure 2). The maternal and paternal grandmothers are included, simply, because they carried the pregnancies during which the maternal and paternal germ cells, reproductive organs, and supporting somatic systems developed.

Value of the Multigenerational View

The multigenerational view of abnormal reproductive outcomes leads us to consider that risk factors may act on either parent or the offspring as far back as the time of conception of either parent. In this context, the factors leading to an abnormal reproductive outcome are:

The specific insult (physiologic damage to an organ, cell type, or biochemical process of the male reproductive system; chromosomal or gene mutation; etc.) that may be caused by an exposure to endogenous or exogenous toxicants or a random error in differentiation;

The individual(s) in whom the insult occurred: either of the grandmothers during her pregnancy, the mother, the father, or during the development of the offspring;

That individual's susceptibilities to the specific insult (pharmacokinetics, metabolism, oxidative status, etc.);

That individual's capacity to repair the prelesion in the specific cell type or stage in which it occurred.

The multigenerational view recognizes that more than one individual is usually involved in each abnormal reproductive outcome. The best example is for postconception toxicity to the pregnant mother that may be detrimental to the health of her developing offspring *in utero*, neonatally, and after birth. Lesions arising in the germ cells of either parent before conception may also result in abnormal reproductive outcomes. For example, numerical aneuploidy in the maternal or paternal germ cells may lead to autosomal trisomy and sex-chromosomal aneuploidy at birth (Epstein, 1986). In addition, the frequency of abnormal outcomes may depend on the capacity of the fertilized egg to repair DNA prelesions in sperm (Matsuda and Tobari, 1989; Genesca *et al.*, 1992). There are well-documented differences in the capacities of fertilized eggs of different strains of mice to repair lesions in sperm of mutagenized mice (Matsuda and Tobari, 1989). Rudak and colleagues (1978) pioneered a technique for analyzing the haploid karyotypes of human sperm after fusion of human sperm with enzymatically prepared hamster eggs (later referred to as the hamster technique). Using the hamster technique, human sperm were found to carry chromosomal aneuploidies and aberrations (Brandriff and Gordon, 1990; Martin and Rademaker, 1990), and the frequency of sperm with aberrations depended on the repair capacity of the hamster egg (Genesca *et al.*, 1992).

In the multigenerational view, the quality of each parent's gametes depends on specific exposures or risk factors experienced between the time they were developing in their respective mother and the time their gametes were released for fertilization. There is epidemiologic evidence for several agents including diethylstilbestrol and ethanol (Eliasson, 1985) suggesting that exposure of male offspring to toxicants in utero, as a child, or as an adolescence can impair their ability to produce fertile gametes and participate in normal reproduction when they are adults.

THE FATHER'S ROLE IN ABNORMAL REPRODUCTIVE OUTCOMES

Viewed in a multigenerational context, a male parent may detrimentally affect the development and health of his offspring in a variety of ways:

1. The father may produce insufficient numbers of functional sperm, resulting in reduced fertility, temporary sterility, or permanent sterility (Committee on Life

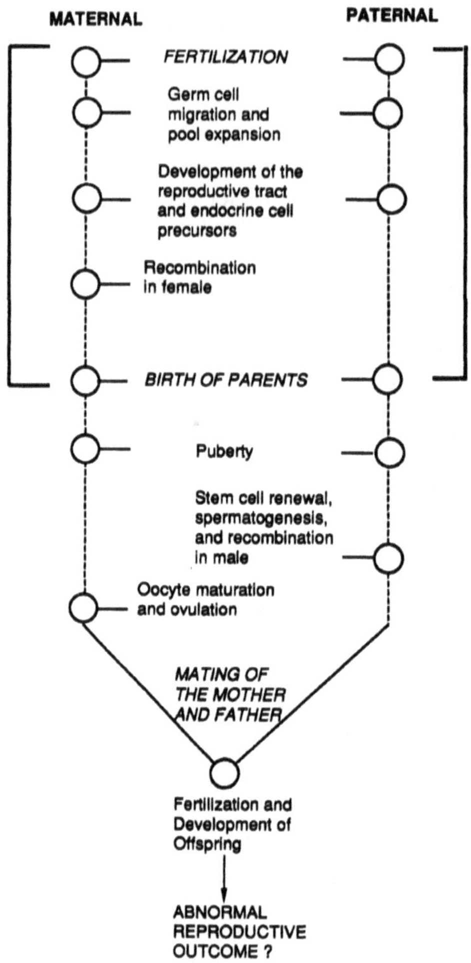

Figure 1. The human reproductive cycle including milestones in development, birth, adolescence, and adulthood of the mother and father and events in the fertilization and development of their offspring. Vertical brackets refer to the development of the mother and father *in utero* in their respective mothers (that is, maternal and paternal grandmothers of the new offspring).

Sciences, 1989). This may be due to genetic factors inherited from his parents or nongenetic factors such as exposure-induced cytotoxicity of cells in the male reproductive system. Exposures may have been to endogenous or exogenous toxicants and may have occurred any time between the time he was *in utero* until he produced the fertilizing sperm (Figure 1).

2. The genetic quality of the father's gametes is a strong determinant of the susceptibility, viability, and health of his offspring *in utero* and after birth (Epstein, 1986). Paternally transmitted genetic defects or mutations can result in early or late fetal loss, malformations, or serious diseases after birth.

3. Certain nonmutational changes in the DNA of the father's gametes are expected to be important for normal development. These are generally referred to as epigenetic changes and are not well defined molecularly. As an example, for certain

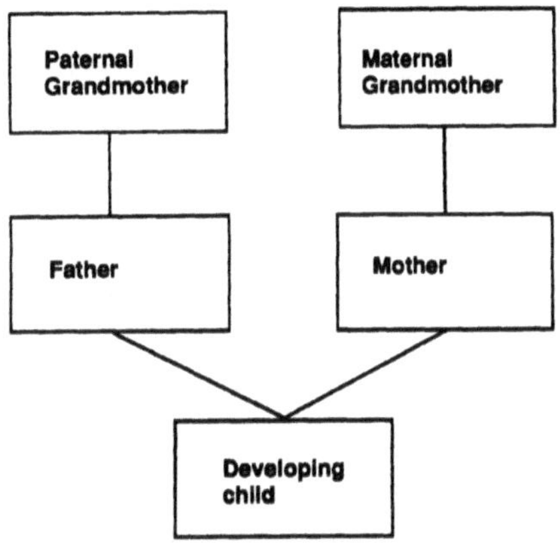

Figure 2. Generations and individuals included in the human reproductive cycle because of their potential relevance for male-mediated reproductive and genetic toxicology. The maternal and paternal grandmothers are included because they carried the pregnancies during which the maternal and paternal germ cells, reproductive organs, and supporting somatic systems developed.

genes only the maternal or paternal alleles are expressed during normal development (an effect referred to as imprinting) (Hall, 1990), and factors or exposures of the male that alter the normal chemistry of imprinting in male gametes may be potentially detrimental for subsequent development *in utero*.

4. Sperm may carry prelesions, the impact of which will depend on the ability of the egg to repair the prelesion (Matsuda and Robari, 1989; Genesca *et al.*, 1992).

5. The father may be a potential source of exposure to toxic agents or infectious agents that negatively affect pregnancy. The relevant exposure to the offspring may be via agents transmitted by semen or via direct exposure of the pregnant mother.

6. The father may also be the source of toxic exposures detrimental to the postnatal development of his child. This may include any factor or agent that interferes with the ability of the male parent to provide for and rear their children into healthy, fertile, young adults.

This broad perspective emphasizes that any comprehensive understanding of the mechanisms leading to male-mediated abnormal reproductive outcomes will require multidisciplinary investigations that include aspects of genetics, epidemiology, toxicology, molecular biology, specific exposure histories (drug taking, history of occupational, environmental, and household exposures), as well as couple and family psychology, sociology, etc.

Direct evidence for abnormal reproductive outcomes of paternal origin

The following lines of evidence indicate that certain abnormal reproductive outcomes are of paternal origin. These include effects on fertility and on the developing offspring:

1. Exposures of the human male to chemical toxicants can diminish the quality and quantity of sperm produced.

2. Male partners are responsible for a substantial fraction of the cases of human infertility, and there is epidemiological evidence that male exposure to a toxicant can diminish fertility.

3. There is evidence that sperm carrying cytogenetic abnormalities can fertilize and that exposure of the human testis to clastogens increases the frequencies of chromosomally abnormal germ cells.

4. There is clinical evidence that chromosomal abnormalities in embryos and newborns are due to abnormalities in the number and structure of paternal chromosomes.

5. There is epidemiological evidence that men with specific occupations or occupational exposures may have an increased risk of fathering children with birth defects or childhood cancers.

The effects of nearly 100 differing exposures have been evaluated for their effects on sperm production in the human male, and ~50 are known to be detrimental to sperm production (Wyrobek *et al.*, 1983). Very few of these agents have been directly evaluated for their effects on human fertility. About 40 to 50% of cases of human infertility are estimated to be due to male factors, yet little is known of the organic, cellular or molecular aspects of the defects. A small but convincing body of epidemiologic studies shows that exposure of the human male to agents such as ionizing radiation and dibromochloropropane (Eliasson, 1985; Mann and Lutwak-Mann, 1983; Nesbit and Karch, 1984) can result in reduced fertility or sterility.

There is now compelling evidence from both human and animal investigations that the impact of the male on reproduction goes well beyond fertilization. Molecular investigations of the parent-of-origin of chromosomal abnormalities in human offspring *in utero* and at birth have identified cytogenetic abnormalities in number (that is, aneuploidies) and structure of paternal chromosomes (Hassold *et al.*, 1984; Jacobs *et al.*, 1989). Aneuploidies involving paternal chromosomes were found to be predominant for children with sex chromosomal abnormalities: XXY (Klinefelter syndrome), XO (Turner syndrome), XXX, and XYY, with paternal chromosomes responsible for ~50, 80, 20, and 100%, respectively, of the cases of each type of abnormality (Hassold *et al.*, 1984). Investigations from several laboratories with the hamster technique have shown conclusively that apparently healthy men produce a small fraction of sperm that are aneuploid or carry chromosomal aberrations and that men who received radiotherapy or chemotherapy for cancer produced increased fractions of chromosomally abnormal sperm (Brandriff and Gordon, 1990; Martin and Rademaker, 1990; Genesca *et al.*, 1990; Martin *et al.*, 1986, 1989).

A new approach has been developed for probing the chromosomal content of human sperm using fluorescence in situ hybridization with chromosome-specific DNA probes to mark the locations of specific chromosomes. The frequencies of aneuploid sperm determined by the hybridization method were consistent with those obtained with the hamster technique, supporting the validity of the more efficient hybridization method (Robbins *et al.*, 1993). Aneuploidy frequencies varied among chromosome and among donors (Robbins *et al.*, 1993). Epidemiologic investigations are in progress to investigate the effects of exposure to tobacco smoke and aneugenic drugs on the frequency of aneuploid sperm.

Numerous epidemiologic investigations have described associations between job descriptions and possible workplace exposures of the father and the likelihood of childhood cancer or birth defects (Savitz and Chen, 1990; Narod *et al.*, 1988). Findings are variable but some occupations have repeatedly shown associations with abnormal reproductive outcomes (for example, paternal exposures in hydrocarbon-related occupations, the petroleum and chemical industries). Paint exposure of the father was linked with brain cancer and leukemias in his offspring (Narod *et al.*, 1988). As yet, however, no conclusive links have been found between specific exposures to the prospective father, specific mechanisms of transmission, and increased frequency of birth defects or childhood cancers. Extensive human investigations are warranted because studies in rodents have demonstrated conclusively that exposure of the male before mating can lead to developmental and behavioral abnormalities in the offspring (Nomura, 1988; Adams *et al.*, 1981) or increases in the frequencies of offspring with cancer (Nomura, 1989). As will be described later, efficient measurements (that is, biomarkers) of internal exposure and early biological effects in the exposed male may be helpful for establishing the link between exposure and abnormal reproductive outcome and for elucidating the mechanism(s) of transmission of defect(s) from the father to his offspring.

ASSESSING THE FUNCTIONAL STATE OF THE MALE REPRODUCTIVE SYSTEM

The pathway between damage induced by exposure of a male to a hypothetical external or endogenous toxicant or randomly occurring damage and abnormal reproductive outcomes is shown schematically in Figure 3. Epidemiologic investigations typically test for associations between the two extremes of the pathway (for example, external exposure versus specific abnormal outcomes). The standard epidemiological studies have inherent difficulties because they have not (a) measured the internal dose of the toxicant to the male, (b) distinguished between maternal and paternal effects, (c) distinguished between somatic and germinal effects, nor (d) distinguished among mechanisms of paternal transmission of damage to the offspring (sperm-mediated, semen-mediated, etc.). An additional complexity is that the relevant exposure may have occurred anytime between conception of the father and his production of the fertilizing sperm.

For recent or ongoing exposures, it may be possible to employ measurements of intermediate biological events along the exposure-response pathway, such as biologically relevant doses and early biological effects. A recent NRC monograph (Committee on Life Sciences, 1989) described biomarkers in reproductive toxicology as "indicators of variation in cellular or biochemical components or processes, structure, or function that are measurable in biological systems or samples.... The interest in biological markers is to identify early stages of health impairment and to understand basic mechanisms of exposure and response." The NRC monograph (Committee on Life Sciences, 1989) classified biomarkers into three categories: exposure, effect, and susceptibility.

A ***biomarker of exposure*** is an exogenous substance or its metabolite(s) or the product of an interaction between a xenobiotic agent and some target molecule or cell that is measured in a compartment within the organism.

A ***biomarker of effect*** is a measurable biochemical, physiological, or other alteration within the organism that, dependent on magnitude, can be recognized as an established or potential health impairment or disease.

A ***biomarker of susceptibility*** is an indicator of an inherent or acquired limitation of an organism's ability to respond to the challenge of exposure to a specific xenobiotic substance.

Biomarker of exposure. The biomarkers of exposure along the exposure-response pathway (Figure 3) range from measures of external exposure, to measures of internal dose, to measures of dose of the critical chemical moiety at the critical biological target (DNA, enzyme, cell membranes, etc.). In studies of exposed human populations, where practicality is an important consideration, biomarkers for internal dose typically

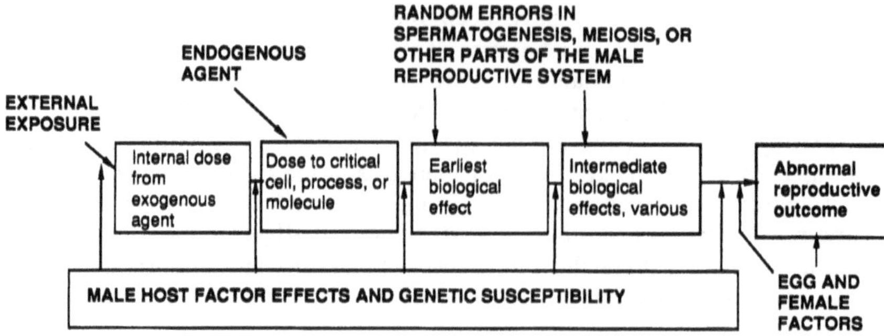

Figure 3. Pathway between exposure of a male to a hypothetical reproductive toxicant or to a random error in differentiation and the abnormal reproductive outcome of his offspring. The progression from insult to response is from left to right. The right-most horizontal arrow may be viewed as the transmission of damage from the male and his mate.

employ easily obtainable body fluids such as saliva and urine. An example for cigarette smokers is the level of cotinine in their saliva (Jarvis *et al.*, 1984). Seminal levels of toxicants or their metabolites may be used for measuring internal dose, but there are difficulties in evaluating semen dose because only a small proportion of the seminal fluid is testicular in origin (Mann and Lutwak-Mann, 1981). Seminal doses may not reflect the relevant doses to the germ cells during spermatogenesis. However, seminal levels may be relevant when considering effects of toxicants on the female partner and her offspring.

Biomarkers of effect. Progress in understanding the male reproductive system has resulted in the development of promising markers of biological effects. These range from measures of the first alteration in the critical biological target to measures of progressively later changes, leading to measures of the ultimate health effect(s) (Figure 3). Biomarkers of effects can be grouped by the types of damage detected: physiologic or genetic (Committee on Life Sciences, 1989). Most of the cell types needed to determine early toxic effects on the male reproductive system are unobtainable without surgery. Thus, semen is a well-studied tissue because of its accessibility, and promising biomarkers of its cellular and chemical components have been developed (Table 1).

Biomarkers of susceptibility. An individual's response to a toxic exposure depends on susceptibilities at the various locations in the exposure-response pathway (Figure 3). Studies with rodents have identified several aspects of susceptibility specific for male germ cells. For example, in mice, susceptibilities to mutations varied as stem cells differentiated into sperm, and pattern of susceptibility depended on the specific genetic endpoint evaluated and the mutagen used (Russell and Shelby, 1985). For specific locus mutations, three patterns of germ cell sensitivity were observed, as measured by the time interval between exposure and the mating week of peak induction of mutations (Russell *et al.*, 1990). Chemicals of pattern one induced mutations in late-step spermatids and sperm; those of pattern two affected early spermatids; and those of pattern three affected stem cells. The importance of genetic variation in response was illustrated by the dramatic increase in the frequencies of postmeiotic mutations found in rats exposed to DBCP compared with mice (Teramoto and Shirasu, 1989). The female-dependent variable of susceptibility for male-mediated mutations involved the capability of the egg to repair prelesions in sperm, as described earlier (Matsuda and Tobari, 1989; Genesca *et al.*, 1992).

Surrogate measurements of susceptibility may be useful until practical molecular biomarkers are developed. These may be differences in responses among individuals or species using biomarkers of exposure and effect at various locations along the exposure-response pathway (Figure 3). For example, a comparison of the change in a specific sperm parameter after equivalent internal doses of a toxicant may serve as a surrogate measure of the relative susceptibilities of two individuals. The biology and genetics of susceptibility are expected to be complex; biomarkers will be needed to assess the effects of pharmacokinetics, metabolism, and repair.

Special features of semen biomarkers

Semen is the most readily obtainable tissue of the human germ line. It contains sperm as well as secretions from numerous male glands (seminal vesicles, prostate, etc.) and products from Sertoli cells, Leydig cells and epididymis (Figure 4) (Desjardins, 1985). Semen is informative for damage to the male reproductive system and may be predictive of what will happen during fertilization or to the embryo during development. Effects on sperm quality and function may be due to either direct or indirect effects on the developing germ cell. Examples of indirect effects are (1) induced hormonal changes that indirectly affect germ cell development, and (2) effects on somatic cells of the testis that alter the microenvironment needed for normal germ cell development. Post-testicular effects are also known to alter sperm function (Desjardins, 1985). In addition, the chemical constitution of the glands that contribute the non-sperm component of semen may be altered by a toxicant exposure (Mann and Lutwak-Mann, 1981), and the semen may contain agents that are directly transmitted to the egg by a fertilizing sperm.

Given the organic complexity of the male reproduction system it is unlikely that any single semen biomarker will be a valid indicator of all mechanisms of male reproductive and genetic toxicity. The NRC report on biomarkers (Committee on Life Sciences, 1989) suggested a battery of biomarkers, but there was little consensus about which measurements to use and in which sequence to apply them. A nonprioritized approach to applying biomarkers is costly and will limit the numbers of chemicals that can be evaluated. As specific mechanisms of action of toxicants are understood, common mechanisms may emerge to guide the selection of biomarkers.

Individual biomarkers are expected to differ in their specificity for mechanisms of action, their sensitivity to toxic exposure and effects, and their predictive value for abnormal reproductive outcomes. Biomarkers may lack specificity because they detect damage induced by several different mechanisms of toxicity. Sperm concentration, for example, may be reduced by direct germ-cell cytotoxicity (for example, destruction of dividing spermatogonia by ionizing radiation), by hormonal down-regulation of spermatogenesis, blockage of efferent ducts, increased frequency of ejaculation, etc. Biomarkers that lack specificity may play an important role as part of a battery of prescreening biomarkers but may be less useful for dissecting specific mechanisms of action. The importance of the predictive value of biomarkers will be discussed later.

BIOMARKERS FOR DETECTING TOXIC EFFECTS ON HUMAN SPERMATOGENESIS AND MALE REPRODUCTIVE SYSTEM

Table 1 lists categories of human biomarkers under development and/or evaluation for assessing physiologic damage of toxic agents on the male reproductive system. These biomarkers are described in the 1989 NRC report (Committee on Life Sciences, 1989), and specific methods for evaluating the human male are reviewed by Comhaire (1993). Biomarkers of genetic damage will be described in the next section.

The most commonly used biomarkers of physiologic damage employ semen and include measurements of (1) numbers of sperm produced, (2) sperm quality (motility, morphology), and (3) capability of sperm to penetrate and fertilize the egg.

The effects of more than 100 agents or mixtures have been evaluated by semen analysis in humans (Wyrobek *et al.*, 1983). Tables 2 and 3 list the agents reported to have detrimental effects on human sperm quantity or quality (Wyrobek *et al.*, 1983). Most chemicals evaluated were experimental and therapeutic drugs (Table 2), but some were recreational drugs and occupational exposures (Table 3). About 50 agents showed detrimental effects on sperm count, motility, or morphology. Since the Genetox report was published in 1983 (Wyrobek *et al.*, 1983), a few additional human epidemiologic investigations of male reproductive toxicity have utilized semen biomarkers. As an example, perchloroethylene showed a small but dose-dependent effect on sperm motion and nuclear shape (Eskenazi *et al.*, 1991). In aggregate, these human studies showed that (1) semen biomarkers are efficient tools for detecting exposure-induced changes on the male reproductive system, (2) the ability to detect induced changes increased using longitudinal instead of cross-sectional sample collection strategies, and (3) there was a need to improve the objectivity of the measurements for sperm motion and structure.

Table 1. Categories of Biologic Markers of Physiologic Toxicity to Human Male Reproduction[a]

Tissue or Data Required	Markers of
Testis (or biopsy)	Histopathology
Seminal sperm	Sperm number Structure[b] Motility[b] Viability Agglutination Penetration and interaction assays: cervical mucus hamster eggs nonliving human eggs Internal and surface domains Chromatin structure
Other seminal parameters	Physical characteristics of semen Categorization of immature germ cells Chemical composition of semen: normal and xenobiotic constituents Seminal measures of the function of: Sertoli cells, Leydig cell, and accessory glands
Blood	Hormone levels
Survey and medical records	Fertility status: standardized fertility ratio, time to conception
Maternal urine	Measures of early pregnancy

[a] Selected categories of biological markers which have been used to characterize the effects of ionizing radiation or chemicals in exposed men and markers for which human baseline data are available (Committee on Life Sciences, 1989).
[b] Automated and computer-assisted methods are under development.

Table 2. Experimental and Therapeutic Drugs with Evidence of Toxic Effects on Human Sperm Quantity and/or Quality[a]

Agent	Agent
Acridinyl anisidide	Metanedienone
Adriamycin	Methotrexate
Aspartic acid	MOPP (Mechlorethamine, vincristine, procarbazine, and prednisone)
Clorambucil	MVPP (Mechlorethamine, vinblastine, prednisolone, and procarbazine)
Clorambucil, mechlorethamine, and azathioprine	Norethandrolone
Clomiphene citrate	Norethindrone
Cyclophosphamide	Norethindrone, norethandrolone, and testosterone
Cyclophosphamide and colchicine	Norgestrel and testosterone enanthate
Cyclophosphamide and prednisone	Norgestrienone and testosterone
Cyclophosphamide, prednisone, and azathioprine	Prednisolone
CVP (cyclophosphamide, vincristine, and prednisone)	Propafenon
CVPP (cyclophosphamide, vincristine, procarbazine, and prednisone)	R-2323 and testosterone
Cyproterone acetate	Sulphasalazine
Danazol and methyl testosterone	Testosterone
Danazol and testosterone enanthate	Testosterone cyclopentyl propionate
Enovid	Testosterone enanthate
Gossypol	Testosterone propionate
Luteinizing hormone releasing factor agonist	VACAM (Vincristine, adriamycin, cyclophosphamide, actinomycin D, and medroxyprogesterone acetate)
Medroxyprogesterone acetate	WIN 13099
Medroxyprogesterone acetate and testosterone enanthate	WIN 13099 and diethylstilbestrol
Medroxyprogesterone acetate and testosterone propionate	WIN 17416
Megestrol acetate and testosterone	WIN 10446

[a]Table entries are based on studies of sperm counts, motility, and morphology. The assignment of individual agents to this category was based on the data provided in the papers reviewed by the Human Sperm Reviewing Committee of the U.S. Environmental Protection Agency (EPA) Gene-Tox Program (Wyrobek *et al.*, 1983). These entries are generally based on few studies and may be expected to change as more data become available.

BIOMARKERS FOR DETECTING GENETIC DAMAGE IN MALE GERM CELLS AND FOR DETECTING MALE-MEDIATED HERITABLE MUTATIONS

Changes in the frequencies of human germ cells carrying genetic lesions are of major concern for several reasons. First, fertilization with sperm carrying numerical and structural abnormalities in chromosomes may affect the viability, development and health of the human embryo and newborn. Second, inherited mutations would affect the individual carrying the defect and may be transmitted to subsequent generations. Third, studies in rodents show convincingly that exposure of the male to ionizing radiation and various chemical mutagens can induce chromosomal and gene mutations in male germ cells and that these can be transmitted to offspring leading to abnormal reproductive outcomes and genetic diseases.

Table 3. Personal Drug Use and Occupational Exposures with Evidence of Adverse Effects on Human Sperm Quality[a]

Personal drug use	Alcoholic beverages (chronic alcoholism) Tobacco smoke Marijuana
Occupational exposures	Carbon disulfide Dibromochloropropane Dibromochloropropane & ethylene dibromide Lead Toluene diamine & dinitrotoluene Carbaryl[b] Kepone[b]

[a] Table entries are based on studies of sperm counts, motility, and morphology. The assignment of individual agents to this category was based on the data provided in the papers reviewed by the Human Sperm Reviewing Committee of the U.S. Environmental Protection Agency (EPA) Gene-Tox Program (Wyrobek *et al.*, 1983). These entries are generally based on few studies and may be expected to change as more data become available.

[b] Agents were judged to have evidence only suggestive of potential adverse effects by the Human Sperm Reviewing Committee of the U.S. Environmental Protection Agency (EPA) Gene-Tox Program.

Biomarkers of genetic damage attempt to detect events which occur less frequently than those detected by biomarkers of physiologic damage. The indicators that have been considered for detecting male-mediated genetic toxicology and mutations were reviewed in the NRC and OTA reports (U.S. Congress, 1986; Committee on Life Sciences, 1989), and examples are listed in Table 4. The categories of indicators include (a) standard epidemiologic studies using reproductive outcome as well as cytogenetic or protein analyses of offspring of exposed parents, (b) analysis of multiple DNA alterations in offspring, and (c) sperm and other germ-cell measurements in the exposed father.

Ionizing radiation is one of the few agents which has been evaluated for the induction of germinal mutations. In a study of men who received testicular irradiation at graded doses up to 6 Gy, a dose-dependent increase was reported in the fraction of

Table 4. Markers of Genetic Damage and Heritable Mutations in the Male Germline[a]

Tissue	Marker
Testis (biopsy)	Cytogenetic analyses of cells in mitosis, meiosis I, and meiosis II
Semen	
Sperm	Sperm cytogenetics Sperm DNA and protein adduction Gene mutations in sperm Sperm aneuploidy
Immature germ cells	Spermatid micronuclei Cytogenetics of ejaculated meiotic I cells
Questionnaire and medical records	Sex ratio Spontaneous abortion Offspring cancer Sentinel phenotypes
Offspring tissue	Cytogenetics DNA sequencing Protein mutations DNA restriction-length polymorphism RNAase digestion Subtractive hybridization of DNA Denaturing gel electrophoresis of DNA Pulse-field electrophoresis of DNA
Mother's urine	Detection of early fetal loss
Somatic cell surrogates:	
In white blood cells	HGPRT mutations
In red blood cells	Hemoglobin mutations Glycophorin A mutations

[a] Specific markers are described and evaluated in NRC 1989 (Committee on Life Sciences, 1989).

chromosomal translocations induced in spermatogonia (Brewen *et al.*, 1975). The shape of the dose-response curve for men was similar with those obtained for laboratory animals. Exposure of human male germ cells to ionizing radiation also increased the frequencies of sperm carrying chromosomal abnormalities, as detected using the hamster technique (Martin *et al.*, 1986, 1989). Epidemiological investigations of survivors of the Japanese atomic bombs, however, did not detect radiation-induced increases in germinal mutations (Schull *et al.*, 1991). The methods employed, however, (heritable chromosomal abnormalities, germinal biochemical mutations, etc.) required large numbers of pregnancies and offspring and were inherently inefficient for detecting mutations induced by low doses of ionizing radiation. Therefore, the apparent discrepancy between the epidemiological investigations and the cytogenetic data for human male germ cells exposed to ionizing radiation may be explained by the relatively low germinal dose received by the Japanese bomb survivors and inefficiency of the specific epidemiological methods used.

Several chemical exposures have also been assessed for effects of paternal exposure on reproductive outcomes and germinal mutations (Narod *et al.*, 1988). The studies of abnormal reproductive outcomes among long-term survivors of cancer therapy (Mulvihill *et al.*, 1987) are unique among these in the certainty that (a) exposure is clearly limited to the male, (b) exposure occurred at a specified time before conception, and (c) there is no additional or low-level post-conception exposure to either the mother or the father. These conditions are rarely met in epidemiologic studies of recreational drug use, environmental agents, or occupational exposures (Narod *et al.*, 1988). Although no detrimental effects were detected among the offspring of men receiving chemotherapy, the studies were limited in (a) the small numbers of pregnancies and offspring evaluated, (b) not all cases included treatment with mutagenic agents, and (c) long duration between exposure and mating means that only stem cell exposures were evaluated. In mice, only a few of the known germ cell mutagens are also mutagenic in spermatogenic stem cells (Russell *et al.*, 1990).

More efficient biomarkers of heritable chromosomal and gene mutations are needed for (a) studies of smaller groups of exposed people and smaller numbers of offspring and (b) distinguishing between mutations that originated in male and female germ cells. DNA-based methods for detecting germinal mutations are under development and these promise increased sensitivity to detecting germinal mutations, because larger number of events will be scored per offspring (U.S. Congress, 1986; Mohrenweiser, in press). Another approach is to develop sperm-based measurements of chromosomal and gene mutations. The hamster technique and the hybridization methods for sperm have already been described (Rudak *et al.*, 1978; Robbins *et al.*, 1993. Other promising biomarkers for exposure levels, chromosomal abnormalities, and gene mutations in human sperm and other seminal constituents are listed in Table 4. There are major advantages for sperm-based biomarkers. Unlike biomarkers based on the analysis of offspring, sperm measurements can be made in individual men and changes in the effects on their germ cells can be monitored over time. Also, similarly exposed individuals can be compared to identify person-to-person variation in mutational response (that is, differences in susceptibilities).

MECHANISM OF ACTION AND IMPORTANCE OF *IN VITRO* APPROACHES

The prediction of potentially harmful effects of drugs, environmental factors, and other factors on the male reproductive system requires an understanding of the fundamental mechanisms regulating spermatozoa production and discharge, chromosomal assortment into gametes, and the production of semen (Desjardins, 1985). This includes processes in the testis, epididymis, and elsewhere in the urogenital duct system. It also involves testicular and nontesticular androgen production, hormonal feedback controls, and the secretions of several glands (seminal vesicle, prostate, etc.). The complexity of the somatic and germ-cell interactions *in vivo* (Figure 4) has impeded the identification of specific target site(s) and mechanism(s) of action of toxicants. Differentiating germ cells interact physically with Sertoli cells and with neighboring germ cells in other steps of spermatogenesis. They also interact through chemical signals with other somatic cells in the testis (for example, Leydig and myoid cells) and other organs. Judging by the diversity of cell types and organ systems that can be affected, the mechanisms of male-mediated abnormal reproductive outcomes are expected to be diverse.

There are three major categories of mechanisms of male reproductive toxicity: nongenetic, genetic, and epigenetic. Agents acting by non-genetic mechanisms of toxicity would be expected to diminish the male's fertility potential. Nongenetic mechanisms of toxicity include any alteration in the normal physiology and morphology of the male reproductive system including abnormalities in (a) spermatogenesis (b) endocrine function, (c) production of semen and (d) delivery of functional sperm into the female reproductive tract. Studies of spermatogenic cytotoxicity have identified several cellular targets of toxicity, depending on the exposure: (Leydig cells, Sertoli cells, epididymal cells, various cell types of spermatogenesis, etc.) (Desjardins, 1985;

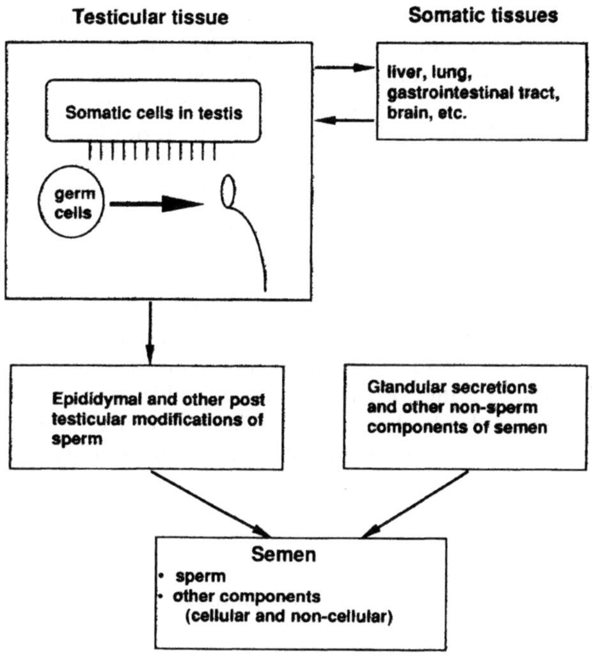

Figure 4. Schematic of the interactions among germinal and somatic tissues required for the production of human semen.

Barlow and Sullivan, 1982). However, specific biochemical and molecular mechanisms of toxicity are not well understood and have been investigated for only a few agents (Zenick, 1984). Table 1 lists the biomarkers thought to detect non-genetic damage to the male reproductive system.

Genetic mechanisms of male reproductive toxicity involve changes in the germ line DNA that may lead to chromosomal abnormalities or gene mutations in the offspring. The sperm may carry either the final lesion (for example, abnormal number of chromosomes) or a prelesion which may be transformed into a chromosomal abnormality or mutation only after fertilization (for example, DNA adducts, breaks in the DNA backbone, labile sites, etc.). Genetic abnormalities in the offspring may result in any of a variety of abnormal reproductive outcomes which may manifest themselves during development (loss, morphological defect, etc.) or after birth (behavioral changes, cancer, disease, etc.). It is well known that exposure of male mice to certain mutagens increases the fraction of germ cells carrying chromosomal abnormalities and/or gene mutations (Russell *et al.*, 1985). Although the evidence for mutagen induction of chromosomal or gene mutations in people is weak, abnormalities in the number of paternal chromosomes have been observed in utero and at birth, and translocations are readily transmitted from father to son (Hassold *et al.*, 1984; Jacobs *et al.*, 1989). Abnormalities in paternal chromosomes of an offspring may have arisen during gametogenesis of its father, or early in its development after fertilization, and more research is needed to understand these mechanisms. Table 4 lists the categories of biomarkers thought to detect genetic damage in male germ cells or male-mediated germinal mutations in offspring.

The hypothetical delivery of toxic agent(s) to the fertilized egg or to the developing offspring via semen may, in concept, be detrimental to fertilization or to embryonic development by genetic or nongenetic mechanisms, depending on the agent.

Epigenetic mechanisms of male reproductive toxicity refer to changes in the process of imprinting which may alter the normal pattern of expressed maternal or a paternal alleles after fertilization (Hall, 1990). How imprinting works during spermatogenesis and how a disturbance in the normal patterning of gene expression during early development leads to abnormal reproductive outcomes are not yet well understood.

The mechanisms of paternally mediated birth defects and childhood cancer remain controversial. These abnormal outcomes may be due to chromosomal abnormalities, gene mutations, alterations in imprinting transmitted via the paternal gametes or they may be due to other causes; arguments can be made for all possible mechanisms of action. Further research is needed to identify specific agents that lead to paternally-mediated birth defects or childhood cancer and to use them to understand mechanisms of action, using both animal models and human epidemiologic studies.

In vitro approaches

The entire complexities of testicular organization and spermatogenesis have not yet been reproduced *in vitro*. Specific aspects of germ-cell differentiation can be maintained in vitro for limited periods of time and methods have also been established for culturing testicular somatic cells and other parts of the male reproductive system (Steinberger and Klinefelter, 1993; Lamb and Chapin, 1993). Single cell type and co-culture methods have been used to investigate the cellular and intercellular features of cytotoxicity for agents such as cadmium (Steinberger and Klinefelter, 1993; Lamb and Chapin, 1993).

The simplicity of the *in vitro* approaches confers notable advantage for the study of mechanisms. By reducing complexity, *in vitro* methods may identify the cell types involved, specific cellular changes, and biochemical mechanisms of toxicity. *In vitro* approaches may be the best means to understanding the central portion of the exposure-response pathway (Figure 3), i.e., the portion dealing with the critical internal dose and the earliest biological effects. At present, however, we generally lack sufficient understanding to (a) relate the dose and response findings from in vitro studies to what is seen in whole animals exposed to the same toxicant and (b) define the specific circumstances under which *in vitro* methods might replace whole animal methods in reproductive risk assessment. A special class of biomarkers may be helpful to relate *in vitro* and *in vivo* studies, as described below.

"BRIDGING" BIOMARKERS IN MALE REPRODUCTIVE AND GENETIC RISK ASSESSMENT

Extrapolating results from animal models to humans and from in vitro studies to the whole animal are complicated problems in male reproductive and genetic risk assessment. The concept of bridging biomarkers may be helpful, and two categories of bridging biomarkers are envisioned:

Biomarkers for comparing the physiologic and genetic responses of human and laboratory animals exposed to the same toxicants.

Biomarkers to compare *in vivo* and *in vitro* findings with the same toxicant.

Bridging biomarkers may provide an efficient way to make valid comparisons among studies of exposed humans, laboratory animals, and *in vitro* models.

Bridging biomarkers for comparisons between exposed men and laboratory species

This category of biomarkers is important for investigating the degree of concordance in response between laboratory species and humans for specific agents as well as for classes of agents. Such comparisons may help identify laboratory species most appropriate for risk characterization. In cases where exposed men are not adequate for study, the laboratory species could be used to determine dose response, persistence of damage, mechanism of toxicity, and other aspects of risk characterization. Bridging biomarkers of this category would be the subset of biomarkers of physiologic and genetic damage (Table 1 and 4) which can be measured directly in exposed males. This category includes biomarkers of exposure and response. Bridging biomarkers of exposure may be measurements of a parent compound or metabolite in a body fluid (saliva, urine, etc.). Bridging biomarkers of response may be any noninvasive procedure to assess alterations in the male reproductive system including ultrasound and measurements of alterations in semen and sperm. Sperm number, motility and morphometry are examples of bridging biomarkers of physiologic damage to sperm while sperm aneuploidy is an example of a bridging biomarker of genetic damage to sperm. To be effective tools for species comparisons, biomarkers must be efficient and practical in human applications and tests of exposed laboratory animals. In addition, to be useful in risk assessment bridging biomarkers must be quantitative and objective measurements.

Identification of laboratory species for the investigation of toxicological and mutagenic mechanisms

It is difficult to predict how well the responses of laboratory species to a toxicant will correspond to the human responses. Using sperm number as a biomarker, Meistrich (1990) found a high degree of concordance between the mouse and human response for several cancer chemotherapeutic drugs. For adriamycin, however, murine spermatogenesis was sensitive while human spermatogenesis appeared to be relatively resistant. A discordance in response was also observed for dibromochloropropane (DBCP), one of the most potent human anti-spermatogenic agents known (Teramoto and Shirasu, 1989). Germ cells of mice were highly resistant to DBCP-induced cytotoxicity, while the rat and human were more sensitive. DBCP's ineffectiveness in inducing specific-locus mutations or dominant lethality in mice was problematic in genetic risk assessment. Was the mouse insensitive to both the toxicological or mutagenic effects of DBCP or (b) was DBCP not mutagenic in mammals? The finding that DBCP induced lesions in rat spermatids leading to dominant lethality among offspring (Teramoto and Shirasu, 1989) argued that the mouse data for genetic toxicity of DBCP may be irrelevant, and that the rat would be a better species for characterizing the mechanisms of cytotoxicity and genetic toxicity. The unpredictable incidence of discordance in the response of the male reproductive system among species underscores the need for a more systematic approach in utilizing animal data in reproductive and genetic risk assessment.

One approach is to utilize the available toxicological data for human sperm to evaluate the responses of laboratory species and to select animal species for investigating the mechanisms of toxicity of agents known to affect human males. The ~100 chemical agents and mixtures of Tables 2 and 3 plus the agents found to be ineffective (data not shown) can be viewed as a reference set of toxicity data for identifying laboratory animals that respond similarly to man, (making allowances for possible classification errors due the lack of repeated human data for some agents). The 1983 Gene-tox report (Wyrobek *et al.*, 1983) noted, however, a major deficiency of corresponding semen data from any laboratory animal for most of these chemicals.

Therefore, the first step will be to prioritize the agents on Tables 2 and 3, and then to evaluate their effects on sperm quantity and quality in various laboratory animals. Rats and mice are candidates, and so are rabbits and other species. Rabbits have several advantages in semen studies of male reproductive toxicity: (1) the morphology of their sperm is much closer to human than that of mice or rats, and (2) unlike rodents which are usually sacrificed to collect sperm samples, repeated semen samples can be collected from a rabbit over time, permitting longitudinal investigations of induction and persistence of damage within the same animals. Disadvantages of rabbits are that they are larger and more costly to house and few baseline data are available for germinal toxicity and mutagenesis. Laboratory animals with human-like responses will be very important for investigations of mechanisms of toxicity *in vivo* and *in vitro*. It is unknown at this time how many different mechanisms of toxicity are represented by the results shown in Table 2 and 3, and how differing animals will respond. This research holds the promise of identifying biomarkers of common mechanisms of toxicity and, possibly, for categorizing laboratory species by their sensitivity to certain classes of human male reproductive toxicants.

Predictive value of bridging biomarkers

It is well recognized that specific numerical relationships between sperm effects and reproductive outcomes are species dependent, as shown with the following example of sperm numbers. In mice, little effect on fertility was seen until sperm numbers were reduced to about 15% of normal and it was unusual for unexposed mice to have numbers that low. In humans, the distribution of sperm number among men is much broader than in mice with a substantial fraction of men in the general population falling below values generally associated with reduced fertility; 1% of men may be azoospermic and 5 to 20% may have sperm concentrations of less than 20 million per mL. Therefore, agents that produce an overall reduction in human sperm numbers would be expected to shift the entire distribution of sperm counts to lower values and thereby increase the proportion of men with fertility problems. An important distinction between the predictive values of murine and human sperm concentration for fertility is that in mice the relationship appears to be thresholded while in the human it probably isn't. To avoid these species-dependent problems, it is suggested that the predictive value of human sperm biomarkers may be established directly in men, and that subjects for study may be from the general population.

Bridging biomarkers for comparing the *in vitro* findings with effects in whole animal

As described earlier, *in vitro* methods typically focus on the portion of the exposure effect pathway in the vicinity of the earliest biological effect, and, therefore, the sperm-based biomarkers are probably not useful. Special bridging biomarkers are needed for comparing the dose of an agent and its biological effects measured in vitro with doses and effects in the same cells of animals treated *in vivo* and the specific biomarkers may depend on the agent used and the mechanism of toxicity being investigated. This category of bridging biomarkers would be of critical importance in evaluating the validity and relevance for the whole animal of mechanisms of toxicity identified *in vitro*.

SUMMARY

Viewed in a multigenerational context, normal reproduction depends on the genetic susceptibilities and exposure histories of the father, as well as those of his spouse and offspring. There is a growing need to identify efficient biomarkers of exposure and biologic response of the male reproductive system to toxic agents, and numerous biomarkers are under consideration. There is a special need for "bridging" biomarkers for comparing responses between humans and laboratory animals, and between *in vitro* and *in vivo* studies. An integrated approach may be helpful for male reproductive and genetic risk assessment which includes (a) practical and efficient biomarkers of exposure and response for use in epidemiological studies of exposed human males, (b) risk characterization using animal species with response patterns similar to those of exposed men, (c) *in vivo* and *in vitro* investigations of the molecular mechanisms of male-mediated reproductive and genetic toxicity for agents known to be active in the human male, and (d) investigations of the predictive value of selected human biomarkers for abnormal reproductive outcomes. The approach is dependent on the development of more efficient biomarkers and on an improved understanding of molecular mechanisms of reproductive and genetic toxicity.

ACKNOWLEDGMENTS

I thank J. Cherniak, J. Kranzler, and W. Robbins for helpful discussions and editing. I also thank Carolin Middleton for editing and formatting.

REFERENCES

Adams, P., Fabricant, J., Legator, M., 1981. Cyclophosphamide-induced spermatogenic effects detected in the F1 generation by behavioral testing. *Science* 211:80–82.

Barlow, S.M., Sullivan, F.M., 1982. Reproductive hazards of industrial chemicals an evaluation of animal and human data. London: Academic Press.

Brandriff, B., Gordon L, 1990. Human sperm cytogenetics and the one-cell zygote. In: Allen, J.W., Bridges, B.A., Lyon, M.F., Moses, M.J., Russell, L.B., Eds. Biology of mammalian germ cell mutagenesis. Banbury #34. Cold Spring Harbor Laboratory Press, 183–194.

Brewen, J.G., Preston, R.J., Gengozian, N., 1975. Analysis of x-ray induced chromosomal translocations in human and marmoset spermatogonial stem cells. *Nature* 253:468–470.

Comhaire, 1993. *Reproductive Toxicology* 7 (in press).

Committee on Life Sciences, National Research Council, 1989. Biologic markers in reproductive toxicology. Washington DC: National Academy Press, 37–146.

Desjardins, C., 1985. Morphological, physiological, and biochemical aspects of male reproduction. In: Dixon, R.L., Ed. Reproductive toxicology. New York: Raven Press, 131–146.

Eliasson, R., 1985. Clinical effects of chemicals on male reproduction. In: Dixon RL, Ed. Reproductive toxicology. New York: Raven Press, 161–172.

Epstein, C.J., 1986. The consequences of chromosomal imbalance: principles, mechanisms and models. Cambridge: Cambridge University Press.

Eskenazi, B., Wyrobek, A.J., Fenster, L., Katz, D., Sadler, M., Lee, J., Hudes, M., Rempel, D., 1991. Perchloroethylene exposure in the dry cleaning industry: I. Effect on semen quality. *Am J Ind Med* 20(5):593–600.

Genesca, A., Caballin, M.R., Miro, R., Benet, J., Bonfill, X., Egozcue, J., 1990. Human sperm chromosomes: long-term effect of cancer treatment. *Cancer Genet Cytogenet* 46:251–260.

Genesca, A., Caballin, M.R., Miro, R., Benet, J., Germa, J.R., Egozcue, J., 1992. Repair of human sperm chromosome aberrations in the hamster egg. *Hum Genet* 89:181–186.

Hall, J.G., 1990. How imprinting is relevant to human disease. Development suppl:141–148.

Hassold, T., Chiu, D., Yamane, J.A., 1984. Parental origin of autosomal trisomies. *Ann Hum Genet* 48:129–44.

Heywood, R., James, R.W., 1985. Current laboratory approaches for assessing reproductive toxicity: testicular toxicity in laboratory animals. In: Dixon, R.L., Ed. Reproductive Toxicology. New York: Raven Press, pp. 147–160.

Jacobs, P., Hassold, T., Harvey, J., May, K., 1989. The origin of sex chromosome aneuploidy. *Prog Clin Biol Res* 311:135–51.

Jarvis, M., Tunstall-Pedoe, H., Feyerabend, C., Vesey, C., Salloojee, Y., 1984. Biochemical markers of smoke absorption and self reported exposure to passive smoking. *J Epidemiol Community Health* 38:335–339.

Lamb, J.C., Chapin, R., 1993. Testicular and germ cell toxicity: *in vitro* approaches. *Reproductive Toxicology* 7 (in press).

Mann, T., Lutwak-Mann, C., 1981. Male reproductive function and semen, themes and trends in physiology, biochemistry, and investigative andrology. Chapter 1. Male reproductive function and the composition of semen. Berlin: Springer-Verlag, 1–38.

Mann, T., Lutwak-Mann, C., 1983. Adverse effects of chemicals on male reproductive function. In: Vouk, V.B., Sheehan, P.J., Eds. Methods for assessing the effects of chemicals on reproductive function. New York: John Wiley and Sons, 135–147.

Martin, R.H., Hildebrand, K., Yamamoto, J., Rademaker, A., Barnes, M., and 4 others, 1986. An increased frequency of human sperm chromosomal abnormalities after radiotherapy. *Mutat Res* 174: 219–225.

Martin, R.H., Rademaker, A., Hildebrand, K., Barnes, M., and 4 others, 1989. A comparison of chromosomal aberrations induced by *in vivo* radiotherapy in human sperm and lymphocytes. *Mutat Res* 226:21–30.

Martin, R.H., Rademaker, A., 1990. The frequency of aneuploidy among individual chromosomes in 6,821 human sperm chromosome complements. *Cytogenet Cell Genet* 53:103–107.

Matsuda, Y., Tobari, I., 1989. Repair capacity of fertilized mouse eggs for x-ray damage induced in sperm and mature oocytes. *Mutat Res* 210:35–47.

Meistrich, M.L., 1990. Comparative male gonadal toxicity from cytotoxic cancer therapies. In: Sherins, R.J., Mulvihill, J.J., Eds. Reproduction and cancer.

Mohrenweiser, H., 1993. Impact of the molecular spectrum of mutational lesions on estimates of germinal gene mutation rates, *Mutat Res*. (in press).

Mulvihill, J.J., McKeen, E.A., Rosner, F., Zarrabi, M.H., 1987. Pregnancy outcome in cancer patients. *Cancer* 60:1143–1150.

Narod, S.A., Douglas, G.R., Nestmann, E.R., Blakey, D.H., 1988. Human mutagens: evidence from paternal exposure. *Environ Mol Mutag* 11:401–415.

Nesbit, I.C.T., Karch, N.J., 1983. Chemical hazards to human reproduction. Park Ridge, New Jersey: Noyes Data Corp.

Nomura, T., 1988. X-ray- and chemically induced germ-line mutation causing phenotypical abnormalities in mice. *Mutat Res* 198:309–320.

Nomura, T., 1989. Role of radiation-induced mutations in multigenerational carcinogenesis. In: Napalov, N.P., Rice, J.M., Tomatis, L., Yamasaki, H., Eds. Perinatal and multigenerational carcinogenesis. Lyon: International Agency for Research on Cancer, 375–387.

Robbins, W.A., Segraves, R., Pinkel, D., Wyrobek, A.J., 1993. Detection of aneuploid human sperm by fluorescence in situ hybridization: evidence for a donor difference in frequency of sperm disomic for chromosomes 1 and Y. *Amer J Hum Genet* (in press).

Rudak, E., Jacobs, P.A., Yanagamachi, R., 1978. Direct analysis of the chromosome constitution of human spermatozoa. *Nature* 274:911.

Russell, L.B., Russell, W.L., Rinchik, E.M., Hunsicker, P.R., 1990. Factors affecting the nature of induced mutations In: Allen, J.W., Bridges, B.A., Lyon, M.F., Moses, M.J., Russell, L.B., Eds. Biology of mammalian germ cell mutagenesis. Banbury Report, Cold Spring Harbor Laboratory Press, 271–289.

Russell, L.B., Shelby, M.D., 1985. Tests for heritable genetic damage and for the evidence of gonadal exposure in mammals. *Mutat Res* 154:69–84.

Savitz, D.A., Chen, J., 1990. Parental occupational and childhood cancer: review of epidemiologic studies. *Environ Health Perspect* 88:325–337.

Schull, W.J., Otake, M., 1991. A review of forty-five years study of Hiroshima and Nagasaki atomic bomb survivors: future studies of the prenatally exposed survivors. *J Radiat Res* (Tokyo) 32 suppl:385–393.

Steinberger A, Klinefelter G, 1993. Sensitivity of Sertoli and Leydig cell to xenobiotics in *vitro* models. *Reproductive Toxicology* 7 (in press).

Teramoto, S., Shirasu, Y., 1989. Genetic toxicity of 1,2-dibromo-3-chloropropane. *Mutat Res* 221:1–9.

U.S. Bureau of the Census, 1992. Statistical abstract of the United States, 112th edition. Washington DC.

U.S. Congress, Office of Technology Assessment, 1986. Technologies for detecting heritable mutations in human beings. Washington DC: U.S. Government Printing Office, OTA-H-298.

Wyrobek, A.J., Gordon, L.A., Burkhart, J.G., Francis, M.C., Kapp, R.W., and 4 others, 1983. An evaluation of human sperm as indicators of chemically induced alterations of spermatogenic function. A report for the U.S. Environmental Protection Agency Gene-ToxProgram. *Mutat Res* 115:73–148.

Zenick, H., 1984. Mechanisms of environmental agents by class associated with adverse male reproductive outcomes. In: Lockey, J.E., LeMasters, G.K., Keye, W.R., Jr, Eds. Reproduction: the new frontier in occupational and environmental health research. New York: Alan R. Liss, 335–361.

SPECIFIC-LOCUS MUTATION TESTS IN GERM CELLS OF THE MOUSE: AN ASSESSMENT OF THE SCREENING PROCEDURES AND THE MUTATIONAL EVENTS DETECTED

Jack Favor

GSF-Institut für Säugetiergenetik
Neuherberg
D-86764 Oberschleissheim
Germany

INTRODUCTION

A detailed characterization of the mutation process in germ cells of mammals includes precise estimations of the spontaneous, radiation- and chemically-induced mutation rates, as well as the identification and quantification of factors which affect the sensitivity to mutation induction. Such studies must rely on experiments based on laboratory animals, of which the choice is the house mouse *Mus musculus*, due to its small size, short generation time, established husbandry procedures, and extensive genetic characterization. In comparison with other laboratory genetic organisms, the use of the mouse for mutation studies is relatively slow, labor intensive and expensive. These disadvantages underline the importance of efficient test methods when using the mouse for mutagenicity studies.

The attributes required of an acceptable or ideal germ cell mutagenicity test in the mouse include that the screening procedures be simple, fast and unambiguous, such that the large populations of animals may be examined which are required to estimate infrequent mutagenic events. To this end a predetermined definition of the traits to be screened as well as the phenotypic variants indicative of mutation to be expected are essential. In the ideal situation the exact number of loci at which mutations are to be screened is defined. In so specifying the loci, the traits and the phenotypic variants to screen for newly occurring mutations, the precision of the mutagenicity test as well as the accuracy of the mutation rate estimate are increased.

The method developed by Russell (1951) to screen for newly occurring recessive visible mutations in the mouse meets the above criteria and has become synonymous with the "mouse specific locus test". Essentially, untreated or treated homozygous wildtype animals are mated to untreated tester animals. The tester stock is homozygous recessive at 7 marker loci (*a*, non-agouti; *b*, brown; c^{ch}, chinchilla; *d*, dilute; *se*, short-ear; *p*, pink-eyed dilution; *s*, piebald), which control coat pigmentation color and pattern as well as the size of the external ear. Offspring resulting from the cross are expected to be heterozygous at all 7 marker loci and to express the wildtype phenotype. In the event of a mutational event affecting the wildtype allele, animals

Male-Mediated Developmental Toxicity, Edited by D.R. Mattison
and A.F. Olshan, Plenum Press, New York, 1994

will express the characteristic homozygous recessive phenotype of the locus at which the mutation occurred. Recovered variants are genetically confirmed by testing for allelism at the locus in question, and the newly occurring mutations are often tested for dominance effects as well as for homozygous viability. The method is the most efficient to test for transmitted and genetically confirmed mutations in germ cells of the mouse and has provided virtually all experimental data on factors affecting the mutation rate in germ cells of mammals. The reader is referred to previous, detailed reviews of the specific locus test methods and results (Ehling and Favor, 1984; Russell *et al.*, 1981; Searle, 1974; Selby, 1981). One aim of mammalian germ cell mutagenicity testing is to provide data with which to base an estimation of the human genetic risk associated with an increased mutation rate due to radiation or mutagen exposure. In this context, recessive visible specific locus test data have been important. However, the recessive visible specific locus test results represent estimations of the mutation rate at a limited number of loci and the method screens only for recessive mutant alleles. In view of these considerations, a number of other mouse germ cell mutagenicity testing procedures have been developed. In the following, I will review the test methods developed to screen for newly occurring mutations in germ cells of the mouse, indicate the specificity of the screening procedures, and compare some representative mutagenicity results.

MOUSE *IN VIVO* GERM CELL MUTAGENICITY TESTS

Table 1 lists those assays developed to systematically screen for transmitted mutations in the mouse. The genetic endpoints screened range from recessive to dominant mutations as well as rearrangements of the chromosomal organization of the mouse genome. Further, the procedures may screen for mutations at a precise set of loci as in the recessive specific locus mutation test, the enzyme electrophoresis and enzyme activity mutation tests as well as the histocompatibility mutation test. At the other extreme are assays which are not locus specific, such as the recessive lethal mutation test, the assay to detect mutagenic effects on fitness parameters as well as the assays to detect chromosomal rearrangements. The dominant visible, dominant cataract and dominant skeletal mutation tests strive for a degree of locus specificity by focusing on a specific set of phenotypes for which variants are identified as presumptive mutants.

The alteration of gene function due to mutation may be grossly categorized into two classes, loss of gene function or gain of gene function. The ability to detect such alterations of gene function is dependent upon the assay employed to screen for mutations. For example, the specific locus test screens for mutations at a set of marker loci in an experimental design in which F_1 offspring are expected to be heterozygous at all marker loci and to express the wild-type phenotype. In the event of a mutation of the wild-type allele at a marker locus, the affected F_1 animal will express the recessive phenotype characteristic for the locus. Mutations resulting in total loss of functional gene product as well as mutations which lead to an intermediate phenotype as the result of an alteration of the functional gene product are identified by this test method.

A similar experimental design has been developed for a sub-set of the enzyme loci screened for electrophoretic variants as well as for a sub-set of the histocompatibility loci (Class I), in which the parental strains chosen are genetically different at the loci screened. For these three mutation assays, mutations resulting in loss or gain of gene function are detectable. By contrast, in those assays developed which do not follow a protocol similar to the specific locus method and in which the parental genotypes are identical, the F_1 animals are expected to be homozygous at the loci screened. Mutations resulting in loss of functional gene product are generally expected not to result in an altered phenotype since as a heterozygote a single copy of the functional gene product usually suffices for the wild-type phenotype. For those assays in which the loci screened do not differ genetically between the parental genotypes employed (dominant visibles, dominant skeletal, dominant cataract, fitness, and the alternative sub-set of electrophoretic or histocompatibility loci [Class II]) it is generally expected that only gain-type of mutation events will be detected. The enzyme activity mutation assay is unique in that the method measures directly the gene product and is able to identify both gain and loss mutational variants although the parental strains employed do not differ genetically at the loci screened.

Table 1. Tests Developed to Systematically Screen for Transmitted Mutations in Germ Cells of the Mouse, *Mus musculus*

Test	Reference	Methods/Endpoints
Specific locus	Russell, 1951 Lyon & Morris, 1966 Searle, 1983	F_1, external visible traits
Recessive lethals	Lüning, 1971	F_2 backcross, embryonic lethals
Dominant visibles	Searle, 1974	F_1, external visible traits
Dominant skeletal	Ehling, 1966 Selby & Selby, 1977	F_1, skeletal defects
Dominant cataract	Kratochvilova & Ehling, 1979	F_1, ophthalmological examination for lens opacity
Dominant fitness	Green, 1968	F_2, litter size effects
Electrophoretic variants	Soares, 1979 Johnson & Lewis, 1981 Pretsch *et al.*, 1982, 1986 Peters *et al.*, 1986	F_1, variant electrophoretic pattern of proteins
Enzyme activity	Charles & Pretsch, 1982, 1986	F_1, specific enzyme activity changes
Histocompatibility	Bailey & Kohn, 1965	F_1, skin graft rejection by day 80
Inversions	Roderick, 1971	Anaphase bridges
Translocations	Koller & Auerbach, 1941 Snell, 1935	Chromosome analysis Semi-sterility

SPONTANEOUS MUTATION FREQUENCIES IN GERM CELLS OF THE MOUSE

Table 2 lists the number of loci screened, the number of mutants recovered and the number of genes tested for the various mouse in vivo germ cell mutation tests to estimate the spontaneous mutation rates. It is evident that the information available is heavily biased towards the specific locus methods owing to the fact that the test has been established for more than 40 years and has been conducted in 3 laboratories. The results systematically collected in the production stocks of the Jackson Laboratory (Schlager and Dickie, 1971) also represent a large survey. A control group has not been conducted for the dominant skeletal mutation test and, therefore, an estimate of the spontaneous mutation frequency for this assay is not available. The number of loci tested for the individual test methods is precisely known with the exception of the dominant visible, dominant cataract and histocompatibility Class II methodologies. For the dominant visible results of Schlager and Dickie (1971) the number of loci stated to have been screened was estimated to be 60 (Schlager and Dickie, 1967). In fact, mutations were recovered at 12 loci (*W*, *Sp*, *To*, *Sl*, *Ta*, *Xt*, *Mo*, *Re*, *Bn*, *Hx*, *Rw* and *Lm*), which affect obvious traits such as pigment distribution, fur texture, morphology of extremities and tail, etc. More recently, Lyon (1983) has estimated as many as 100 loci for which mutations to dominant visible alleles would be observed. The number of loci estimated to be screened in the dominant cataract mutation test is based upon human genetic data as outlined by Ehling (1985) and it is at present not known if this is an over- or an underestimate. For the Class II histocompatibility loci screened, the authors have arbitrarily assumed the number of loci screened to be 50.

Table 2. The Frequency of Spontaneous Mutations Observed *in vivo* for Various Genetic Endpoints in the Mouse, *Mus musculus*

Test	Loci	Mutants	Genes Tested[a]	Reference
Specific locus (♂)	7	28	3,720,500	Russell, 1965
"	7	11	1,101,947	Searle, 1974
"	7	19	1,594,635	Ehling *et al.*, 1985
Specific locus (♀)	7	3[b]	1,432,473	Russell, 1977
Recessive visible	5	21	1,976,932	Schlager & Dickie, 1971
Dominant visible	~60	54	not given	"
	?	3	375,224	Searle, 1974
Dominant cataract	~30	1	1,355,640	Ehling *et al.*, 1985
Protein charge	23	0	133,676	Ehling *et al.*, 1985
	32	1	1,289,500	Neel & Lewis, 1990
	4	0	1,452	Peters, unpubl. (cited in Neel & Lewis, 1990)
Enzyme activity	12	0	86,640	Pretsch & Charles, 1984
Histocompatibility				
Class I	30	24	not given	Melvold & Kohn, 1975
Class II	~50	18	not given	"
Inversions	--	0	44	Roderick & Hawes, 1973
Translocations	--	6	16,163	Adler, 1990; Generoso *et al.*, 1985, 1990

[a] The value is derived from the number of animals examined for a locus times the number of loci times the number of gametes in which a mutation could occur. For the specific locus and recessive visible assays, the number of gametes in which a mutation could occur is 1, while for the dominant visible, dominant cataract, protein charge and enzyme activity mutation assays the number of gametes in which a mutation could occur is 2.

[b] A total of 8 mutant animals were recovered, of which a large cluster of 6 mutants were derived in the offspring of a single female parent. The results represent 3 independent mutational events. See Russell (1977), Lyon, Phillips and Fisher (1979) and Ehling and Neuhäuser-Klaus (1988) for a discussion of the rationale to estimate the spontaneous mutation frequency based upon the number of independent mutational events as opposed to the total number of mutants recovered.

The observed spontaneous mutation frequencies (per locus, per gamete) are presented in Table 3 for the various classes of genetic endpoints. The 3 independent results for the specific locus test in male germ cells are similar with mutation frequency estimates which range from 0.75 to 1.2 x 10^{-5} and were combined (0.9 x 10^{-5}). The spontaneous specific locus mutation frequency in female germ cells (0.2 x 10^{-5}) appears to be lower than in male germ cells, which may be a reflection of the difference in the number of cell divisions required to produce a mature gamete in the two sexes (Lyon, 1981). It should also be noted that female germ cells remain for a major portion of their time in the arrested oocyte stage, which is a stationary phase and would allow a long time for DNA repair between the time of occurrence of a DNA lesion and the next round of DNA replication. The remaining mutagenicity assays represent combined results with the possibility of mutation in either the male or female gamete.

Table 3. Spontaneous Mutation Frequencies per Locus per Gamete (x10^{-5}) for Different Classes of Genetic Endpoint in the Mouse, *Mus musculus*

Class	Frequency[a]	95% Conf. Limits[b]	References
Specific locus (♂)	0.9	0.7 - 1.2	Russell, 1979; Searle, 1974; Ehling *et al.*,1985
Specific locus (♀)	0.2	0.06 - 0.6	Russell, 1977
Recessive visible	1.0	0.7 - 1.6	Schlager & Dickie, 1971
Dominant visible	0.008	0.006 - 0.01	"
Dominant cataract	0.07	0.003 - 0.4	Ehling *et al.*, 1985
Electrophoretic	0.07	0.003 - 0.4	Ehling *et al.*, 1985; Neel & Lewis, 1990; Peters, unpubl., cited in Neel & Lewis, 1990
Enzyme Activity	0	0 - 3.8	Pretsch & Charles, 1984
Histocompatibility			
Class I	1.8	1.1 - 2.6	Melvold & Kohn, 1975
Class II	1.0	0.6 - 1.5	"

[a] Since the number of genes tested was not given for the dominant visible and histcompatibility results, spontaneous mutation frequency estimates were taken directly from the references as cited.
[b] Calculated for the expectation of a Poisson variable according to Crow and Gardner (1959).

For dominant mutations, the spontaneous mutation frequency of Schlager and Dickie (1971), 0.008 x 10^{-5} and the results for the dominant cataract mutation test, 0.07 x 10^{-5}, both indicate a per locus per gamete mutation frequency for dominant mutations at least an order of magnitude lower than the mutation frequency to recessive alleles. This discrepancy has been previously discussed (Favor, 1983, 1986, 1990) and may be the result of the inherent stability of the loci screened and more importantly due to the broader spectrum of DNA alterations which would result in a recessive mutant allele as compared to the DNA alterations which result in a dominant mutant allele. The observed mutation frequencies for the histocompatibility test should also be considered. The experimental protocol for Class I loci is similar to the specific locus method and allows the recovery of both gain and loss mutational events. However, the observed mutation frequency is twice that observed for recessive alleles. Further, there is a discrepancy within the Class I histocompatibility loci screened. All mutations recovered at the Class I H-2 locus (major histocompatibility complex) were simultaneous loss and gain events, whereas at the Class I non-H-2 loci the mutational events were exclusively losses.

As expected, at the Class II loci the mutational events recovered were exclusively gains. The Class II histocompatibility protocol is most comparable to the methods to screen for dominant mutational events. A comparison of the Class II histocompatibility mutation frequency to that for dominant alleles indicates the histocompatibility loci to be three times more mutable. Certainly the screening methods are more sensitive to identify mutational alterations of histocompatibility loci than the screening methods to identify dominant visibles or dominant cataract mutations. It is also plausible that there is a selective advantage to a higher instability of the histocompatibility loci resulting in higher antigen diversity, which is reflected in a higher mutation frequency. The estimates of the spontaneous mutation frequencies for electrophoretic variants are most comparable to the mutation frequency to dominant alleles and may reflect the predominance of loci screened at which only alterations of gene product could be detected as opposed to loci at which alterations as well as loss of gene product could be identified.

INDUCED MUTATION FREQUENCIES IN GERM CELLS OF THE MOUSE

Representative mutagenicity results have been compiled for the various genetic endpoints following irradiation with an accumulated total dose of approximately 6 Gy (Table 4) as well as for approximately 250 mg ENU per kg body weight (Table 5). Differences in the number of offspring screened in the various mutagenicity tests reflect, to a large degree, differences in the ease of the screening procedures. For the electrophoretic genetic endpoint, the results of Peters *et al.* (1986) could be distinguished between those loci at which the parental genotypes differed, and only those results have been included. Mutations have been recovered in both radiation and ENU mutagenicity experiments for all genetic endpoints for which an adequate number of offspring was screened. The per locus mutation rates have been calculated, and are presented in Table 6. The highest per locus mutation frequencies are observed for the recessive visible specific locus test followed by the results for electrophoretic and enzyme activity mutations.

These observations are consistent with the fact that all 3 tests represent results for screening systems in which both gain and loss mutational events are detectable. Results for the dominant genetic endpoints indicate consistently lower per locus mutation frequencies, which parallel the observations for spontaneous mutations and have been previously interpreted to be due to the narrower spectrum of mutational events which would result in an observable dominant mutant allele (Favor, 1983, 1989; Kohn, 1979, 1983; Kohn and Melvold, 1976; Lyon and Morris, 1966, 1969; Lyon and Phillips, 1975; Lyon *et al.*, 1964, 1972, 1979; Melvold and Kohn, 1975; Muller, 1950).

The results for the histocompatibility mutation test are the most divergent and may be due to the lower number of offspring which were screened. Alternatively, it has been proposed that newly occurring mutations at histocompatibility loci are often due to gene conversion events (Egorov and Egorov, 1988; Hanson *et al.*, 1984) and this mechanism of mutational alteration may not respond to radiation or ENU treatment to as great an extent as the induction of deletions or base pair substitutions.

Table 4. Observed Mutation Frequencies in Various Mutation Assays Following Irradiation in Mouse Spermatogonia

Test	Loci	Dose (Gy)	Mutants/Offspring	References
Specific locus	7	6	111/ 119,326	Russell, 1965
Dom. visible	~60	6	11/ 35,972	Searle, 1984
Dom. skeletal	?	6	5/ 754	Ehling, 1966
Dom. cataract	~30	6	3/ 11,095	Ehling, *et al.*, 1982
Electrophoretic	4	3+3	3/ 10,000	Peters, unpubl., in Neel & Lewis, 1990
Enzyme activity	12	3+3	1/ 3,388	Charles & Pretsch, 1986
Histocompatibility				
Class I	~30	6.5	0/ 1,285	Kohn & Melvold, 1976
Class II	~50	"	4/ 1,285	Dunn & Kohn, 1981

Table 5. Observed Mutation Frequencies in Various Mutation Assays Following Ethylnitrosourea Treatment in Mouse Spermatogonia

Test	Loci	Dose (Gy)	Mutants/Offspring	References
Specific locus	7	250	35/ 7,584	Russell, 1965
Dom. visible	~60	250	24/ 10,246	Lyon, 1983
Dom. skeletal	?	3x100	10/ 243	Selby & Niemann, 1984
Dom. cataract	~30	250	17/ 9,352	Favor, 1983
Electrophoretic	5-6	250	12/	Peters, Ball & Andrews, 1986
Enzyme activity	12	250	9/ 1,402	Charles & Pretsch, 1987
Histocompatibility				
Class I	not tested			
Class II	~80	250	0/ 2,484	Egorov & Egorov, 1988

Table 6. Per Locus Mutation Rates Observed for Various Genetic Endpoints Following Radiation for Ethylnitrosourea Treatments§

Test	Radiation*	Ethylnitrosourea*
Specific locus	13.3 (10.8-15.6)	65.9 (44.7-89.8)
Dom. visible	0.5 (0.2-0.9)	3.9 (2.4-5.6)
Dom. skeletal	?	?
Dom. cataract	0.9 (0.2-2.4)	6.1 (3.4-9.4)
Electrophoretic	7.5 (2.0-20.3)	53.8 (27.3-91.2)
Enzyme activity	2.5 (01.1-13.3)	53.5 (26.5-99.7)
Histocompatibility		
Class I	0 (0-8.5)	nt
Class II	6.2 (2.1-14.9)	0 (0-1.7)

§ Data listed in Tables 4 and 5.

* Mutation rate per locus x 10^{-5}. The lower and upper 95% confidence limits of the estimates are given in parentheses, and were calculated for the expectation of a Poisson variable according to Crow and Gardner (1959).

LOCUS SPECIFICITY

Following the initial recessive visible specific locus mutation experiments, it was quickly noted that there were differences in the observed mutation rates among the loci studied. Table 7 presents the spectra of mutations recovered in control, irradiation and ethylnitrosourea (ENU) experimental groups. The spontaneous mutation rates appear to be higher at the *b*, *c* and *p* loci, although the total number of mutations recovered is limited. In comparison, following irradiation there is a shift in the spectrum of mutations recovered, the *s* locus exhibiting the highest mutation rate, the *a* and *se* loci had the lowest mutational response, and the *b*, *c*, *d* and *p* loci were intermediate. Employing ENU as a mutagenic agent the spectrum of mutations was much different. The *a*, *se*, and *s* loci exhibited the lowest mutational response, while the *d* and *p* loci had the highest number of mutants. These results imply that the estimations of spontaneous or induced mutation rates are dependent upon the loci studied. Indeed both Lyon and Morris (1966) and Searle (1986) have employed recessive visible specific locus mutation test protocols for which mutations at loci other than those screened by the tester stock of Russell (1951) and the observed mutation rates following radiation differed greatly depending on the loci assayed. The locus specificity of the mutation rate indicates a difference in the inherent stability of the loci in question which may be due to a number of factors including target size, protein-DNA complex structure, whether the loci are normally expressed or not, as well as neighboring genes which if involved in the DNA alteration may render the mutant bearing germ cell or offspring nonviable.

Table 7. Locus Specificity of Spontaneous and Induced Recessive Visible Mutation Rates in the Mouse

	Locus							
	a	b	c	d	se	ps	s	Reference
Control	0	7	3	1	0	4	0	Neuhäuser-Klaus, unpub., cited in Ehling & Favor, 1984
Radiation	2	34	18	25	2	22	71	Russell & Russell, 1959
ENU	10	21	25	37	10	61	9	Ehling & Neuhäuser-Klaus, 1984

This phenomenon of a difference in the inherent stability of loci as well as a difference in the susceptibility to mutation induction at loci may be safely assumed for the various genetic endpoints assayed in different mutation tests in the mouse. Dominant mutations are of special importance since they represent the most relevant genetic endpoints for the human situation. Appropriate experimental data are available for the dominant visible mutations (Table 8).

As for the recessive visible mutations, there are differences in the spontaneous mutation frequencies among the loci at which mutations were recovered as well as a shift in the mutation spectrum following radiation or ENU treatment to increase the mutation rates. As previously indicated, an experienced person would screen for obvious phenotypic variants which could be due to dominant visible mutations at 60 to 100 loci. In fact in the extensive screening program of Schlager and Dickie (1971)

Table 8. Locus Specificity of Spontaneous and Induced Dominant Visible Mutation Rates in the Mouse*

	Locus								
	W	Sp	Mo	Sl	Ta	T	Xt	Re	Other
Control	23	12	8	5	2	1	1	1	4
Radiation	11	2	2	4	8	17	4	2	39
ENU	2	-	-	2	4	2	2	-	10

*Summarized in Lyon, 1983

as well as the results for radiation or ENU treatment, most mutations were recovered at only 12 loci, which would also be consistent with a difference in the inherent mutability of the loci screened.

Table 9 lists and describes the loci at which dominant visible mutations were recovered by Schlager and Dickie, indicates the spontaneous mutation rates as well as the radiation-induced alleles which have been recovered.

Table 9. Loci with Relatively High Spontaneous and Induced Mutation Rates to Dominant Visible Alleles in Germ Cells of the Mouse

Locus	Description	Spontaneous Mutation Rate*	Radiation Induced Alleles
W	Dominant spotting	2.2×10^{-6}	W^{19H}, *Wa*
Sp	Splotch	1.15×10^{-6}	Sp^{r}, Sp^{1H}, Sp^{2H}
Mo	Mottled, X-linked	1.00×10^{-6}	Mo^{dp}
Sl	Steel	0.48×10^{-6}	Sl^{cg}, Sl^{con}
Ta	Tabby, X-linked	0.13×10^{-6}	
Xt	Extra toes	0.07×10^{-6}	Xt^{bph}
Re	Rex	0.07×10^{-6}	Re^{we}
Sey	Small eye	not reported	Sey^{H}
T	Brachyury	not reported	T^{c}, T^{Or}
Bn	Bent tail, X-linked	0.07×10^{-6}	
Hk	Hook (tail)	0.07×10^{-6}	
Hx	Hemimelic extra toes	0.07×10^{-6}	
Rw	Rump white	0.07×10^{-6}	*Rw*
Lm	?	0.07×10^{-6}	

* Reported in Schlager & Dickie, 1971

Information pertaining to the question of mutational locus specificity for dominant visibles may be obtained from the molecular characterization of mutant alleles. Table 10 lists the characterization of spontaneous alleles at those loci which would be screened for phenotypic variants in a dominant visible mutation assay and for which molecular analyses are possible. It is evident that essentially all types of

DNA alterations have lead to dominant visible mutant alleles, ranging from base pair substitutions to insertions, deletions, duplications, rearrangements as well as retroposon insertions. The high frequency of dominant visible mutations shown to be due to gene loss events is unexpected. The phenomenon of haploid insufficiency is known and per definition describes loci at which 2 functional gene copies are required for the normal phenotype.

The results in Table 10 indicate the *Sey*, *Sl*, *T* and *W* loci all to be haploid insufficient. The *Sey* gene contains paired-like and homeobox domains, and functions as a transcriptional regulatory gene during the process of differentiation (Hill *et al.*, 1991; Walther and Gruss, 1991). The *W* locus codes for the *c-kit* proto-oncogene (Geissler *et al.*, 1988), a tyrosine kinase receptor, and interestingly the product of the *Sl* locus is the ligand of the *c-kit* receptor (Flanagan and Leder, 1990; Huang *et al.*, 1990; Williams *et al.*, 1990; Zsebo *et al.*, 1990). The *T* gene is important for the differentiation of the embryonic primitive streak and the subsequent formation of a sufficient mesodermal layer (Bennett, 1975; Frischauf, 1985; Sherman and Wudl, 1977; Silver, 1985). Due to the fact that the loci are haploid insufficient, a wider spectrum of DNA alterations leading to gain or loss of functional gene product are detectable as mutations. This correlates with the higher mutation frequencies observed at the 4 loci. The *Sp* locus has also been shown to have a relatively high mutation frequency to dominant alleles and is likely haploid insufficient. The *Sp* locus contains paired-like and homeobox domains (Epstein *et al.*, 1991) and it my be assumed that the *Sp* gene product functions in a manner similar to the product of the *Sey* locus. Further, radiation-induced *Sp* alleles have been shown to be associated with deletions.

Table 10. Molecular Characterization of Dominant Visible Spontaneous Germ Cell Mutations of the Mouse

Mutation	Mutational event	Reference
Sey^{Dey}	deletion	Glaser *et al.*, 1990
Sey	G to T	Hill *et al.*, 1991
Sl	deletion	Huang *et al.*, 1990
Sl^{d}	deletion (242 bp)	Flanagan *et al.*, 1991
Sl^{J}	deletion	Copeland *et al.*, 1990
Sl^{gb}	deletion	"
W^{x}	rearrangement	Geissler *et al.*, 1988
W^{44}	insertion (4-5 kb)	"
W^{42J}	G to A	Tan *et al.*, 1990
W^{v}	C to T	Nocka *et al.*, 1990
W^{37J}	G to A	"
W^{41J}	G to A	"
W	deletion (234 bp)	"
W^{55J}	C to T	Reith *et al.*, 1990
T	deletion (160-200 kb)	Herrmann *et al.*, 1990
T^{2J}	deletion (80-110 kb)	"
T^{Wis}	retroposon insertion	"
T^{hp}	deletion (very large)	"
t^{Tu3}	duplication	"

CONCLUSIONS

The use of efficient mouse germ cell mutagenicity tests is essential for a characterization of the mutation process in mammals. The recessive visible specific locus test protocol of Russell (1951) is the most sensitive mouse in vivo germ cell mutagenicity assay and is the method of choice to assess in mammalian germ cells the mutagenic activity of various treatments as well as the physical and biological factors which may influence the degree of mutagenic response. Further, genetic and molecular methods are available to characterize the mutational events (see Russell, 1986). However, an extrapolation of results based on mouse mutagenicity assays to humans in an attempt to quantify the expected genetic risks associated with an increased mutation rate following mutagenic response requires that the mouse experimental results accurately represent the mutation rate of the relevant genetic endpoints in humans.

As shown above there is a marked locus specificity in the degree of mutational response which would emphasize the recommendation that the genetic endpoints chosen for study in mouse mutagenicity tests be similar in function to the human genetic endpoints to which the experimental results are to be extrapolated. This necessitates a characterization of both the mouse and human genetic endpoints. Mulvihill and Czeizel (1983) have identified a group of candidate dominant traits appropriate for epidemiological studies to demonstrate an increase in the mutation frequency following mutagenic exposure. Based on the mouse results as reviewed above, those traits controlled by haploid insufficient loci will be the most sensitive dominant traits. As the genetic and molecular characterization of the mouse and human genome progresses we may anticipate the identification of such loci as well as homologous loci in mouse and man. For example, the *Sey* gene of the mouse, which is a haploid insufficient gene, is the mouse homolog of the human aniridia gene (Hill *et al.*, 1991). In this way one may progress to a most direct comparison and extrapolation of the mutational process in mouse and man.

REFERENCES

Adler, I-D. (1990) Clastogenic effects of acrylamide in different germ-cell stages of male mice, in: "Banbury Report 34: Biology of Mammalian Germ Cell Mutagenesis," (J. Allen, B.A. Bridges, M.F. Lyon, M.J. Montrose and L.B. Russell, eds.), Cold Spring Harbor Laboratory Press, Cold Spring Harbor, NY pp. 115-131.

Bailey, D.W., and H.I. Kohn (1965) Inherited histocompatibility changes in progeny of irradiated and unirradiated inbred mice. *Genet. Res.*, Camb., 6, 330-340.

Bennett, D. (1975) The T-locus of the mouse. *Cell* 6, 441-454.

Charles, D.J., and W. Pretsch (1982) Activity measurements of erythrocyte enzymes in mice: Detection of a new class of gene mutations. *Mutation Res.*, 97, 177-178.

Charles, D.J., and W. Pretsch (1986) Enzyme activity mutations detected in mice after paternal fractionated irradiation. *Mutation Res.*, 160, 243-248.

Charles, D.J., and W. Pretsch (1987) Linear dose-response relationship of erythrocyte enzyme-activity mutations in offspring of ethylnitrosourea-treated mice. *Mutation Res.* 176, 81-91.

Copeland, N.G., D.J. Gilbert, B.C. Cho, P.J. Donovan, N.A. Jenkins, D. Cosman, D. Anderson, S.D. Lyman and D.E. Williams (1990) Mast cell growth factor maps near the Steel locus on mouse chromosome 10 and is deleted in a number of Steel alleles. *Cell* 63, 175-183.

Crow, E.L., and R.S. Gardner (1959) Confidence intervals for the expectation of a Poisson variable. *Biometrika* 46, 441-453.

Dunn, R., and H.I. Kohn (1981) Some comparisons between induced and spontaneous mutation rates in mouse sperm and spermatogonia. *Mutation Res.* 80, 159-164.

Egorov, I.K., and O.S. Egorov (1988) Detection of new MHC mutations in mice by skin grafting, tumor transplantation and monoclonal antibodies: A comparison. *Genetics* 118, 287-298.

Ehling, U.H. (1966) Dominant mutations affecting the skeleton in offspring of X-irradiated male mice. *Genetics*, 54, 1381-1389.

Ehling, U.H. (1985) Induction and manifestation of hereditary cataracts, in: "Assessment of Risk from Low-Level Exposure to Radiation and Chemicals" (A.D. Woodhead, C.J. Shellabarger, V. Pond and A. Hollaender, eds.), Plenum, New York, pp. 345-367.

Ehling, U.H., and J. Favor (1984) Recessive and dominant mutations in mice, in: "Mutations, Cancer and Malformation" (E.H.Y. Chu and W.M. Generoso, eds.), Plenum, New York, pp. 389-428.

Ehling, U.H., and A. Neuhäuser-Klaus (1984) Dose-effect relationships of germ-cell mutations in mice, in: "Problems of Threshold in Chemical Mutagenesis," (Y. Kazima, S. Kondo and Y. Kuroda, eds.), Kokusai-bunken, Tokyo, pp. 15-25.

Ehling, U.H., and A. Neuhäuser-Klaus (1988) Induction of specific-locus mutations in female mice by 1-ethyl- 1-nitrosourea and procarbazine. *Mutation Res.* 202, 139-146.

Ehling, U.H., J. Favor, J. Kratochvilova and A. Neuhäuser-Klaus (1982) Dominant cataract mutations and specific-locus mutations in mice induced by radiation or ethylnitrosourea. *Mutation Res.* 92, 181-192.

Ehling, U.H., D.J. Charles, J. Favor, J. Graw, J. Kratochvilova, A. Neuhäuser-Klaus, and W. Pretsch (1985) Induction of gene mutations in mice: The multiple endpoint approach. *Mutation Res.*, 150, 393-401.

Epstein, D.J., M. Vekemans and P. Gros (1991) Splotch (Sp^{2H}), a mutation affecting development of the mouse neural tube, shows a deletion within the paired homeodomain of Pax-3. *Cell* 67, 767-774.

Favor, J. (1983) A comparison of the dominant cataract and recessive specific-locus mutation rates induced by treatment of male mice with ethylnitrosourea. *Mutation Res.*, 110, 367-382.

Favor, J. (1986) A comparison of the mutation rates to dominant and recessive alleles in germ cells of the mouse. *Prog. Clin. Biol. Res.* 209B, 519-526.

Favor, J. (1989) Risk estimation based on germ-cell mutations in animals. *Genome* 31, 844-852.

Favor, J. (1990) Multiple endpoint mutational analysis. *Prog. Clin. Biol. Res.* 340C, 115-124.

Flanagan, J.G., and P. Leder (1990) The kit ligand: A cell surface molecule altered in Steel mutant fibroblasts. *Cell* 63, 185-194.

Flanagan, J.G., D.C. Chan and P. Leder (1991) Transmembrane form of the kit ligand growth factor is determined by alternative splicing and is missing in the Sl^d mutant. *Cell* 64, 1025-1035.

Frischauf, A-M (1985) The T/t complex of the mouse. *Trends Genet.* 1, 100-103.

Geissler, E.N., M.A. Ryan and D.E. Housman (1988) The dominant-white spotting (W) locus of the mouse encodes the c-kit proto-oncogene. *Cell* 55, 185-192.

Generoso, W.M., K.T. Cain, N.L.A. Cacheiro and C.V. Cornett (1985) ^{239}Plutonium-induced heritable translocations in male mice. *Mutation Res.* 152, 49-52.

Generoso, W.M., K.T. Cain, C.V. Cornett, N.L.A. Cacheiro and L.A. Hughes (1990) Concentration-response curves for ethylene-oxide-induced heritable translocations and dominant lethal mutations. *Environ. Mol. Mutagen.* 16, 126-131.

Glaser, T., J. Lane and D. Houseman (1990) A mouse model for the Aniridia-Wilm's Tumor deletion syndrome. *Science* 250, 823-827.

Green, E.L. (1968) Genetic effects of radiation on mammalian populations. *Ann. Rev. Genet.* 2, 87-120.

Green, M.C. (1989) Catalog of mutant genes and polymorphic loci, in: "Genetic Variants and Strains of the Laboratory Mouse, 2nd Edition," (M.F. Lyon and A.G. Searle, eds.), Oxford University Press, Oxford-New York-Tokyo, pp. 12-407.

Hansen, T.H., D.G. Spinella, D.R. Lee, and D.C Shreffer (1984) The immunogenetics of the mouse major histocompatibility complex. *Ann. Rev. Genet.* 18, 99-129.

Herrmann, B.G., S. Labeit, A. Poustka, T.R. King and H. Lehrach (1990) Cloning of the T gene required in mesoderm formation in the mouse. *Nature* 343, 617-622.

Hill, R.E., J. Favor, B.L.M. Hogan, C.C.T. Ton, G.F. Saunders, I.M. Hanson, J. Prosser, T. Jordan, N.D. Hastie and V. van Heyningen (1991) Mouse Small eye results from mutations in a paired-like homeobox-containing gene. *Nature* 354, 522-525.

Huang, E., K. Nocka, D.R. Beier, T-Y. Chu, J. Buck, H-W. Lahm, D. Wellner, P. Leder and P. Besmer (1990) The hematopoietic growth factor KL is encoded by the steel locus and is the ligand of the c-kit receptor, the gene product of the W locus. *Cell* 63, 225-233.

Johnson, F.M., and S.E. Lewis (1981) Electrophoretically detected germinal mutations induced in the mouse by ethylnitrosourea. *Proc. Nat. Acad. Sci.*, USA 78, 3138-3141.

Kohn, H.I. (1979) X-ray mutagenesis: Results with the H-test compared with others and the importance of selection and/or repair. *Genetics* 92, s63-s66.

Kohn, H.I. (1983) Radiation genetics: The mouse's view. *Rad. Res.* 94, 1-9.

Kohn, H.I., and R.W. Melvold (1976) Divergent x-ray-induced mutation rates in the mouse for H and "7-locus" groups of loci. *Nature* 259, 209-210.

Koller, P.C., and C. Auerbach (1941) Chromosome breakage and sterility in the mouse. *Nature* 148, 501-502.

Kratochvilova, J., and U.H. Ehling (1979) Dominant cataract mutations induced by gamma-irradiation of male mice. *Mutation Res.*, 63, 221-223.

Lüning, K.G. (1971) Testing for recessive lethals in mice. *Mutation Res.*, 11, 125-132.

Lyon, M.F. (1981) Sensitivity of various germ-cell stages to environmental mutagens. *Mutation Res.* 87, 323-345.

Lyon, M.F. (1983) Comparison of the dominant visible and other mutation tests in the mouse. INSERM 119, 153-164.

Lyon, M.F., and T. Morris (1966) Mutation rates at a new set of specific loci in the mouse. *Genet. Res.*, Camb., 7, 12-17.

Lyon, M.F., and T. Morris (1969) Gene and chromosome mutation after large fractionated or unfractionated radiation doses to mouse spermatogonia. *Mutation Res.* 8, 191-198.

Lyon, M.F., and R.J.S. Phillips (1975) Specific locus mutation rates after repeated small radiation doses to mouse oocytes. *Mutation Res.* 30, 375-382.

Lyon, M.F., R.J.S. Phillips and A.G. Searle (1964) The overall rates of dominant and recessive lethal and visible mutation induced by spermatogonial X-irradiation of mice. *Genet. Res.* 5, 48-467.

Lyon, M.F., R.J.S. Phillips and H.J. Bailey (1972) Mutagenic effects of repeated small doses to mouse spermatogonia. I. Specific-locus mutation rates. *Mutation Res.* 15, 185-190.

Lyon, M.F., R.J.S. Phillips, and G. Fisher (1979) Dose-response curves for radiation-induced gene mutations in mouse oocytes and their interpretation. *Mutation Res.* 63, 161-173.

Melvold, R.W. and H.I. Kohn (1975) Histocompatibility mutation rates: H-2 and non-H-2. *Mutation Res.* 27, 415-418.

Muller, H.J. (1950) Radiation damage to the genetic material. Part I. Effects manifested mainly in the descendants. *Am. Sci.* 38, 33-59.

Mulvihill, J.J., and A. Czeizel (1983) Perspectives in mutation epidemiology, 6: A 1983 view of sentinel phenotypes. *Mutation Res.* 123, 345-261.

Neel, J.V. and S.E. Lewis (1990) The comparative radiation genetics of humans and mice. *Ann. Rev. Genet.* 24, 327-362.

Nocka, K., J.C. Tan, E. Chiu, T.Y. Chu, P. Ray, P. Traktman and P. Besmer (1990) Molecular bases of dominant negative and loss of function mutations at the murine c-kit/white spotting locus: W^{37}, W^{v}, W^{41} and *W*. *EMBO J.* 9, 1805-1813.

Peters, J., S.T. Ball, and S.J. Andrews (1986) The detection of gene mutations by electrophoresis, and their analysis. *Prog. Clin. Biol. Res.*, 209B, 367-374.

Pretsch, W. (1986) Protein-charge mutations in mice. *Prog. Clin. Biol.* Res., 209B, 383-388.

Pretsch, W. and D.J. Charles (1984) Detection of dominant enzyme mutants in mice: model studies for mutations in man, in: "Monitoring Human Exposure to Carcinogenic and Mutagenic Agents: IARC Scientific Publications No. 59," (A. Berlin, M. Draper, K. Hemminki, and H. Vainio, eds.), pp. 361-369.

Pretsch, W., D.J. Charles and K.R. Narayanan (1982) The agar contact replica technique after isoelectric focusing as a screening method for the detection of enzyme variants. *Electrophoresis* 3, 142-145.

Reith, A.D., R. Rottapel, E. Giddens, C. Brady, L. Forrester and A. Bernstein (1990) W mutant mice with mild or severe developmental defects contain distinct point mutations in the kinase domain of the c-kit receptor. *Genes Dev.* 4, 390-400.

Roderick, T.H. (1971) Producing and detecting paracentric chromosomal inversions in mice. *Mutation Res.* 11, 59-69.

Roderick, T.H., and N.L. Hawes (1973) Nineteen paracentric chromosomal inversions in mice. *Genetics* 76, 109-117.

Russell, W.L. (1951) X-ray-induced mutations in mice. Cold Spring Harbor Symposia on Quantitative Biology, Vol. XVI. pp. 327-356.

Russell, W.L. (1965) The nature of the dose-rate effect of radiation on mutation in mice. *Jap. J. Genet.* 40, 128-140.

Russell, W.L. (1977) Mutation frequencies in female mice and the estimation of genetic hazards of radiation in women. *Proc. Nat. Acad. Science* USA 74, 3523-3527.

Russell, L.B. (1986) Information from specific-locus mutants on the nature of induced and spontaneous mutations in the mouse. *Prog. Clin. Biol. Res.* 209B, 437-447.

Russell, W.L., and L.B. Russell (1959) The genetic and phenotypic characterization of radiation induced mutations in mice. *Rad. Res.* Suppl. 1, 296-305.

Russell, L.B., P.B. Selby, E. von Halle, W. Sheridan and L. Valcovic (1981) The mouse specific-locus test with agents other than radiations: Interpretation of data and recommendations for future work. *Mutation Res.* 86, 329-354.

Russell, W.L., E.M. Kelly, P.R. Hunsicker, J.W. Bangham, S.C. Maddux and E.L. Phipps (1979) Specific-locus test shows ethylnitrosourea to be the most potent mutagen in the mouse. *Proc. Nat. Acad. Sci.*, USA 76, 5818-5819.

Schlager, G., and M.M. Dickie (1967) Spontaneous mutations and mutation rates in the house mouse. *Genetics* 57, 319-330.

Schlager, G. and M.M. Dickie (1971) Natural mutation rates in the house mouse: Estimates for five specific loci and dominant mutations. *Mutation Res.* 11, 89-96.

Searle, A.G. (1974) Mutation induction in mice, in: "Advances in Radiation Biology, Vol. 4," (J.T. Lett, H.I. Adler, and M. Zelle, eds.), Academic Press, New York. pp. 131-207.

Searle, A.G. (1983) Some ideas on future test systems, in: Utilization of Mammalian Specific Locus Studies, in: Hazard Evaluation and Estimation of Genetic Risk, (F.J. de Serres and W. Sheridan, eds.), Plenum, New York, pp. 279-288.

Searle, A.G. (1986) The role of dominant visibles in mutagenicity testing. *Prog. Clin. Biol. Res.* 209B, 511-518.

Selby, P.B. (1981) Radiation Genetics, in: The Mouse in Biomedical Research, Vol. 1., (H.L. Foster, J.D. Small, and J.G. Fox, eds.), Academic Press, New York, pp. 263-283.

Selby, P.B., and S.L. Niemann (1984) Non-breeding methods for dominant skeletal mutations shown by ethylnitrosourea to be easily applicable to offspring examined in specific-locus experiments. *Mutation Res.* 127, 93-105.

Selby, P.B., and P.R. Selby (1977) Gamma-ray-induced dominant mutations that cause skeletal abnormalities in mice. I. Plan, summary of results and discussion. *Mutation Res.* 43, 357-375.

Sherman, M.I. and L.R. Wudl (1977) T-complex mutations and their effects. In: "Concepts in Mammalian Embryogenesis" (M.I. Sherman, Ed.), MIT Press, Cambridge, Mass., pp. 136-234.

Silver, L.M. (1985) Mouse t haplotypes. *Ann. Rev. Genet.* 19, 179-208.

Snell, G.D. (1934) The production of translocations and mutations in mice by means of X-rays. *Am. Naturalist* 68, 178.

Snell, G.D. (1935) The induction by X-rays of hereditary changes in mice. *Genetics* 20, 545-567.

Soares, E.R. (1979) TEM-induced gene mutations at enzyme loci in the mouse. *Environ. Mut.* 1, 19-25.

Tan, J.C., K. Nocka, P. Ray, P. Traaktman and P. Besmer (1990) The dominant W^{42} spotting phenotype results from a missense mutation in the c-kit receptor kinase. *Science* 247, 209-212.

Walther, C., and P. Gruss (1991) *Pax-6*, a murine paired box gene, is expressed in the developing CNS. *Devel.* 113, 1435-1449.

Williams, D.E., J. Eisenman, A. Baird, C. Rauch, K. van Ness, K.J. March, L.S. Park, U. Martin, D.Y. Mochizuki, H.S. Boswell, G.S. Burgess, D. Cosman and S.D. Lyman (1990) Identification of a ligand for the c-kit proto-oncogene. *Cell* 63, 167-174.

Zsebo, K.M., D.A. Williams, E.N. Geissler, V.C. Broudy, F.H. Martin, H.L. Atkins, R-Y. Hsu, N.C. Birkett, K.H. Okino, D.C. Murdock, F.W. Jacobsen, K.E. Langley, K.A. Smith, T. Takeishi, B.M. Cattanach, S.J. Galli and S.V. Suggs (1990) Stem cell factor is encoded at the Sl locus of the mouse and is the ligand for the c-kit tyrosine kinase receptor. *Cell* 63, 213-224.

EFFECTS OF SPERMATOGENIC CELL TYPE ON QUANTITY AND QUALITY OF MUTATIONS

Liane B. Russell

Biology Division
Oak Ridge National Laboratory
Oak Ridge, TN 37831-8077

INTRODUCTION

It is probable that most instances of male-mediated developmental toxicity will be traceable to damage induced in male germ cells. With respect to the action of environmental agents, germ cells obviously constitute a specialized system. Their response mechanisms are very likely to be different from those of cells used for other mutagenicity test systems; and, since the organism at risk is not the father but his offspring, not only the induction but the transmission of a genetic lesion must be considered.

Not only must germ cells as a whole be considered a specialized response system, but the heterogeneity *within* the germ-cell populations of both sexes must be explored. Differences between the responses of male and female germ cells have been well demonstrated for radiations; for chemical agents, the paucity of data on females hinders comparative analyses (L.B. Russell and W.L. Russell, 1992). In the male, considerable evidence has been amassed that indicates major differences among the spermatogenic cell types with regard both to the quantitative yield of mutations and the structural nature of the mutations. The results of earlier analyses are summarized here.

The length of time required for specific spermatogenic cells to develop and appear in the ejaculate have been delineated in classical studies (Oakberg, 1984). Thus, offspring that trace to paternal exposures during specific spermatogenic stages can readily be identified on the basis of the interval between exposure and conception. Offspring conceived in the first seven post-treatment weeks are, successively, derived from exposed spermatozoa, late spermatids, early spermatids, diplotene and late pachytene spermatocytes, early pachytene and leptotene spermatocytes, late differentiating spermatogonia, and early differentiating spermatogonia. Starting with the eighth post-treatment week, all offspring are derived from exposed spermatogonial stem cells.

Male-Mediated Developmental Toxicity, Edited by D.R. Mattison and A.F. Olshan, Plenum Press, New York, 1994

Both the quantitative and structural mutation data that will be used to illustrate response differences between spermatogenic stages are derived by means of the mouse morphological specific-locus test, SLT (W.L. Russell, 1951). This efficient and reliable test has been used for decades to measure frequencies of heritable mutations induced in germ cells of various types by radiations and chemicals. In recent years, molecular-genetics advances have facilitated the structural characterizations of specific-locus mutations, adding outstanding new qualitative capabilities to the SLT's quantitative ones (L.B. Russell, 1991). Though the SLT's indicators of mutation -- specific changes in coat color and pattern and in ear length associated with 7 marker loci -- are deceptively simple, such mutations have now facilitated molecular entry to most of the marker loci and their surrounding regions, which are becoming characterized in greater molecular detail than most of the mouse genome (Rinchik and Russell, 1990).

DIFFERENTIAL GERM-CELL-STAGE EFFECTS AS REVEALED BY PRODUCTIVITY DATA

Differential responses of the various spermatogenic stages are revealed even by the productivity data that are obtained as ancillary results in SLT studies. Reductions in average litter size are generally indicative of the induction of dominant-lethal mutations (i. e., of chromosome breakage in germ cells that results in the early death of the conceptus). Dominant-lethal induction is often verified by independent studies. When reductions in total numbers of offspring are due primarily to depressions in numbers of litters (as opposed to offspring per litter), a cytotoxic effect on germ cells is indicated. Thus, information of several types--additional to mutation-rate data --is revealed in the course of a SLT.

Different patterns of effect are found with different chemicals. For example, acrylamide monomer (L.B. Russell *et al.*, 1991) (typical of several other chemicals) produces reductions in the size of litters conceived week-1 and week-2 postexposure, indicating dominant-lethal induction in spermatozoa and late spermatids; while treatment with 6-mercaptopurine (L.B. Russell and Hunsicker, 1987) leads to a sharp decline in the size of litters conceived 32-38 days later, suggesting (and confirmed by independent findings) that dominant lethals are induced in preleptotene spermatocytes. A complex productivity pattern such as that following exposure to chlorambucil is illustrated in Figure 1 (a similar pattern is obtained with melphalan.). Dominant-lethal induction peaks sharply in week 3 (exposure to early spermatids), but occurs also in week 1 (exposure to spermatozoa). The reduction in offspring during weeks 5 and 6, which is largely due to reduction in number of litters, results from the killing of differentiating spermatogonia. Killing of spermatogonial stem cells, as would be indicated by productivity reductions subsequent to week 7, is observed following higher doses of chlorambucil, as well as of several other chemicals and of radiations. Such doses are avoided in the conduct of SLT studies because they might result in spermatogonial cell selection which would complicate the interpretation of mutation rates.

QUANTITATIVE YIELD OF SPECIFIC-LOCUS MUTATIONS AS A FUNCTION OF EXPOSED GERM-CELL STAGE

In analyzing specific-locus mutation yields for different spermatogenic stages, the first broad comparison that can be made is between spermatogonial stem cells and poststem-cell stages as a whole. Table 1 lists all agents for which studies were

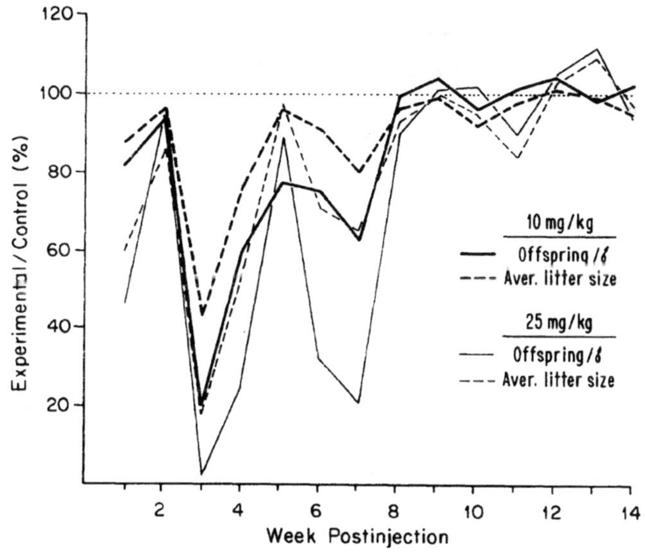

Figure 1. Number of offspring per male and average size of litters sired in successive weeks after exposure to chlorambucil. Each value is expressed as a percentage of the corresponding number in concurrent controls (From L.B. Russell *et al.*, 1989).

conducted with *both* sets of stages. Only 6 agents have to date been found to be clearly mutagenic in stem-cell spermatogonia, and for 4 of these, the mutation yield from stem cells is smaller than that from a subsequent spermatogenic stage.

Further, the poststem-cell population is not a homogeneous one with respect to specific-locus-mutation yield; major differences are found among these stages, with the pattern dependent on the agent. Two examples are illustrated in Table 2. Following exposure to chlorambucil, CHL, a peak in mutation yield is observed in

Table 1. Specific-Locus-Test Outcomes for Agents that have been Investigated in Both Stem-Cell Spermatogonia and Poststem-cell Stages[a]

Agent[b]	Stem-cell Spermatogonia	Relation	Poststem-cell stages
Radiation[c]	+	<	+
ENU	+	≈	+
MNU	+	<	+
PRC	+	≈	+
TEM	+	<	+
MLP	+	<	+
CHL	+?		+
AA	-		+
CPP	-		+
DES	-		+
EMS	-		+
6MP	-		-
ADR	-		-
PLA	-		-
UR	-		-

[a] Modified from L.B. Russell *et al.*, 1990. + and - indicate positive and negative results as determined by Genetox criteria (L.B. Russell, *et al.*, 1981).
[b] Abbreviations as follows: ENU, ethylnitrosourea; MNU, methylnitrosourea; PRC, procarbazine hydrochloride; TEM, triethylenemelamine; MLP, melphalan; CHL, chlorambucil; AA, acrylamide monomer; CPP, cyclophosphamide; DES, diethyl sulfate; EMS, ethyl methane sulfonate; 6MP, 6-mercapto purine; ADR, adriamycin; PLA, platinol; UR, urethan. For references to original publications, see L.B. Russell *et al.*, 1990.
[c] W.L. Russell *et al.*, 1958.

offspring conceived during week 3, deriving from exposed early spermatids. Similar peak periods are found with melphalan and with radiation. By contrast, following exposure to methyl-nitrosourea, MNU, a sharp peak in mutation yield is seen in offspring conceived during week 6, deriving from exposed Type-B spermatogonia or preleptotene spermatocytes. A similar pattern for postspermatogonial stages is observed with ethylnitrosourea, ENU; however, ENU is considerably more mutagenic than MNU in stem-cell spermatogonia.

Table 2. The Germ-Cell Stage Yielding Maximum Mutational Response Varies with the Chemical: Comparison of Patterns Produced by Chlorambucil (CHL) and Methylnitrosourea (MNU)[a]

	Treatment			
Post injection week	10 mg CHL/kg		75 mg MNU/kg	
	Offspring	Mutation rate per locus (x 10^{-5})	Offspring	Mutation rate per locus (x 10^{-5})
1	4921	11.6	1868	
2	5879	4.9	1092	
3	1660	77.5	1210	4.2
4	4815	8.9	2319	
5	3191		3787	
6	2841	1.6	2827	91.0
7	2915		2933	4.9

[a]Data for CHL from L.B. Russell *et al.*,1989; data for MNU from W.L. Russell, unpublished.

Figure 2 summarizes the three patterns of specific-locus mutation yield that have been observed to date. Patterns 2 and 3 are the ones illustrated by examples in Table 2. The largest numbers of chemicals tested to date fall into Pattern 1, where peak yield is obtained following exposure of spermatozoa and late spermatids.

To summarize the quantitative information derived from specific-locus tests: it is clear that specific-locus-mutation yield, as well as dominant-lethal incidence (reflected in the productivity patterns derived as ancillary information) are highly dependent on the spermatogenic stage that is exposed. The pattern of peak yields depends on the nature of the mutagenic agent; to date, three different patterns have emerged for postspermatogonial stages.

STRUCTURAL NATURE OF MUTATIONS AS A FUNCTION OF EXPOSED GERM-CELL STAGE

Direct Characterization of Specific-Locus Mutations

Owing to the design of the SLT, mutations--rather than being merely counted--can be propagated in breeding stocks for subsequent analysis by whatever techniques have become available. Extensive genetic studies (using such stocks) have provided framework maps for the regions surrounding several of the SLT markers, and these have constituted highly favorable resources for more detailed molecular characterizations. Thus, the past products of SLT studies, i.e., the mutations generated over the years, have considerably facilitated the development of high-density structural information for several regions of the mouse genome.

Reciprocally, the new information generated by the intensive molecular mapping is making possible the structural characterizations of the products of mutagenesis (i.e., newly induced specific-locus mutations or old mutations that are being propagated in breeding stocks or have been preserved in frozen embryos). As a result, the SLT is now a favorable test not only for measuring mutation rates but for analyzing structural

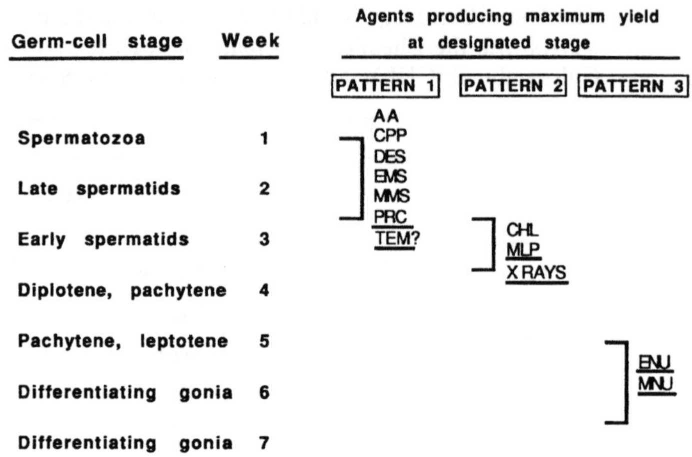

Figure 2. Germ-cell stage sensitivity patterns for agents that produce a positive response in poststem-cell stages of spermatogenesis. Agents are listed opposite the stage(s) that yield maximum frequencies of specific-locus mutations. Those agents that also elicit a positive response in stem-cell spermatogonia are underlined. For abbreviations, see Table 1. For references to original publications, see L.B. Russell et al., 1990, from which this figure is modified.

changes induced in indigenous genes of a mammal. Some of the steps here outlined are discussed in slightly greater detail below.

The SLT yields multiple independent mutations of, or encompassing, a given locus. Many of the radiation-induced mutations have turned out to be deletions that involve not only the marker locus but different extents of the surrounding regions of the chromosome. The first step in the genetic analysis defines the limits of each deletion relative to standard genetic markers that may be available in the region. Subsequently, intercrosses of the various mutations of a region and the study of resulting phenotypes define segments of overlap and non-overlap between the deletions and generate information about the normal functions of gene(s) residing in different segments of the region (e.g., L.B. Russell *et al.*, 1982).

Derivation of a structural map builds on the information of the genetic map. Molecular entry to the region may be obtained in various ways, one of which is to map cloned DNA sequences (from enriched libraries or human homology regions) first to one of the largest deletions of a region, and then to sub-regions defined on the basis of the previously derived genetic map. Subsequently, by means of chromosome walking, long-range mapping techniques, and the identification of fusion fragments (i.e., fragments that encompass DNA flanking both ends of a deletion), which permit direct jumps from one end of a deletion to the other, a region becomes quickly populated with molecular tools for further high-intensity mapping. Some of the marker loci themselves have been cloned and partially sequenced. Thus, the regions surrounding the SLT markers are already among the most highly characterized regions of the mouse genome (Rinchik and Russell 1990), and further characterization is progressing at a rapid rate (e.g., Klebig *et al.*, 1992).

Because the products of SLT mutagenesis experiments have thus led (and are leading) to the development of high-density structural information for several genomic regions, investigators are now in an increasingly favorable position to avail themselves of the other half of the reciprocal relation--to structurally characterize the products of mutagenesis. The simplest application is to use DNA probes from the characterized regions to determine (by means of Southern blots) whether a mutation involves a deletion (Rinchik *et al.*, 1990; L.B. Russell *et al.*, 1992). To date, this has been done for mutations at any of three of the SLT markers, but the probes are now available to expand the studies to three additional markers. Also, until recently, we have used only one or two probes at each tested locus to determine whether the gene is deleted or not; but the molecular tools already exist (and are rapidly multiplying) for characterizing the breakpoints of the deletions with some accuracy.

Retroactive Classification of Specific-Locus Mutations

While the newly available molecular tools provide a direct approach to the qualitative analysis of specific-locus mutations being (and to be) generated, it is clearly desirable to obtain some measure of information from the results of the many SLTs that were carried out in the past, by other laboratories as well as our own. Fortunately, these data can now be *retroactively* analyzed -- at least crudely -- with regard to lesion size. The criteria for this retrospective analysis are based on the information gained from complementation maps and molecular data about the phenotype of the null condition for the markers.

The null condition occurs when overlapping complementing deletions, in combination, totally ablate the marker (i.e., it is not present in either of the two chromosomes) without removing neighboring genes that might affect viability or other properties of the animal. The phenotype for the null condition at 6 of the 7 loci has been determined (Rinchik and Russell, 1990). At five of these loci, the null combination is fully viable, while at one it is a juvenile lethal; at all six, the nulls resemble, in color or ear length, animals that are homozygous either for the marker allele or for a more extreme one.

If a mutation is found that has a phenotype *additional* to that characteristic of the null, e.g., if homozygotes are prenatally lethal or have some morphological or physiological defect, it can be concluded that additional genetic material has been deleted, a condition defined as a large lesion, LL. Another classification is also possible. Since it is known that the nulls resemble, in color or ear length, animals that are homozygous for the marker allele or a more extreme one, it may be concluded that any mutation which produces a color or ear length that is intermediate between those characteristic of the null and the wild type is the result of an *intra*genic change, IG. Mutations that cannot be retroactively classified as either LL or IG are designated "other" lesions, OL. The criteria for the three types are summarized in Table 3.

The IG/LL/OL classification can thus be made for 6 of the 7 loci from published records that provide information on the homozygous lethality/viability of the

Table 3. Criteria for Retrospective Structural Classification of Specific-Locus Mutations

Expression		Designation
Marker phenotype	Additional phenotypes	
Null	Yes	LL
Null	No	OL
Altered	No	IG

mutations and on whether intermediate phenotypes were observed. (The former information is more often provided than the latter.) It should be noted that retrospective classification yields minimum estimates for the IG and LL categories; more refined analysis of an OL mutation would result in its reclassification as either IG or LL. An example of such direct analysis of OL mutations that arose in the past, but are still being propagated in breeding stocks, is described in a subsequent section.

Structural Comparisons of Mutations Induced in Different Germ-Cell Stages

Direct methods (genetic, molecular, and cytogenetic) for characterizing the structure of mutations have been applied in recent chemical mutagenesis experiments. Of particular interest are the large mutation sets derived from two very powerful mutagens, CHL and MLP. Retrospective classification was used to analyze data accumulated earlier for other chemicals and for radiation. The frequencies of LLs in each of these sets of data are summarized in Table 4 (note: only low-LET, i.e., X- or gamma-, radiation data are included here), both for spermatogonia and for postspermatogonial stages.

Table 4. Frequency of LLs Among Mutations Induced in Spermatogonia or in Postspermatogonial Stages

Mutagen	Type of analysis[a]	Mutations induced in:			
		Spermatogonia[b]		Postspermatogonial stages	
		No.	%LL	No.	%LL
CHL and MLP	Direct[c]	11	9.1	29	82.3
Other chemicals[d]	Retrospective	320	3.8	45	60.0
Low-LET radiation	Retrospective[e]	102[f]	20.6	25	48.0

[a] Data for all seven loci are included, except in the case of the radiation data, which are for three loci only.
[b] Includes differentiating spermatogonia where these were scored (CHL, MLP, and some of the other chemicals).
[c] About one-half of the mutations were analyzed with molecular probes (Rinchik *et al.*, 1990; L.B. Russell *et al.*, 1992).
[d] Exposure to both spermatogonia and postspermatogonial stages involved the following chemicals: ethylnitrosourea, methylnitrosourea, procarbazine hydrochloride, triethylenemelamine. Postspermatogonial stages were also exposed to methyl methane sulfonate, ethyl methane sulfonate, cyclophosphamide, diethyl sulfate, and acrylamide monomer; and spermatogonia were also exposed to mitomycin-C. For references to original publications, see L. B. Russell et al, 1990.
[e] Data from L. B. Russell, 1971; L. B. Russell *et al.*, 1979.
[f] Does not include mutations induced in "sensitized" spermatogonia, for which see Table 5.

For each of the three data sets, it is clear that exposure of spermatogonia (stem-cell or differentiating) yields relatively few LLs, while LLs are common products of exposure of postspermatogonial stages. It should be recalled that in the retrospectively analyzed data sets, the LL category represents a minimum estimate, because direct analysis would have reclassified many of the mutations presently contained in the "default" OL category. Consistent with this argument, the postspermatogonial frequency of LLs is greatest in the case of the CHL and MLP mutations, about one-half of which were analyzed with molecular probes.

A molecular analysis of mutations that had earlier been classified only by retrospective criteria was recently carried out for one data set, namely radiation-induced mutations involving the *c* (albino, or tyrosinase) locus (Rinchik *et al.*, 1992). Of 31 mutations analyzed with molecular probes for the region, 28 were viable albinos, i.e., null with respect to the marker phenotype, and with no additional phenotypes (therefore, OL by the criteria summarized in Table 3); the remainder were mutations to intermediate alleles, and thus presumed to be IG mutations. DNA deletions were detectable in 14 of the 31 mutations. Four of these 14 were found to break in the tyrosinase gene itself, and their sizes ranged from <36 kb to about 480 kb. Some of the viable deletions extended through flanking proviral loci, suggesting that they could be at least 2000 kb in length. As predicted from the retrospective criteria, none of the three intermediate alleles analyzed (presumed IGs) was found to be associated with a detectable deletion.

Structural Comparisons of Mutations Induced in Different Types of Stem-Cell Spermatogonia

The reclassification of a large set of *c*-locus OL mutations as either LL or IG has added more precision to structural comparisons that may be made for mutations induced in different types of cells. We have recently examined the structural nature of mutations induced in "sensitized" spermatogonia, a term applied to spermatogonia that received a second dose of X rays 24 hours following a preliminary dose (L.B. Russell and Rinchik, in press). Table 5 presents the distribution of such mutations, along with those for irradiated normal stem-cell spermatogonia and poststem-cell stages. There is a significant difference in the frequency of LLs not only between normal stem-cell spermatogonia and postspermatogonial stages (P=0.002), but also between "sensitized" and normal stem-cell spermatogonia (P=0.016). The distribution for "sensitized" spermatogonia approaches that for postspermatogonial stages.

Table 5. Distribution of Lesion Sizes Among Mutations Induced by Low-LET Radiations in Male Germ Cells

Germ-cell stage	No. of mutations	Distribution of mutational lesions[a] (in percent)		
		IG	OL	LL
Spontaneous	48	20.8	75.0	4.2
Normal stem-cell spermatogonia	103	20.4	58.3	21.3
"Sensitized" stem-cell spermatogonia	30	6.7	50.0	43.3
Poststem-cell stages	25	4.0	44.0	52.0

[a] IG=intragenic lesion; LL=large lesion; OL=other lesion (see text).

It was first shown by W.L. Russell (1962) that a 24-hour fractionation regime elevates the mutation rate considerably above the linear extrapolation made from the dose-response curve, and it was postulated that the first dose induces changes in the composition of the spermatogonial population that lead to a predominance of mutagenically-sensitive cells 24 hours later. As discussed in more detail by L.B. Russell and Rinchik (in press), the structural analyses of mutations support the conclusions, (a) that the "extra" mutations that are produced when the second fraction of the dose encounters an altered cell population are primarily large DNA lesions, and

(b) that the nuclear state of "sensitized" stem-cell spermatogonia may be different from the state of those normal stem cells that respond to a single dose of radiation. If some special nuclear state(s) is indeed conducive to the formation of LLs this state may be common to postspermatogonial stages and "sensitized" spermatogonial stem cells.

It has sometimes been suggested that the low frequency of LLs that generally characterizes mutations induced in spermatogonia is the result of selection against LL mutations during the course of subsequent germ-cell development. However, the finding, here summarized, of structural differences between mutations induced in two different populations of spermatogonial stem cells argues against this explanation. Thus, LLs can readily be derived from "sensitized" spermatogonia, from which they would have to face the same selection problems as do LLs induced in normal spermatogonial stem cells. Another argument against the selection hypothesis is that, with few exceptions, LLs that are induced in postspermatogonial stages, once they are transmitted to offspring, have no problems getting through spermatogenesis of the F_1 or later generations (W.L. Russell, 1964). The findings for "sensitized" spermatogonia also argue against the possibility that diploidy might be a protection against LL induction.

Rather, it appears that some special chromosomal state(s) that is found in postspermatogonial stages, "sensitized" spermatogonia, and oocytes (L.B. Russell and W.L. Russell, 1992) is conducive to the formation of LLs. Alternatively, or in addition, it may be that these cell types lack an LL-repair system that is present in normal spermatogonial stem cells.

CONCLUSIONS AND IMPLICATIONS FOR HUMAN HEALTH

Clearly, major differences exist among various types of male germ cells in their responses to mutagens. The differential mutation frequency is agent-specific, i.e., different agents produce peak yields in different germ-cell stages (with three patterns identified to date). This is probably a function of the ways in which different chemicals interact with different chromatin constituents. An even more intriguing problem is posed by the finding that the structural nature of mutations appears to be governed primarily by the type of germ cell in which they are induced, with the same agent producing different mutational products in different germ-cell types. Elucidating the underlying mechanisms for this phenomenon is a major future challenge in mutagenesis research.

The findings have implications for human health effects. Thus, if, on the basis of epidemiological data, an agent is suspected of being a male-mediated environmental toxicant, it would seem advisable to determine experimentally not only that this agent is a germ-cell mutagen, but also what type of differential mutation yield it produces in the various germ-cell stages. Such information would help to define what segments of the human study population are likely to yield the most informative results, and/or could suggest procedures for preventing or limiting future damage (e.g., if mutation yield turned out to be at a peak following exposure of one or more postspermatogonial stages, men could be advised to refrain from procreation for limited periods after acute or semi-acute exposures, a procedure earlier recommended for radiation--W.L. Russell, 1956). Information about the structural nature of mutational lesions induced in germ cells will eventually (after more research has been done on the relation between arrays of lesion sizes and phenotypes) allow one to estimate the likelihood that genetic effects will be dominant (i.e., that damage will be found in an exposed male's immediate offspring), and may even point to certain classes of likely phenotypes.

ACKNOWLEDGEMENTS

Research jointly sponsored by the Office of Health and Environmental Research, U.S. Department of Energy under contract DE-AC05-84OR21400 with Martin Marietta Energy Systems, Inc., and by the National Institute of Environmental Health Sciences under IAG No. 222Y01-ES-10067.

REFERENCES

Klebig, M.L., Russell, L.B., and Rinchik, E.M., 1992, Murine fumarylacetoacetate hydrolase (*Fah*) gene is disrupted by a neonatally lethal albino deletion that defines the hepatocyte-specific developmental regulation-1 (*hsdr-1*) locus, *Proc. Natl. Acad. Sci. USA*, 89:1363-1367.

Oakberg, E.F., 1984, Germ cell toxicity: significance in genetic and fertility effects of radiation and chemicals, *Environ. Sci. Res.*, 31, 549-590.

Rinchik, E.M. and Russell, L.B., 1990, Germ-Line Deletion Mutations in the Mouse: Tools for Intensive Functional and Physical Mapping of Regions for the Mammalian Genome, *in:* Genome Analysis, K. Davies and S. Tilghman, eds., Cold Spring Harbor Laboratory Press, N. Y., Vol. I, Genetic and Physical Mapping, 121-158.

Rinchik, E.M., Stoye, J.P., Frankel, W.N., Coffin, J., Kwon, B.S., and Russell, L.B., 1992, Molecular analysis of viable spontaneous and radiation-induced albino (*c*)-locus mutations in the mouse, *Mutat. Res.*, in press.

Rinchik, E.M., Bangham, J.W., Hunsicker, P.R., Cacheiro, N.L.A., Kwon, B.S., Jackson, I.J., and Russell, L.B., 1990, Genetic and Molecular Analysis of Chlorambucil-Induced Germline Mutations in the Mouse, *Proc. Natl. Acad. Sci., USA,* 87:1416-1420.

Russell, L.B., 1971, Definition of functional units in a small chromosomal segment of the mouse and its use in interpreting the nature of radiation-induced mutations, *Mutat. Res.* 11:107-123.

Russell, L.B., 1991, Factors that affect the molecular nature of germ-line mutations recovered in the mouse specific-locus test, *Environ. Molec. Mutagen.*, 18:298-302.

Russell, L.B., and Hunsicker, P.R., 1987, Study of the base analog 6 - mercaptopurine in the mouse specific-locus test, *Mutat. Res.* 176:47-52.

Russell, L.B. and Russell, W.L., 1992, Frequency and nature of specific-locus mutations induced in female mice by radiations and chemicals: a review, *Mutat. Res.*, in press.

Russell, L.B., and Rinchik, E.M., (1993), Structural differences between specific-locus mutations induced in two different populations of spermatogonial stem cells of the mouse, *Mutat. Res.* 288:187-195.

Russell, L.B., Russell, W.L., and Kelly, E.M., 1979, Analysis of the albino-locus region of the mouse. I. Origin and viability, *Genet.* 91:127-139.

Russell, L.B., Selby, P.B., von Halle, E., Sheridan, W., and Valcovic, L., 1981, The mouse specific-locus test with agents other than radiation: Interpretation of data and recommendations for future work, *Mutat. Res.* 86:329-354.

Russell, L.B., Montgomery, C.S., and Raymer, G.D., 1982, Analysis of the albino-locus region of the mouse: IV. Characterization of 34 deficiencies, *Genet.* 100:427-453.

Russell, L.B., Hunsicker, P.R., Cacheiro, N.L.A., Bangham, J.W., Russell, W.L., and Shelby, M.D., 1989, Chlorambucil effectively induces deletion mutations in mouse germ cells, *Proc. Natl. Acad. Sci., USA,* 86:3704-3708.

Russell, L.B., Russell, W.L., Rinchik, E.M., and Hunsicker, P.R., 1990, Factors Affecting the Nature of Induced Mutations, in: *Banbury Report 34: Biology of Mammalian Germ Cell Mutagenesis*, Cold Spring Harbor Laboratory Press, pp 271-289.

Russell, L.B., and Hunsicker, P.R., Cacheiro, N.L.A., and Generoso, W.M., 1991, Induction of specific-locus mutations in male germ cells of the mouse by acrylamide monomer, *Mutat. Res.*, 262:101-107.

Russell, L.B., Hunsicker, P.R., Cacheiro, N.L.A., and Rinchik, E.M., 1992, Genetic, cytogenetic, and molecular analyses of mutations induced by melphalan demonstrate high frequencies of heritable deletions and other rearrangements from exposure of postspermatogonial stages of the mouse, *Proc. Natl. Acad. Sci. USA*, 89:6182-6186.

Russell, W.L., 1951, X-ray-induced mutations in mice. *Cold Spring Harbor Symposia on Quant. Biol.* 16:317-336.

Russell, W.L., 1956, Genetic effects of radiation in mice and their bearing on the estimation of human hazards. *in:* "Peaceful Uses of Atomic Energy," Proc. Internatl. Conf., Geneva, August 1955, Vol. 11. United Nations, New York, pp. 382-383, 401-402.

Russell, W.L., 1962, An augmenting effect of dose fractionation on radiation-induced mutation rate in mice, *Proc. Natl. Acad. Sci. USA*, 48:1724-1727.

Russell, W.L., 1964, Evidence from mice concerning the nature of the mutation process. *in:* "Genetics Today," Proc. Intern. Congr. Genet. XI, The Hague, The Netherlands, 1963, ed., S.J. Geerts, Pergamon Press (Oxford), Vol. 2, 257-264.

Russell, W.L., Russell, L.B., Gower, J.S., and Sheppard, C.W., 1953, Neutron-induced dominant lethals in the mouse, *Genet.* 38:688.

DOMINANT MUTATIONS IN MICE

Udo H. Ehling
GSF - Institut für Säugetiergenetik, Neuherberg
D-85758 Oberschleissheim
Germany

INTRODUCTION

Methods for the detection of mutations in mammals have been discussed by Hertwig (1932; 1935), Snell (1935), Catcheside (1947), and Falconer (1949). The pioneering studies of Hertwig (1935), Brenneke (1937), and Schaefer (1939) had already indicated that the litters sired during the pre-sterile period of irradiated mice were of reduced size. Since there was no effect on sperm mobility and since the number of fertilized eggs was normal, it was concluded that the reduced litter size was due to death of embryos after fertilization. The observation of various nuclear and chromosomal abnormalities in fertilized ova led to the conclusion that embryonic death was caused by chromosomal abnormalities, induced by irradiation of spermatozoa. These earlier studies on dominant lethal mutations and translocations by Hertwig, Snell and Strandskov have been reviewed by Russell (1954).

An effort to investigate dominant mutations was conducted during a period extending from the late 1943 until about 1950, at the University of Rochester School of Medicine, under the auspices of the wartime Manhattan Project and, later, its successor, the U. S. Atomic Energy Commission. It was concluded that the incidence of mortality, rare morphological anomalies, visible mutations and mutations affecting fertility taken together are definitely increased by radiation at the rate of at least 1.16×10^{-4} per R (Charles *et al.*, 1960).

Dominant mutations can be measured by comparing first generation descendants from treated and untreated populations, but, for many characteristics, it is difficult to distinguish between the effects of newly occurring genetic changes and the within-strain variations. These problems have been solved for dominant mutations affecting the skeleton and the lens of the mouse. Systematic investigations of induced skeletal mutations were initiated by Ehling in 1959 (Ehling and Randolph, 1962) and for dominant cataract mutations by Ehling in 1977 (Kratochvilova and Ehling, 1979). Detailed accounts of these methods and the results obtained in these studies were published by Ehling and Favor (1984) and Selby (1990).

Male-Mediated Developmental Toxicity, Edited by D.R. Mattison
and A.F. Olshan, Plenum Press, New York, 1994

DOMINANT SKELETAL MUTATIONS

Dominant skeletal mutation experiments represent the earliest attempt at systematically screening for mutations affecting one body system. In the initial experiments, phenotypic variants of cleared and stained skeletons of the F_1 were classified according to whether the variant phenotype was unique or occurred more than once. The variants were further subdivided into those that were unilateral or bilateral, and whether additional skeletal variations were simultaneously present. A class of phenotypic variants identified as highly likely to be mutations represented variants occurring only once in the experiments (Ehling, 1966). Three dominant mutations affecting the skeleton were found by Ehling (1970) to be transmitted to the second and third generation. One of these mutations, a disproportionate micromelia (*Dmm*), was described in detail by Brown *et al.* (1981).

In a subsequent experiment the inheritance of presumed mutations was tested by Selby and Selby (1977). F_1 offspring derived from irradiated males were allowed to produce an F_2 generation before skeletons were prepared and classified by the criteria developed by Ehling (1966). Selby and Selby (1977) confirmed the conclusion of Ehling (1966) that the presumed mutations are true mutations. Later, Selby and Niemann (1984) developed an indicator method for the detection of skeletal variants. An excess of variants in the experimental group over the control group can be an indication of a mutational component, but it cannot be used for the determination of the mutation rate and, therefore, cannot be used for the estimation of the genetic risk due to radiation or chemical mutagens (Ehling, 1984a).

The difficulty with the skeletal system is in proving that the variant is due to a mutation. To test the transmission of the variant, F_2 offspring must be obtained before the skeletal preparation of the F_1 generation. Because this procedure is time- and space-consuming, we developed another method by screening for dominant cataracts. When compared with screening for dominant skeletal mutations, screening for phenotypic variants of the lens has the advantage of a more rapid examination that can be performed on living animals.

DOMINANT CATARACT MUTATIONS IN MICE

A cataract is an opacity of the lens causing a reduction of visual function. The organogenesis of the lens is similar in various mammals. Therefore, a gene that disturbs the same process in the normal development of the lens in different species allows the same cataract type to be seen in these different species. Ehling (1964) pointed out that morphologically comparable cataracts in humans and other mammals frequently have the same mode of inheritance. The comparability of the genetic endpoint in mice and man is one advantage of this system. A detailed description of radiation-induced dominant cataracts in mice (Kratochvilova, 1981; Kratochvilova and Favor, 1988), drawings (Kratochvilova, 1981) and photographs of these dominant lens-opacity mutants (Ehling, 1980; Favor and Pretsch, 1990; Kratochvilova, 1981) were published. Similar types of cataracts were described in man (Waardenburg *et al.*, 1961). A genetic homology between the human X-linked cataract-dental syndrome, the Nance-Horan Syndrome, and the mouse X-linked cataract was reported by Favor and Pretsch (1990).

The systematic investigation of induced dominant cataracts in mice was initiated by Ehling in 1977 and subsequently reported in a series of papers (Ehling, 1980; Ehling, 1983; Ehling, 1984b).

The detection of dominant cataract mutations can be combined with the scoring of specific-locus mutations (Ehling *et al.*, 1982). This combination is the basis for the multiple endpoint approach.

THE MULTIPLE ENDPOINT APPROACH

The problem with specific-locus experiments is that only one class of mutations (recessive visibles) can be detected. This disadvantage can be overcome by combining the scoring of specific-locus alleles at 7 loci with the screening of approximately 30 loci coding for dominant cataract mutations (Ehling, 1985), 23 loci controlling protein-charge changes (Pretsch *et al.*, 1982), and 12 loci for enzyme-activity alterations (Charles and Pretsch, 1987). The multiple endpoint approach was developed in Neuherberg (Ehling *et al.* 1982, 1985). It allows the comparison of induced mutation rates to recessive visible alleles and to dominant cataract alleles, as well as mutation rates to protein-charge or enzyme-activity alterations in the same animals.

Experiments to screen for the approximately 70 loci in the same offspring of treated male mice were performed with ethylnitrosourea (ENU), procarbazine, and X-ray exposure. Mutations were recovered for each genetic endpoint in all treatment groups where a sufficient number of offspring was scored. The mutations were confirmed by breeding tests (Table 1).

Table 1. Mutation Frequencies of Different Genetic Endpoints Following Spermatogonial Treatment of (101/ElxC3H/El)F_1-hybrid male mice*

Treatment	Specific-locus mutations[a]	Dominant-cataract mutations	Protein-charge mutations	Enzyme-activity mutations
Historical control	19(13)/277805	1/22594	0/5812	0/3610
ENU				
160 mg/kg	35(32)/8658	14/6435	nt	nt
	52(50)/13018	nt	1/1892	14/3093
250 mg/kg	64(58)/9766	17/9352	3/4254	4(3)[b]/505
Procarbazine				
600 mg/kg	4/13071	1/12056	1/7506[c]	1/2974
X-Irradiation	18/10054	3(2)[b]/8889	nt	nt
3 + 3 Gy	11/8085	0/6662	0/1003	1/3388

[a]Number of independent mutational events in parentheses.
[b]Presumed cluster (2 mutations were recovered in the same litter and express the same phenotype. However, a test for allelism has not yet been conducted).
[c]Includes results of the present multiple endpoint experiment (0/5725) as well as results of an independent experiment (1/1781).
nt = not tested.
*Data from Ehling *et al.*, 1985.

The advantage of a combined investigation is at least threefold:

1st The number of scorable mutations is increased in comparison with a simple specific locus experiment by a factor of ten.

2nd The combined investigation allows the comparison of the mutation frequency of selected and unselected loci.

3rd In the same experiment the frequency of mutations with dominant and recessive modes of inheritance can be compared.

QUANTIFICATION OF GENETIC RISK

In using the mice data to arrive at quantitative estimates of the genetic risks from radiation and environmental mutagens for humans, the following assumptions have been made: (*a*) The various biological and exposure factors affect the magnitude of the induced mutation frequency in a similar way and to a similar extent in mice and in humans (Ehling, 1991). (*b*) There is a non-thresholded linear or linear quadratic dose-response. In the simplest case, response depends linearly on dose: $y-y_o=bD$, where y is the frequency of observed mutations per gamete, y_o is the background or spontaneous frequency of mutations per gamete, D is the dose and b is the slope of the dose-response line (Ehling *et al.*, 1983).

Risk estimation is based on the results of induced mutations in stem-cell spermatogonia of mice (BEIR Report, 1990, UNSCEAR Report, 1982). The reason is that only A_s(stem-cell) spermatogonia continue to divide throughout reproductive life. Cells that go through meiotic divisions and through successive differentiation stages are known as spermatocytes, spermatids, and spermatozoa. These postspermatogonial germ-cell stages have only a transitory existence, approximately five weeks in mice (Oakberg, 1956) and almost twice as long in man (Heller and Clermont, 1964).

Following the advice of EPA (1986) there are two main approaches in making genetic risk estimates. One of these, termed the direct method, expresses risks in terms of expected frequencies of genetic changes induced per unit dose, the other, referred to as the doubling dose method or the indirect method, expresses risks in relation to the observed incidence of genetic disorders now present in man.

Indirect or Doubling Dose Method

This approach uses data mainly from human genetics. Only the doubling dose is determined in animal experiments. For the quantification of the genetic risk of chemical mutagens we have to take into account the differences of the metabolism of mice and man. Quantification is only necessary for compounds where the benefit is more important than the risk. Only few mutagens are belonging to this group of chemicals. The doubling doses of chemical mutagens in mice are given in Table 2.

The doubling dose is defined as the dose necessary to induce as many mutations as occur spontaneously in one generation. The determination of the doubling dose is based exclusively on results obtained in our laboratory, except for triethylenemelamine, which is based on results obtained by Cattanach (1966, 1982).

The doubling dose is a useful indicator for the evaluation of the potential human hazard for a given exposure. The ratio between the individual dose and the doubling dose can be used to determine the individual risk. The doubling dose for a male patient for procarbazine is 110 mg/kg body weight, the therapeutic dose is 215 mg/kg. From this comparison one would conclude that the procarbazine treatment of a patient with Hodgkin's disease would induce two times as many mutations as arise spontaneously. For a female patient the risk is four times the spontaneous frequency. The information about the individual risk is the basis for human genetic counseling.

More important, the doubling dose can be used for the estimation of the population risk. The ratio between the population dose and the doubling dose gives an indication of the genetic risk of the exposed population.

However, in the latest UNSCEAR-Report (1988) and BEIR-Report (1990) the doubling dose method is mainly used to estimate the population effect of dominant

Table 2. Doubling Doses of Chemical Mutagens for Specific-Locus Mutations in Mice*

Compound	Germ cell stage affected	Doubling dose (mg/kg)
Cyclophosphamide	spermatozoa	6
Diethyl sulfate	spermatozoa	17
Cisplatin	spermatids	2
Ethyl methanesulfonate	spermatids	9
Methyl methanesulfonate	spermatids	2
Busulfan	spermatids	2
Chlormethine	spermatids	0.1
Ethylnitrosourea	spermatogonia	4
Isoniazid	spermatogonia	219
Mitomycin C	spermatogonia	1
Procarbazine	spermatogonia	110
Triethylenemelamine	spermatogonia	0.3
Ethylnitrosourea	oocytes	7
Procarbazine	oocytes	53
Triethylenemelamine	oocytes	0.2

* Data from Ehling, 1989.

mutations. The rational of this procedure is very questionable, because the doubling dose is estimated for recessive specific locus mutations in male mice. The per locus ratio of radiation-induced dominant to recessive mutations in spermatogonia of the mouse is approximately 1:10 (Ehling, 1985). The estimated radiation doubling doses for dominant cataract alleles ranged from 0.96 to 2.14 Gy with a weighted mean of 1.57 Gy (Favor, 1989). The mean doubling dose for dominant mutations under equal radiation condition in the mouse is 5 times higher than for recessive mutations. Since the estimation for the doubling dose of dominant cataracts are based on a single mutation in the controls, they have wide confidence limits (Ehling, 1991).

A detailed analysis of the problems of using the indirect or doubling dose method for the estimation of the population risk was recently published by Ehling (1991). To overcome these difficulties we developed the concept of the direct estimation of the genetic risk for dominant mutations.

Direct method

For the quantification with the direct method, we used the following equation by Ehling (1987):

Expected number of induced dominant mutations in man	=	Induced mutation frequency in mice	x	Multiplication factor for the overall dominant mutation frequency in man	x	Genetically significant dose in man

Using this equation one can calculate the number of dominant mutations expected to be manifested in the offspring of survivors of the atomic bombings in Hiroshima and Nagasaki (Table 3).

Table 3. Direct Estimation of the Genetic Risk in 19,000 Offspring After Parental Exposure in Hiroshima and Nagasaki Based on Dominant Cataract Mutations in Mice*

	(1984)	(1991)
Mutations/Gamete/Gy	0.45 - 0.55 x 10^{-4}	
Multiplication factor for the overall dominant mutation rate	41(a)	47(c)
Sv (genetically significant population dose)	1.1 x 10^{4}(b)	0.9 x 10^{4}(d)
Total number of expected dominant mutations	20 - 25	19 - 23

(a) McKusick (1983); (b) Oftedal (1984), (c) McKusick (1988); (d) Neel *et al.* (1988)
*Data from Ehling, 1984, 1991.

The induced mutation rate of dominant cataracts for single exposure with a dose rate of 0.5 Gy/min is 0.45-0.55x10^{-4} mutations/gamete/Gy. The mutation rates are based on two independent experiments.

The change in the multiplication factor reflects our increasing knowledge about dominant genes in man. According to McKusick (1983) the ratio of all dominant mutations to dominant cataracts was 934 to 23. In 1988 this ratio according to McKusick is 1443 to 31.

The genetically significant dose of the survivors of the atomic bombings at Hiroshima and Nagasaki was taken from an estimation of Oftedal (1984). Since then the dosimetry was reevaluated (Neel *et al.*, 1988). This leads to a reduction in the estimated neutron yield coupled with a less marked increase, in parts of the cities, in the estimated gamma yield. In addition, the estimates of the shielding from radiation provided by the roof tiles of Japanese homes and by concrete walls have been revised, as well as the amount of radiation reaching the gonads.

These estimations can be compared with the data of Schull, Otake and Neel (1981) for four indicators of genetic effects from studies of children born to survivors of atomic bombings at Hiroshima and Nagasaki. These indicators are:

- frequency of untoward pregnancy outcomes (stillbirth, major congenital defects, death during first postnatal week);
- occurrence of death in live children through an average life expectancy of 17 years;
- frequency of children with sex chromosome aneuploidy; and
- frequency of children with mutation resulting in an electrophoretic variant.

In a recent publication Neel and coworkers (1990) extended their observation and included four more indicators:

- malignancies in the F_1;
- frequency of balanced structural rearrangements of chromosomes;
- sex ratio among children of exposed mothers;
- growth and development of the F_1.

Using these indicators one would expect to observe only a fraction of the 20 dominant mutations predicted by our equation. Therefore, the conclusion of the authors, "in no instance is there a statistically significant effect of parental exposure; but for all indicators the observed effect is in the direction suggested by the hypothesis that genetic damage resulted from the exposure," is in agreement with the direct estimation of genetic risk for children of Hiroshima and Nagasaki (Schull *et al.*, 1981) based on mouse experiments. The important message is not the actual number of our estimations, but the estimation of the order of magnitude.

The same concept can be used for the quantification of chemically induced mutations (Ehling, 1988, 1989; Ehling and Neuhäuser, 1979).

CONCLUSION

From a practical point of view, at present the direct method for the estimation of the first generation effects, supplemented by the doubling dose method for human genetic counseling are the best-founded concepts for estimating risk. Regulatory decisions concerning risk to the individual or the population at large must be based on one of these concepts.

To improve our estimates, we cannot rely on an overall doubling dose. Different doubling doses must be determined for distinct genetic endpoints. Furthermore, a better experimental data base must be compiled for the direct estimation of genetic risk. Our model systems for risk estimation of chromosomal diseases need improving and experimental approaches must be developed to estimate risk for the complex constitutional and degenerative diseases. Refinements in mutation recognition at the DNA level should permit the analysis of new mutational endpoints in Hiroshima and Nagasaki. This analysis must be combined with controlled experimentation in animals . Last, but not least, estimation of the incidence of genetic disorders in man must be improved.

We have the knowledge to protect the human genome against mutagenic agents, but our resources, especially for animal experimentation, are insufficient. The hazard from chemical mutagens and the interaction of ionizing radiation and chemical mutagens are probably more important for the genetic health of future generations than the hazard from radiation alone.

ACKNOWLEDGMENT

Experiments were supported by contract BI-6-0156-D(B) and EV5V-CT-91-0012(MNLA) by the Commissions of the European Communities.

REFERENCES

BEIR Report (Biological Effects of Ionizing Radiations), 1990, Health Effects of Exposure to Low Levels of Ionizing Radiation, BEIR V., National Academy Press, Washington, D.C.

Brennecke, H., 1937, Strahlenschäden von Mäuse- und Rattensperma, beobachtet an der Frühentwicklung der Eier, *Strahlentherapie*, 60:214-238.

Brown, K. S., Cranley, R. E., Greene, R., Kleinmann, H. K., and Pennypacker, J. P., 1981, Disproportionate micromelia (*Dmm*): An incomplete dominant mouse dwarfism with abnormal cartilage matrix, *J. Embryol. exp. Morph.*, 62:165-182.

Catcheside, D.G., 1947, Genetic effects of radiations, *Br. J. Radiol. 1 (Suppl)*:109-116.

Cattanach, B.M., 1966 , Chemically induced mutations in mice, *Mutation Res.*, 3:346-353.

Cattanach, B.M., 1982 , Induction of specific-locus mutations in female mice by triethylenemelamine (TEM), *Mutation Res.* 104:173-176.

Charles, D.R., Tihen, J.A., Otis, E.M., and Grobman, A.B., 1960, Genetic effects of chronic X-irradiation exposure in mice, UR-565 AEC Research Development Report, Univ. Rochester Atomic Energy Project, Rochester, NY.
Charles, D. J., and Pretsch, W., 1987, Linear dose-response relationship of erythrocyte enzyme-activity mutations in offspring of ethylnitrosourea-treated mice, *Mutation Res.* 176:81-91.
Ehling, U., 1964, Vererbung von Augenleiden im Tierreich, in: "Bericht über die" 65. Zusammenkunft der Deutschen Ophthalmologischen Gesellschaft in Heidelberg 1963," ed. W. Jaeger, pp. 228-238, Bergmann, München.
Ehling, U.H., 1966, Dominant mutations affecting the skeleton in offspring of X-irradiated male mice, *Genetics* 54:1381-1389.
Ehling, U.H., 1970, Evaluation of presumed dominant skeletal mutations, in "Chemical Mutagenesis in Mammals and Man," ed. F. Vogel, G. Röhrborn, pp. 162-166, Springer, Berlin.
Ehling, U.H., 1980, Comparison of the mutagenic effect of chemicals and ionizing radiation in germ cells of the mouse, in "Progress in Environmental Mutagenesis," ed. M. Alacevic, pp. 47-58, Elsevier/North Holland Biomedical Press, Amsterdam.
Ehling, U.H., 1983, Cataracts - Indicators for dominant mutations in mice and man, in "Utilization of Mammalian Specific Locus Studies in Hazard Evaluation and Estimation of Genetic Risk," ed. F.J. de Serres, W. Sheridan, pp. 169-190, Plenum, New York.
Ehling, U.H., 1984 a, Variants and mutants, *Mutation Res.* 127:189-190.
Ehling, U.H., 1984 b, Methods to estimate the genetic risk, in: "Mutations in Man," ed. G. Obe, pp. 292-318, Springer, Berlin.
Ehling, U.H., 1985, Induction and manifestation of hereditary cataracts, in: "Assessment of Risk from Low-Level Exposure to Radiation and Chemicals," eds. A.D. Woodhead, C.J. Shellabarger, V. Pond and A. Hollaender, pp. 345-367, Plenum Publishing Corporate, New York.
Ehling, U.H., 1987, Quantifizierung des strahlengenetischen Risikos, *Strahlentherapie und Onkologie,* 163:283-291.
Ehling, U.H., 1988, Quantification of the genetic risk of environmental mutagens, *Risk Anal.,* 8:45-57.
Ehling, U.H., 1989, Quantifizierung des chemogenetischen Risikos, *Naturwissenschaften* 76: 194-199.
Ehling, U. H., 1991, Genetic risk assessment, *Annual Review of Genetics,* 25:255-280.
Ehling, U.H., and Favor, J., 1984, Recessive and dominant mutations in mice, in: "Mutation, Cancer, and Malformation," ed. H.E.Y. Chu and W.M. Generoso, pp. 389-428, Plenum, New York.
Ehling, U.H., and Neuhäuser, A., 1979, Procarbazine induced specific locus mutations in male mice, *Mutation Res.* 59:245-256.
Ehling, U.H., and Randolph, M.L., 1962, Skeletal abnormalities in the F_1 generation of mice exposed to ionizing radiations, *Genetics* 47:1543-1555.
Ehling, U.H., Favor, J., Kratochvilova, J., and Neuhäuser-Klaus, A., 1982, Dominant cataract mutations in mice induced by radiation or ethylnitrosourea, *Mutation Res.* 92:181-192.
Ehling, U.H., Averbeck, D., Cerutti, P.A., Friedman, J., Greim, H., Kolbye, Jr., and Mendelsohn, M.L., 1983, Review of the evidence for the presence or absence of thresholds in the induction of genetic effects by genotoxic chemicals, *Mutation Res.* 123:281-341.
Ehling, U.H., Charles, D.J., Favor, J., Graw, J., Kratochvilova, J., Neuhäuser-Klaus, A., and Pretsch., W., 1985, Induction of gene mutations in mice: The multiple endpoint approach, *Mutation Res.* 150:393-401.
EPA (Environmental Protection Agency), 1986, Guidelines for Mutagenicity Risk Assessment, *Fed. Regist.,* 51 (185):34006-34012.
Falconer, D.S., 1949, The estimation of mutation rates from incompletely tested gametes, and the detection of mutations in mammals, *J. Genet.* 49:226-234.
Favor, J., 1989, Risk estimation based on germ-cell mutations in animals, *Genome* 31:844-852.
Favor, J., and Pretsch, W., 1990, Genetic localization and phenotypic expression of X-linked cataract *(Xcat)* in *Mus musculus, Genet. Res.* 56:157-162.
Heller, C. G., and Clermont, Y., 1964, Kinetics of the germinal epithelium in man, in "Recent Progress in Hormone Research," ed. G.Pincus, 20:545-575, Academic, New York.
Hertwig, P., 1932, Wie muß man züchten, um bei Säugetieren die natürliche oder experimentelle Mutationsrate festzustellen? *Arch. Rassen. Gesellschaftsbiolog.* 27:1-12.
Hertwig, P., 1935, Sterilitätserscheinungen bei röntgenbestrahlten Mäusen, *Z. indukt. Abstamm. Vererbungsl.* 70:517-523.
Kratochvilova, J., 1981, Dominant cataract mutations detected in offspring of gamma-irradiated male mice, *J. Hered.* 72:302-307.
Kratochvilova, J., and Ehling, U.H., 1979, Dominant cataract mutations induced by γ-irradiation of male mice, *Mutation Res.* 63:221-223.

Kratochvilova, J., and Favor, J., 1988, Phenotypic characterization and genetic analysis of twenty dominant cataract mutations detected in offspring of irradiated male mice, *Genet. Res.* 52:125-134.
McKusick, V.A., 1983, "Mendelian Inheritance in Man," *The Johns Hopkins Univ. Press*, Baltimore, London, 6th ed.
McKusick, V.A., 1988, "Medelian Inheritance in Man," *The Johns Hopkins Univ. Press,* Baltimore, London, 8th ed.
Neel, J.V., Satoh, C., Goriki, K., Asakawa, J., Fujita, M.,Takeshi K., and Ryuji H., 1988, Search for mutations altering protein charge and/or function in children of atomic bomb survivors: Final report, *Am. J. Hum. Genet.* 42:663-676.
Neel, J.V., Schull, W.J., Awa, A. A., Satoh, C., Kato H., Otake, M., and Yoshimoto, Y., 1990, The children of parents exposed to atomic bombs: Estimates of the genetic doubling dose of radiation for humans, *Am. J. Hum. Genet.* 46:1053-1072.
Oakberg, E.F., 1956, Duration of spermatogenesis in the mouse and timing of stages of the cycle of the seminiferous epithelium, *Am. J. Anat.* 99:507-516.
Oftedal, P., 1984, Genetic damage following nuclear war, in "Effects of Nuclear War on Health and Health Services," pp. 163-174, *WHO,* Geneva.
Pretsch, W., Charles, D.J., and Narayanan, K.R., 1982, The agar contact replica technique after isoelectric focusing as a screening method for the detection of enzyme variants, *Electrophoresis* 3:142-145,
Russell, W.L., 1954, Genetic effects of radiation in mammals, in "Radiation Biology," Vol.1, ed., Hollaender, A., pp. 825-859, McGraw-Hill, New York.
Schaefer, H., 1939, Die Fertilität von Mäusemännchen nach Bestrahlung mit 200 r, *Z. Mikrosk. Anat. Forsch*. 46:121-152.
Schull, W.J., Otake, M., and Neel, J.V., 1981, Genetic effects of the atomic bombs: A reappraisal, *Science* 213:1220-1227, Quotation p. 1220.
Selby, P.B., 1990, Experimental induction of dominant mutations in mammals by ionizing radiations and chemicals, *Issues Rev. Teratol.* 5:181-253.
Selby, P.B., and Niemann, S.L., 1984, Non-breeding-test methods for dominant skeletal mutations shown by ethylnitrosourea to be easily applicable to offspring examined in specific-locus experiments, *Mutation Res. 127:93-105.*
Selby, P.B., and Selby, P.R., 1977, Gamma-ray-induced dominant mutations that cause skeletal abnormalities in mice. I. Plan, summary of results and discussion. *Mutation Res*. 43:357-375.
Snell, G.D., 1935, The induction by X-rays of hereditary changes in mice, *Genetics* 20:545-567.
Strandskov, H.H., 1932, Effect of X-rays in an inbred strain of guinea-pigs, *J. Exptl. Zool.* 63:175-202.
UNSCEAR Report (United Nations Scientific Committee on the Effects of Atomic Radiation), 1982, Ionizing Radiation: Sources and Biological Effects.
UNSCEAR Report (United Nations Scientific Committee on the Effects of Atomic Radiation), 1988, Sources, Effects and Risks of Ionizing Radiation.
Waardenburg, P.J., Franceschetti, A., Klein, D., 1961, "Genetics and ophthalmology," Vol. 1. Assen: Van Gorcum.

ANEUPLOIDY TESTS: CYTOGENETIC ANALYSES OF MAMMALIAN MALE GERM CELLS*

James W. Allen[1], Barbara W. Collins[1],
Ronald E. Cannon[1], Pamela W. McGregor[2],
Arash Afshari[2], and James C. Fuscoe[2]

[1]Genetic Toxicology Division
Health Effects Research Laboratory
U.S. Environmental Protection Agency
Research Triangle Park, NC 27711
[2]Environmental Health Research and Testing, Inc.
P.O. Box 12199
Research Triangle Park, NC 27709

INTRODUCTION

Human gametes are frequently aneuploid. It is estimated that approximately 2% of sperm, 18% of eggs and 20% of conceptuses have extra or missing chromosomes (Burgoyne *et al.*, 1991). This condition of genetic imbalance accounts for much of the embryonic loss during development, and for various disabling syndromes in liveborn. Paternal contributions to the aneuploidy load are less than those which are maternally-mediated; nevertheless, they are substantial. Roughly 6% of individuals with Trisomy 21, the most common aneuploidy condition, are thought to have inherited the extra chromosome from their father (Sherman *et al.*, 1991). In regard to sex chromosome aneuploidies, all of the XYY trisomies, most of the XO monosomies, and half of the XXY trisomies are considered to be of paternal origin (reviewed by May *et al.*, 1990).

The high levels of aneuploidy in gametes are generally attributed to malsegregation of meiotic chromosomes during cell division. Such errors might arise through physiological changes influenced by genetic background or aging processes. However, the view is also widely held that exposures to certain environmental agents pose

* This document has been reviewed in accordance with United States Environmental Protection Agency policy and approved for publication. Mention of trade names or commercial products does not constitute endorsement or recommendation for use.

Male-Mediated Developmental Toxicity, Edited by D.R. Mattison
and A.F. Olshan, Plenum Press, New York, 1994

significant risks for induction of aneuploidy. Toxic effects directly on chromosomes or spindle structures, or on physiological factors governing chromosomal behavior in meiosis, could be implicated in an aneuploidy outcome. Some potential targets are unique to meiosis, eg. chromosomal structures and metabolic activities functioning in homologue synapsis and recombination, while others are common to all dividing cells, eg. microtubules and associated proteins. Lesions in post-meiotic cells may also mediate chromosome malsegregation at the first cleavage division, although this pathway has not been well studied.

To our knowledge, no environmental factor has been clearly shown to induce heritable aneuploidy in human reproductive cells. However, a number of chemical and physical agents have consistently tested positive in mammalian and non-mammalian test systems. Uncertainties regarding the extent of public health hazard associated with exposure to various environmental chemicals has led to several reviews of the major aneuploidy test systems and data (Dellarco *et al.*, 1985; Parry and Parry, 1987), as well as those more specifically concerned with mammalian male germ cells (Russell and Shelby, 1985; Russell, 1985; Allen *et al.*, 1986). Some general conclusions were that: 1) existing assays need to be standardized; 2) new assays need to be developed; and 3) relatively little data on the induction of aneuploidy in well-defined tests exist; extensive chemical testing is needed for risk assessment. The Commission of the European Communities (CEC) has established research programs to develop and validate an appropriate battery of aneuploidy tests (Parry and Parry, 1987).

With regard to male germ cell assays, the analysis of hyperploidy at metaphase II in spermatocytes or at the first cleavage division following fertilization, and genetic analysis of offspring following parental treatment, eg. the numerical sex chromosome anomalies test (Russell, 1985), were considered promising for further development and/or applications. Metaphase II hyperploidy analysis has been widely used for chemical and radiation testing in rodents, and some recent results are summarized below. Chromosome counts in pronuclei have been assessed in the human sperm-hamster egg fusion test (Martin *et al.*, 1989; Martin and Rademaker, 1990; Brandriff and Gordon, 1990), and have provided direct information on aneuploidy levels in sperm of normal and irradiated individuals. However, there has been little work using this method of analysis after treatment of experimental animals. The data available from genetic analyses of offspring after paternal treatments are also quite limited. Autosomal aneuploidies in mice are not generally compatible with survival to term; these analyses are based upon genetic complementation tests or phenotypic expression of X-linked genetic markers (Russell, 1985; Parry and Parry, 1987). Principles of these and other tests used to measure aneuploidy in male germ cells are discussed in the aforementioned reviews.

The present discussion of aneuploidy tests in male germ cells is intended as a survey to update progress in certain research areas concerning assay development and chemical test results. As this brief account must be selective, it is limited to cytogenetic and molecular cytogenetic analyses of meiotic and post-meiotic germ cells. Recently, these areas of investigation have been especially active, resulting in test refinement and applications as well as in the development of new technical assay procedures. Specifically, a brief review of some results from metaphase II hyperploidy tests, and discussions of the potential usefulness of spermatid micronucleus and fluorescence in situ hybridization (FISH) methods in aneuploidy testing are provided below.

ANEUPLOIDY TESTS

Treatments and Mechanisms

Aneuploidy testing is complicated by the diversity of targets and mechanisms by which environmental agents may cause chromosome distribution processes to fail. It is important that aneuploidy tests accommodate a variety of treatment times since potential mechanisms of chemical interaction with complex meiotic "machinery" are typically not very well understood. Exposures close to the time of meiotic cell divisions, for example, may interfere with spindle formation, or with motor molecules (Hoffman, 1992) which effect chromosome movement along the spindle. A number of agents are known to interfere with different components or specific metabolic steps in the developing spindle apparatus (Liang and Brinkley, 1985), and many of these substances have tested positive in diverse aneuploidy test systems (Waters *et al.*, 1985).

Exposures of pre-meiotic or early meiotic cells may result in disruption of homologue pairing/synapsis, and inhibition of crossing-over. This genetic exchange process is believed to regulate disjunction through achieving proper chromosome alignment and chiasma formation; irregular synapsis/recombination may result in abnormalities of chromosome segregation. Recent studies of individuals with Trisomy 21 (Warren *et al.*, 1987) and 47,XXY (Hassold *et al.*, 1991) conditions support the view that aberrant meiotic recombination frequencies are implicated in aneuploidy. DNA polymorphism analyses of the nondisjoined chromosomes revealed that, in many cases, frequencies of meiotic recombination in parental germ cells appeared reduced; indeed, nondisjoined X,Y chromosomes often showed no evidence that crossing-over (expected in the pseudoautosomal region) had occurred. Studies of synaptonemal complex formation in mutagen-treated rodents have revealed that homologue synaptic processes are subject to many forms of perturbation, some types of damage being correlated with subsequent hyperploidy (Allen *et al.*, 1988; Backer *et al.*, 1988).

Metaphase II Hyperploidy

Counting extra chromosomes at meiotic metaphase II provides a direct and reliable measure of aneuploidy. The analysis of hypoploidy at this stage is equivocal due to the potential for chromosomes to be missing as a result of technical artifact. In some laboratories C-banding stains have been used to distinguish between whole chromosomes (or chromatids) and chromosomal acentric fragments, thereby aiding in the distinction between clastogenic and aneuploidy mechanisms (Miller and Adler, 1992; Liang and Pacchierotti, 1988). In the last several years, a number of studies to determine chemical and radiation induction of metaphase II hyperploidy in rodents have been reported. Table 1 summarizes some study results, along with experimental details concerning treatments. Effective and ineffective doses listed in the table provide some guidance; however, it is notable that hyperploidy induction often does not follow clear patterns of dose-dependent increase. That treatments of various spermatogonial and meiotic cell stages give rise to positive hyperploidy results indicates that different target structures/functions are susceptible to disruption. Some agents, eg. chloral hydrate, exert effects at multiple cell stages to result in nondisjunction, while other agents, eg. *p*-fluorophenylalanine and vinblastine, appear more specific in their mechanisms of action that are limited to certain stages. A species difference in results is evident for ethanol. Of the several known spindle poisons acting at diakinesis-metaphase I to give hyperploidy results, most were not tested at other cell stages. Miller and Adler (1992) have made the interesting observation that meiotic delay tends to accompany positive hyperploidy results, and have suggested that such delay might serve as a prescreen endpoint for aneuploidy.

Table 1. **Meiotic Metaphase II Hyperploidy Results from Chemical and Radiation Treatments in Rodents**

Treatment[1]	Treatment Time (Stage)[2]	Aneuploidy result[3]	LEDT/ HIDT[4]	Reference
Adriamycin	3d (diplotene) 5d (pachytene)	+	24 mg/kg 12 mg/kg	Liang *et al.* (1986)
Cytosine Arabinoside	2d (diplotene) 7d (pachytene) 10d (zygotene) 12d (preleptotene) 14 & 21d (spermatogonia)	+	2000 mg/kg 2000 mg/kg 2000 mg/kg 500 mg/kg 2000 mg/kg	Liang *et al.* (1986)
Nitrilotriacetic Acid	6h (diakinesis/MI)	+	275 mg/kg	Zordan *et al.* (1990)
"	6h (diakinesis/MI)	+	275 mg/kg	Costa *et al.* (1988)
p-Fluorophenylalanine	4h (MI)	+	100 mg/kg	Brook & Chandley (1986)
"	8d+10h (zygotene) 11d+4h (preleptotene)	-	100 mg/kg	Brook & Chandley (1986)
Diethylstilbestrol	6h (diakinesis/MI) 10d (zygotene)	+	100 mg/kg	Liang & Pacchierotti (1988)
Chloral Hydrate	5d (pachytene) 12d (preleptotene) 21 & 42d (spermatogonia)	+	82.7 mg/kg	Russo *et al.* (1984)
"	6h (diakinesis/MI) 14 & 22h (diplotene)	+	200 mg/kg	Miller & Adler (1992)
"	5d (pachytene) 12d (preleptotene) 21d (spermatogonia)	+	165 mg/kg	Liang & Pacchierotti (1988)
Cyclophosphamide	6h (diakinesis/MI) 2d (diplotene) 5d (pachytene) 10d (zygotene) 12d (preleptotene) 21d (spermatogonia)	+	30 mg/kg	Pacchierotti *et al.* (1983)
"	13d (preleptotene/ spermatogonia)	-	100 mg/kg	Backer *et al.* (1988)
Demecolcine	4h (MI)	+	0.1 mg/kg	Risley *et al.* (1990)
Colchicine	6h (diakinesis/MI) 14h (diplotene) 22h (diplotene)	+	3 mg/kg 1.5 mg/kg	Miller & Adler (1992)
Ethanol	2-6h (diakinesis/MI)	+	12.5%	Hunt (1987)
"	0.5-5h (diakinesis/MI) [13 & 16d of daily dosing before sacrifice]	-	12.5%	Daniel & Roane (1987)
Econazole	6h (diakinesis/MI) 14 & 22h (diplotene)	+	80 mg/kg	Miller & Adler (1992)
Hydroquinone	6h (diakinesis/MI)	+	100 mg/kg	Miller & Adler (1992)
Diazepam	6h (diakinesis/MI)	+	150 mg/kg	Miller & Adler (1992)
Cadmium Chloride	22h (diplotene)	+	1 mg/kg	Miller & Adler (1992)

Table 1. (continued)

Treatment[1]	Treatment Time (Stage)[2]	Aneu-ploidy result[3]	LEDT/ HIDT[4]	Reference
Vinblastine	14h (diplotene) 22h (diplotene)	+	2 mg/kg 1 mg/kg	Miller & Adler (1992)
"	10d zygotene) 21d (spermatogonia) 56d (stem cell)	-	2.7 mg/kg	Liang *et al.* (1986)
Cobalt 60 Radiation	2d (diplotene)	+	1 Gy	Liang *et al.* (1986)
Neutrons	5d (pachytene) 12d (preleptotene)	+	0.22 Gy 0.56 Gy	Liang & Pacchierotti (1988)
X-rays	5d (pachytene) 12d (preleptotene)	+	1 Gy 2 Gy	Liang & Pacchierotti (1988)
EDTA	6h (diakinesis/MI) 5d (pachytene)	-	186 mg/kg	Zordan *et al.* (1990)
Thiabendazole	6h (diakinesis/MI) 14d & 22h (diplotene)	-	400 mg/kg	Miller & Adler (1992)
Pyrimethamine	6h (diakinesis/MI) 22h (diplotene)	-	20 mg/kg	Miller & Adler (1992)
Thimerosal	6h (diakinesis/MI) 14 & 22h (diplotene)	-	20 mg/kg	Miller & Adler (1992)
Phenylalanine	4h (MI) 8d+10h (zygotene) 11d+4h (preleptotene)	-	100 mg/kg	Brook & Chandley (1986)
6-Mercaptopurine	4h (MI) 8d+10h (zygotene) 11d+4h (preleptotene)	-	150 mg/kg	Brook & Chandley (1986)
Acrylamide	50d (stem cell)	-	125 mg/kg	Backer *et al.* (1989)

[1]Mice were treated in all studies except that of Daniel & Roane (1987), who used Chinese hamsters.
[2]Treatment times are the days/hours prior to harvest of metaphase II cells. Stages treated were provided in the referenced literature or were calculated according to Oakberg (1957).
[3]Aneuploidy results shown are the conclusions of the referenced authors; positive results are not necessarily statistically significant.
[4]LEDT/HIDT: Lowest effective dose tested/highest ineffective dose tested. In some cases the dose indicated was the only dose tested. In other cases, the doses/times tested were too numerous or complicated to list, and representative test conditions and results were selected for the table.

Spermatid Micronuclei

The rodent spermatid micronucleus assay (Lähdetie and Parvinen, 1981; Tates *et al.*, 1983) has also been used to detect radiation and chemical effects on various germ cell stages. However, spermatid micronuclei are more ambiguous than chromosome counts at metaphase II with regard to their interpretation as acentric fragments, or whole chromosomes, which failed to segregate properly at anaphase. In order to distinguish between clastogenic and aneuploidy mechanisms, kinetochore staining may be used in conjunction with scoring of micronuclei (Moroi *et al.*, 1980; Eastmond and Tucker, 1989). Anti-kinetochore antibodies and immunofluorescence highlight the presence of kinetochore proteins; micronuclei with kinetochore signals are presumed to represent whole chromosomes, while those without signals are more interpretable as fragments. This technique has recently been used to reveal significant increases in kinetochore-positive micronuclei in spermatids of mice treated with x-rays or acrylamide (Collins *et al.*, 1992; Figure 1).

There are additional uncertainties in the interpretation of spermatid micronuclei as mediating events in the development of aneuploidy. In particular, the fate of micronucleated cells in subsequent stages of spermiogenesis, or in pronuclei, is not known. Are there selection factors operating to inhibit the differentiation of these abnormal cells to mature gametes? And, if such cells reach the pronuclear stage, are there opportunities during reorganization of the nucleus for earlier expelled cytoplasmic chromosomes to become reincorporated? Despite the questions surrounding micronuclei observed in spermatids, this endpoint has particular advantages for aneuploidy studies. Importantly, it can signify chromosomal loss occurring at either or both meiotic divisions. Also, there is evidence indicating that spermatid micronuclei can result from chromosome lagging as a mechanism separate from nondisjunction. Russo and Levis (1992) have noted that EDTA exposure to preleptotene cells in mice does not result in metaphase hyperploidy or spermatid micronuclei; however, exposure at the time of meiotic metaphase divisions does lead to elevated spermatid micronuclei, while metaphase hyperploidy remains negative.

FISH Signals of Specific Chromosomes/Chromosome Regions

Fluorescence in situ hybridization provides a powerful methodology for detecting aneuploidy in metaphase or interphase cells (Pinkel *et al.*, 1986). In addition to determining the aneuploid state, the particular chromosomes that are extra or missing can be identified. FISH exploits the property of complementary DNA strands to hybridize and form stable sequence-specific double-strand molecules. Probe DNA is prepared by chemical or enzymatic fluorochrome labeling of sequences that are specific for certain chromosomes or chromosome regions. Next, the probe is hybridized to the denatured target DNA, which is typically cells or metaphase chromosomes fixed to slides. Specifically bound probe is visualized in situ by fluorescence microscopy. To date, most applications of FISH to detect chromosomal imbalance have involved studies of translocations and aneuploidies in clinical conditions. However, there is considerable interest in its potential use for toxicological studies, in both humans and experimental animals.

Overall, the technical advantages of FISH for cytogenetic analysis coupled with the vast number of cDNA and genomic sequences that have been isolated and characterized should provide considerable potential for aneuploidy (in particular, hyperploidy) studies. Variations and refinements of these methods now enable the detection and enumeration of individual chromosomes, chromosome subregions, total genomes, or unique sequences in both metaphase and interphase cells. In human cells, single copy sequences with lengths of 1-10 kb (Lawrence *et al.*, 1988; Bhatt *et al.*, 1988; Lichter and Ward, 1990) can be revealed, as can specific centromere repeats (Meyne *et al.*, 1989), or whole chromosomes by "painting" probes (Cremer *et al.*, 1988; Lichter *et al.*, 1988; Pinkel *et al.*, 1988). Some relative advantages and disadvantages of different size probes for certain aneuploidy detection studies have been recognized (Kuo *et al.*, 1991). The development of fluorescently tagged nucleotides capable of emitting red, green, or blue now provide for multiple color labeling to elucidate several different chromosomes or chromosomal regions simultaneously. A multicolor FISH detecting probe kit has been developed for studies in human uncultured amniocytes to detect the most common aneuploidies (Ried *et al.*, 1992).

Recently, FISH procedures have been used to detect aneuploidy in human sperm (Wyrobek *et al.*, 1990; Wyrobek *et al.*, 1992a,b). It was shown that baseline frequencies of disomy for chromosomes 1 and Y approximate those measured by the

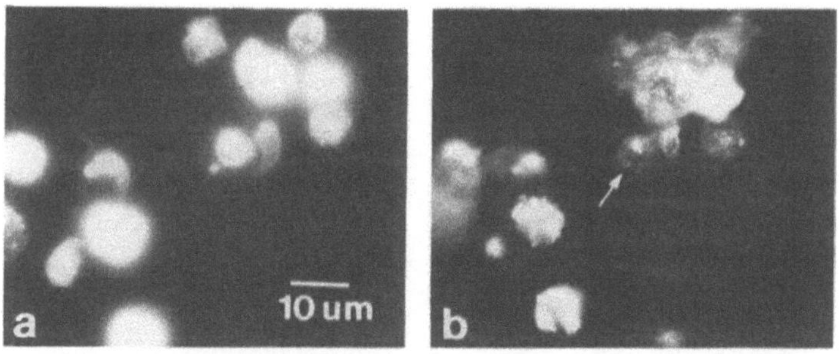

Figure 1. (a) Micronucleated round spermatid from a mouse following intraperitoneal exposure of acrylamide (100 mg/kg body weight) during meiosis, as seen with simultaneous phase contrast microscopy and DAPI fluorescence. (b) The same cell with fluorescein illumination showing the kinetochore-positive micronucleus (arrow) (From: Collins et al., 1992).

direct cytogenetic analyses of human sperm pronuclei in the human sperm-hamster egg fusion test (Wyrobek *et al.*, 1992a,b). With increasing availability of chromosome-specific probes in mice, similar studies oriented to toxicological assessments of induced aneuploidy should, in principle, be feasible. FISH analysis of mouse spermatids or sperm should allow the detection of specific chromosome aneuploidies resulting from chemical-induced chromosome malsegregation during meiotic divisions. This methodology has the important advantage of providing very large numbers of cells for quantitative analysis (Robbins *et al.*, 1992).

FISH methods may also be used in conjunction with micronucleus analysis for the identification of centric chromosomes lost from the nucleus (Becker *et al.*, 1990; Miller *et al.*, 1991). This application is similar in purpose to determining the presence or absence of kinetochores in micronuclei; however, DNA sequences in the centromeric region rather than kinetochore proteins are detected. Such studies have been carried out employing mouse bone marrow cells for analysis (Miller *et al.*, 1991) and using probes to gamma satellite DNA which signal those centromeric sequences in all mouse chromosomes except the Y (Weier *et al.*, 1991). Interestingly, the results from kinetochore vs. gamma satellite DNA detection in micronuclei were not always the same; questions of chemical-specific activities to damage kinetochores and diminish their immunological detection were raised (Miller *et al.*, 1991). We have recently employed a DNA probe that efficiently highlights centromeric regions in spermatid micronuclei (Figure 2), and studies are underway to compare the use of this technique with kinetochore analysis for aneuploidy induction studies involving chemical and radiation effects on pre-meiotic or meiotic cells.

SUMMARY

As we continue to learn more about the role of chromosome malsegregation, not only in fetal loss and aneuploidy syndromes, but also in cancer and uniparental disomy disorders, the need for detecting hazardous agents which may induce these chromosomal anomalies is increasingly important. Testing capabilities in the male germ line have been steadily improved by the further validation of existing assays and the development of new test methods. Metaphase II analysis has been used to detect nondisjunction at the first meiotic division caused by various chemicals and radiation

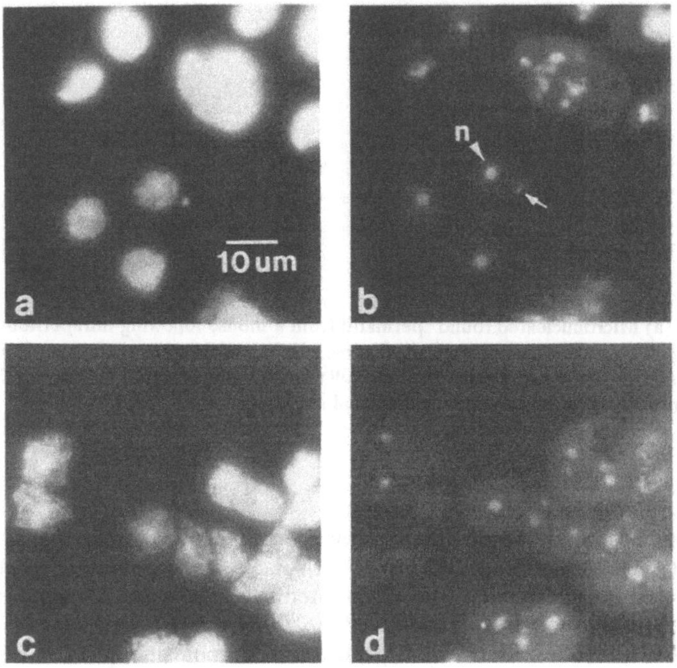

Figure 2. (a,c) Mouse micronucleated early spermatid following radiation exposure (4 Gy/mouse) during meiosis, as seen with simultaneous phase contrast microscopy and DAPI fluorescence.(b) The same cell as in (a) with fluorescein illumination showing the dense arrangement of highly repeated centromere specific DNA sequences in the nucleus (n) and in the micronucleus (arrow). (d) The same cell as in (c) with fluorescein illumination demonstrating the lack of centromere specific DNA sequences in the micronucleus. (Color photographs are available upon request from authors.)

acting on different germ cell stages. The development of spermatid and sperm assays addresses the important need to detect both chromosome loss and gain which may have occurred at either or both meiotic divisions. Major advantages of using DNA probes in these assays would seem to be just beginning. For mechanistic studies, probe hybridization should be useful for identifying and perhaps tracking induced changes in specific chromosomes through consecutive stages of development. As new probes continue to be developed and combined for simultaneous detection of multiple chromosomes and/or chromosome regions, these advances should convey increasing sensitivity and versatility for detecting aneuploidy in the germ line.

ACKNOWLEDGMENT

The authors are grateful to Dr. Sally Perreault Darney for helpful discussions.

REFERENCES

Allen, J.W., Gibson, J.B., Poorman, P.A., Backer, L.C., and Moses, M.J., 1988, Synaptonemal complex damage induced by clastogenic and anti-mitotic chemicals: implications for non-disjunction and aneuploidy, *Mutation Res.*, 201:313.

Allen, J.W., Liang, J.C., Carrano, A.V., and Preston, R.J., 1986, Review of literature on chemical-induced aneuploidy in mammalian male germ cells, *Mutation Res.*, 167:123.

Backer, L.C., Dearfield, K.L., Erexson, G.L., Campbell, J.A., Westbrook-Collins, B., and Allen, J.W., 1989, The effects of acrylamide on mouse germ-line and somatic cell chromosomes, *Environ. Molec. Mutagen.*, 13:218.

Backer, L.C., Gibson, J.B., Moses, M.J., and Allen, J.W., 1988, Synaptonemal complex damage in relation to meiotic chromosome aberrations after exposure of male mice to cyclophosphamide, *Mutation Res.*, 203:317.

Becker, P., Scherthan, H., and Zankl, H., 1990, Use of a centromere-specific DNA probe (p82H) in nonisotopic in situ hybridization for classification of micronuclei, *Genes, Chromosomes, & Cancer*, 2:59.

Bhatt, B., Burns, J., Flannery, D., and McGee, J., 1988, Direct visualization of single copy genes on banded metaphase chromosomes by nonisotopic *in situ* hybridization, *Nucleic Acids Res.*, 16:3951.

Brandriff, B.F., and Gordon, L.A., 1990, Human sperm cytogenetics and the one-cell zygote, *in*: "Banbury Report 34: Biology of Mammalian Germ Cell Mutagenesis," J.W. Allen, B.A. Bridges, M.F. Lyon, M.J. Moses, and L.B. Russell, eds., Cold Spring Harbor Laboratory Press, Cold Spring Harbor.

Brook, J.D., and Chandley, A.C., 1986, Testing for the chemical induction of aneuploidy in the male mouse, *Mutation Res.*, 164:117.

Burgoyne, P.S., Holland, K., and Stephens, R., 1991, Incidence of numerical chromosome anomalies in human pregnancy estimation from induced and spontaneous abortion data, *Hum. Repro.*, 6:555.

Collins, B.W., Howard, D.R., and Allen, J.W., 1992, Kinetochore-staining of spermatid micronuclei: Studies of mice treated with X-radiation or acrylamide, *Mutation Res.*, 281:287.

Costa, R., Russo, A., Zordan, M., Pacchierotti, F., Tavella, A., and Levis, A.G., 1988, Nitrilotriacetic acid (NTA) induces aneuploidy in *Drosophila* and mouse germ-line cells, *Environ. Molec. Mutagen.*, 12:397.

Cremer, T., Lichter, P., Borden, J., Ward, D.C., and Manuelidis, L., 1988, Detection of chromosome aberrations in metaphase and interphase tumor cells by *in situ* hybridization using chromosome-specific library probes, *Hum. Genet.*, 80:235.

Daniel, A., and Roane, D., 1987, Aneuploidy is not induced by ethanol during spermatogenesis in the Chinese hamster, *Cytogenet. Cell Genet.*, 44:43.

Dellarco, V.L., Voytek, P.E., and Hollaender, A., 1985, "Aneuploidy: Etiology and Mechanisms," Plenum Press, New York.

Eastmond, D.A., and Tucker, J.D., 1989, Identification of aneuploidy-inducing agents using cytokinesis-blocked human lymphocytes and an antikinetochore antibody, *Environ. Molec. Mutagen.*, 13:34.

Hassold, T.J., Sherman, S.L., Pettay, D., Page, D.C., and Jacobs, P.A., 1991, XY chromosome nondisjunction in man associated with diminished recombination in the pseudoautosomal region, *Am. J. Hum. Genet.*, 49:253.

Hoffman, M., 1992, Motor molecules on the move, *Science*, 256:1758.

Hunt, P.A., 1987, Ethanol-induced aneuploidy in male germ cells of the mouse, *Cytogenet. Cell Genet.*, 44:7.

Kuo, W.-L., Tenjin, H., Segraves, R., Pinkel, D., Golbus, M.S., and Gray, J., 1991, Detection of aneuploidy involving chromosomes 13, 18, or 21, by fluorescence in situ hybridization (FISH) to interphase and metaphase amniocytes, *Am. J. Hum. Genet.*, 49:112.

Lähdetie, J., and Parvinen, M., 1981, Meiotic micronuclei induced by X-rays in early spermatids of the rat, *Mutation Res.*, 81:103.

Lawrence, J.B., Villnave, C.A., and Singer, R.H., 1988, Sensitive, high-resolution chromatin and chromosome mapping *in situ*: presence and orientation of two closely integrated copies of EBV in a lymphoma line, *Cell*, 52:51.

Liang, J.C., and Brinkley, B.R., 1985, Chemical probes and possible targets for the induction of aneuploidy, *in*: "Aneuploidy: Etiology and Mechanisms," V.L. Dellarcoo, P.E. Voytek, and A. Hollaender, eds., Plenum Press, New York.

Liang, J.C., and Pacchierotti, F., 1988, Cytogenetic investigation of chemically-induced aneuploidy in mouse spermatocytes, *Mutation Res.*, 201:325.

Liang, J.C., Sherron, D.A., and Johnston, D., 1986, Lack of correlation between mutagen-induced chromosomal univalency and aneuploidy in mouse spermatocytes, *Mutation Res.* 163:285.

Lichter, P., Cremer, T., Borden, J., Manuelidis, L., and Ward, D.C., 1988, Delineation of individual human chromosomes in metaphase and interphase cells by *in situ* suppression hybridization using recombinant DNA libraries, *Hum. Genet.*, 80:224.

Lichter, P., and Ward, D.C., 1990, High resolution mapping of human chromosome 11 by *in situ* hybridization with cosmid clones, *Science*, 247:64.

Martin, R.H., and Rademaker, A., 1990, The frequency of aneuploidy among individual chromosomes in 6,821 human sperm chromosome complements, *Cytogenet. Cell Genet.*, 53:103.

Martin, R.H., Rademaker, A., Hildebrand, K., Barnes, M., Arthur, K., Ringrose, T., Brown, I.S., and Douglas, G., 1989, A comparison of chromosomal aberrations induced by in vivo radiotherapy in human sperm and lymphocytes, *Mutation Res.*, 226:21.

May, K.M., Jacobs, P.A., Lee, M., Ratcliffe, S., Robinson, A., Nielsen, J., and Hassold, T.J., 1990, The parental origin of the extra X chromosome in 47,XXX females, *Am. J. Hum. Genet.*, 46:754.

Meyne, J., Littlefield, L.G., and Moyzis, R.K., 1989, Labeling of human centromeres using an alphoid DNA consensus sequence: application to the scoring of chromosome aberrations, *Mutation Res.*, 226:75.

Miller, B.M., and Adler, I.-D, 1992, Aneuploidy induction in mouse spermatocytes, *Mutagenesis*, 7:69.

Miller, B.M., Zitzelsberger, H.F., Weier, H.-Ul.G., and Adler, I.-D., 1991, Classification of micronuclei in murine erythrocytes: immunofluorescent staining using CREST antibodies compared to *in situ* hybridization with biotinylated gamma satellite DNA, *Mutagenesis*, 6:297.

Moroi, Y., Peebles, C., Fritzler, M.J., Steigerwald, J., and Tan, E.M., 1980, Autoantibody to centromere (kinetochore) in scleroderma sera, *Proc. Natl. Acad. Sci. (U.S.A.)*, 77:1627.

Oakberg, E.F., 1957, Duration of spermatogenesis in the mouse, *Nature*, 180:1137.

Pacchierotti, F., Bellincampi, D., and Civitareale, D., 1983, Cytogenetic observations, in mouse secondary spermatocytes, on numerical and structural chromosome aberrations induced by cyclophosphamide in various stages of spermatogenesis, *Mutation Res.*, 119:177.

Parry, J.M., and Parry, E.M., 1987, Comparisons of tests for aneuploidy, *Mutation Res.*, 181:267.

Pinkel, D., Landegent, J., Collins, C., Fuscoe, J., Segraves, R., Lucas, J., and Gray, J., 1988, Fluorescence *in situ* hybridization with human chromosome-specific libraries: detection of trisomy 21 and translocations of chromosome 4, *Proc. Natl. Acad. Sci. (U.S.A.)*, 85:9138.

Pinkel, D., Straume, T., and Gray, J.W., 1986, Cytogenetic analysis using quantitative, high-sensitivity, fluorescence hybridization, *Proc. Natl. Acad. Sci. (U.S.A.)*, 83:2934.

Ried, T., Landes, G., Dackowski, W., Klinger, K., and Ward, D.C., 1992, Multicolor fluorescence *in situ* hybridization for the simultaneous detection of probe sets for chromosomes 13, 18, 21, X, and Y in uncultured amniotic fluid cells, *Hum. Molec. Genet.*, 1:307.

Risley, M.S., Saccomanno, C.F., and Pohorenec, G.M., 1990, An improved method for cytogenetic analysis of meiotic aneuploidy in rodent and frog spermatocytes, *Mutation Res.*, 234:361.

Robbins, W., Segraves, R., Pinkel, D., and Wyrobek, A.J., 1992, Frequencies of aneuploid sperm in healthy men vary with donor and chromosome, *Environ. Molec. Mutagen.*, 19 (Suppl. 20):53.

Russell, L.B., 1985, Experimental approaches for the detection of chromosomal malsegregation occurring in the germline of mammals, *in*: "Aneuploidy: Etiology and Mechanisms," V.L. Dellarco, P.E. Voytek, and A. Hollaender, eds., Plenum Press, New York.

Russell, L.B., and Shelby, M.D., 1985, Tests for heritable genetic damage and for evidence of gonadal exposure in mammals, *Mutation Res.*, 154:69.

Russo, A., and Levis, A.G., 1992, Further evidence for the aneuploidogenic properties of chelating agents: Induction of micronuclei in mouse male germ cells by EDTA, *Environ. Molec. Mutagen.*, 19:125.

Russo, A., Pacchierotti, F., and Metalli, P., 1984, Nondisjunction induced in mouse spermatogenesis by chloral hydrate, a metabolite of trichloroethylene, *Environ. Mutagen.*, 6:695.

Sherman, S.L., Takaesu, N., Freeman, S.B., Grantham, M., Phillips, C., Blackston, R.D., Jacobs, P.A., Cockwell, A.E., Freeman, V., Uchida, I., Mikkelsen, M., Kurnit, D.M., Buraczynska, M., Keats, B.J.B., and Hassold, T.J., 1991, Trisomy 21: Association between reduced recombination and nondisjunction, *Am. J. Hum. Genet.*, 49:608.

Tates, A.D., Dietrich, A.J.J., de Vogel, N., Neuteboom, I., and Bos, A., 1983, A micronucleus method for detection of meiotic micronuclei in male germ cells of mammals, *Mutation Res.*, 121:131.

Warren, A.C., Chakravarti, A., Wong, C., Slaugenhaupt, S.A., Halloran, S.L., Watkins, P.C., Metaxotou, C., and Antonarakis, S.E., 1987, Evidence for reduced recombination on the nondisjoined chromosomes 21 in Down syndrome, *Science*, 237:652.

Waters, M.D., Stack, H.F., Mavournin, K.H., and Dellarco, V.L., 1985, Special committee report, part II: Quantitative evaluation of chemicals that induce aneuploidy using the genetic activity profile method, *in*: "Aneuploidy: Etiology and Mechanisms," V.L. Dellarco, P.E. Voytek, and A. Hollaender, eds., Plenum Press, New York.

Weier, H.-U., Zitzelsberger, H.F., and Gray, J.W., 1991, Non-isotopical labeling of murine heterochromatin *in situ* by hybridization with *in vitro*-synthesized biotinylated gamma (major) satellite DNA, *BioTechniques*, 10:498.

Wyrobek, A.J., Ahlborn, T., Balhorn, R., Stanker, L., and Pinkel, D., 1990, Fluorescence in situ hybridization to Y chromosomes in decondensed human sperm nuclei, *Mol. Reprod. Dev.*, 27:200.

Wyrobek, A., Weier, H.-U., and Pinkel, D., 1992a, Molecular detection of aneuploidy in human sperm, *Environ. Molec. Mutagen.*, 19 (Suppl. 20):72.

Wyrobek, A.J., Weier, H.-U., Robbins, W., Mehraein, Y., and Pinkel, D., 1992b, Detection of sex-chromosomal aneuploidies in human sperm using two-color fluorescence *in situ* hybridization, *Environ. Molec. Mutagen.*, 19 (Suppl. 20):72.

Zordan, M., Russo, A., Costa, R., Bianco, N., Beltrame, C., and Levis, A.G., 1990, A concerted approach to the study of the aneuploidogenic properties of two chelating agents (EDTA and NTA) in the germ and somatic cell lines of *Drosophila* and the mouse, *Environ. Molec. Mutagen.*, 15:205.

STRATEGIES FOR THE USE OF A MULTIPLE-ENDPOINT SYSTEM FOR MAMMALIAN GERM CELL MUTATION TESTING

S.E. Lewis, L.B. Barnett, and L.S. Niedziela

Chemistry and Life Sciences
Research Triangle Institute
PO Box 12194
Research Triangle Park, NC 27709-2194

There are specific advantages to the use of a multiple-endpoint system, as opposed to any individual *in vivo* mutation test system. First of all, a larger sampling of the genome becomes possible than with any individual system. It has become clear that some loci (both within and between systems) appear to be more sensitive to mutagens (Favor, 1989; Neel and Lewis, 1990). By examining a broad spectrum of loci, as is possible in a multiple-endpoint system, a better idea of differential sensitivities of regions of the genome becomes possible. Furthermore, it permits more effective use of resources by performing multiple assays on the same animals. This capability is especially valuable in experiments on postgonial stages and those experiments in which fertility is affected by the test compound. Finally, a multiple-endpoint system allows for flexibility. It is clearly an option to add other assays which may become available in the future and to allow selective use of individual endpoints, according to the requirements of a given experiment.

The multiple-endpoint system we are devising includes: (1) the recessive visible specific-locus test, (2) the electrophoretic specific-locus test, (3) dominant visible mutation screening, (4) the dominant cataract assay, and (5) molecular screening using RFLP analysis.

The seven-locus visible test has several advantages for mutation testing (Russell *et al.*, 1981; Searle, 1974, 1975). First of all, the test is relatively easy to perform, involving visible examination of animals at weaning. Furthermore, the largest data base in vivo mouse mutagenesis has already been collected on both ionizing radiations (Searle, 1975) and chemicals (Russell *et al.*, 1981). Use of this assay in the multiple-endpoint system will permit extrapolation of results obtained in the multiple-endpoint system to this enormous data base. In addition, this system may be more responsive to certain mutagens than many other mutation-detection systems (Favor, 1989; Neel and Lewis, 1990). The seven-locus visible system has been successfully utilized in combination with electrophoretic screening by Dr. J. Peters of Harwell (Peters *et al.*, 1986; data cited in Neel and Lewis, 1990).

Male-Mediated Developmental Toxicity, Edited by D.R. Mattison and A.F. Olshan, Plenum Press, New York, 1994

Mutations causing alterations in externally-detectable, dominant phenotypes are found in mouse populations treated with mutagens, as well as unexposed mouse populations (Schlager and Dickie, 1967). These are important endpoints for assessment of mutagenic risk. First of all, dominant mutations can be observed as F_1 effects, often with deleterious effects on the heterozygous carriers. There is a current interest in evaluating F_1 effects of mutagens in the performance of human risk assessment.

Dominant visible mutations have been found at many loci in the mouse (Green, 1981). Many of these have recently been shown to be models of known heritable human dysmorphic syndromes using the human/mouse homology map (Erickson, 1990). Significant dominant effects have been found among the progeny of males exposed to environmental chemicals (Lewis *et al.*, 1986). Finally, dominant visible effects are relatively easy to assess at weaning when scrutiny of the animals for recessive visibles is done. Only a small additional amount of time is needed, in addition to that required to screen for recessive visible mutations, to observe each animal for dominant visible mutations.

The dominant cataract test, as developed by Dr. J. Favor (1989), has several specific merits which justify its inclusion in the multiple-endpoint system. It has been well developed and validated using ENU and radiation in the Neuherberg multiple-endpoint system (Ehling *et al.*, 1985; Favor, 1989). It has the advantage of detecting genetic damage expressed in the F_1 progeny. Finally, there is obvious medical relevance of this system to human health.

The advantages of the electrophoretic system for detection of specific-locus mutations and extrapolation to human risk are as follows. First, the loci screened in this assay are among those that have considerable conservation of linkage in mice and humans (Johnson *et al.*, 1981; Searle *et al.*, 1987), thus indicating significant similarities between the genomes of these two organisms. Second, at least 19 loci, which are presently evaluated in the mouse electrophoretic system (Johnson *et al.*, 1981; Lewis *et al.*, 1985; Lewis, unpublished), have been used in screening of humans for detection of mutations (Neel *et al.*, 1980a,b).

In addition, some of the loci in the electrophoretic system are associated with animal models of human disease. These models have great potential importance, not only for human risk assessment but also for their experimental value to medical research. Two human disease models have been identified in our program. These are an α-thalassemia of spontaneous origin (Skow *et al.*, 1983) and a carbonic anhydrase deficiency induced by ENU (Lewis *et al.*, 1988) analogous to the human disease. Additional interesting human disease models have been found at electrophoretically-studied loci by other investigators (Peters *et al.*, 1985).

The I/D/R system was chosen for inclusion in the multiple-endpoint system because it approximates the system of Mohrenweiser *et al.* (1989) which was designed for human mutation screening. This molecular screening system is targeted at detection of major DNA lesions, such as insertions, deletions, and rearrangements. In addition, it is able to detect base-pair mutations when these occur at restriction sites.

The I/D/R system described here uses Southern blot analysis of mouse genomic DNA to determine frequencies of insertions, deletions, and rearrangements that occur spontaneously and that are induced by chemical exposure. For germinal mutation studies, factors such as the number of subjects, multigenerational families, controlled exposure, and variability make human studies difficult to conduct. Therefore, the I/D/R system in mice has been proposed as a pilot study for use of these types of analyses in humans.

The efficacy of each of the assays proposed for the multiple-endpoint system is being determined using two model mutagens which have been shown to be particularly effective in inducing mutations in the germ cells of mice. The first, chlorambucil (CHL), induces large numbers of germinal mutations in postmeiotic germ cell stages (Russell *et al.*, 1989). It produces large molecular lesions (Rinchik *et al.* 1990). The second mutagen used to test the system is ethylnitrosourea (ENU), which is a powerful in vivo mutagen, especially active in spermatogonia (Russell *et al.*, 1979; Lewis *et al.*, 1991). In contrast to CHL, ENU induces a high proportion of point mutations in mouse germ cells (Popp *et al.*, 1983; Lewis *et al.*, 1985; Peters *et al.*, 1985). Thus, using these two mutagens, the sensitivity of the various endpoints can be evaluated.

ACKNOWLEDGEMENTS

This program is supported in part by NIEHS Contract No. N01-ES-95273. The authors are grateful to Dr. Michael Shelby for helpful comments on the manuscript.

REFERENCES

Ehling, U.H., D.J. Charles, J. Favor, J. Graw, J. Kratochvilova, A. Neuhauser-Klaus, and W. Pretsch (1985) Induction of gene mutations in mice: The multiple-endpoint approach, *Mutation Res.*, 150, 393-401.

Erickson, R.P. (1990) Invited Editorial: Mapping dysmorphic syndromes with the aid of the human/ mouse homology map, *Am. J. Hum. Genet.*, 46, 1013-1016.

Favor, J. (1989) Risk estimation based on germ-cell mutations in animals, in: D.D. Moens (ed.) Proceedings of the XVIth International Congress of Genetics, August 20-27, 1988, Toronto, Canada, pp. 844-852, The National Research Council of Canada.

Green, M.C. (ed.) (1981) Genetic Variants and Strains of the Laboratory Mouse, Gustav Fischer Verlag, Stuttgart, New York.

Johnson, F.M., G.T. Roberts, R.K. Sharma, F. Chasalow, R. Zweidinger, A. Morgan, R.W. Hendren, and S.E. Lewis (1981) The detection of mutants in mice by electrophoresis: Results of a model induction experiment with procarbazine, *Genetics*, 97, 113-124.

Lewis, S.E., F.M. Johnson, L.C. Skow, L.B. Barnett, and R.A. Popp (1985) A mutation in the mouse β-globulin gene detected in the progeny of a female treated with ethyl nitrosourea, *Proc. Natl. Acad. Sci.* USA, 82, 5829-5831.

Lewis, S.E., C. Felton, L.B. Barnett, W. Generoso, N. Cacheiro, and M.D. Shelby (1986) Dominant visible and electrophoretically expressed mutations induced in male mice exposed to ethylene oxide by inhalation, *Env. Mutagen.*, 8, 867-872.

Lewis, S.E., R.P. Erickson, L.B. Barnett, P. Venta, and R. Tashian (1988) Ethylnitrosourea-induced null mutation at the mouse *Car-2* locus: An animal model for human carbonic anhydrase II deficiency syndrome, *Proc. Natl. Acad. Sci.* USA, 85, 1962-1966.

Lewis, S.E., L.B. Barnett, B.M. Sadler, and M.D. Shelby (1991) ENU mutagenesis in the mouse electrophoretic specific locus test: 1. dose-response relationship of electrophoretically-detected mutations arising from mouse spermatogonia treated with ethylnitrosourea, *Mutation Res.*, 249, 311-315.

Mohrenweiser, H.W., R.D. Larsen, and J.V. Neel (1989) Development of molecular approaches to estimating germinal mutation rates, I. Detection of insertion/deletion/rearrangementvariants in the human genome, *Mutation Res.*, 212, 241-252.

Neel, J.V., C. Satoh, H.B. Hamilton, M. Otake, K. Goriki, T. Kageoka, M. Fujita, S. Nerishi, and J. Asakowa (1980a) Search for mutations affecting protein structure in children of atomic bomb survivors: Preliminary report, *Proc. Natl. Acad. Sci.* USA, 71, 4222-4225.

Neel, J.V., H. Mohrenweiser, and M. Meisler (1980b) Rate of spontaneous mutation at human loci encoding protein structure, *Proc. Natl. Acad. Sci.* USA, 77, 6037-1041.

Neel, J.V., and S.E. Lewis (1990) The comparative radiation genetics of humans and mice, *Ann. Rev. Genet.*, 24, 327-362.

Peters, J., S.J. Andrews, J.F. Loutit, and J.B. Clegg (1985) A mouse β-globin mutant that is an exact model of hemoglobin Rainier in man, *Genetics*, 110, 709-721.

Peters, J., S.T. Ball, and S.J. Andrews (1986) The detection of gene mutations by electrophoresis and their analysis, in: C. Ramel, B. Lambert, J. Magnusson (eds.), Genetic Toxicology of Environmental Chemicals, Part B: Genetic Effects and Applied Mutagenesis, Alan R. Liss, Inc., New York, pp. 367-374.

Rinchik, E.M., J.W. Bangham, P.R. Hunsicker, N.L. A. Cacheiro, B.S. Kwon, I.J. Jackson, and L.B. Russell (1990) Genetic and molecular analysis of chlorambucil-induced germline mutations in the mouse, *Proc. Natl. Acad.* Sci. USA, 87, 1416-1420.

Russell, L.B., P.B. Selby, E. von Halle, W. Sheridan, and L. Valcovic (1981) The mouse specific-locus test with agents other than radiation. Interpretation of data and recommendations for future work, *Mutation Res.*, 86, 329-354.

Russell, L.B., P.R. Hunsicker, N.L. A. Cacheiro, J.W. Bangham, and W.L. Russell (1989) Chlorambucil effectively induces deletion mutations in mouse germ cells, *Proc. Natl. Acad. Sci.* USA, 86, 3704.

Russell, W.L., E.M. Kelly, P.R. Hunsicker, J.W. Bangham, S.C. Maddux, and E.L. Phipps (1979) Specific-locus test shows ethylnitrosourea to be the most potent mutagen in the mouse, *Proc. Natl. Acad. Sci.*, 76, 5818-5819.

Schlager, G., and M.M. Dickie (1967) Spontaneous mutations and mutation rates in the house mouse, *Genetics*, 57, 319-330.

Searle, A.G. (1974) Mutation induction in mice, in: J.T. Lett, H. Adler, and M. Zeble (eds.), Advances in Radiation Biology, Vol. 4, Academic Press, New York, pp. 131-207.

Searle, A.G. (1975) The specific locus test in the mouse, *Mutation Res.*, 31, 277-290.

Searle, A.G., J. Peters, M.F. Lyon, E.P. Evans, J.H. Edwards, and V.J. Buckle (1987) Chromosome maps of man and mouse, III, *Genomics*, 1, 3-18.

Skow, L.C., B.A. Burkhart, F.M. Johnson, R.A. Popp, D.M. Popp, S.Z. Goldberg, W.F. Anderson, L.B. Barnett, and S.E. Lewis (1983) A mouse model for α-thalassemia, *Cell*, 34, 1043-1052.

TRANSGENIC MICE IN DEVELOPMENTAL TOXICOLOGY

R. P. Woychik

Biology Division
Oak Ridge National Laboratory
P.O. Box 2009, MS 8077
Oak Ridge, TN 37831-8077

INTRODUCTION

Recent advances in molecular biology and embryology are being utilized for experiments in mammalian genetics and toxicology. Some of the most important developments are based on the generation of transgenic mice, which are animals that contain specific additions, deletions, or modifications of genes or sequences in their DNA. Originally, pronuclear microinjection or retroviral infection of early-stage embryos was used to generate transgenic mice, and the integration of the exogenous DNA constructs occurred randomly throughout the host genome (reviewed by Palmiter and Brinster, 1986). In the past several years, mouse embryonic stem cells and homologous recombination procedures have made it possible to target specific DNA structural alterations to highly localized region in the host chromosomes (reviewed by Capecchi, 1989; Frohman and Martin, 1989). Most important, the majority of the DNA structural rearrangements in transgenic mice can be passed through the germ line and used to establish new genetic traits in the carrier animals. Since the use of transgenic mice is having such an enormous impact on so many areas of mammalian biological research, including developmental toxicology, it will be the objective of this review to briefly describe the fundamental methodologies for generating transgenic mice and to describe one particular application that has direct relevance to the field of genetic toxicology.

THE PRONUCLEAR MICROINJECTION PROCEDURE FOR GENERATING TRANSGENIC MICE

The most widely used transgenic technology for adding genes or sequences to the mouse germ line is based on the direct pronuclear microinjection procedure originally developed by Gordon and Ruddle (Gordon and Ruddle, 1981). With this method, about a picoliter of solution containing several hundred linearized molecules

Male-Mediated Developmental Toxicity, Edited by D.R. Mattison
and A.F. Olshan, Plenum Press, New York, 1994

of DNA are microinjected directly into the male pronucleus of the fertilized egg. Any DNA fragment can be used for the generation of transgenic mice, including prokaryotic sequences that will not express in mammalian cells. The embryos that survive the microinjection are then transplanted into a pseudopregnant host mother which carries them to term. Typically about 20% of the animals that develop from the microinjected embryos are found to have integrated one or more copies of the exogenously added DNA fragment into one, or a limited number of sites, on their chromosomes. They are classified as transgenic "founder" animals. The integrated sequences, called the "transgene," can then be passed through the germ line and used to establish a new line of transgenic mice. In most cases the inheritance of the transgene follows simple Mendelian principles.

The precise mechanism by which the transgene integrates following pronuclear microinjection is presently unclear. Moreover, the means through which sites on the host DNA are selected for integration is also unknown, although it is generally assumed that integration can occur randomly throughout the genome. One important feature of the integration reaction is that when multiple copies of the exogenously added DNA integrate into the host chromosomes, they usually do so at a single site to form a multiple-copy head-to-tail concatemer. Remarkably, in the vast majority of cases, these concatemer transgenes, which can be over hundreds of kilobases in length, are perfectly stable when passed through the germ line.

Overall, the pronuclear microinjection procedure has been used with great success for generating transgenic lines that express cloned mammalian gene constructs. Additionally, since the integration site appears to be random, the procedure can be applied for generating insertional mutations (for example, see Woychik *et al.*, 1985, Woychik *et al.*, 1990). It is becoming clear that this form of germ line mutagenesis is particularly useful for generating an entire spectrum of mutations, including variable-sized deletions and translocations. The most significant feature of these insertional mutations is that the mutant locus is "tagged" with the transgene and can, in most cases, be immediately accessed by standard DNA cloning procedures. This has allowed several investigators to establish a molecular relationship between individual genes and the phenotypes observed in the corresponding mutant animals (see Meisler, 1992).

GENERATION OF TARGETED MUTATIONS BY HOMOLOGOUS RECOMBINATION IN EMBRYONIC STEM CELLS

In the mid-1980's it was demonstrated that homologous recombination could occur between exogenously added DNA fragments and homologous genes present within the chromosomes of cultured cells (Smithies *et al.*, 1985). Unfortunately, however, since it appeared that only one out of a thousand integration events occurred by homologous recombination (as opposed to random) in cultured cells, it was estimated that the generation of transgenic mice using homologous recombination would be impractical, if not impossible, with the existing labor-intensive methods for introducing exogenous DNA into embryos. For this reason, attention was then directed to mouse embryonic-stem-cell (ES) lines that had the capability of passing their chromosomes through the mouse germ line. With these cultured cells, it was possible to start with thousands of cells and to select for the rare homologous integration events in individual cells by utilizing powerful PCR-based molecular tools or a drug-based selection scheme referred to as positive/negative selection (PNS) (Mansour *et al.*, 1988). ES cell lines, each containing the products of a specific homologous recombination event, could then be injected into a wild-type host blastocyst, which was introduced into a host mother and carried to term to form a germ line chimeric founder animal. These chimeric animals could then be used to establish mutant lines of mice each of which contained a DNA structural rearrangement that was originally introduced by homologous recombination in ES cells.

This methodology for introducing mutations into the mouse germ line by homologous recombination is now becoming routine in several laboratories and allows one to mutate at will any gene or DNA sequence that has been cloned from the mouse. Not unexpectedly, this technology is revolutionizing mouse molecular genetics and is making an enormous contribution to our understanding of the functional role of many developmental genes that have been characterized previously. Therefore, it would not be unreasonable to expect that mutant mice engineered by targeted mutagenesis will also contribute to our understanding of basic mechanisms associated with developmental toxicology.

DESCRIPTION OF THE TRANSGENIC MOUSE MODEL DESIGNED SPECIFICALLY FOR GENETIC AND DEVELOPMENTAL TOXICOLOGY EXPERIMENTS

The most direct application of the transgenic mouse technology to the fields of genetic and developmental toxicology involves lines of transgenic mice that are presently commercially available and are being marketed as "kits" for *in vivo* mutagenicity testing (Kohler *et al.*, 1990; Provost *et al.*, 1990). Both of these lines were generated by the pronuclear microinjection procedure and contain multi-copy transgene constructs. These lines are technically not the same, but do rely on the same fundamental principle. Therefore, since it is not the purpose of this review to advocate the use of one or the other of these lines, a generic description of the lines will be presented here.

Each of these transgenic lines contains a transgene that is comprised of many DNA fragments, each of which is referred to as a shuttle vector. Shuttle vectors are DNA constructs that have the ability to replicate in both bacteria and mammalian cells and can be readily "shuttled" between different kinds of cells. This ability to "shuttle" is desirable for mammalian mutagenesis experiments because growth in bacteria allows the exploitation of genetic tools for the detection of mutations in the DNA, while growth in mammalian cells provides the opportunity to mutagenize DNA fragments while they are in the context of a mammalian cellular environment. In a typical experiment utilizing shuttle vectors, the shuttle vector DNA is introduced into mammalian cells, exposed to an exogenous treatment, and then shuttled back into bacteria where it can be quickly analyzed to evaluate whether a mutation had occurred in the DNA.

There are several types of shuttle vectors, some of which are based on the bacteriophage lambda genome. As a group, the lambda-based vectors are 40-50 kb in length and contain all of the genes that are necessary for lytic growth in bacteria (which means that during growth in bacteria, they have the capability of forming bacteriophage plaques on a bacterial lawn). Additionally, these vectors contain two other important sequences: the lambda COS site, which is the cis-acting regulatory element on the lambda genome that allows the virus to be packaged into an infectious viral particle, and the lac-z or lac-i gene, which are "reporter" genes that indicate the presence of a mutation within the shuttle vector construct. With the lac-z gene, the wild-type color is blue (and the mutant white) in the presence of the X-gal indicator compound, while under the same conditions with lac-i, the mutant color is blue (and the wild-type white).

In a typical shuttle experiment, these vectors are introduced into mammalian cells and are allowed to integrate into the host chromosomes. After treatment with a putative mutagen, the shuttle vector is selectively removed from the chromosomal DNA through a single-step *in vitro* packaging of the shuttle-vector DNA into an infectious bacteriophage particle (utilizing the COS regulatory element on the vector sequences). The bacteriophage particle is then used to infect bacteria and gives rise to a plaque on a bacterial lawn. The color formed in the presence of the X-gal indicator compound indicates which plaques contain a mutated reporter gene within the shuttle-vector construct. This system is simple, efficient, and works well. From just a few micrograms of genomic DNA, hundreds of thousands of plaques can be generated, each of which corresponds to a single shuttle-vector DNA molecule that was originally integrated in the mammalian cell.

The commercially available transgenic lines each contain a bacteriophage lambda shuttle vector integrated as a multi-copy head-to-tail concatemer. Therefore, every cell of every tissue in these animals contains the integrated shuttle vector sequences. To perform an experiment, a transgenic animal carrying the shuttle-vector transgene is first treated with an exogenous agent. If this agent causes structural changes in the DNA, the "reporter" gene within the shuttle vector sequences will become mutated in some cells. To detect the presence of the mutated sequence within the shuttle vector, a small amount of genomic DNA is prepared from any or all of the tissues of the mouse, and the shuttle-vector sequences are selectively extracted from the host genomic DNA and packaged into infectious bacteriophage particles through a single-step *in vitro* packaging reaction. Most often, hundreds of thousands of plaques can be generated, and the ratio of white/blue plaques with the lac-z gene, or blue/white plaques with the lac-i gene, would give an indication of the relative mutagenicity of a particular exogenous agent.

There are several advantages of the commercially available transgenic shuttle vector systems. First, they do represent a cost-effective test for *in vivo* mutagenicity, and may eventually prove to be an extremely reliable system for mutagenicity testing. Second, they can be utilized for detecting mutations in any tissue of the animal, including sperm, provided that several micrograms of genomic DNA can be extracted. Therefore, in a single experiment, mutagenicity can be evaluated in many somatic tissues like the liver, brain, heart, lung, intestine, etc., as well as in sperm. Third, these transgenic lines can be purchased commercially, and the "kits" that are available make it possible for investigators with only minimal molecular biology experience to conduct these experiments.

On the other hand, it is also important to keep in mind that these commercially available systems have not been totally validated with respect to their reliability as mutagenicity tests. Some investigators in the genetic toxicology community are concerned by the fact that these transgenic models are incapable of detecting mutagenic events that cause large megabase deletions and other structural changes like chromosomal translocations. Other concerns relate to the fact that mutations need one round of DNA replication to become fixed in the DNA, and it is presently unclear whether this will occur in the bacterial host cell for mutations that occur in nonreplicating somatic tissues. Both systems are currently being extensively tested with a variety of existing mutagens, and the appropriate data should be available in the published literature in the near future.

REFERENCES

Capecchi, M.R., The new mouse genetics: altering the genome by gene targeting, *TIG* 5:70-76 (1989).

Frohman, M.A., and Martin, G.R., Cut, paste, and save: New approaches to altering specific genes in mice, *Cell* 56:145-147 (1989).

Gordon, J.W., and Ruddle, F.H., Integration and stable germ line transmission of genes injected into mouse pronuclei, *Science* 214:1244-1246 (1981).

Kohler, S.W., Provost, G.S., Kretz, P.L., Dycaico, M.J., Sorge, J.A., and Short, J.M., Development of a short-term, *in vivo* mutagenesis assay: the effects of methylation on the recovery of a lambda phage shuttle vector from transgenic mice, *Nucl. Acids Res.* 18:3007-3013 (1990).

Mansour, S.L., Thomas, K.R., and Capecchi, M.R., Disruption of the proto-oncogene *int-2* in mouse embryo-derived stem cells: a general strategy for targeting mutations to non-selectable genes, *Nature* 336:348-352 (1988).

Meisler, M.H., Insertional mutation of "classical" and novel genes in transgenic mice, TIG 8:341-344 (1992).

Palmiter, R.D., and Brinster, R.L., Germ-line transformation of mice, *Ann. Rev. Genet.* 20:465-499 (1986).

Provost, G.S., Kohler, S.W., Fieck, A., Kretz, P.L., Monlina, T., and Short, J.M., Strategies 4:55-56 (1990).

Smithies, O., Gregg, R.G., Boggs, S.S., Koralewski, M.A., and Kucherlapati, R.S., Insertion of DNA sequences into the human chromosomal B-globin locus by homologous recombination, *Nature* 317:230-234 (1985).

Woychik, R.P., Stewart, T.A., Davis, L.G., D'Eustachio, P. and Leder, P.: An inherited limb deformity created by insertional mutagenesis in a transgenic mouse. *Nature* 318: 36-40 (1985).

Woychik, R. P., D. Maas, R. Zeller, T. F. Vogt, and P. Leder. The formins: a novel class of proteins deduced from the variable transcripts of the *limb deformity* gene. *Nature* 346: 850-853 (1990).

MALE MICE RECEIVING VERY LOW DOSES OF IONIZING RADIATION TRANSMIT AN EMBRYONIC CELL PROLIFERATION DISADVANTAGE TO THEIR PROGENY EMBRYOS

Lynn M. Wiley

Division of Reproductive Biology and Medicine
Department of Obstetrics and Gynecology
School of Medicine
University of California, Davis
Davis, CA 95616

INTRODUCTION

Over the past eight years we have studied male transmitted effects with a novel assay system that uses mouse embryo aggregation chimeras. The hallmark of this 'chimera assay' is its exquisite sensitivity for detecting male germ cell exposure to ionizing radiation. Its published detection limit for male germ cell exposure is presently 0.01 Gy exposure of low LET radiation (e.g., x-radiation and gamma radiation). An important feature of the chimera assay is that it can provide us with information about the cellular identity and heritability of the radiosensitive target(s) of the male germ cell to very low doses of ionizing radiation. The following paragraphs will describe the chimera assay and summarize the results it has produced over the past eight years. The relevance of these results to human health issues will then be discussed in light of what has been published on male transmission of health effects to progeny. Finally, hypotheses will be presented on the radiosensitive target of the male germ cell that is expressed in this assay system.

THE CHIMERA ASSAY

Chimeras are composite individuals comprised of cells originating from more than one fertilized egg (McLaren, 1976). There are two starting conditions that determine the relative proportions of the different cell types comprising the chimera: 1) the starting proportions of the different cell types, and 2) the relative cell proliferation rates of the different cell types. Chimeric mice are readily produced by

Male-Mediated Developmental Toxicity, Edited by D.R. Mattison and A.F. Olshan, Plenum Press, New York, 1994

aggregating together two or more cleavage stage embryos and transferring the aggregate to a foster mother for development to term.

The chimera assay is an extension of *in vitro* assays with preimplantation embryos, which are used to detect genetic and reproductive effects (Spielmann and Eibs, 1978). Preimplantation embryo cultures are more sensitive to these outcomes than fertility tests (Goldstein and Spindle, 1976) and are capable of providing information regarding induction of sister chromatid exchange (Bennett and Pedersen, 1984), DNA repair (Spielmann and Eibs, 1978), and assessment of early postimplantation losses (Goldstein, 1984a,b). In addition, these assays have been used to detect germ cell damage *in vivo* (Goldstein, 1984a,b).

Most endpoints for estimating reproductive risk have involved infertility, embryo lethality, or gross malformations. Subtle changes, which may become revealed under suboptimal conditions or with additional exposures of the parent or of the progeny from exposed parents have proven difficult to assess. In particular, there are few reliable assays that detect environmentally induced abnormalities during the preimplantation period, when embryonic loss for humans may be 50% of conceptions under natural conditions (Biggers, 1981; Kline and Stein, 1985).

The chimera assay is intended to detect changes that may not normally be expressed in the embryo, but which may become expressed when the embryo is additionally exposed later in embryonic development or during postnatal life. This is especially important within the context of cancer induction.

A male mouse can be exposed to doses of ionizing radiation that are by themselves weakly carcinogenic to its progeny. However, if these progeny are subsequently exposed to tumor promoters postnatally, they develop tumors at an incidence that is 20-250 times greater than those produced by the radiation dose alone (Nomura, 1989).

How the Assay is Performed

The assay exploits two fundamental observations of preimplantation embryos. First, that direct cell-cell contact with a normal embryo is a competitive situation that provides a very stringent test of an experimental embryo's viability and vigor (Kelly and Rossant, 1976). The second observation is that embryo cell number is one of the most sensitive endpoints for detecting impaired embryo viability (Spielmann and Eibs, 1978; Eibs and Spielmann, 1977).

The chimera assay is initiated with 4-cell mouse embryos of the CD1 strain (Figure 1). Each chimera consists of two embryos of which one is pre-labelled with the viable dye fluorescein isothiocyanate (FITC; Overstreet, 1973) so that its progeny cells can be distinguished from progeny cells of its partner embryo. Zonae pellucidae

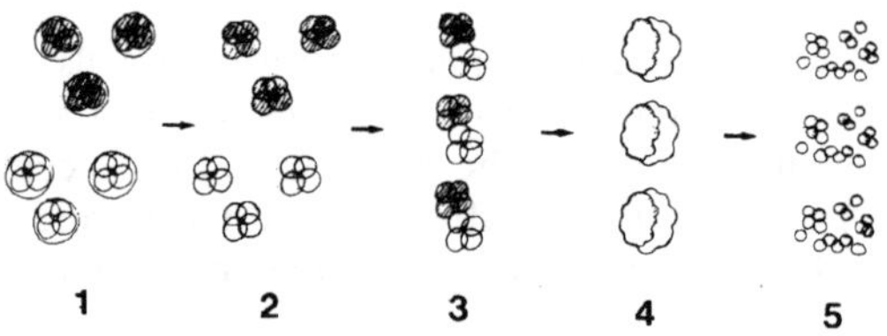

Figure 1. Diagram of the Chimera Assay. Four-cell embryos are labelled with FITC as desired (1), treated with acidified buffer to remove zonae pellucidae (2), aggregated into pairs using PHA (3), and cultured 35-40h (4). Chimeras are then partially dissociated using calcium-free medium and examined to obtain proliferation ratios (5).

are removed by a brief treatment with acidified buffer and phytohemagglutinin (PHA; Mintz *et al.*, 1973) is used to aggregate the embryos into pairs to form two types of chimeras: 1) "Control chimeras," containing one FITC-labelled normal embryo and one non-labelled normal embryo; these control for any effect of the FITC and provide a standard deviation for statistical analyses. And, 2) "Heterologous chimeras," containing one FITC-labelled normal embryo and one non-labelled 'experimental' embryo (progeny embryo from an irradiated parent).

The chimeras are cultured in 20 μl drops of culture medium for 35-40 hours to allow time for 2-3 cell cycles. Chimeric embryos are then partially dissociated with calcium-free medium and examined under phase contrast optics to obtain total chimera cell number and then examined under epifluorescence to obtain the number of progeny cells from the unlabelled (non-fluorescent) partner embryo. The ratio of unlabelled cell number: total chimera cell number provides a **proliferation ratio** for a given chimera and these ratios are averaged for a dose group and the **mean proliferation ratios** are compared using a variety of parametric and non-parametric statistical tests (Obasaju *et al.*, 1989; Warner *et al.*, 1991). A ratio of 0.5 occurs when both 4-cell embryos comprising a chimera contributed equal numbers of progeny cells to the chimera by the time of dissociation. A ratio less than 0.5 occurs when the experimental 4-cell partner embryo expressed a cell proliferation disadvantage and contributed fewer progeny cells to the chimera than did its partner control embryo.

From 10 to 15 animals are included per dose group and produce from 5 to 15 chimeras per animal. Coefficient of variation for mean proliferation ratios for a given dose group is typically 10-15%, which is very low for reproductive assessments using in vivo exposure of intact, genetically outbred animals such as the CD1 strain used in the chimera assay.

The CD1 random-bred strain has been a popular strain in the carcinogenicity studies conducted by the NCI and NTP programs (Maita *et al.*, 1988). As a result, there exists an enormous databank on the baseline incidence of a wide range of tumor types in this strain (Charles River Digest, vol. 22, January issue, 1983; Jerry M. Rice, Chief of the Laboratory of Comparative Carcinogenesis, NIH/NCI, personal communication).

Rationale for Using Chimeras as Radiation Biodosimeters for Embryonic Effects

This rationale rests on two observations from the literature on the effects of low-dose irradiation. The first observation is that by culturing 2-cell mouse embryos to the blastocyst stage in very low concentrations of ^{3}H-thymidine (5 μCi/ml) has no effect on embryo cell number of blastocysts, but prevents completely any such embryos from developing to term following transfer to foster mothers (Horner and McLaren, 1974). This observation shows that 'asymptomatic' radiation exposure during preimplantation development can still produce embryonic death after implantation.

The second observation was obtained using dissociated 8-cell blastomeres that were treated for 2h with ^{3}H-thymidine (0.25 μCi/ml). When such blastomeres are self-aggregated to reconstitute 8-cell embryos, their preimplantation vigor and cell proliferation rate are unaffected. However, when treated blastomeres are aggregated with non-treated blastomeres to reconstitute 8-cell embryos, the treated blastomeres become significantly under represented after 2-3 cell cycles. This observation shows that 'asymptomatic' irradiation can become expressed as a cell proliferation disadvantage when irradiated blastomeres are challenged by direct contact with unirradiated blastomeres in aggregation chimeras (Kelly and Rossant, 1976). Together, these two observations led to the hypothesis that this embryonic cell proliferation disadvantage, as detected in aggregation chimeras prior to implantation, is predictive of effects that can persist after postimplantation to impair embryo viability.

RESULTS OF THE CHIMERA ASSAY

In Vitro Irradiation of Preimplantation Embryos

The long-term objective is to use chimeras as biodosimeters of embryonic effects stemming from parental exposure to environmentally encountered ionizing radiation, particularly the more common photon radiation modalities such as X-rays and gamma-rays. It was therefore necessary to confirm that photon radiation would also produce the same embryonic cell proliferation disadvantage as radiation produced by beta decay of tritium concentrated in the DNA. This was shown to be the case using X-rays with mean proliferation ratios from irradiated<–>non-irradiated 4-cell chimeras exhibiting a dose-response from 0.05 Gy to 2.0 Gy. (Table 1; Obasaju *et al.*, 1988).

Table 1. Proliferation Ratios Produced by 4-cell Embryos Irradiated With X-rays *in vitro*

Type of Chimera	Mean Proliferation ratio (no. chimeras)
0 Gy<–>0 Gy (control chimeras)	0.49 ± 0.03 (32)
0 Gy<–>0.05 Gy	0.43 ± 0.01 (30)
0 Gy<–>0.25 Gy	0.41 ± 0.08 (23)
0 Gy<–>0.50 Gy	0.40 ± 0.04 (21)
0 Gy<–>1.0 Gy	0.37 ± 0.06 (23)
0 Gy<–>2.0 Gy	0.33 ± 0.06 (17)

All ratios are given as the mean ± s.d. All mean proliferation ratios from chimeras containing irradiated embryos are significantly different (ANOVA) from the mean proliferation ratio from control chimeras.

How the Chimera Assay is Used to Study Embryonic Effects From In Vivo Irradiation of the Male Germ Cell

The following experimental design is a traditional one that has been used extensively to study male-mediated developmental toxicity. Cohorts of males are irradiated as desired and then bred weekly to superovulated females over a 9-week post-irradiation period. The 4-cell embryos obtained from these breedings are tested in the chimera assay for a transmitted embryonic cell proliferation disadvantage expressed as decreases in mean proliferation ratios (Figure 2). This experimental design allows progressively younger stages of germ cells to be tested each week, with epididymal sperm being tested during post-irradiation week one and type A spermatogonia being tested during post-irradiation week 8. These different stages of spermatogenesis differ in radiosensitivity with respect to cell-killing, with mature sperm being most radioresistant (LD_{50} of 200 Gy) and the type B spermatogonium being most radiosensitive (LD_{50} of 0.25-0.40 Gy, post-irradiation weeks 6/7).

With doses of 1.7 Gy or more, significant decreases in mean proliferation ratios begin to appear during post-irradiation week 3 (Figure 3). Mean proliferation ratios continue to fall until post-irradiation week 7 and return to the control value of 0.5 by post-irradiation week 9. When the much lower dose of 0.01 Gy is used, two spermatogenic stages continue to transmit decreases in mean proliferation ratios in the

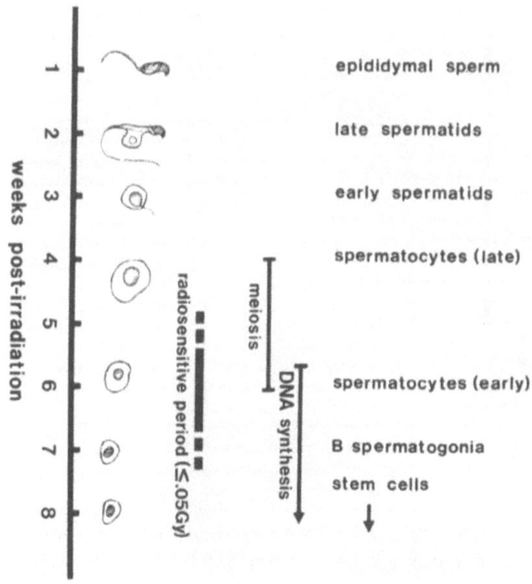

Figure 2. Correlation between transmitted decreases in mean proliferation ratios with spermatogenic stage at the time of irradiation.

chimera assay (Figure 4; Warner *et al.*, 1991). The earliest of these occurs at post-irradiation week 4, which corresponds to the pachytene spermatocyte while the later one occurs at post-irradiation weeks 6/7, which corresponds to the type B spermatogonium and likely also the late type A spermatogonium.

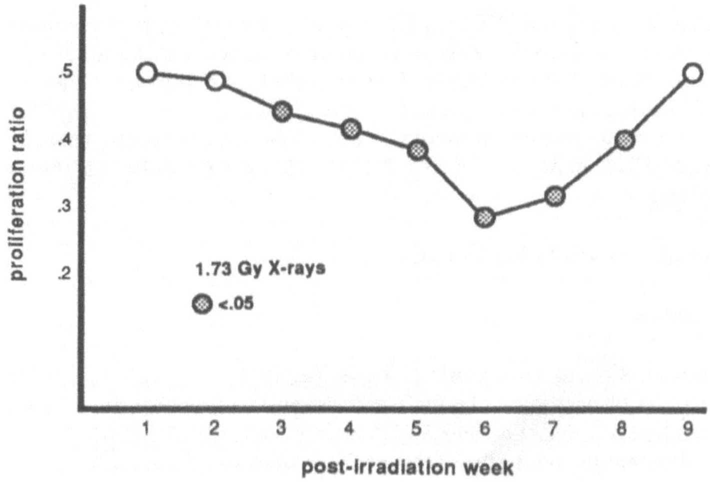

Figure 3. Time course of decreases in proliferation ratios obtained with male germ cells irradiated *in vivo*.

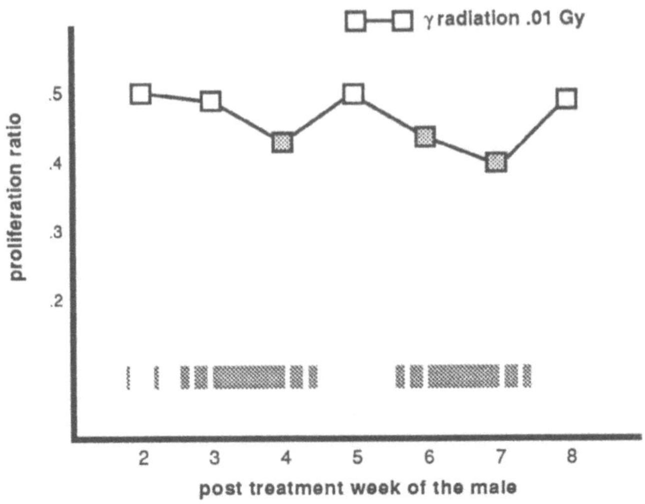

Figure 4. Correlations between Low Proliferation Ratios and Embryonic Effects. Shaded horizontal bars indicate the post-irradiation weeks when sperm from treated males transmitted significant increases in the incidences of birth anomalies (week 3, Kirk and Lyon, 1984), cancer (week 3, Nomura, 1989; week 6, Nomura, 1988) and decreases *in vitro* fertilization capacity (week 4 through weeks 6/7, Matsuda *et al.*, 1985). Shaded symbols represent significantly decreased mean proliferation ratios.

It is important to appreciate that these two time periods and stages of spermatogenesis differ in two fundamental ways. Pachytene spermatocytes are in meiosis and are not undergoing DNA synthesis while type B and type A spermatogonia are still mitotic. Consequently spermatogonia are fully repair-competent because fixation of DNA damage can occur during DNA replication and/or by unscheduled DNA synthesis. Although DNA repair may take place in the pachytene spermatocytes no repair via unscheduled DNA synthesis appears to occur with radiation doses as high as 1.0 Gy (of X-irradiation; Meistrich *et al.*, 1975). Therefore, at the very low doses examined in the chimera assay, the two times when decreases in proliferation ratios are observed--post-irradiation weeks 4 and 6/7--may represent sperm transmitting non-repaired DNA changes and 'fixed' DNA changes (mutations), respectively.

RELEVANCE TO HEALTH ISSUES

Animal Studies

Transmission of embryonic cell proliferation disadvantage has an interesting correlation with transmission of increases in several embryonic effects in mice including birth anomalies, age-onset cancer and decreases in fertilizing capacity *in vitro* (Figure 4). The correlations are not perfect, perhaps because these embryonic effects were obtained using radiation doses one to three orders of magnitude greater than 0.01 Gy and using slightly different post-irradiation breeding protocols and different mouse strains. These correlation, however rough, are consistent with the idea that the decreases in proliferation ratios transmitted by exposed pachytene spermatocytes and type B spermatogonia are predictive of an increased transmission of embryonic effects.

Human Studies

Although highly controversial, the epidemiological studies published by Gardner *et al.* (1990) recorded a significant increased risk of childhood leukemias (mostly acute lymphocytic leukemias) in children whose fathers working at the Sellefield Nuclear Power Plant had received 0.01 Gy or more of radiation within 6 months of conception. In these cases, the involved sperm would have been irradiated in the postgonial stages of spermatogenesis. At odds with this study are the Hiroshima/Nagasaki data analysis (see Evans, 1990), which has not paralleled the Sellefield data analysis of Gardner *et al.* (1990). One possibility is that the final outcome from the Hiroshima/Nagasaki bombings is yet to emerge, and that it will do so as an increased risk for age-onset cancer as the survivors reach advanced age (Nomura, 1990). Other possibilities accounting for this discrepancy include race-related differences in sensitivities in germline transmissible mutations and/or in subsequent exposure of progeny to environmental tumor promotors (Nomura, 1990). These possibilities are supported by animal studies using different strains of mice (Nomura, 1986) and a combination of male parental exposure to radiation followed by postnatal exposure to tumor promoters (Nomura, 1988, 1989).

NATURE OF THE RADIOSENSITIVE TARGET EXPRESSED IN THE CHIMERA ASSAY

Argument for a Nuclear DNA Target

Straightforward calculations show that almost 1,000 ion pairs will be generated by 0.01 Gy of photon-type radiation to the type B spermatogonium (average cell diameter, 10 microns) and around 100 of these will appear in the nucleus. This would in theory deposit enough energy to provide an opportunity for lesions within the DNA. Since the type B spermatogonia undergo DNA repair (during mitosis and by additional mechanisms), these lesions, if they occurred, would have the possibility of being repaired, or, fixed as heritable mutations. Added to these two considerations is the polygenic nature of the regulation of cell proliferation. In theory there are hundreds of genes whose products affect cell proliferation including cytoskeletal proteins, growth factors and their ligands, cellular proto-oncogenes and gene products participating in signal transduction cascades and so forth. From the cancer literature, there is substantial evidence that photon radiation induces the expression of several cellular proto-oncogenes (Anderson *et al.*, 1992), tubulin and protein kinase C (Woloshak *et al.*, 1990). Some of these gene products, c-fos and protein kinase C in particular, are up regulated during expression of receptors encoding epidermal growth factor (EGFR), for example, and EGFR is expressed and functional in mouse preimplantation embryos (Wood and Kaye, 1989; Wiley *et al.*, 1992).

Taken together, these observations are consistent with the idea that in the chimera assay, the experimental embryo expresses a dysfunctional mutated gene product involved in cell proliferation that confers a competitive cell proliferation disadvantage when challenged by cell-cell contact with a normal embryo. Every embryo exhibiting this disadvantage could be expressing one dysfunctional gene product out of the several hundred (thousand?) gene products that can affect cell proliferation. Collectively, all of these genes could present a huge genetic target sharing a common endpoint, embryonic cell proliferation disadvantage in the competitive setting within an aggregation chimera.

Arguments for a Non-Nuclear DNA Target

The classic Oak Ridge specific-locus assays using seven visible recessive marker genes required hundreds of thousands of progeny to be analyzed to detect transmission frequencies of 10^{-5} for embryonic effects (birth anomalies) using radiation doses of 0.3 Gy or higher (Russell and Kelly, 1982). However, roughly 20% of experimental 4-cell embryos from a cohort of males irradiated with 0.01 Gy express proliferation ratios less than 0.5. This is a difference of at least four orders of magnitude between the Oak Ridge and chimera assay data that begs explanation. Although the above argument for a large polygenic target could account for some of this discrepancy, the shape of the dose-response curve for the chimera assay suggests that additional factors are involved.

Hypothesis for the Radiosensitive Target

This hypothesis combines the polygenic target idea discussed above with facts regarding the biology of spermatogenesis. An interesting observation emerges when a dose-response curves is generated for 4-cell embryos obtained from males during post-irradiation week 6 for doses ranging from 0.01 Gy to 1.8 Gy (air doses; Figure 5). The slope of this curve is much steeper at doses below 0.2 Gy than at higher doses.

During spermatogenesis, the developing sperm cells that are mitotic progeny of one spermatogonium can remain connected by cytoplasmic bridges. By the end of spermiogenesis, hundreds of spermatids can be linked into a syncytium. These bridges can be one micron in diameter (Clermont and Rambourg, 1978) and can transmit mRNA and proteins (Braune *et al.*, 1989; De Boer *et al.*, 1990). The result of such transmission is that even though they will be genetically distinct, the interconnected spermatids can be phenotypically equivalent.

If irradiation caused some genetic alteration early in spermatogenesis to cause altered plasma membrane components, for example, this could conceivably affect the

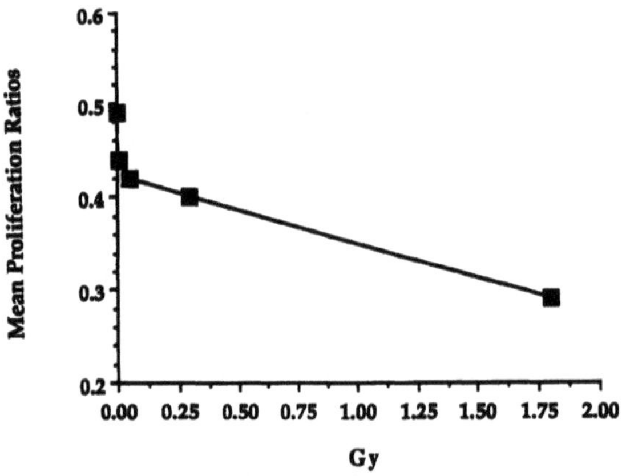

Figure 5. Dose-response curve from the chimera assay of embryos obtained from males during post-irradiation week 6.

viability of the early embryo. This is because the plasma membrane of the newly formed embryo is a mosaic of both sperm and oocyte plasma membrane. A sufficient amount of sperm-originated antigens remain on the surface of the embryo to mediate complement-dependent cell lysis of embryonic stages when proliferation ratios are determined in the chimera assay (Gaunt, 1983; Menge and Rajesh, 1988). This observation suggests that enough plasma membrane is inherited from the sperm to affect embryonic health. As a result, a mutagenic event leading to altered plasma membranes in a single spermatogonial germ cell could be amplified to result in a substantially greater transmission frequency of sperm capable of impairing embryo viability than that expected from a single mutation's effects remaining confined to a single sperm.

This argument becomes more convincing in light of the fact that experimental and control embryos conventionally cultured within their respective zonae pellucidae do not produce decreased proliferation ratios even for doses of 1.8 Gy (Obasaju *et al.*, 1989). In other words, direct cell-cell contact between experimental and control embryos is required for decreases in proliferation ratios. In other words, decreased proliferation ratios are (in part?) a result of altered plasma membrane function (transport? gap function communication?). There might be other sperm-originating non-nuclear entities that produce an (additive?) effect on viability of the early embryo.

It is therefore hypothesized that the increased sensitivity of the chimera assay for very low radiation doses results from non-nuclear targets whose effects predominate over those from genetic effects. At higher doses, these non-nuclear effects might be overshadowed by genetic ones to produce the straight negative slope that would be expected from a stochastic increase in affected targets of similar position (nuclear) and composition (DNA).

CONCLUSIONS

The chimera assay is characterized by an extraordinary sensitivity for detecting transmitted embryonic effects from irradiation of the male germ cell *in vivo*. It is hypothesized that the radiosensitive target that becomes expressed in this assay has two components. One of these components may be genetic, based on the polygenic regulation of cell proliferation, which is the endpoint of this assay. The other component may be non-genetic based on the phenotypic inheritance the embryo receives from the gametes and the persistence of cytoplasmic bridges linking male germ cells during spermatogenesis. At higher doses (>0.2 Gy) it is hypothesized that the genetic effects may predominate while the non-genetic effects become more predominant at lower doses and confer the greater than expected sensitivity of the assay at very low doses. To my knowledge, this is the first documented situation where exposure of the male to a known mutagen may be able to induce transmitted, non-DNA alterations in the sperm that affects the viability of the F1 embryo.

Experiments are now in progress to determine the heritability and existence of the hypothesized two component-radiosensitive target whose endpoint is embryonic cell proliferation disadvantage. In addition, it will be important to determine whether subsequent postnatal exposure of these F1 progeny to radiation or promotors will produce further decreases in proliferation ratios of F2 embryos.

REFERENCES

Anderson, A. and Woloschak, G.E., 1992, Cellular proto-oncogene expression following exposure of mice to gamma rays, *Radiat. Res.* 130:340.

Bennett, J. and Pedersen R.A., 1984, Early mouse embryos exhibit strain variation in radiation-induced sister-chromatid exchange: Relationship with DNA repair. *Mutat. Res.* 126:153.

Biggers, J.D., 1981, *In vitro* fertilization and embryo transfer in human beings, *N. Engl. J. Med.* 304:336.

Braune, R.E., Behringer, R.R., Peschon, J.J., Brinster, R.L. and Palmiter, R.D., 1989, Genotypically haploid spermatids are phenotypically diploid, *Nature* 337:373.

Clermont, Y. and Rambourg, A., 1978, Evolution of endoplasmic reticulum during rat spermiogenesis, *Am. J. Anat.* 151:191.

De Boer, P., Redi, C.A., Garagna, S. and Winking, H., 1990, Protamine amount and cross linking in mouse teratospermatozoa and aneuploid spermatozoa, *Mol. Reprod. Dev.* 25:297.

Eibs, H.-G. and Spielmann, H., 1977, Inhibition of post-implantation development of mouse blastocysts *in vitro* after cyclophosphamide treatment *in vivo*. *Nature* 270:52.

Evans, H.J., 1990, Leukaemia and radiation, *Nature* 345:16.

Gardner, M.J,, Snee, M.P., Hall, A.J., Powell, C.A., Downes, S. and Terrell, J.D., 1990, Results of case-control study of leukaemia and lymphoma among young people near Sellafield nuclear plant in West Cumbria, *BMJ* 300:423.

Gaunt, S.J., 1983, Spreading of a sperm surface antigen within the plasma membrane of the egg after fertilization, *J. Embryol. Exp. Morph.* 75:259.

Goldstein, L.S. and Spindle, A.I., 1976, Detection of X-ray induced dominant lethal mutations in mice: An *in vivo* approach, *Mutat. Res.* 41:289.

Goldstein, L.S., 1984a, Dominant lethal mutations induced in mouse spermatozoa by antineoplastic drugs, *Mutat. Res.* 140:193.

Goldstein, L.S., 1984b, Use of an *in vitro* technique to detect mutations induced by antineoplastic drugs in mouse germ cells, *Cancer Treat. Rep.* 68:855.

Horner, D. and McLaren, A., 1974, The effect of low concentrations of [3H]thymidine on pre- and postimplantation mouse embryos. *Biol. Reprod.* 11:553.

Kelly, S.J. and Rossant, J., 1976, The effect of short-term labelling in [3H]thymidine on the viability of mouse blastomeres: Alone and in combination with unlabelled blastomeres, *J. Embryol. Exp. Morph.* 35:95.

Kitazawa, T., Enomoto, A., Inui, K. and Shirasu,Y., 1988, Mortality, major cause of moribundity, and spontaneous tumors in CD-1 mice. *Toxicol Pathol.*, 16:340.

Kline, J. and Stein, Z., 1985, Very early pregnancy, in "Reproductive Toxicology," R.L. Dixon (ed), Raven Press, New York, NY, pp. 251.

Maita, K., Hirano, M., Harada, T., Mitsumori, K., Yoshida, A., Takahashi, K., Nakashima, N., Kirk, M. and Lyon, M.F, 1984, Induction of congenital malformations in the offspring of male mice treated with X-rays at pre-meiotic and post-meiotic stages. *Mutat. Res.* 125:75.

Matsuda, Y., Tobari, I. and Yamada, T., 1985, *In vitro* fertilization of mouse eggs with sperm after X-irradiation at various spermatogenetic stages, *Mutat. Res.* 142:59.

Meistrich, M.L., Reid, B.O. and Barcellona, W.J., 1975, Meiotic DNA synthesis during mouse spermatogenesis. *J. Cell. Biol.* 64:211.

Menge, A.C. and Rajesh, K.N., 1988, Immunologic reactions involving sperm cells and preimplantation embryos, *Am. J. Reprod. Immunol. Microbiol.* 18:17.

Mintz, B., Gearhart, J.D. and Guymont, A.O., 1973, Phytohaemagglutinin-mediated blastomere aggregation and development of allophanic mice, *Dev. Biol.* 31:195.

Nomura, T., 1986, Further studies on X-ray and chemically induced germ-line alterations causing tumors and malformations in mice, in "Genetic Toxicology of Environmental Chemicals," C. Ramel, B. Lambert and J. Magnusson, eds., Liss, New York, NY, pp. 13-20.

Nomura, T., 1988, X-ray and chemically-induced germ-line mutation causing phenotypical anomalies in mice, *Mutat. Res.* 198:309.

Nomura, T., 1989, Role of radiation-induced mutations in multigeneration carcinogenesis, in "Perinatal and Multigeneration Carcinogenesis," N.P. Napaikov, J.M. Rice, L. Tomatis and H. Yamasaki, eds., International Agency for Research on Cancer, Lyon, France, pp. 375.

Nomura, T., 1990, Of mice and men?, *Nature* 345:671.

Obasaju, M.F., Wiley, L.M., Oudiz, D.J., Miller, L., Samuels, S.J., Chang, R.J. and Overstreet, J.W., 1988, An assay using embryo aggregation chimeras for the detection of nonlethal changes in x-irradiated mouse preimplantation embryos, *Radiat. Res.* 113:289.

Obasaju, M.F., Wiley, L.M., Oudiz, D.J., Raabe, O.G. and Overstreet, J.W., 1989, A chimera assay reveals a decrease in embryonic cellular proliferation induced by sperm from x-irradiated male mice, *Radiat. Res.* 118:246.

Overstreet, J.W., 1973, The labelling of live rabbit ova with fluorescent dyes, *J. Reprod. Fertil.* 32:291.

Russell, W.L. and Kelly, E.M., 1982, Mutation frequencies in male mice and the estimation of genetic hazards of radiation in men, *Proc. Nat . Acad. Sci.*, USA, 79:542.

Spielmann, H. and Eibs, H.-G., 1978, Recent progress in teratology: A survey of methods for the study of drug actions during the preimplantation period. *Arzneim-Forsch./Drug Res.* 28:1733.

Warner, P., Wiley, L.M., Oudiz, D.J., Overstreet, J.W. and Raabe, O.G., 1991, Paternally inherited effects of gamma radiation on mouse preimplantation development detected by the chimera assay, *Radiat. Res.* 128:48.

Wiley, L.M,. Wu, J.-X., Harari, I. and Adamson, E.D., 1992, Initiation of epidermal growth factor receptor gene activity occurs after the four cell preimplantation stage of murine development, *Develop. Biol.* 149:247.

Woloschak, E.G., Lie, C.-M. and Shearin-Jones, P., 1990, Regulation of protein Kinase C by ionizing radiation, *Cancer Res.* 50:3963.

Wood, S.A. and Kaye, P.L., 1989, Effects of epidermal growth factor on preimplantation mouse embryos, *J. Reprod. Fert.* 85:575.

POST-TESTICULAR MECHANISMS OF MALE-MEDIATED DEVELOPMENTAL TOXICITY

Bernard Robaire and Barbara F. Hales

Department of Pharmacology and Therapeutics
Centre for the Study of Reproduction
McGill University
3655 Drummond Street
Montreal, Quebec, Canada, H3G 1Y6

INTRODUCTION

The cause of most (as much as 70%) of the congenital malformations in man is not known (Wilson, 1977). There is increasing evidence and concern that male-mediated effects on development have a major importance in adverse pregnancy outcome (this conference; Robaire *et al.*, 1985; Robaire and Hales, 1993). The contribution of the father to chemically-mediated adverse progeny outcome may be subdivided into three categories based on the site of action of the chemical in the male reproductive tract. The first category is that of direct exposure of the fertilized egg, blastocyst, embryo or fetus to drugs, or their metabolites, which are present in semen. The second category focuses on changes in spermatozoa after they leave the testis, i.e. changes that may take place in spermatozoa in the excurrent duct system or at the time secretions from the sex accessory glands (prostate, seminal vesicles, bulbourethral gland) are mixed with spermatozoa during ejaculation. The final category is that of drugs directly affecting spermatozoa as they are being synthesized in the seminiferous tubules. This last category is dealt with in detail in another chapter in this volume (Hales and Robaire, 1994).

DRUGS IN SEMEN

In most mammals the overwhelming majority of the fluid constituting semen arises from the secretions of sex accessory glands at the time of ejaculation. In man, approximately 60% of seminal fluid arises from the seminal vesicles, 30% from the prostate and only about 10% of seminal fluid volume arises from spermatozoa and the fluid bathing them in the excurrent duct system (Polakoski and Zaneveld, 1977). Consequently, the access of drugs to the secretions of the accessory glands largely

Male-Mediated Developmental Toxicity, Edited by D.R. Mattison
and A.F. Olshan, Plenum Press, New York, 1994

determines their presence in semen. A wide range of compounds has been shown to enter semen (Mann and Lutwak-Mann, 1982). It has long been established that drugs are well absorbed after intravaginal administration (Benziger and Edelson, 1983). Thus, drugs in seminal fluid may be absorbed by the mother and may potentially affect the conceptus.

While most mammals will not allow intercourse after the initiation of pregnancy, humans continue to have intercourse throughout pregnancy; hence, the conceptus may be exposed to a drug in the seminal fluid not only at the time of conception, but at any subsequent time during development. As a consequence, all animal experimentation will underestimate the potential risk of drugs in semen to progeny outcome.

To demonstrate that a drug or its metabolite given to the father can affect progeny outcome, it is essential to establish first that the drug can enter the male reproductive tract, second that it can be transmitted to the female, and finally that the transmitted drug can produce an effect on progeny outcome.

Entry of Drugs into the Male Reproductive Tract

Though it has been hypothesized for many years that drugs taken by a male can potentially enter his semen, it is only in the last twenty years that the ability of drugs and endogenous chemicals to cross different regions of the male reproductive tract has been investigated. There have been very few studies designed to test the ability of specific drugs to cross the blood-testis or blood-epididymis barriers. The ability of the male reproductive system to concentrate drugs varies greatly. Some compounds such as the antimicrobial, thiamphenicol, are concentrated in semen relative to serum twenty-four hours after injection (Plomp *et al.*, 1978). Others, such as amphetamine (Smith, 1981) or cyclophosphamide (Hales *et al.*, 1986), attain concentrations similar to those in serum while others still, such as diphenylhydantoin (Swanson *et al.*, 1978), are found in semen in concentrations consistently lower than those in serum, even after prolonged exposure.

In micropuncture studies from Howards' laboratory (Forrest *et al.*, 1981), it was shown that cytotoxic agents such as cyclophosphamide and vincristine will penetrate the blood-testis barrier and enter the lumen of the seminiferous epithelium. However, the entry of drugs into semen is postulated to occur primarily via passage from the capillaries into ducts in the prostate and the seminal vesicles (Eliasson and Dornbusch, 1980; Malmborg, 1978). Indeed, we have demonstrated that cyclophosphamide and/or its metabolites can enter all tissues of the male reproductive tract, including the seminal vesicle fluid (Hales *et al.*, 1986). In this experiment, ^{14}C-cyclophosphamide was given intravenously to adult male Sprague-Dawley rats. Radiolabel (greater than 96% cyclophosphamide) was first found in seminal vesicle fluid within 10 minutes and equilibrated with plasma radioactivity within 30 minutes (Figure 1). The ratio of seminal vesicle fluid to plasma radiolabel remained close to 1 for at least two hours after drug administration. Moreover, more than 96% of the radiolabel in the seminal vesicle fluid co-chromatographed with authentic cyclophosphamide on thin-layer plates (Hales *et al.*, 1986). Thus, radiolabel from cyclophosphamide entered the male reproductive tract and reached equilibrium with blood concentrations. These concentrations were maintained for at least 2 hours following administration of the drug.

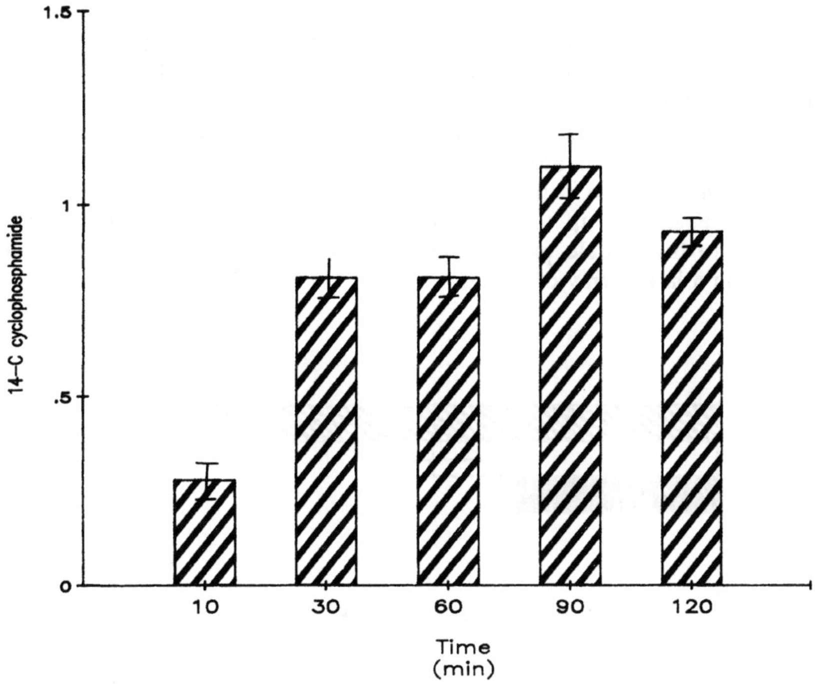

Figure 1. Ratio of radiolabel in the seminal vesicle fluid to that in plasma in adult male rats within two hours after a single intravenous injection of cyclophosphamide (10 mg/kg unlabelled cyclophosphamide, 50 μCi/rat ^{14}C-cyclophosphamide). Values represent mean ± SEM (n=4).

Transmission of Drugs in the Semen to the Female during Intercourse

Many drugs that are found in semen can be readily absorbed by the vagina and some can affect both the female partner and the progeny. Vinblastine in semen has been associated with vulvovaginitis in women (Paladine *et al.*, 1975). Pretreatment of male rats with high doses of estradiol results in uterine hypertrophy and hyperplasia in the female partner (Ericsson and Baker, 1966).

Evidence has accumulated which indicates that short-term paternal exposure to methadone results in an increased perinatal mortality and decreased birth weight of the offspring (Soyka *et al.*, 1978).

In our studies with cyclophosphamide, the next goal was to determine whether the radiolabel present in the seminal fluid of the male could be transmitted to the female during mating. Male rats received an intraperitoneal injection of cyclophosphamide (10 mg/kg unlabelled drug, 50 μCi/rat ^{14}C-cyclophosphamide); 1, 3 and 5 hours later they were each exposed to one ovariectomized, estrogen and progesterone primed female rat (Hales *et al.*, 1986).

Decreasing amounts of radiolabel were recovered in the seminal plugs in the females mated to males 1, 3 and 5 hours after the injection of the labelled drug; by 5 hours the amount of radiolabel had decreased to nearly one-third of that found after 1 hour (Figure 2). Radioactivity was also found in a variety of tissues in the female rats 50 minutes after mating. Radiolabel was found in decreasing amounts after the 1 to 5 hour matings in tissues of the female reproductive tract such as the vagina and cervix, albeit at a level approximately one order of magnitude below that found in the seminal plugs. Interestingly, the presence of radioactivity was not limited to tissues of the female reproductive tract, or tissues of drug metabolism and excretion such as the liver and kidney, but was also found in brain and muscle. Thus, radiolabel from cyclophosphamide and/or its metabolites was excreted in the seminal fluid of the male in sufficient quantities up to 5 hours post-drug treatment such that it could be transmitted to the female partner where it was absorbed across the vagina and distributed in a large array of tissues.

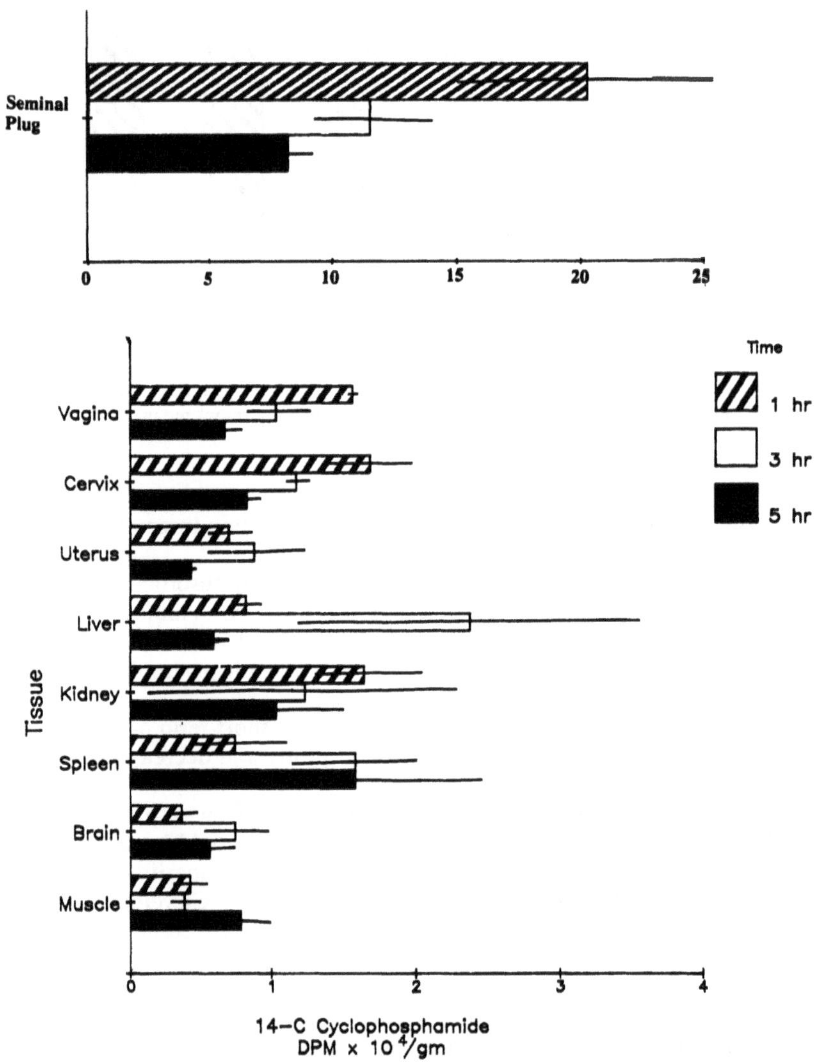

Figure 2. Amount of ^{14}C-cyclophosphamide in seminal plugs (upper panel) and in tissues of female rats mated to males 1, 3 and 5 hours after treatment of the male with an intraperitoneal injection of cyclophosphamide (10 mg/kg unlabelled drug, 50 μCi/rat ^{14}C-radiolabelled cyclophosphamide). Females were killed 50 minutes after mating and tissues were prepared for the assessment of radiolabel by liquid scintillation counting.

Effects of Drugs in Semen on Progeny Outcome

To determine the ability of paternally-administered cyclophosphamide to alter progeny outcome, male rats were given vehicle or a single dose of cyclophosphamide of 10, 30 or 100 mg/kg immediately prior to co-habitation with females in proestrus. The females were killed on day 20 of gestation and pregnancy outcome was evaluated. Preimplantation loss was assessed by counting the number of corpora lutea in both ovaries and subtracting this from the number of resorption sites in the uterus. Postimplantation loss represented the total number of resorbed or dead fetuses per pregnant female. There was no effect of cyclophosphamide treatment on the number of pregnant females per sperm positive females but there was a significant dose -

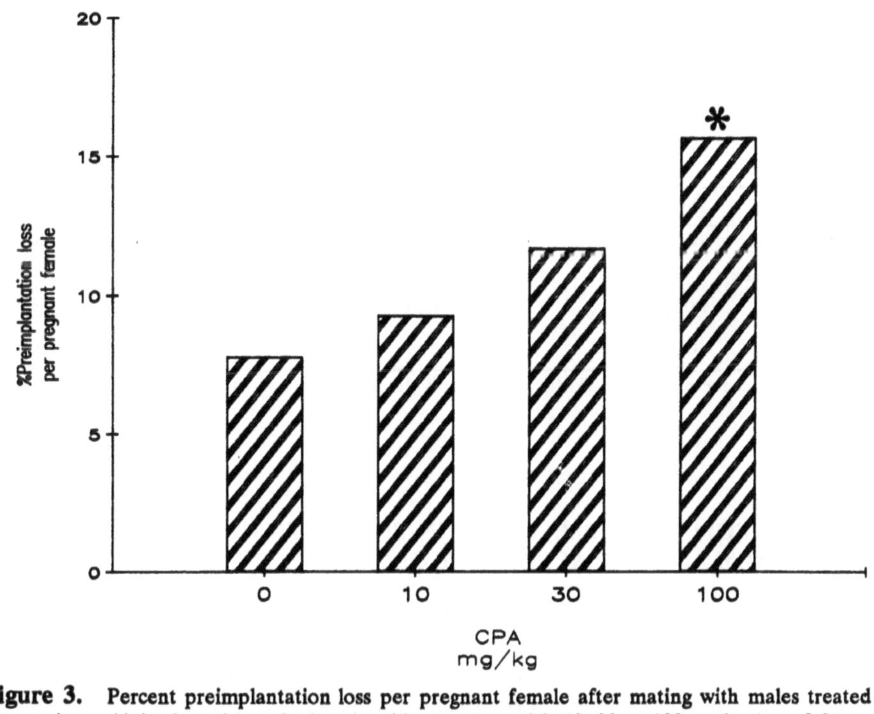

Figure 3. Percent preimplantation loss per pregnant female after mating with males treated by intraperitoneal injection with cyclophosphamide at a dose of 0, 10, 30 or 100 mg/kg (n = 8-9/group). * Significantly different from control, $p \leq 0.05$.

dependent increase in preimplantation loss (Figure 3). There was no increase in postimplantation loss or the number of abnormal fetuses (Hales *et al.*, 1986).

The effect of treatment of males with cyclophosphamide on preimplantation loss could be due either to an effect of the presence of the drug itself in the semen or to an effect of the drug on spermatozoa in the testicular excurrent duct system. To distinguish between these possibilities, a second experiment was done. In this study, females in proestrus mated to control males were remated within two hours to vasectomized males treated acutely with cyclophosphamide (50 or 100 mg/kg) or saline.

Once again, preimplantation loss (number of corpora lutea minus number of implantation sites divided by number of corpora lutea) was increased significantly in both drug treated groups (Table 1). The magnitude of the preimplantation loss in this experiment was nearly identical to that observed in the previous study. Thus, this increase in preimplantation loss appears to be due to the presence of the drug and or its metabolites in the seminal fluid, rather than to an effect on spermatozoa stored in the testicular excurrent duct system.

It is thus clear that the presence of drugs in semen can modify progeny outcome. The effects may be quantitatively important but occur early during gestation. If similar effects were to occur in humans, they would most likely be interpreted as an increased rate of idiopathic infertility in the couple.

Table 1. Numbers of Corpora Lutea and Implantation Sites in Females Mated to Control Males and Subsequently to Vasectomized Males Treated Acutely with Saline or Cyclophosphamide

Treatment Group	Pregnant/sperm positive females		Corpora lutea/ pregnant female	Implantations/ pregnant female	Corpora lutea-implantations/total corpora lutea	
	No.	%			No.	%
Saline	10/12	83	14.1±0.6[a]	13.6±0.5	5/141	3.6
Cyclophosphamide						
50 mg/kg	9/12	75	15.3±0.4	13.4±1.3	17/138	12.3*
100 mg/kg	10/12	83	13.6±0.5	11.5±1.2	21/136	15.5*

[a] Values represent the means ± S.E.M. * Significantly different from control, $p \leq 0.05$.

DRUGS AFFECTING SPERMATOZOA IN THE TESTICULAR EXCURRENT DUCT SYSTEM AND SEX ACCESSORY GLANDS

Structure and Functions of the Epididymis

When spermatozoa leave the testis they are neither motile nor do they have the ability to fertilize eggs. These functions are acquired during their lengthy transit (approximately 10 days in most mammals) through the epididymal duct. Depending on the species, the ability to fertilize eggs is acquired in the distal caput epididymidis (e.g. marmoset), in the corpus epididymidis (e.g. rabbit), or in the proximal cauda epididymidis (e.g. rat, man) (Robaire and Hermo, 1988). Storage of spermatozoa takes place in the cauda epididymidis (Robaire and Hermo, 1988). The epididymal duct, measuring several metres, is a single convoluted tubule that begins at the point where the ductuli efferentes converge to a single tube, and ends as the vas deferens. The environment seen by spermatozoa as they traverse this long duct undergoes dramatic changes with respect to its ionic make-up as well as the composition of small organic molecules and proteins (Robaire and Hermo, 1988). The entry of chemicals into the epididymal lumen is controlled by the blood-epididymal barrier which is analogous to the blood-testis barrier found between adjacent Sertoli cells in the seminiferous tubules (Hinton, 1985).

Drugs Affecting Progeny Outcome by Acting on Spermatozoa in the Testicular Excurrent Duct System

Remarkably few studies have been designed to test whether drugs that act selectively on the male excurrent duct system, or drugs that are administered in such a way so that their action is limited to the excurrent duct system, can affect progeny outcome. A wide range of chemicals will result in reduced fertility due to an androgen antagonist action or a direct inhibition of spermatogenesis. Compounds that have been shown to clearly act on the excurrent duct system are few in number and include α-chlorohydrin, 6-chloro-6-deoxy glucose, ethyl dimethanesulfonate, methyl chloride and cyclophosphamide. The first two agents act by forming spermatoceles in the caput epididymidis and hence blocking transport of spermatozoa (Jones, 1978; Ford and Waites, 1982). At lower doses, α-chlorohydrin can also act on spermatozoa residing in the cauda epididymidis by blocking the glyceraldehyde 3-phosphate dehydrogenase found in spermatozoa, and thus rendering them incapable of fertilizing oocytes (Jones, 1978).

Ethyl dimethanesulfonate (EDS) is an alkylating agent that selectively kills Leydig cells of the testis, thus inhibiting androgen production (Bartlett *et al.*, 1986, Morris *et al.*, 1986). This agent had previously been shown to affect the epididymis (Morris *et al.*, 1986), but it was not until recently that Klinefelter and his colleagues established that EDS could directly act on the epididymis, as opposed to having an indirect action via suppressing androgen production (Klinefelter *et al.*, 1990). The range of actions of this drug include a dramatic disruption of normal epididymal epithelial histology, decreased motility of spermatozoa, and a change in the membrane proteins of spermatozoa. Indeed, evidence has been presented that EDS prevents the maturation of spermatozoa in the epididymis by altering the normal changes in sperm membrane proteins (Klinefelter *et al.*, 1992). Though these investigators did not study the effect of treatment with this agent on progeny outcome, the effects they have found would indicate a high likelihood that treatment with this agent, using a protocol that would establish an epididymal site of action, would affect progeny outcome.

Methyl chloride can have a wide range of actions on the male reproductive system. However, Working and his colleagues (Working and Chellman, 1989; Chellman *et al.*, 1986a) have demonstrated that the administration of methyl chloride can cause an increase in dominant lethal mutations as a consequence of a selective inflammatory activity of this agent on the epididymis; this increase in embryo loss could be reversed by administering an anti-inflammatory agent (Chellman *et al.*, 1986b).

Previous studies from our laboratories have suggested that, in addition to an effect on spermatozoa in the testis, cyclophosphamide may have an adverse effect on spermatozoa after they leave the testis, during epididymal transit (Trasler *et al.*, 1986, 1987). To elaborate on this post-testicular effect on germ cells, and to determine at which site(s) in the epididymis germ cells is(are) most sensitive to cyclophosphamide treatment, three experiments were undertaken (Qiu *et al.*, 1992). In the first experiment, the time course of the effect of treatment of male rats with cyclophosphamide on their progeny outcome was studied in order to assess whether these were specific effects on germ cells residing in the caput (7 days), corpus (4 days) or cauda (1 day) epididymidis. Male rats were treated daily by gavage with saline or one of two doses of cyclophosphamide (6.8 mg/kg or 10.0 mg/kg) for 1, 4 or 7 days. At the end of each treatment period, males were mated to females to assess the effect on pregnancy outcome. No effect was observed on preimplantation loss at any time among any of the groups (Figure 4, upper panel); in contrast there was a time dependent and dose related increase in postimplantation loss (Figure 4, lower panel). Postimplantation loss was significantly increased after 4 days of treatment and reached

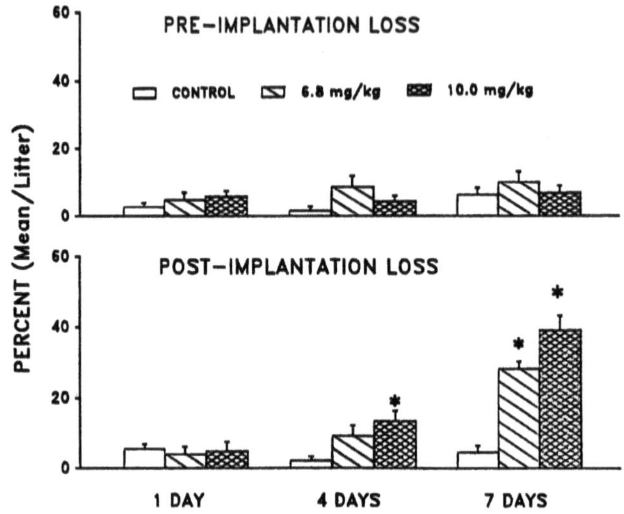

Figure 4. Effects of treatment of adult male rats with saline or cyclophosphamide (6.8 or 10.0 mg/kg/day) for 1, 4 or 7 days on progeny outcome. The upper panel indicates the percentage, calculated on a mean per litter basis, of preimplantation loss (number of corpora lutea minus number of implantation sites); the lower panel indicates the percentage of postimplantation loss (the number of implantation sites per pregnant female minus the number of resorptions per pregnant female). Values represent the mean ± SEM. * Significantly different from control, $p \leq 0.05$.

nearly 40% after 7 days of drug exposure in the 10 mg/kg cyclophosphamide treatment group. These data suggest that cyclophosphamide is acting on spermatozoa in the caput and corpus but not the cauda epididymidis.

To further test the hypothesis that cyclophosphamide is affecting progeny outcome by acting on spermatozoa in the excurrent duct system, a second experiment was undertaken. The objective of this study was to determine whether treatment of male rats with single higher doses of cyclophosphamide could affect spermatozoa in the cauda epididymidis as well as in the caput and corpus epididymidis. Male rats were treated with a single dose of cyclophosphamide (10, 30 or 70 mg/kg) and bred 1 and 4 days post-treatment. No significant change in preimplantation loss was observed at either time point; no change in postimplantation loss was found in the 1 day post-treatment groups. However, a significant increase in postimplantation loss was observed 4 days post-treatment in the 30 and 70 mg/kg cyclophosphamide treatment groups (25 to 30%, compared to controls of less than 5%). Hence, it would appear that spermatozoa residing in the cauda epididymidis cannot be affected by cyclophosphamide, whereas those residing in the corpus can be modified in such a way that a major increase in postimplantation loss occurs. Furthermore, the lack of an effect on postimplantation loss after 1 day exposure to a high dose of cyclophosphamide clearly indicates the "functional" absence of this drug from the semen at the time of mating.

In order to obtain unequivocal proof that cyclophosphamide was acting on spermatozoa residing in the excurrent duct system, and not on those still in the testis, a third experiment was undertaken. In this study the effect of cyclophosphamide treatment on animals after efferent duct ligation was determined. Bilateral ligation of the efferent ducts for 7 days did not affect male progeny outcome; however, ligation followed by the administration of cyclophosphamide, at a dose of 10 mg/kg/day for 7 days produced the same percentage of postimplantation loss as that produced by cyclophosphamide treatment alone (Figure 5).

Together these results demonstrate that short-term paternal treatment with cyclophosphamide (4 or 7 days) exerts an adverse effect on progeny outcome which is mediated by a post-testicular mechanism. That four days are required for the adverse effect of cyclophosphamide on post-testicular spermatozoa to be manifested suggests that spermatozoa in transit through the caput and corpus epididymides are affected by cyclophosphamide, while those in the cauda epididymidis are resistant.

Role of the Secretions of Sex Accessory Tissues in Progeny Outcome

The exact role played by the secretions of sex accessory tissues in fertility and progeny outcome has been a subject of ongoing controversy. However, a study by O *et al.* (1988) has clearly established that, in the golden hamster, removal of any of the sex accessory glands (ventral prostate, ampullary glands, or all of the accessory glands) did not affect the rate of fertilization or cleavage in the early embryo in the first 48 hrs, but did cause a dramatic decrease in the rate of cleavage at 72 hrs and in the number of implantation sites at 122 hrs post-conception. Such results would indicate that secretions from accessory glands can affect preimplantation embryonic development and have an impact on embryo survival.

SUMMARY AND CONCLUSIONS

It is thus clear that male-mediated adverse effects on progeny outcome can result from both direct effects of a drug in the semen and indirect effects via alterations of post-testicular spermatozoa or effects on the function of sex accessory tissues. There is a major need to initiate additional studies using an array of drugs to which men are exposed during their reproductive years. It is evident from the animal studies above that the effects of exposure of the father to a drug may include an

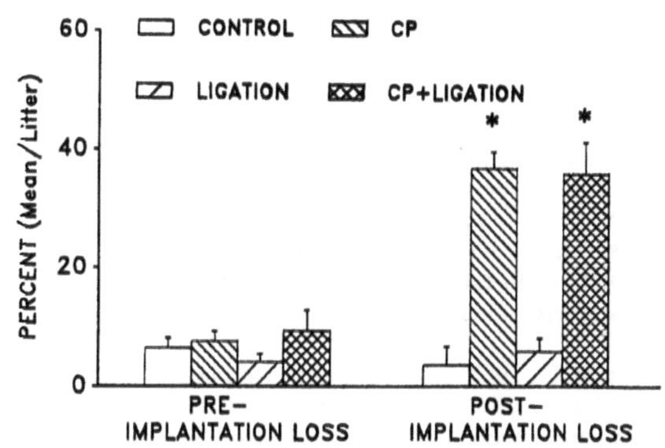

Figure 5. Effects of efferent duct ligation with or without cyclophosphamide treatment (10.0 mg/kg/day) for 7 days on progeny outcome. The left panel indicates the percentage, calculated on a mean per litter basis, of preimplantation loss (number of corpora lutea minus number of implantation sites per pregnant female); the right panel indicates the percentage of postimplantation loss (the number of implantation sites minus the number of resorptions per pregnant female). Values represent the mean ± SEM. * Significantly different from control, $p \leq 0.05$.

increased incidence of early spontaneous abortions that will likely go unnoticed; such an effect will be perceived as infertility and be diagnosed as an increase in idiopathic infertility.

ACKNOWLEDGMENTS

The studies from our laboratories that are reported in this chapter were supported by the Medical Research Council of Canada and by a National Foundation March of Dimes Reproductive Hazards in the Workplace Grant. We thank J. Qiu and S. Smith for their dedication to this project.

REFERENCES

Bartlett, J.M.S., Kerr, J.B., and Sharpe, R.M., 1986, The effect of selective destruction and regeneration of rat Leydig cells on the intratesticular distribution of T and the morphology of the seminiferous epithelium, *J. Androl.,* 7:240-253.

Benziger, D.P., and Edelson, J., 1983, Absorption from the vagina, *Drug Metabolism Reviews* 14(2):137-168.

Chellman, G.J., Bus, J.S., and Working, P.K., 1986a, Role of epididymal inflammation in the induction of dominant lethal mutations in Fischer 344 rat sperm by methyl chloride, *Proc. Natl. Acad. Sci. USA,* 83:8087-8091.

Chellman, G.J., Morgan, K.T., Bus, J.S., and Working, P.G., 1986b, Inhibition of methyl chloride toxicity in male F-344 rats by the anti-inflammatory agent BW755C, *Toxicol. Appl. Pharmacol.* 85:367-379.

Eliasson, R., and Dornbusch, K, 1980, Secretion of metronidazole into the human semen. *Int. J. Androl.* 3:236-242.

Ericsson, R.J., and Baker, V.F., 1966, Transport of oestrogens in semen to the female rat during mating and its effect on fertility. *J. Reprod. Fertil.* 12:381-384.

Ford, W.C.L., and Waites, G.M.H., 1982, Activities of various 6-chloro-6-deoxysugars and (S)α-chlorohydrin in producing spermatocoeles in rats and paralysis in mice and in inhibiting glucose metabolism in bull spermatozoa in vitro, *J. Reprod. Fertil.* 65:177-183.

Forrest, J.B., Turner, T.T., and Howards, S.S., 1981, Cyclophosphamide, vincristine, and the blood testis barrier, *Invest. Urol.* 18:443-444.

Hales, B.F., Smith, S., and Robaire, B., 1986, Cyclophosphamide in the seminal fluid of treated males: transmission to the female during mating and effect on pregnancy outcome. *Toxicol. Appl. Pharmacol.* 84: 423-430.

Hales, B.F. and Robaire B., 1993, The male-mediated developmental toxicity of cyclophosphamide. Andrew F. Olshan and Donald R. Mattison, eds., Plenum Press, New York (this volume).

Hinton, B.T., 1985, The blood-epididymis barrier, *in*: "Male Fertility and its Regulation," T. Lobl and E.S.E. Hafez, eds., MTP Press Ltd., Lancaster, pp. 371-393.

Jones, A.R., 1978, The antifertility actions of α-chlorohydrin in the male, *Life Sciences* 23:1625-1646.

Klinefelter, G.R., Laskey, J.W., Roberts, N.R., Slott, V., and Suarez, J.D., 1990, Multiple effects of ethane dimethylsulfonate on the epididymis of adult rats, *Toxicol. Appl. Pharmacol.* 105:271-287.

Klinefelter, G.R., Roberts, N.L., and Suarez, J.D., 1992, Direct effects of ethane dimethylsulfonate on epididymal function in adult rats: an in vitro demonstration, *J. Androl.* 13:409-421.

Malmborg, A.S., 1978, Antimicrobial drugs in human seminal plasma, *J. Antimicrobe Chemother.* 6:483-485.

Mann, T., and Lutwak-Mann, C., 1982, Passage of chemicals into human and animal semen: mechanisms and significance, *CRC Crit. Rev. Toxicol.,* 11:1-14.

Morris, I.D., Philips, D.M., and Bardin, C.W., 1986, Ethylene dimethanesulphonate destroys Leydig cells in the rat testis, *Endocrinology* 118:709-719.

O, W.S., Chen, H.Q., and Chow, P.H., 1988, Effects of male accessory sex gland secretions on early embryonic development in the golden hamster, *J. Reprod. Fertil.* 84:341-344.

Paladine, W.J., Cunningham, T.J., Donavan, M.A., and Dumper, C.W., 1975, Possible sensitivity to vinblastine in prostatic or seminal fluid. *N. Engl. J. Med.* 292:52.

Plomp, T.A., Mattelaer, J.J., and Maes, R.A.A., 1978, The concentration of thiamphenicol in seminal fluid and prostatic tissue. *J. Antimicrobe Chemother.* 4:65-71.

Polakoski, K.L., and Zaneveld, L.J.D., 1977, Biochemical examination of the human ejaculate, *in*: "Techniques of Human Andrology", E.S.E. Hafez, ed., Elsevier/North-Holland Biomedical Press, Amsterdam, pp.265-286.

Qiu, J., Hales, B.F. and Robaire, B., 1992, Adverse effects of cyclophosphamide on progeny outcome can be mediated through post-testicular mechanisms in the rat, *Biol. Reprod.* 46:926-931.

Robaire, B., and Hales, B., 1993, Paternal exposure to chemicals before conception. Some children may be at risk. *Brit. Med. J.* 307:341-342.

Robaire, B., and Hermo, L., 1988, Efferent ducts, epididymis and vas deferens: Structure, functions and their regulation, *in*: "Physiology of Reproduction," E. Knobil and J. Neill, *et al.*, eds, Raven Press, New York, pp. 999-1080.

Robaire, B., Trasler, J.M., and Hales, B.F., 1985, Consequences to the progeny of paternal drug exposure, *in*: "Male Fertility and its Regulation," T. Lobl and E.S.E. Hafez, eds., MTP Press Ltd., Lancaster, pp. 225-243.

Smith, F.P., 1981, Detection of amphetamine in bloodstains, semen, seminal stains, saliva and saliva stains, *Forensic Science* 17:225-228.

Soyka, L.F., Peterson, J.M., and Joffe, J.M., 1978, Lethal and sublethal effects on the progeny of male rats treated with methadone. *Toxicol. Appl. Pharmacol.* 45:797-807.

Swanson, B.N., Leger, R.M., Gordon, W.P., Lynn, R.K., and Gerber, N., 1978, Excretion of phenytoin into semen of rabbits and man. Comparison with plasma levels. *Drug Metabol. Dispos.* 6:70-74.

Trasler, J.M., Hales, B.F., and Robaire, B., 1986, Chronic low dose cyclophosphamide treatment of adult male rats: effect on fertility, pregnancy outcome and progeny, *Biol. Reprod.* 34: 275-283.

Trasler, J.M., Hales, B.F., and Robaire, B., 1987, A time course study of chronic paternal cyclophosphamide treatment in rats: effects on pregnancy outcome and the male reproductive and hematologic systems, *Biol. Reprod.* 37:317-326.

Wilson, J.G.,1977, Embryotoxicity of drugs in man, *in*: "Handbook of Teratology," Volume 1, J.G. Wilson and F.C. Fraser, eds., Plenum Press, New York, pp. 309-355.

Working, P.K., and Chellman, G.J., 1989, The use of multiple endpoints to define the mechanism of action of reproductive toxicants and germ cell mutagens, *in*: "Sperm Measures and Reproductive Success: Institute for Health Policy Analysis Forum on Science, Health, and Environmental Risk Assessment," Alan R. Liss, Inc, New York, pp. 211-227.

THE MALE-MEDIATED DEVELOPMENTAL TOXICITY OF CYCLOPHOSPHAMIDE

Barbara F. Hales and Bernard Robaire

Department of Pharmacology & Therapeutics
Centre for the Study of Reproduction
McGill University
3655 Drummond Street
Montréal, Québec, Canada H3G 1Y6

INTRODUCTION

Over the past thirty years there has been increased concern over the potential teratogenic effects of drugs or chemicals. This concern has been focused primarily on the consequences of maternal exposure to teratogens; much less attention has been given to the possible adverse effects on the progeny of paternal exposure to drugs or environmental chemicals. However, a wide range of agents, including environmental pollutants (e.g. lead) and therapeutic agents (e.g. cyclophosphamide) (Lindbohm *et al.*, 1991; Robaire *et al.*, 1985; Robaire and Hales, 1993), have been reported to produce an abnormal progeny outcome after paternal exposure. Drugs or environmental chemicals to which the male is exposed may be present in his seminal fluid, and thus may have direct effects on the ovulated egg, on the process of fertilization or on embryo development itself. Alternatively, radiation, drugs or other chemicals may have adverse effects on the fetus by "functionally" altering male germ cells. Paternally-mediated adverse effects on progeny outcome may be manifested as decreased fertility, embryolethality, external or internal malformations evident at the time of birth, or more subtle changes such as modification of post-natal development or behavior, or a subsequent increased incidence of cancer. Such effects have been found in experimental animal models after exposure of the male to X-rays (Kirk and Lyon, 1984) and to drugs such as cyclophosphamide (Trasler *et al.*, 1986, 1987) or to environmental chemicals such as lead or dibromochloropropane (Uzych, 1985; Whorton *et al.*, 1979). In man, several studies (Fried *et al.*, 1987; Potashnik and Yanai-Inbar, 1987; Sentura *et al.*, 1985) have failed to show an increase in fetal malformations in children fathered by men exposed to such chemicals, while other studies (Gulati *et al.*, 1986) have indicated such effects.

Much of the experimental evidence showing the existence of drug-induced alterations in the germ cell in mammalian systems has come from studies with antineoplastic agents (Jackson, 1964; Lu and Meistrich, 1979). Though one might expect

Male-Mediated Developmental Toxicity, Edited by D.R. Mattison
and A.F. Olshan, Plenum Press, New York, 1994

that the maximum susceptibility to the effects of these drugs would be seen during mitosis or meiosis, a number of studies have demonstrated that each phase of spermatogenesis and sperm maturation can be affected. For example, procarbazine, after treatment of the male mouse, caused a dominant lethal effect when spermatids and spermatocytes were first exposed and a failure of fertilization when spermatogonia were first exposed (Ehling, 1974). Cytogenetic analyses revealed chromosomal abnormalities after exposure of spermatogonial stages to this drug (Adler, 1982); that such abnormalities can be found in the progeny was shown using the specific locus mutation test (Ehling and Neuhäuser, 1979). Using this test, it has also been demonstrated that procarbazine caused chromosomal alterations in the progeny after exposure of post-spermatogonial stages (Ehling and Neuhäuser, 1979).

CYCLOPHOSPHAMIDE AS A MODEL DRUG FOR STUDIES OF MALE-MEDIATED DEVELOPMENTAL TOXICITY

Studies from our laboratories (Trasler *et al.*, 1985; 1986; 1987; Hales *et al.*, 1986; Hales and Robaire, 1990; Hales *et al.*, 1992; Kelly *et al.*, 1992; Qiu *et al.*, 1992), as well as from a number of others (Adams *et al.*, 1981; Auroux *et al.*, 1986; Jenkinson and Anderson, 1990; Sotomayor and Cumming, 1975), have focused on cyclophosphamide as a model drug to elucidate the mechanisms of male-mediated developmental toxicity. Cyclophosphamide is a nitrogen mustard analog which requires metabolic activation to be an alkylating agent; two reactive metabolites, phosphoramide mustard and acrolein, are formed (Foley *et al.*, 1961). Both of these reactive metabolites alkylate DNA; this alkylation leads to base substitution mutations, DNA-DNA crosslinks and DNA-protein crosslinks as well as DNA strand breaks (Crook *et al.*, 1986). Exposure to activated metabolites of cyclophosphamide causes base substitution mutations as well as chromosomal translocations and tangles (Hales, 1982; Benedict *et al.*, 1978). Cyclophosphamide has been reported to induce heritable translocations after exposure of the male mouse during the development of spermatocytes, spermatids and spermatozoa (Sotomayor and Cumming, 1975).

In elucidating the mechanism(s) by which a drug given to the father can affect his progeny outcome, the first step is to distinguish between the action of a drug due to its presence in the semen and its effects on the male germ cells. Both of these mechanisms were operative for cyclophosphamide. Indeed, single high doses of cyclophosphamide given to the father shortly before mating resulted in a significant increase in preimplantation loss (see chapter by Robaire and Hales, this volume).

Low dose chronic treatment of the male rat with cyclophosphamide had marked dose-dependent and time-specific effects on progeny outcome (Trasler *et al.*, 1985, 1986, 1987); these appeared to be mediated by stage-specific effects on the male germ cell. Such effects on progeny outcome were not accompanied by any apparent effects on parameters of male reproductive function including serum LH, FSH or testosterone, epididymal or testicular weights or sperm reserves or sex accessory tissue weights (Trasler *et al.*, 1986), although some changes in the histologic appearance of the epididymal epithelium and of spermatozoa in the epididymis (Trasler *et al.*, 1988), as well as some biochemical markers of epididymal function (Trasler and Robaire, 1988) were observed. The percent sperm positive females, the number of seminal plugs present after mating, and the number of pregnant females per sperm positive females did not differ in any of the treatment groups (Trasler *et al.*, 1986).

Despite the apparently minimal effects of 9 weeks of treatment with cyclophosphamide (doses of 1.4, 3.4 or 5.1 mg/kg/day) on the male reproductive system and on fertility, there were a number of effects on pregnancy outcome (Trasler *et al.*,

1986). The results for the control (top) and high dose cyclophosphamide (bottom) groups are presented in Figure 1.

There was an increase in preimplantation loss at 4-6 weeks that was not evident at other times. The timing of this effect suggested that it was mediated by the action of cyclophosphamide on early spermatids and spermatocytes. A progressive and dose dependent increase in postimplantation loss was observed, starting at 2 weeks and increasing dramatically to plateau at 80% in the high dose group by 4 weeks of treatment. This clear rise in postimplantation loss was associated with a phase of spermiogenesis. When the resorptions were defined as early (<100 mg) or late (>100 mg), it became apparent that this dramatic dose-dependent increase in postimplantation loss was due to an increase in early or peri-implantation resorptions.

An increase in malformed and growth retarded (weighing less than 75% of the weight of their litter mates) fetuses was observed after 7-9 weeks of treatment (Figure 1). The external malformations observed were principally hydrocephaly, edema and micrognathia. Thus, external malformations and growth retardation are produced in progeny resulting from paternal cyclophosphamide treatment of germ cells exposed as spermatogonia. Francis *et al.* (1990) observed abnormal karyotypes in 2/21 abnormal offspring from cyclophosphamide-treated male rats.

Thus, depending on the stage of spermatozoal development when germ cells were first exposed to cyclophosphamide, different rates of preimplantation loss, postimplantation loss and congenital malformations were found (Trasler *et al.*, 1985, 1986, 1987). The stages of spermatogenesis that are sensitive to the effects of a toxic chemical must be dependent on the accessibility of germ cells at that stage to damage, as well as on the likelihood of cell death compared to the ability of repair mechanisms to correct the damage. A drug which alters or blocks the mitotic and meiotic divisions, which germ cells undergo before differentiation, would be expected to result in the arrest of spermatogenesis and hence infertility (Meistrich, 1986). This was not found after low dose chronic exposure of male rats to cyclophosphamide; this treatment did not affect fertility or spermatozoal reserves.

A drug may also act on post-meiotic germ cells (spermatids and spermatozoa) as they differentiate (spermiogenesis) or mature (Jackson, 1964). At this stage of development, germ cells are haploid and hence have lost the ability to repair drug-induced DNA damage. Treatments which work via this mechanism include cyclophosphamide, as well as other alkylating agents; these agents often cause dominant lethal mutations (pre- and/or post-implantation loss) and fetal malformations (Batemena and Epstein, 1971; Nagao, 1987). In the studies of Fabricant *et al.* (1983), the stage of spermatogenesis most susceptible to the induction of learning abnormalities in the progeny after paternal exposure to cyclophosphamide was the spermatid. The absence of unscheduled DNA synthesis or DNA repair in the late spermatids and spermatozoa may contribute to their increased susceptibility to alkylating agents (Sega, 1974). The susceptibility of the more mature germ cell stages to damage with alkylating agents is also correlated with those stages where protamine has replaced the chromosomal histones (Pogany *et al.*, 1981).

The observations that pre- and postimplantation effects as well as growth retardation and malformations can occur after paternal treatment with such drugs as cyclophosphamide suggest that more than one mechanism may be involved. The importance of "specific" genes or chromosomes as targets and the "mechanism" by which the adverse effects of chemicals on the male germ cells produce embryolethality or embryo malformations are not known. The entire male genome cannot be "shattered" because paternal genes are expressed as early as the two cell embryo (Sawicki *et al.*, 1981), and most of the embryos survive until implantation. The types of genetic abnormalities that have been associated with dominant lethality are chromosomal breaks and rearrangement and mis-segregation (Brewen *et al.*, 1975).

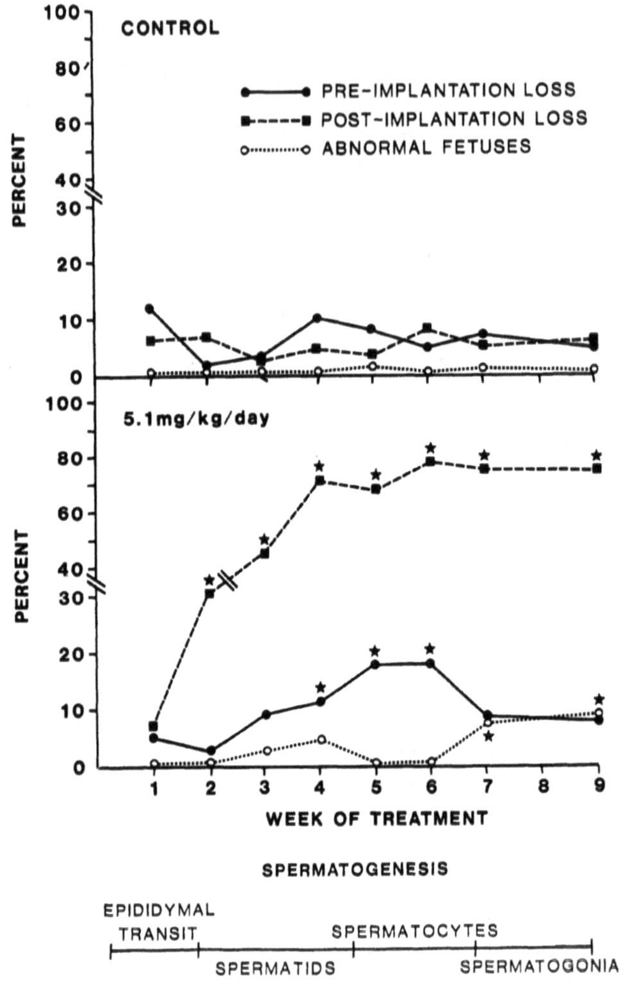

Figure 1. Relative incidence of pre- and post-implantation loss as well as malformations in the progeny of male Sprague Dawley rats treated with vehicle (top) or cyclophosphamide, 5.1 mg/kg/day (bottom) for periods of time from one to nine weeks. * indicates a significant difference from vehicle control, P≤0.05.

Studies from several laboratories (Covarrubias *et al.*, 1985; Mahon *et al.*, 1988) have revealed that the integration of exogenous DNA into the mouse genome caused DNA rearrangements, leading to postimplantation embryo deaths.

The presence of both the maternal and paternal genome is essential for normal development in the mouse. There is accumulating evidence that the two parental genomes are not equivalent (Barton *et al.*, 1984; McGrath and Solter, 1984). McGrath and Solter (1984) and Surani and his co-workers (Barton *et al.*, 1984; Surani *et al.*, 1984, 1987) have suggested that during gametogenesis chromosomes are imprinted in a germ-line specific manner, and that this "imprinting" introduces chromosomal determinants of development. We hypothesize that cyclophosphamide treatment of the male rat is affecting specific "non-complementary" (imprinted?) genes in the male genome, and that this may be involved in the early post-implantation death, growth retardation and malformations which are observed.

HERITABILITY OF THE MALE-MEDIATED DEVELOPMENTAL TOXICITY OF CYCLOPHOSPHAMIDE

Although litter size is drastically reduced after exposure of male rats to cyclophosphamide, some "apparently normal" pups survive. The question which must be addressed, then, is whether these offspring have in fact inherited some adverse effect(s) of the paternal drug exposure which can, in turn, be transmitted to their progeny. Previous studies by Auroux and his co-workers (1988; 1990) have reported that some of the behavioral abnormalities caused by paternal cyclophosphamide treatment persist in subsequent generations; no increase in postimplantation loss or malformations was observed.

Male rats were treated with saline or with 3.4 or 5.1 mg/kg/day of cyclophosphamide for 4 or 18 weeks, and then mated to untreated females. These females were allowed to deliver normally and males and females from each treatment group (F_1 generation) were randomly mated. The offspring of males treated with saline or with a dose of cyclophosphamide of 3.4 mg/kg/day for 4 or 18 weeks were not significantly different from control with respect to their F_2 generation progeny outcome (Figure 2). In contrast, there was a dramatic and significant increase in postimplantation loss among the offspring of the group whose fathers had been exposed to a dose of cyclophosphamide of 5.1 mg/kg/day. Exposure to this dose of cyclophosphamide also resulted in an F_2 generation with a significant increase in malformed fetuses. The malformations observed among the F_2 generation included open eyes, omphalocele, generalized edema, syndactyly, gigantism and dwarfism (Hales *et al.*, 1992). Furthermore, the F_2 progeny of male rats exposed to this dose of cyclophosphamide had a significantly decreased mean fetal weight per litter (Hales *et al.*, 1992). It is interesting that the adverse consequences observed in the F_2 generation (increased postimplantation loss and increased malformations) were similar to those previously reported for the F_1 progeny (Trasler *et al.*, 1985; 1986; 1987). The increase in postimplantation loss and malformations observed in the F_2 fetuses may reflect genes of lower penetrance that are transmissible to a later generation, or the presence of balanced translocations. Thus, it is clear that exposure of the male rat to cyclophosphamide did result in a specific and heritable alteration in his surviving "apparently normal" F_1 progeny (Hales *et al.*, 1992). The possibility of second generation consequences after treatment of men with cyclophosphamide and other anticancer drugs needs to be considered.

REVERSIBILITY OF THE MALE-MEDIATED DEVELOPMENTAL TOXICITY OF CYCLOPHOSPHAMIDE

A major question that arises from our studies on the consequences of chronic treatment of adult male rats with cyclophosphamide is the degree to which these effects are reversible. To resolve this question, adult male rats were treated with saline or one of three doses of cyclophosphamide, 1.4, 3.4 or 5.1 mg/kg/day, 6 days/week, for 9 weeks. After the 9 week treatment period, each male was exposed

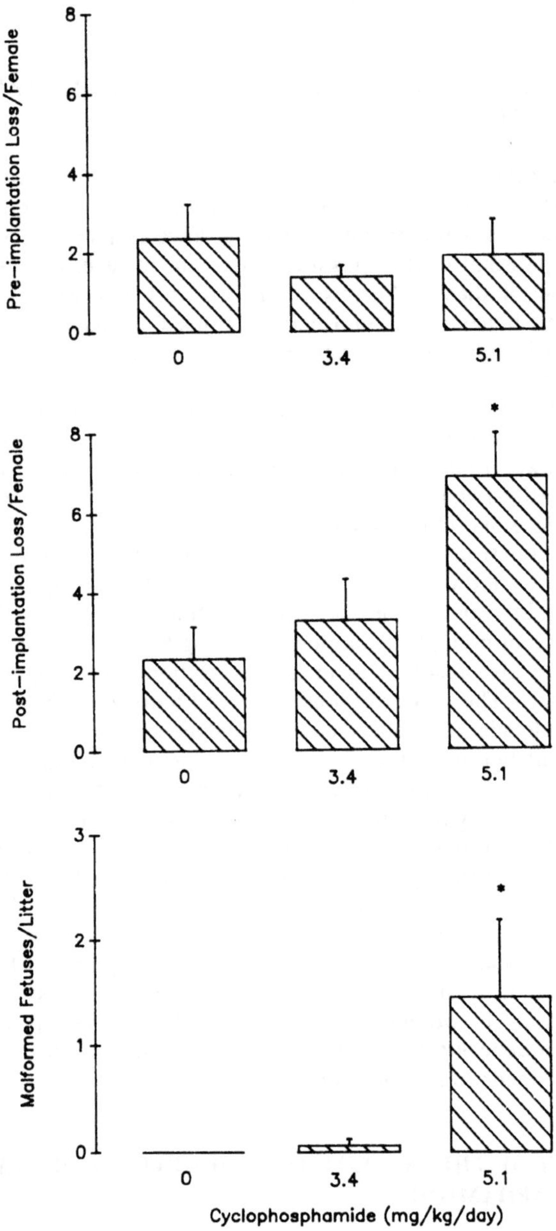

Figure 2. Effects of treatment of male rats with vehicle or cyclophosphamide (3.4 or 5.1 mg/kg/day) on the rate of preimplantation loss (top), postimplantation loss (middle) and the incidence of malformations (bottom) in the F_2 progeny. Both the F_1 male and female rats were derived from the offspring of the males from the designated treatment groups. * indicates a significant difference from vehicle control, $P \leq 0.05$.

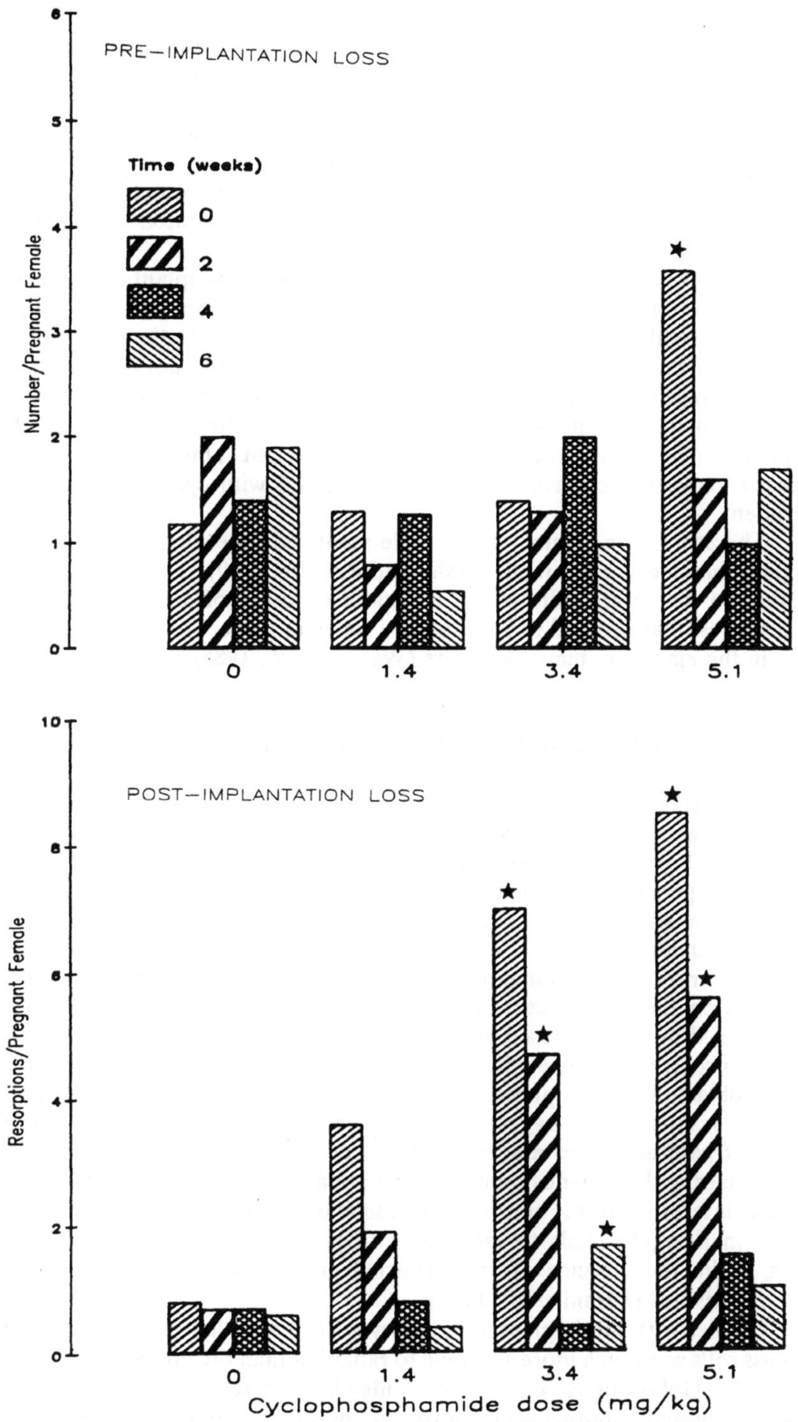

Figure 3. Incidence of preimplantation loss (top panel) and postimplantation loss (bottom panel) in litters derived from males that had been treated daily for 9 weeks with vehicle or cyclophosphamide (1.4, 3.4 or 5.1 mg/kg/day) and mated at various periods (0, 2, 4 and 6 weeks) after discontinuing treatment. * indicates a significant difference from vehicle control, $P \leq 0.05$.

to two females in proestrus. These matings were repeated every 2 weeks for a period equivalent to two cycles of the seminiferous epithelium. At the end of the 9 week treatment period, preimplantation loss was significantly increased from 6% in the control group to 21% in the 5.1 mg/kg/day cyclophosphamide treatment group (Figure 3). Preimplantation loss rapidly decreased upon cessation of treatment, returning to the control level by the first two week post-treatment mating period. This observation would suggest that the effect of cyclophosphamide administered to the male on preimplantation loss was due to an effect of the drug on spermatozoa in the epididymis or on the final stages of spermatogenesis in the testis.

After 9 weeks of exposure to cyclophosphamide, postimplantation loss increased dramatically and in a dose-dependent manner (from 5% in the control group to 74% in the 5.1 mg/kg/day cyclophosphamide treatment group). Postimplantation loss was significantly decreased by two weeks post-treatment, but was not decreased to within the control range until four weeks post-treatment. Thus, postimplantation loss was due to an additional effect of cyclophosphamide on germ cells in the testis, probably on spermatids. The number of malformed or low weight fetuses was not significantly increased above the control in the groups treated with cyclophosphamide in this experiment.

Therefore, in rats, the male-mediated developmental toxicity of cyclophosphamide was reversible (Hales and Robaire, 1990). The limited data that are available on the reversibility of the effects of exposure of men to anticancer drugs would suggest that reversal in humans, at least with respect to effects on numbers of sperm in the ejaculate, may take years (Watson *et al.*, 1985).

MECHANISMS OF PERI-IMPLANTATION EMBRYO DEATH

To further approach the mechanism by which paternal cyclophosphamide exposure triggers such a dramatic increase in postimplantation loss, the timing and manner of embryo death were examined (Kelly *et al.*, 1992). To determine which cells were dying in peri-implantation embryos sired by cyclophosphamide treated males, day 7 implantation sites from embryos sired by control males and males treated with a cyclophosphamide dose of 6 mg/kg/day for 4 weeks were dissected, fixed, and embedded for serial sectioning; these sections were examined at the light microscope for effects on the tissues derived from inner cell mass and trophoblast cells. In embryos sired by drug-exposed male rats, we made the striking observation that the cell death occurred selectively in those tissues derived from the inner cell mass, while the trophoblast-derived trophectoderm cells appeared morphologically normal (Figure 4). These results would suggest that exposure of the male rat to cyclophosphamide may affect paternal genes essential for the development of inner cell mass in the embryo, sparing those genes required for normal trophectoderm development. This finding is interesting in the context of genomic imprinting as the absence of the male genome in a zygote leads to embryos with poor development of extra-embryonic tissues, while the absence of the maternal genome leads to embryos with little embryonic tissues (Surani *et al.*, 1986).

Studies with pre-implantation embryos have revealed that growth of the inner cell mass cells was much more sensitive to radiation than the growth of the trophoblast giant cells (Goldstein *et al.*, 1975). Indeed, irradiated morulae and blastocysts frequently formed outgrowths in culture with normal numbers of trophoblast cells but which lacked or were deficient with respect to inner cell mass components. The growth requirements of inner cell mass cells are clearly different from those of trophoblast cells. It is not evident how much such early differences in the regulation

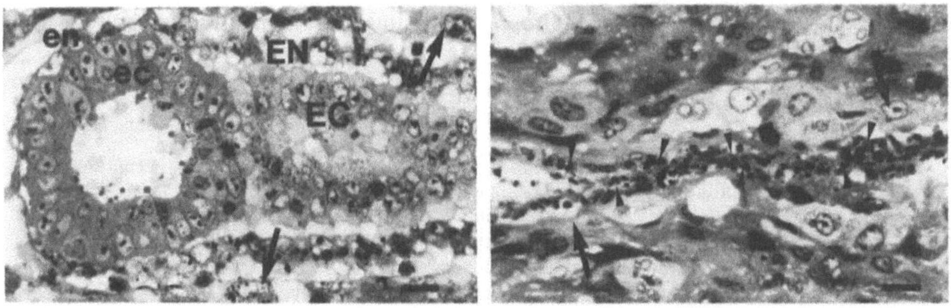

Figure 4. Day 7 implantation sites sired by males treated with vehicle (left) or cyclophosphamide (6 mg/kg/day) for 4 weeks. en, embryonic endoderm; ec, embryonic ectoderm; EN, extraembryonic endoderm; EC, extraembryonic ectoderm; arrows indicate trophectoderm; arrowheads indicate debris. The bar represents 1 μm.

of growth are dependent on specific factors or genes. Clearly, future studies must investigate the relationship of this cell lineage-specific cell death after paternal cyclophosphamide exposure, or after irradiation, to the phenomenon of genome imprinting. In this regard growth factors important in early embryo development, such as the insulin like growth factors (IGFs), are interesting (Telford *et al.*, 1990). Recently, knockout experiments have permitted the identification of IGF II as a paternally imprinted gene (DeChiara *et al.*, 1991), while the IGF II receptor gene is maternally imprinted (Barlow *et al.*, 1991). Disruption of the paternal allele of IGF II resulted in growth retardation of the offspring, while disruption of the allele inherited from the mother had no phenotypic effect (DeChiara *et al.*, 1991). IGF II and its receptor, as well as other imprinted genes, may serve as selective targets for mutagens with parental specificity in their effects, as they act as "functionally recessive," or "non-complementary" genes.

SUMMARY AND CONCLUSIONS

Over the past twenty years it has become clear that drugs given to the father can and do affect his progeny outcome. The range of effects that can occur encompasses infertility and reduced fertility, as well as malformations or growth retardation, and behavioral alterations in the progeny. Furthermore, it is also apparent that the germ cell line of the progeny may be affected. The molecular mechanisms mediating any of these consequences to the progeny of paternal drug exposure remain a mystery. Although few of these observations have been extended to the human, it seems essential that carefully designed epidemiological studies be undertaken to establish the extent to which such effects may take place in man.

ACKNOWLEDGEMENTS

The studies from our laboratories were done with the support of grants from the Medical Research Council of Canada and the National Foundation March of Dimes. J. M. Trasler, S. Kelly, J. Qiu, S. Smith and K. Crosman are thanked for their dedication to this project.

Dimes. J. M. Trasler, S. Kelly, J. Qiu, S. Smith and K. Crosman are thanked for their dedication to this project.

REFERENCES

Adams, P.M., Fabricant, J.D., and Legator, M.S., 1981, Cyclophosphamide-induced spermatogenic effects detected in the F_1 generation by behavioral testing, *Science* 211:80-82.

Adler, I.-D., 1982, Male germ cell cytogenetics, *in*: "Cytogenetic Assays of Environmental Chemicals," T.C. Hsu, ed., Allanheld, Osmun and Co., Totowa, N.J. pp. 249-276.

Auroux, M.R., Dulioust, E.M., Nawar, N.Y., and Yacoub, S.G., 1986, Antimitotic drugs (cyclophosphamide and vinblastine) in the male rat. Deaths and behavioral abnormalities in the offspring, *J. Androl.* 7:378-386.

Auroux, M., Dulioust, E.J.B., Nawar, N.N.Y., Yacoub, S.G., Mayaux, M.J., Schwartz, D., and David, G., 1988, Antimitotic drugs in male rat. Behavioral abnormalities in the second generation, *J. Androl.* 9:153-159.

Auroux, M., Dulioust, E., Selva, J., and Rince, P., 1990, Cyclophosphamide in the F_0 male rat: Physical and behavioral changes in three successive adult generations, *Mutation Res.* 229:189-200.

Barlow, D.P., Stoger, R., Germann, B.G., Saito, K., and Schweifer, N., 1991, The mouse insulin-like growth factor type-2 receptor is imprinted and closely linked to the Tme locus, *Nature* 349:84-87.

Barton, S.C., Surani, M.A.H., and Norris, M.L., 1984, Role of paternal and maternal genomes in mouse development, *Nature* 311:374-376.

Batemena, A.J., and Epstein, S.A., 1971, Dominant lethal mutations in mammals, *in*: "Chemical Mutagens -Principles and Methods for their Detection", Vol. 2, A. Hollaender, ed., Plenum Press, New York-London, pp. 541-568.

Benedict, W.F., Banerjee, A., and Venkatesan, N., 1978, Cyclophosphamide-induced oncogenic transformation, chromosomal breakage and sister chromatid exchange following metabolic activation, *Cancer Res.* 38:2922-2924.

Brewen, J.G., Payne, H.S., Jones, K.P., and Preston, R.J., 1975, Studies on chemically induced dominant lethality, I. The cytogenetic basis of MMS-induced dominant lethality in post-meiotic germ cells, *Mutation Res.* 33:239-250.

Covarrubias, L., Nishida, Y., and Mintz, B., 1985, Early developmental mutations due to DNA rearrangements in transgenic mouse embryos, *CSH Symp. Quant. Biol.* L: 447-452.

Crook, T.R., Souhami, R.L., and McLean, A.E.M., 1986, Cytotoxicity, DNA cross-linking and single strand breaks induced by activated cyclophosphamide and acrolein in human leukemia cells, *Cancer Res.* 46:5029-5034.

DeChiara, T.M., Robertson, E.J., and Efstratiadis, A., 1991, Parental imprinting of the mouse insulin-like growth factor II gene, *Cell* 64:849-859.

Ehling, U.H., 1974, Differential spermatogenic response of mice to the induction of mutations by antineoplastic drugs, *Mutation Res.*, 26:285-295.

Ehling, U.H., and Neuhäuser, A., 1979, Procarbazine-induced specific-locus mutations in male mice, *Mutation Res.* 59:245-256.

Fabricant, J.D., Legator, M.S., and Adams, P.M., 1983, Post-meiotic cell mediation of behavior in progeny of male rats treated with cyclophosphamide, *Mutation Res.* 119:185-190.

Foley, G.E., Friedman, O.M., and Drolet, B.P., 1961, Studies on the mechanism of action of cytoxan --evidence of activation in vivo and in vitro, *Cancer Res.* 21:57-63.

Francis, A.J., Anderson, D., Evans, J.G., Jenkinson, P.C., and Godbert, P., 1990, Tumours and malformations in the adult offspring of cyclophosphamide-treated and control male rats - preliminary communication, *Mutation Res.* 229:239-246.

Fried, P., Steinfeld, R., Casileth, B., and Steinfeld, A., 1987, Incidence of developmental handicaps among the offspring of men treated for testicular seminoma, *Int. J. Androl.* 10:385-387.

Goldstein, L.S., Spindle, A.I., and Pedersen, R.A., 1975, X-ray sensitivity of the preimplantation mouse embryo in vitro, *Radiation Res.* 62:276-287.

Gulati, S.C., Vega, R., Gee, T., Koziner, B., and Clarkson, B., 1986, Growth and development of children born to patients after cancer therapy, *Cancer Invest.* 4(3):197-205.

Hales, B.F., 1982, The mutagenicity and teratogenicity of cyclophosphamide and its active metabolites, 4-hydroxycyclophosphamide, phosphoramide mustard and acrolein, *Cancer Res.* 42:3016-3021.

Hales, B.F., Smith, S., and Robaire, B., 1986, Cyclophosphamide in the seminal fluid of treated males: transmission to females by mating and effects on progeny outcome, *Toxicol. Appl. Pharmacol.* 84:423-430.

Hales, B.F. and Robaire, B., 1990, Reversibility of the effects of chronic paternal exposure to cyclophosphamide on pregnancy outcome in rats, *Mutation Res.* 229:129-134.
Hales, B.F., Crosman, K., and Robaire, B., 1992, Increased postimplantation loss and malformations among the F_2 progeny of male rats chronically treated with cyclophosphamide, *Teratology* 45:671-678.
Jackson, H., 1964, The effects of alkylating agents on fertility, *Brit. Med. J.* 20:107-114.
Jenkinson, P.C., and Anderson, D., 1990, Malformed foetuses and karyotype abnormalities in the offspring of cyclophosphamide and allyl alcohol-treated male rats, *Mutation Res.* 229:173-184.
Kelly, S.M., Robaire, B. and Hales, B.F., 1992, Paternal cyclophosphamide treatment causes postimplantation loss via inner cell mass-specific cell death, *Teratology* 45:313-318.
Kirk, K.M., and Lyon, M.F., 1984, Induction of congenital malformations in the offspring of male mice treated with X-rays at pre-meiotic and post-meiotic stages, *Mutation Res.* 125:75-85.
Lindbohm, M.-L., Sallmén, M., Anttila, A., Taskinen, H., and Hemminki, K., 1991, Paternal occupational lead exposure and spontaneous abortion, *Scand. J. Work Environ. Health* 37: 95-103.
Lu, C.C., and Meistrich, M.L., 1979, Cytotoxic effects of chemotherapeutic drugs on mouse testis cells, *Cancer Res.* 39:3575-3582.
Mahon, K.A., Overbeek, P.A., and Westphal, H., 1988, Prenatal lethality in a transgenic mouse line is the result of a chromosomal translocation, *Proc. Natl. Acad. Sci. USA* 85:1165-1168.
McGrath, J., and Solter, D., 1984, Completion of mouse embryogenesis requires both the maternal and paternal genomes, *Cell* 37:170-183.
Meistrich, M.L., 1986, Critical components of testicular function and sensitivity to disruption, *Biol. Reprod.* 34:17-28.
Nagao, T., 1987, Frequency of congenital defects and dominant lethals in the offspring of male mice treated with methylnitrosourea, *Mutation Res.* 177:171-178.
Pogany, G.C., Corzett, M., Weston, S., and Balhorn, R., 1981, DNA and protein content of mouse sperm: implications regarding sperm chromatin structure, *Exp. Cell Res.* 136:127-136.
Potashnik, G., and Yanai Inbar, I., 1987, Dibromochloropropane (DBCP): an 8 year reevaluation of testicular function and reproductive performance, *Fertil. Steril.* 47:317-323.
Qiu, J., Hales, B.F. and Robaire, B., 1992, Adverse effects of cyclophosphamide on progeny outcome can be mediated through post-testicular mechanisms in the rat, *Biol. Reprod.* 46:926-931.
Robaire, B., and Hales, B., 1993, Paternal exposures to chemicals before conception. Some children may be at risk, *Brit. Med. J.* 307:341-342.
Robaire, B., Trasler, J.M., and Hales, B.F., 1985, Consequences to the progeny of paternal drug exposure, *in*: "Male Fertility and its Regulation," Lobl,T., and Hafez,E.S.E., eds. MTP Press Ltd., Lancaster. pp. 225-243.
Robaire, B. and Hales, B.F., 1993, Post testicular mechanisms of male-mediated developmental toxicity, *in* "Male-Mediated Developmental Toxicity," Andrew F. Olshan and Donald R. Mattison, eds., Plenum Press, New York (this volume).
Sawicki, J., Magnuson, T. and Epstein, C., 1981, Evidence for expression of the paternal genome in the two cell mouse embyro, *Nature* 294:450-451.
Sega, G.A., 1974, Unscheduled DNA synthesis in the germ cells of male mice exposed in vivo to the chemical mutagen ethyl methanesulfonate, *Proc. Natl. Acad. Sci. USA* 71:4955-4959.
Sentura, Y.D., Peckman, C.S., and Peckman, M.J., 1985, Children fathered by men treated for testicular cancer, *Lancet* 2:766-769.
Sotomayor, R.E., and Cumming, R.B., 1975, Induction of translocations by cyclophosphamide in different germ cell stages of male mice: cytological characterization and transmission, *Mutation Res.* 27:375-388.
Surani, M.A.H., Barton, S.C, and Norris, M.L., 1984, Development of reconstituted mouse eggs suggests imprinting of the genome during gametogenesis, *Nature* 308:548-550.
Surani, M.A.H., 1986, Evidences and consequences of differences between maternal and paternal genomes during embryogenesis in the mouse, *in:* "Experimental Approaches to Mammalian Embryonic Development." Rossant, J., and Pedersen, R.A., eds. Cambridge University Press, New York, pp. 401-435.
Surani, M.A.H., Barton, S.C. and Norris, M.L., 1987, Influence of parental chromosomes on spatial specificity in androgenetic-parthenogenetic chimeras in the mouse, *Nature* 326:395-397.
Telford, N.A., Hogan, A., Franz, C.R., and Schultz, G.A., 1990, Expression of genes for insulin and insulin-like growth factors in early postimplantation mouse embryos and embryonal carcinoma cells, *Mol. Reprod. Dev.* 27:81-92.
Trasler, J.M., Hales, B.F., and Robaire, B., 1985, Paternal cyclophosphamide treatment of rats causes fetal loss and malformations without affecting male fertility, *Nature* 316: 144-146.

Trasler, J.M., Hales, B.F., and Robaire, B., 1986, Chronic low dose cyclophosphamide treatment of adult male rats: effect on fertility, pregnancy outcome and progeny, *Biol. Reprod.* 34: 275-283.
Trasler, J.M., Hales, B.F., and Robaire, B., 1987, A time course study of chronic paternal cyclophosphamide treatment in rats: effects on pregnancy outcome and the male reproductive and hematologic systems, *Biol. Reprod.* 37:317-326.
Trasler, J.M., Hermo, L., and Robaire, B., 1988, Morphological changes in the testis and epididymis of rats treated with cyclophosphamide: a quantitative analysis, *Biol. Reprod.* 38:463-479.
Trasler, J.M., and Robaire, B., 1988, Effects of cyclophosphamide on selected cytosolic and mitochondrial enzymes in the epididymis of the rat, *J. Androl.* 9:143-152.
Uzych, L., 1985, Teratogenesis and mutagenesis associated with the exposure of human males to lead: a review, *Yale J. Biol. Med.* 58:9-17.
Watson, A.R., Rance, C.P., and Bain, J., 1985, Long term effects of cyclophosphamide on testicular function, *Brit. Med. J.* 291:1457-1460.
Whorton, M.D., Milby, T.H., Krauss, R.M., and Stubbs, H.A., 1979, Testicular function in DBCP exposed pesticide workers, *J. Occup. Med.* 21:161-166.

MALE-MEDIATED TERATOGENESIS: IONIZING RADIATION/ETHYLNITROSOUREA STUDIES

Taisei Nomura

Department of Radiation Biology
Faculty of Medicine
Osaka University
Suita, Osaka 565
Japan

INTRODUCTION

If the father's exposure to radiation and chemicals can induce birth defects in their offspring, this should clearly indicate that the germ cell alterations result in these defects. Animal experiments anticipated that parental exposure to radiation and chemicals would induce varieties of defects in the offspring, including embryonic deaths, tumors and malformations (Nomura, 1975; 1978; 1982; 1986; Tomatis, 1981). However, these findings have not been supported by a large scale epidemiological survey of the children of atom-bomb survivors in Hiroshima and Nagasaki until the age of 20 (Yoshimoto *et al.*, 1990), although the case control studies in the United States (Graham et al, 1966; Shiono *et al.*, 1980) and China (Shu *et al.*, 1988) and the Cohort study in the Vietnam (Ton, 1981, Cân, 1984) suggested a higher risk of these defects in the children of fathers who were preconceptionally exposed to radiation and chemicals. Recently, Gardner *et al.* (1990) documented a 6 to 7 times higher risk of leukemia (mostly acute lymphatic leukemia) in the children of fathers who were employed at the Sellafield nuclear reprocessing plant (in the United Kingdom) and exposed to more than 100 mSv as external doses before conception, especially more than 10 mSv during 6 months before conception, while the estimated external doses were about one-fourth or one-fortieth, respectively, of that in Hiroshima and Nagasaki (435 mSv) (Nomura, 1990; Yoshimoto *et al.*, 1990) and radiation exposure at Sellafield was protracted for long period at low dose rate.

In this paper, I will review previous animal studies on embryonic deaths and congenital malformations in the offspring who were preconceptionally exposed to ionizing radiation and ethylnitrosourea, in order to reconcile the problems that lie at the basis in the human studies.

Male-Mediated Developmental Toxicity, Edited by D.R. Mattison and A.F. Olshan, Plenum Press, New York, 1994

DIFFERENTIAL SENSITIVITY OF DEVELOPING GERM CELLS

Development of germ cells and experimental procedures are shown in the scheme (Figure 1). In the testis, stem germ cells (spermatogonia) undergo meiosis (primary and secondary spermatocytes) after a period of mitotic proliferation, and then differentiate into mature gametes from spermatids to spermatozoa. The development goes as if they are on the moving side-walk. Adult male mice were exposed to X-rays 1 to 7 days before conception for the treatment of spermatozoa stage, 15 to 21 days for the spermatid stage, and 64 to 80 days or more for the spermatogonial stage. In the ovary, meiosis finished in part during fetal age, and oocytes and follicles undergo maturation. Meiosis starts again a few hours before ovulation. Adult females were also exposed to X-rays at various intervals before conception to treat oocytes at various follicular stages (Nomura, 1978; 1982; 1988).

Mouse oocytes at early follicular stages are extremely sensitive to radiation. One Gy of X-rays were enough to kill these oocytes, resulting in sterility in 5 weeks after irradiation (Nomura, 1978; 1988). Especially, immature oocytes at neonatal and young ages were much more sensitive to radiation killing (Oakberg, 1962; Dobson and Kwan, 1977; Nomura, 1978; 1988). Histologically, there were no oocytes in the irradiated mouse ovary, while human oocytes were highly resistant to 2 Gy and more

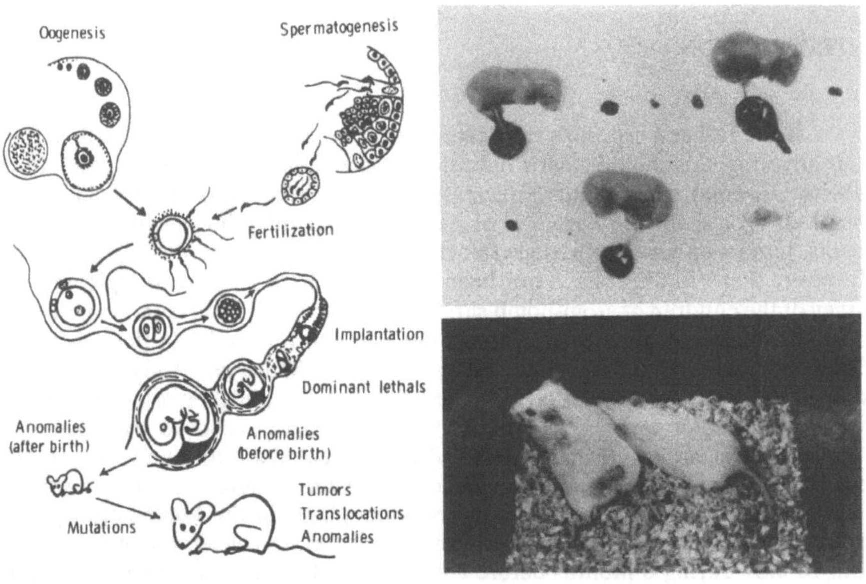

Figure 1. Scheme for the experimental procedures to detect fertility, embryonic deaths (dominant lethals), congenital anomalies, tumors etc. in the F_1 offspring after parental exposure to X-rays and chemicals (Nomura, T. 1988 with permission). An estrous female mouse was mated with males in the evening, and next morning vaginal plug was examined to determine the day of conception. Conception (fertilization) occurs at the time of ovulation (about 2:00 on the plug day), when the mouse room is lit from 4:00 to 18:00 (Nomura et al, 1987). Male mice had been treated with X-rays or chemicals at various intervals before conception, i.e., 1-7 days before conception for the treatment of spermatozoa, 15-21 days for spermatids, more than 64 days for spermatogonia. Right upper: embryonic deaths (dominant lethals) and living F_1 fetuses of male ICR mice exposed to 5.04 Gy of X-rays at spermatozoa stage (examined on the 18th day of gestation). Right lower: malformed (kinky and short tail, left) and normal (right) offspring resuscitated and foster-nursed after the cesarian operation.

doses of radiation (Blot and Sawada, 1971). These differences in the radiation sensitivity between mouse and human oocytes may be caused by the 10 fold difference in the cytoplasm (much repair enzymes), suggesting the difficulty to extrapolate radiation risk to human oocytes from mouse oocyte data.

For the male treatment, germ cells at a spermatogonial stage are very sensitive to radiation in both mouse and human. Consequently, reduction of the sperm number occurs 7 weeks or 4 months after irradiation (2-5 Gy) in mice and men, respectively, resulting in temporary or permanent sterility (depending on gonadal doses) and also morphological abnormality of the surviving sperm (Searle and Beechey, 1974; Kumatori *et al.*, 1980) (Figure 2).

The F_1 offspring derived from these irradiated germ cells developed embryonic deaths, congenital malformations, tumors, etc. (Nomura, 1978; 1982; 1983; 1986; 1988; 1989a).

EMBRYONIC DEATHS

Preconceptional irradiation of ICR mice induced high incidence of dead embryos (dominant lethals) which correspond to abortion or miscarriage in human (Figure 1). As shown in Figure 3, dominant lethals scored as embryonic deaths per implant increased almost linearly with parental doses of X-rays at the postmeiotic stages of male germ cells (spermatozoa and spermatids).

Thus, the father's exposure to radiation can kill the offspring. X-irradiation of mouse oocytes at late follicular stages also increased the incidence of embryonic deaths. However, there was no increase of embryonic deaths at 2.16 Gy but a little increase at 5.04 Gy, if at all, after the spermatogonial irradiation. Doubling dose (doses double the spontaneous incidence) was extremely large (>3.2 Gy) for

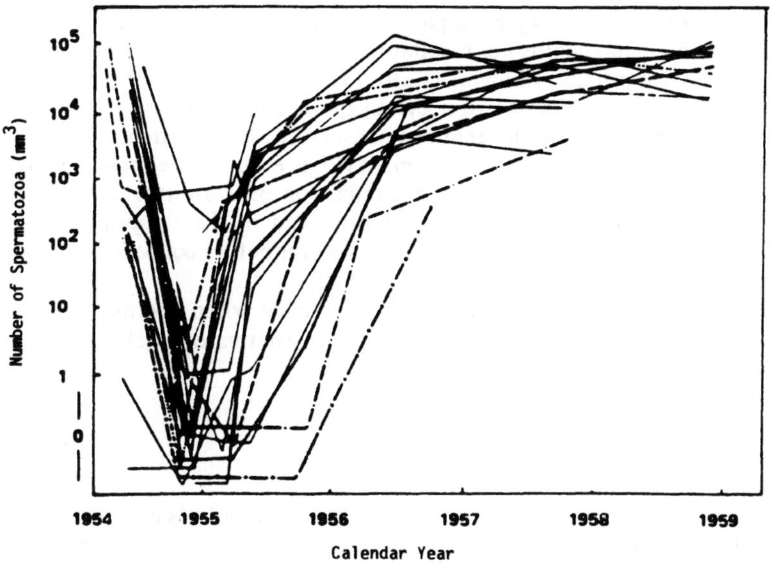

Figure 2. Changes in the number of spermatozoa in Japanese fishermen exposed to radioactive fallout at Bikini in 1954. Total whole body γ-ray doses were 1.7 to 6.9 Gy (Kumatori et. al. 1980). From Kumatori et. al. 1980 with permission.

spermatogonia, while these were approximately 0.38 Gy, 0.24 Gy and 0.39 Gy for spermatozoa, spermatids, and oocytes at late follicular stages, respectively (Table 1).

Table 1. Induced Rate and Doubling Dose of Embryonic Deaths (dominant lethals) in F_1 Offspring of ICR Mice Exposed to X-rays before Conception

	Induced Rate/Gy ($x 10^2$)	Doubling dose (Gy)
Spermatozoa	6.7 (7.8)	0.38 (0.33)
Spermatids	10.9 (11.7)	0.24 (0.22)
Spermatogonia	0.24 ~ 0.8 (0)	3.2 ~ 10 (∞)
Oocytes	6.5 (6.0)	0.39 (0.43)

Induced rates and doubling doses were calculated and averaged from the values at doses of 2.16 and 5.04 Gy for spermatozoa and spermatids, and at doses of 2.16 and 3.6 Gy for oocytes. Doubling doses were calculated as uniparental exposure following Lüning and Searle (1971). As for spermatogonia, two independent data at 5.04 Gy were used for calculation, since there was no increase of embryonic deaths at a dose of 2.16 Gy. Figures in parentheses are for the protracted irradiation by dose fractionation. Experimental procedures are given in the legend to Figure 3. A majority of data was derived from Nomura (1978, 1982).

Spermatogonia with large chromosomal changes, which cause dominant lethals, died during meiosis, and only the sperm without such damages could be ejaculated, resulting in no increase of embryonic deaths. There were no reduction in the incidence of embryonic deaths by the protracted irradiation of postmeiotic sperm. However, slight reduction was observed by the protracted irradiation of mature oocytes (Figure 3). Human oocytes must be much more resistant to radiation for embryonic deaths, i. e., dominant lethals, than mouse oocytes (Nomura, 1988), because of more repair activity in human oocytes in parallel with their volume of cytoplasm. Neither prenatal and neonatal deaths nor resultant decrease of live-births were detected in the F_1 offspring of atom-bomb survivors (Kato, 1975) which involved both spermatogonial exposure (in males) and immature oocytes exposure (of females). Negative results in human must be caused by the reduced sensitivities to radiation of these germ cell stages as indicated by the mouse experiment (Figure 3). Consequently, genetic risk of abortions and stillbirths and consequent reduction of live-births should be carefully estimated in humans, because these are produced in the offspring only when the fathers are exposed to radiation during a few months before conception (at postmeiotic stages of male germ cells) (Nomura *et al.*, 1990).

ENU also induced dominant lethals in mice when spermatozoa stage were treated but little increase was observed when spermatogonial stage was treated (Nomura, 1988; Nagao, 1990).

CONGENITAL MALFORMATIONS

Parental exposure of ICR mice to X-rays and chemicals increased the incidence of prenatal and postnatal malformations which were detected just before birth (on the 18th day of gestation) or 7 days after birth, respectively (Nomura, 1975; 1978; 1982;

1988; 1989b). Morphological malformations (dwarf, cleft palate, tail anomalies, open eyelid, exencephalus, hydrocephalus, gastroschisis, etc.) increased with paternal doses of X-rays up to 2.16 Gy, but the incidence leveled off at a very high dose (5.04 Gy).

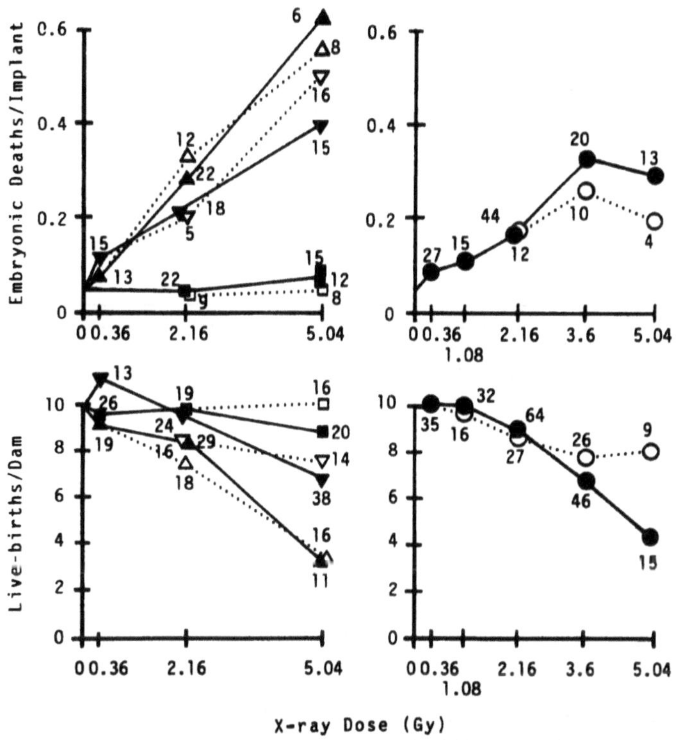

Figure 3. Embryonic deaths and birth-rate in the F_1 offspring of ICR mice preconceptionally exposed to X-rays. An estrous ICR female (63-65 days old) was mated with adult male ICR mice (63-65 days old) exposed to a single dose (closed symbols and solid line) or protracted dose (open symbols and dotted lines) of X-rays at spermatozoa (1-7 days before conception; ▼, ▽), spermatid (15-21 days; ▲, △), or spermatogonia (64-80 days; ■, □) stage. Adult female ICR mice (63-65 days old) were also treated with a single (●) or protracted (○) dose of X-rays at late follicular stage (1-14 days before conception). The upper panel shows the average frequency of embryonic deaths (early deaths) per implant as an indicator of dominant lethals, and the lower panel shows the average number of live-birth per dam. Figures around the symbols indicate the number of parents examined. Doses of X-rays to parental ICR mice were given on the abscissa. A Toshiba KC-18-2A (Toshiba Medical Co., Ltd., Tokyo, Japan) was used for irradiation, operating at 20 mA and 180 KVp with a filter of 0.5 mm of aluminum and 0.5 mm of copper. Dose rate was 0.72 Gy/min at 45 cm distance (measured by Fricke dosimetry). For protracted irradiation, 0.36 Gy of X-rays was given at 2hr intervals.

In general, higher incidence of malformations was observed prenatally than postnatally, simply because many malformations were lethal shortly after birth (Figure 4). However, it is noted that most lethal malformations in mice are viable in human after surgical operation (see review by Nomura, 1988). The induced rate of congenital malformations per Gy by the postmeiotic exposure was 2-3 times higher than that by the spermatogonial exposure (Table 2).

Table 2. Incidence (%) and Induced Rate per Gy ($\times 10^3$) of Congenital Malformations in the F_1 Offspring (before and after birth) of ICR Mice Exposed to X-rays before Conception in Comparison with those of Dominant Skeletal Mutations

	Congenital malformation (Dose: 2.16 Gy)		Skeletal mutation (Dose: 6 Gy)	
	Before Birth[1]	After Birth[2]	Before Birth[3]	After Birth[4]
Post-gonia	3.0% (11.6)	1.13 % (4.5)	2.4 % (3.3)	1.8% (2.4)
Gonia	1.8 % (6.6)	0 % 1.1 % (1.9)[5]	1.1 % (1.2)	1.3 % (2.1)
Oocytes	2.4 % (10.4)	1.7% (6.2)	---	---
Control	0.41 %	0.12 %	0.42 %	0.06 %

Figures in parentheses show the induced rate of congenital malformations and dominant mutations affecting skeletal bones which were detected before or after birth.
[1] [2] Nomura (1978, 1982, 1988); detected on the 18th day (Day 19) of gestation or 7 days after birth.
[3] Bartsch-Sandhoff (1974); detected in Day 19 fetuses.
[4] Ehling (1966, 1984) detected 4 weeks after birth.
[5] 5.04 Gy.

High incidence of malformations was also detected by oocyte exposure. Malformation incidence was slightly reduced by the protracted irradiation of spermatogonia and oocytes but not of postmeiotic sperm (Nomura, 1982). As shown in Table 2, induced rate of malformation per Gy was about 4 times higher than that of dominant mutation affecting skeletal bones, because overall anomalies were detected in my experiment (Nomura, 1978; 1988). Doubling doses were 0.12, 0.27 and 0.19 Gy for acute irradiation of spermatozoa, spermatogonia and mature oocytes, respectively (Nomura, 1988). However, acute 1 Gy exposure of spermatogonia increases only 0.2 % of malformations. This indicates that congenital malformation will not be an adequate marker to estimate genetic risk of radiation in Hiroshima and Nagasaki, i.e., negligible increase of congenital malformations (about 0.1 %) must be detected amongst of high back-ground incidence (4.7 %) in the children of atom-bomb survivors who were exposed to the average dose of 0.435 Sv at the spermatogonial stage. ENU also induced varieties of congenital malformations in the offspring (Figure 4) (Nomura, 1988; Nagao, 1990). However, spermatogonial stages were more sensitive to ENU than postmeiotic sperm for congenital malformation in contrast to X-rays. These results were similar to these for specific locus mutations (Russell *et al.*, 1982).

Recently, I found that preconceptional treatment of male ICR mice with ethylnitrosourea (ENU) induced respiratory distress syndrome (RDS), i.e., asphyxia, in the offspring (Nomura *et al.*, 1990). Although about half of RDS fetuses had specific morphological malformations (dwarfism and gigantic thymus), the remainder showed no morphological changes (Figure 5). In our preliminary results, preconceptional X-ray exposure of male mice also induces RDS. Since functional defects like RDS are more commonly observed in human neonates than major morphological defects (Avery, 1987), the study should be extended both experimentally and epidemiologically to variety of physiological and biochemical defects.

HERITABILITY OF CONGENITAL MALFORMATIONS

Open eyelid, dwarf, and tail anomalies were predominant types of viable anomalies, but some dwarfs were lethal before weaning and about half of the survivors were sterile. To confirm the heritability of congenital malformations, the F_1 progeny of treated parents were mated with untreated mice and their progeny were examined before or just after birth. The results were confirmed by continuing the urethane-treated groups into the F_3 generation. The total incidence of these anomalies in F_3 generation was 9.9 % (14 tail anomalies, 6 dwarfs, and 7 open eyelids in 274 F_3 progeny). Among viable anomalies in F_3 generation, 1 of 4 tail anomalies tested was inherited with low expressivity (7%), and 3 of 6 open eyelids tested were inherited with 11, 5, and 50 % expressivity, when these malformed F_3 mice were crossed with normal ICR mice. Cross of male and female progeny with open eyelid from ENU-treated parents yielded 41.7-100 % expressivity from F_2 to F_{12} generations. Recent studies on X-ray-induced malformations revealed that 6 of 8 dwarfs were heritable with 7-40 % expressivity and 1 of 3 tail anomalies was heritable with 18 % expressivity when mated with normal mice. Thus, some germ-line alterations causing phenotypical malformations seem to transmit to the next generation as if they are dominant

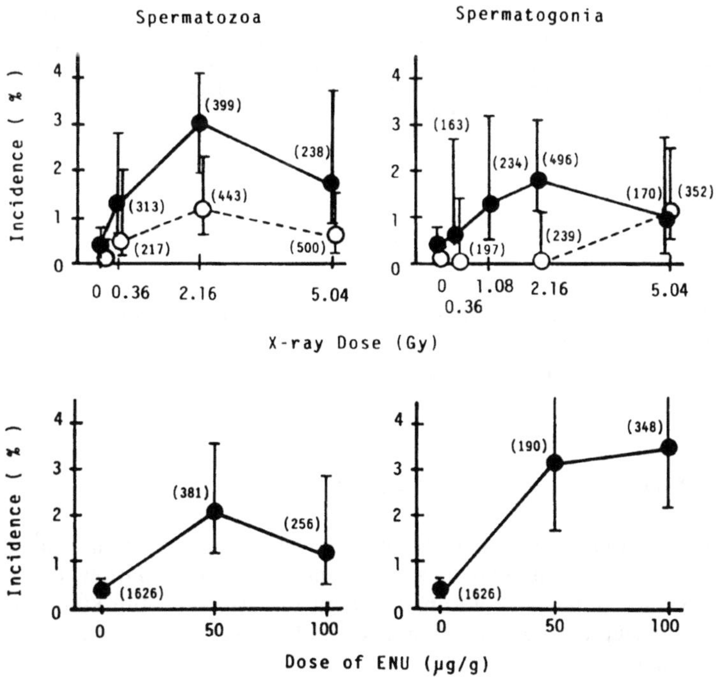

Figure 4. Incidence of congenital malformations in the F_1 offspring of male ICR mice treated with a single dose of X-rays or ENU at spermatozoa (left) or spermatogonial (right) stage. Adult males (63-65 days old) were treated at spermatozoa (1-14 days before conception) or spermatogonial (64-180 days before conception) stage. Morphological malformations were examined prenatally (●) or postnatally (○). Figures in parentheses show numbers of F_1 fetuses or live-born F_1 mice examined. Vertical bars indicate 90 % binomial confidence limits. Details for experimental procedures and types of induced malformations were given in the previous report (Nomura, 1988). From Nomura, 1988 with permission.

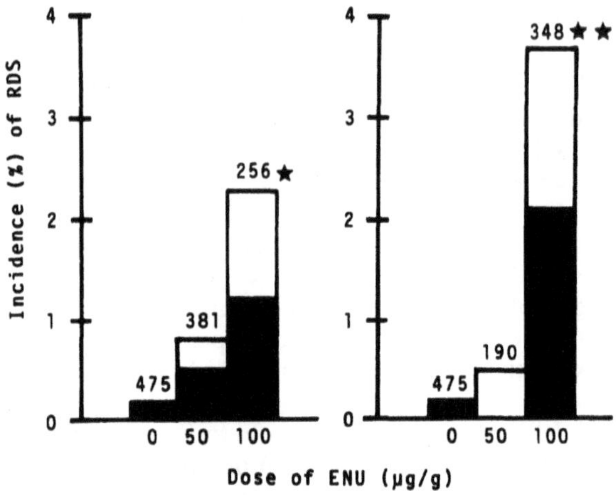

Figure 5. Offsprings showing functional defects (respiratory distress) after paternal exposure to ethylnitrosourea. An estrous ICR female was mated with adult male ICR mice (63-65 days old) treated with ethylnitrosourea (ENU) 1-21 days or 90-180 days before conception so that postmeiotic stages or spermatogonial stage of male germ cells were treated. Pregnant mice were killed on the 18th day of gestation by cervical dislocation, and then fetuses were taken out by cesarian operation. Fetuses were resuscitated immediately after the cesarian operation by patting very gently with tissue paper and wiping amniotic fluid from nose, mouth and body surface. Fetuses were then classified as apneic or pneic. Incidence of apneic fetuses showing respiratory distress syndrome (RDS) was given on the ordinate. Left: post-meiotic treatment. Right: spermatogonial treatment. Figures upon the histogram indicate the number of fetuses examined. ■ ; RDS with morphological malformations (dwarf, cleft palate, gigantic thymus, and exencephalus), □ ; RDS without morphological malformation. * $p<0.02$, ** $p<0.001$ See Nomura *et al.* (1990) for details.

mutations with reduced penetrance. Although some chromosomal changes (deletion, inversion and trisomy) were found in malformed offspring, no correlation has been detected between congenital malformation and specific chromosomal abnormality (Nomura, 1988).

COMPARISON WITH HUMAN STUDIES

Information obtained from the work on germ-line alterations causing congenital malformations are invaluable to assess genetic risk of radiation and chemicals to humans, because a majority of human genetic diseases show this kind of irregular and uncertain inheritance and most of induced malformations in mice are similar to those found in human.

Epidemiological study in Vietnam revealed that congenital malformations were induced in the children as a consequence of herbicides exposure to fathers (Ton, 1981, Cân, 1984). Significantly higher incidence of congenital malformations was observed in the children of the North Vietnam soldiers who fought in the sprayed area of the South Vietnam and married with the North Vietnam women after returning to their homes than the children of the North Vietnam soldiers who had never been in the South Vietnam and married with the North Vietnam women (Table 3). This suggests that paternal exposure to the herbicides (dioxines) induced malformations in their offspring by germ-cell alterations, as in the case of mouse experiments.

Table 3. Congenital Malformations in the Children of Fathers Exposed to Herbicides Before Conception in Vietnam

Fathers	Ton (1981)	Cân (1984)
Exposed to herbicides	47/1496 (3.1 %)	28/1801 (1.6 %)
Not exposed to herbicides	3/1438 (0.2 %)	118/16570 (0.7 %)

Details are given in the text.

Although there were no increase of abortion/stillbirth and congenital malformations in the children of the atomic bomb survivors in Hiroshima and Nagasaki who were exposed at the spermatogonial stage, this was the case in my mouse experiments when the spermatogonial stage was treated, as described in this paper. Furthermore, genetic markers which had been examined in Hiroshima and Nagasaki were proven to be inappropriate from the mouse experiments (Nomura, 1988; 1989a, b; 1990; this paper). Especially, electrophoretic mutants of serum proteins examined in the children of atomic bomb survivors are caused by base substitution which is rarely induced by radiation. Although the same gene loci were examined in mice, electrophoretic mutations have never been induced in the offspring of mice exposed to X-rays (Charles and Pretsch, 1986; Neel *et al.*, 1988; Nomura, 1989a). It is noted that one enzyme activity mutation caused by deletion has been detected in about 5000 children of atomic bomb survivors (Sato and Neel, 1988). Although numbers of cases are extremely small, there is no substantial difference in the mutation rate per Gy per locus of enzyme activity mutation between man and mice (Nomura, 1989). The value is also same in specific locus mutation in mice (Nomura, 1989). Consequently, further examination on the deletion mutations which are commonly induced by radiation must be important to estimate genetic risk of radiation to human. Such studies are in progress in Japan.

In conclusion, we must note that not only gestational exposure to mothers but also father's exposure to environmental or industrial toxic agents can increase the incidence of congenital malformations in the offspring. Although incidence of congenital malformations caused by parental exposure is very low when compared with gestational exposure, critical stage continues almost through life, while it is a few week for gestational exposure. Not only mothers but also fathers should be taken into account for the risk estimation of birth defects.

REFERENCES

Avery, G. B., 1987, "Neonatology," 3rd Edn., J. B. Lippincott Co., Philadelphia Bartsch-Sandhoff, M., 1974, Skeletal abnormalities in mouse embryos after irradiation of the sire, *Humangenetik* 25: 93-100.

Blot, W. J., and Sawada, H., 1971, Fertility among female survivors of the atomic bombs Hiroshima and Nagasaki, *Atomic Bomb Casualty Commission Technical Report* 26: 1-21.

Cân, N., 1984, Effects on offspring of parental exposure to herbicides, in "Herbicides in War - the Long Term Ecological and Human Consequences," A. H. Westing, ed., pp. 137-14, Pailor and Francis, Philadelphia.

Charles, D. J., and Pretsch, W., 1986, Enzyme-activity mutations detected in mice after paternal fractionated irradiation, *Mutat. Res.* 160: 243-248.

Dobson, R. L., and Kwan, J. C., 1977, The tritium RBE at low-level exposure-variation with dose, dose rate, and exposure duration, Curr. Top. *Radiat. Res.* 12: 44-62.

Ehling, U. H., 1966, Dominant mutations affecting the skeleton in offspring of X-irradiated male mice, *Genetics* 54: 1381-1389.

Ehling, U. H., 1984, Methods to estimate the genetic risk, in mutations in man, *G. Obe,* ed., pp. 292-318, Springen-Verlag, Berlin.

Gardner, M. J., Snee, M. P., Hall, A. J., Powell, C. A., Downes, S., and Terrell, J. D., 1990, Results of case-control study of leukaemia and lymphoma among young people near Sellafield nuclear plant in West Cumbria, *Br. Med. J.* 300: 423-429.

Graham, S., Levin, M. L., Lilienfield, A. M., Schuman, L. M., Gibson, R., and Dowd, J. E., 1966, Preconception, intrauterine, and postnatal irradiation as related to leukemia, *Natl. Cancer Inst. Monogr.* 19: 347-371.

Kato, H., 1975, Early genetic survey and mortality study, *J. Radiat. Res.* 16: 67-74.

Kumatori, T., Ishihara, T., Hirashima, K., Sugiyama, H., Ishii, S., and Miyoshi, K., 1980, Follow-up studies over a 25-year period on the Japanese fishermen exposed to radioactive fallout in 1954, in "The Medical Basis for Radiation Accident Preparedness", K. F. Hübner and S. A. Fry, eds., pp. 33-54, Elsevier/North-Holland, New York.

Lüning, K. G., and Searle, A. G., 1971, Estimates of the genetic risks from ionizing irradiation, *Mutat. Res.* 12: 291-304.

Nagao, T., and Fujikawa, K., 1990, Genotoxic potency in mouse spermatogonial stem cells of triethylenemelamine, mitomycin C, ethylnitrosourea, procarbazine, and propyl methanesulfonate as measured by F1 congenital defects, *Mutat. Res.* 229: 123-128.

Neel, J. V., Satoh, C., Goriki, K., Asakawa, J., Fujita, M., Takahashi, N., Kageoka, T., and Hazama, R., 1988, Search for mutations altering protein charge and/or function in children of atomic bomb survivors: final report, *Am. J. Hum. Genet.* 42: 663-676.

Nomura, T., 1975, Transmission of tumors and malformations to the next generation of mice subsequent to urethan treatment, Cancer Res. 35: 264-266.

Nomura, T., 1976, Comparison of tumour susceptibility among various organs of foetal, young, and adult ICR/Jcl mice, *Br. J. Cancer* 33: 521-534.

Nomura, T., 1978, Changed urethan and radiation response of the mouse germ cell to tumour induction, in "Tumours of Early Life in Man and Animals", L. Severi, ed., pp. 873-891, Pergia Univ. Press, Perugia.

Nomura, T., 1982, Parental exposure to X rays and chemicals induces heritable tumours and anomalies in mice, *Nature*. 296: 575-577.

Nomura, T., 1986, Further studies on X-ray and chemically induced germ-line alterations causing tumors and malformations in mice, in "Genetic Toxicology of Environmental Chemicals, Part B: Genetic Effects and Applied Mutagenesis," C. Ramel, ed., pp. 13-20, Alan R. Liss, New York.

Nomura, T., 1988, X-ray-and chemically induced germ-line mutation causing phenotypical anomalies in mice, *Mutat. Res.* 198: 301-320.

Nomura, T., 1989a, Role of radiation-induced mutations in multigeneration carcinogenesis, *IARC Sci. Publ.* 96: 375-387.

Nomura, T., 1989b, Congenital malformations as a consequence of parental exposure to radiations and chemicals in mice, *J. Univ. Occupat. Environ. Health* 11: 406-415.

Nomura, T., 1990, Of mice and men? *Nature* 345: 671.

Nomura, T., 1991, Paternal exposure to radiation and offspring cancer in mice: reanalysis and new evidences, *J. Radiat. Res.*, Suppl. 2: 64-72.

Nomura, T., Hata, S., Shibata, K., and Kusafuka, T., 1987, Killing of preimplantation mouse embryos by main ingredients of cleansers AS and LAS, *Mutat. Res.* 190: 25-29.

Nomura, T., Gotoh, H., and Namba, T., 1990, An examination of respiratory distress and chromosomal abnormalities in the offspring of male mice treated with ethylnitrosourea, *Mutat. Res.* 229: 115-122.

Oakberg, E. F., 1962, Gamma-ray sensitivity of oocytes of immature mice, *Proc. Soc. Exptl. Biol. Med.* 109: 763-767.

Russell, W. L., Hunsicker, P. R., Raymer, G. D., Steele, M. H., Stelzner, K. F., and Thompson, H. M., 1982, Dose-response curve for ethylnitrosourea-induced specific-locus mutations in mouse spermatogonia, *Proc. Natl. Acad. Sci.* (U.S.A.) 79: 3589-3591.

Sato, C., and Neel, J. V., 1988, Biochemical mutations in the children of atomic bomb survivors, Gann Monogr. *Cancer Res.* 35: 191-208.

Searle, A. G., and Beechey, C. V., 1974, Sperm-count, egg-fertilization and dominant lethality after X-irradiation of mice, *Mutat. Res.* 22: 63-72.

Shiono, P. H., Chung, C. S., and Myrlanthopoulos, N. C., 1980, Preconception radiation, intrauterine diagnostic radiation, and childhood neoplasia, *J. Natl. Cancer Inst.* 65: 681-686.

Shu, X. O., Gao, Y. T., Brinton, L. A., Linet, M. S., Tu, J. T., Zheng, W., and Fraumeni, J. F. Jr., 1988, A population-based case-control study of childhood leukemia in Shanghai, *Cancer* 62: 635-644.

Tomatis, L., Cabral, J. R. P., Likhachev, A. J., and Ponomarkov, V., 1981, Increased cancer incidence in the progeny of male rats exposed to ethylnitrosourea, *Int. J. Cancer* 28: 475-478.

Ton, T., 1981, US chemical warfare and its consequences, Vietnam Courier, Hanoi, reviewed by Watanuki, R. in "Kagaku," 51: 577-580 (in Japanese).

Yoshimoto, Y., Neel, J. V., Schull, W. J., Kato, H., Soda, M., Eto, R., and Mabuchi, K., 1990, Malignant tumors during the first 2 decades of life in the offspring of atomic bomb survivors, *Am. J. Hum. Genet.* 46: 1041- 1052.

PRECONCEPTION EXPOSURE OF MALES AND NEOPLASIA IN THEIR PROGENY: EFFECTS OF METALS AND CONSIDERATION OF MECHANISMS

Lucy M. Anderson, Kazimierz S. Kasprzak, and Jerry M. Rice

Laboratory of Comparative Carcinogenesis
National Cancer Institute
Frederick Cancer Research and Development Center
Frederick, MD 21702

INTRODUCTION

Exposures of male rodents, either as fetuses or in the weeks before mating, to X- or neutron radiation, or to an assortment of chemical carcinogens (urethane, ethylnitrosourea, 4-nitroquinoline N-oxide, diethylstilbestrol, or cyclophosphamide), have resulted in significant increases in the incidences of tumors in their progeny and sometimes later generations (review in Tomatis *et al.*, 1992; see also Francis *et al.*, 1990; Loktionov *et al.*; 1992, Turusov *et al.*, 1992; Takahashi *et al.*, 1992). Target organs have included the nervous system and uterus of rats, and lung, lymphoid tissue, ovary, uterus, liver, and intestine of mice. Exposures of females have also revealed multigenerational carcinogenesis effects, with implication of additional compounds (methylnitrosourea, *o*-aminoazotoluene, 3-methylcholanthrene, 7,12-dimethylbenz(*a*)anthracene, and benzo(*a*)pyrene) and target organs (skin, mammary gland, and pituitary). Broad significance of the phenomenon is suggested by the diversity of tumorigenic agents causing it and the wide range of target organs.

For the human, numerous epidemiological studies have revealed significant correlations between paternal exposures and incidences of childhood cancers (reviewed in Olsen *et al.*, 1991; Bunin *et al.*, 1992). One of the strongest associations, in terms of consistency of findings and magnitude of apparent effect, has been between occupations involving exposures to metals (welder, machinist, auto body repair, etc.) and childhood cancers at a variety of sites, including kidney (Wilm's tumor), lymphoid tissue (leukemia), central nervous system, soft tissues (sarcoma), liver (hepatoblastoma), and eye (retinoblastoma) (Olshan *et al.*, 1990; Bunin *et al.*, 1992). Some common metals, including chromium and nickel, are known to be tumorigenic in adult humans and rodents (Sunderman, 1986; Magos, 1991), and nickel has recently been demonstrated to be a transplacental carcinogen in rats (Diwan *et al.*, 1992).

Male-Mediated Developmental Toxicity, Edited by D.R. Mattison
and A.F. Olshan, Plenum Press, New York, 1994

been demonstrated to be a transplacental carcinogen in rats (Diwan *et al.*, 1992). However, the biological plausibility of a putative preconception tumorigenic effect of metals in the human was uncertain, since metals had not been tested in this regard in an animal model.

We therefore addressed experimentally the question of whether preconception exposure of males to metals may result in increased tumor incidence in their offspring. The experimental system chosen was the mouse lung tumor model developed by Nomura (1975, 1982, 1989), which is well defined, has given consistent results, and is relatively rapid. Attention was focused first on constituents of welding fumes, based on the epidemiology. Seven metal compounds as well as sodium fluoride, all prominent components of welding fumes, were tested.

ASSAY OF PRECONCEPTION TUMORIGENIC EFFECTS OF METALS

Methods

Swiss Cr:NIH(s) mice were obtained from the Animal Production Area of the Frederick Cancer Research and Development Center and maintained under standardized pathogen-free conditions as described previously (Anderson *et al.*, 1989). Male mice, 6 weeks old, were treated once i.p. with solutions of the compounds in saline and at the doses listed in Table 1. Initial doses were chosen on the basis of published LD-50 values for rodents, as those likely to be biologically effective. Two weeks later, so that sperm utilized in fertilization would have been exposed post-meiotically, each male was offered five 8-week old females for 3 weeks. The males were then killed for histological examination of kidneys and testes. The females were held for an additional 3 weeks and housed separately when visibly pregnant.

Offspring were weaned at 4 weeks and held without further treatment until kill at 20 weeks with complete necropsy. Lungs were fixed in Bouin's solution and tumors identified under a dissecting microscope as described previously (Anderson *et al.*, 1989); the latter were confirmed in paraffin sections with hematoxylin and eosin staining. All other masses and lesions were examined histologically. Results were evaluated by the Fisher Exact Test or Student's t test as appropriate. To take into account any possible litter effects, incidences of tumor bearers were in addition analyzed by Wilcoxon's rank-sum test on the raw proportions of tumor-bearing animals in each litter; by analysis of variance (ANOVA) of the arcsin transformations of these proportions, weighted proportionally to the litter size; and by ANOVA of the Freeman-Tukey binomial transformations of the proportions, similarly weighted (Haseman and Kupper, 1979).

Toxicity

Doses of 500 μmol/kg of $FeCl_2$, $MnCl_2$, and NaF, 600 μmol/kg of $FeCl_3$, or 1000 μmol/kg $CrCl_3$ were well tolerated by the male mice, with no overt signs of toxicity. $Na_2Cr_2O_7$, $NiCl_2$, and $CoCl_2$ were more toxic, with death 1-2 days after treatment at doses of 300 (as Cr), 500, and 500 μmol/kg respectively (Table 1). The dichromate salt was tolerated only at 75 μmol Cr/kg; death occurred 10-12 days after treatment with 150 μmol Cr/kg. The nickel and cobalt chlorides were tolerated at 250 μmol/kg.

Kidneys of the treated males were evaluated for mineralization of the medulla, glomerulonephritis, cortical tubular degeneration or regeneration, proteinaceous casts, tubular dilatation, cortical tubular hyaline droplets, cortical mineralization, proximal tubular karyomegaly, and interstitial chronic inflammation. Changes were minimal in all groups and did not differ from controls, except for those receiving the toxic doses of sodium dichromate, in which we observed mild to moderate multifocal alterations in all categories except for mineralization of the medulla.

Testes were evaluated for tubular degeneration, cysts, and interstitial cell hyperplasia. Again changes were minimal and not different in treated vs. control mice, except for minimal to mild interstitial hyperplasia in one male each from the $Na_2Cr_2O_7$, $NiCl_2$, and $CoCl_2$ treatments.

Table 1. Toxicity after Exposure of Male Mice to Metal Salts

Compound[1]	Dose[2] μmol/kg	Av. # Female Impregnated ± S.D. [3]	Av. # Offspring Weaned Per Litter ± S.D.	
			Females	Males
$FeCl_2$	500	4.25 ± 0.5	4.1 ± 1.1	4.3 ± 1.4
$FeCl_3$	600	3.4 ± 1.3	3.6 ± 1.3	3.5 ± 1.0
$MnCl_2$	500	4.25 ± 1.0	3.0 ± 0.3	3.8 ± 0.6
$CrCl_3$	1000	3.6 ± 1.0	4.1 ± 0.6	2.8 ± 1.9
$Na_2Cr_2O_7$	300	males died		
	150	males died		
	75	3.6 ± 1.1	2.6 ± 1.1	2.7 ± 1.0
$NiCl_2$	500	males died		
	250	3.5 ± 1.7	2.7 + 1.8	4.0 ± 3.0
$CoCl_2$	500	males died		
	250	3.6 ± 1.3	3.6 ± 1.0	2.5 ± 1.5
NaF	500	4.4 ± 0.9	4.2 ± 1.3	3.7 ± 1.6
Saline		3.8 ± 0.8	3.1 ± 1.1	3.9 ± 1.1

[1] Aldrich Co., Milwaukee, WI; highest purity available
[2] Calculated per metal or fluoride
[3] In each group, 0-3 litters died before weaning.

With regard to reproductive parameters, each of the males impregnated an average of 3.4 to 4.4 of the 5 females offered, with no significant differences among the groups (Table 1). All of these females gave birth within 4-5 weeks after introduction of the males. In each group regardless of treatment 0-3 of the litters born died before weaning. The average numbers of male and female offspring weaned are given in Table 1. Litter sizes were 7-8; there were slightly fewer on average for the dichromate treated males (average 5.3) and those given $CoCl_2$ (average 6.1), but these differences were not of statistical significance when evaluated with the treated male as the experimental unit. Sex ratios were not significantly altered in any group, although again there was a suggestion of an effect with the cobalt salt, in favor of female offspring.

In sum, at sublethal doses none of the compounds applied had major toxic effects on the internal organs of the males or on their reproductive performances.

Tumorigenicity

Results are summarized in Table 2. In the saline control mice, lung tumors occurred with an incidence of 0.9%, in good agreement with the low incidence of spontaneous lung tumors expected in this strain at 20 weeks of age. No other neoplasms occurred.

Lung tumor incidence in the offspring of the NaF-treated males was similar, 0.7%. However, 1 ovarian teratoma was found.

All of the offspring of males exposed to metal salts other than NaF had lung tumor incidences higher than those in the control and NaF groups. The incidence after $CrCl_3$ was 7%, a difference of statistical significance compared with controls, including statistical tests that would correct for litter effect. Each of the 5 treated males contributed to the fathering of the 8 tumor-bearing mice; 2 litters each included 2 tumor bearers. A disseminated lymphoblastic lymphoma was found in one male offspring.

Lung tumor incidences after the remaining metal salts were 2-3%. Although 2- to 3-fold greater than experimental and historical controls, the numbers were too low for an effect of statistical significance to be demonstrated. A lymphoblastic lymphoma occurred in a male offspring after $FeCl_2$.

Conclusions

These results indicate that some metal compounds, like organic carcinogens and radiation, can, after exposure of males, increase tumor incidence in their offspring, even in the absence of notable morphological effects or immediate reproductive toxicity.

The statistically-significant effect of $CrCl_3$ on lung tumor incidence seems quite convincing but of course requires confirmation. Effects of the other metal salts on lung tumors, and the single occurrence of other types of neoplasms, are at present only suggestive but will be explored with additional experiments. The dose of $CrCl_3$ applied was greater than that of the other compounds and may have contributed to its particularly striking effect. It is of interest that Cr(III) was found to accumulate in the interstitial tissue of testes of rats to a greater extent than Cr(VI) (Danielsson et al., 1984). Also, Cr(III) caused degeneration of spermatogenic epithelium and interruption of spermatogenesis in rabbits to a greater extent than Cr(VI)(Behari et al., 1978). Transitional suppression of spermatogenesis and disintegration of spermatozoa was also observed in rats given nickel sulfate (Hoey, 1966). Preimplantation loss of embryos after exposure of male mice to nickel nitrate was associated with toxic effect of nickel on spermatids and spermatogonia (Jacquet and Mayence, 1982). The effects of chromium, nickel, and welding fumes on reproduction and prenatal toxicity have been recently reviewed by IARC (1990). No relevant data are available for the other metals studied.

Complete understanding and management of risks entailed in human parental preconception exposure to metals and other carcinogens will ultimately depend in part on elucidation of the mechanisms of this remarkable effect. To facilitate design of meaningful experiments on mechanism, we have examined the literature on rodent multigenerational carcinogenesis for clues as to the types of phenomena that might be occurring. Some thoughts generated by this analysis are given below. Data from both

Table 2. Tumors in Offspring after Exposure of Fathers to Metal Salts

Compound	Lung Tumors			Other Neoplasms
	No. With Tumor/ Total (Percent)	# Fathers with Tumor-Bearing Offspring	# Litters with Tumor-Bearing Offspring	
$FeCl_3$	3/121/ (2.5%) 1 M, 2F	2	2	none
$FeCl_2$	3/108 (2.8%) 1 M, 2 F (1 with 2)	2	2	lymphoblastic lympoma (M)
$MnCl_2$	3/109 (2.8%) 1 M, 2 F	3	3	none
$CrCl_3$	8/119 (7%)[1] 2 M, 6 F	5	6	lymphoblastic lumphoma (M)
$Na_2Cr_2O_7$	3/96 (3%) 3 F	2	3	none
$NiCl_2$	3/110 (2.7%) 1 M, 2 F	2	3	none
$CoCl_2$	3/114 (2.6%) 1 M, 2 F	3	3	none
NaF	1/147 (0.7%) 1 F	1	1	ovarian teratoma
Control	1/114 (0.9%) 1 F	1	1	none

[1] P = 0.021 vs controls, Fisher exact test; P = 0.029, Wilcoxon's rank-sum test of unweighted raw proportions; P = 0.018, ANOVA of the arcsin transformed proportions, weighted by litter size; P = 0.014, ANOVA of the Freeman-Tukey transformed proportions, weighted by litter size. All P values are one-tailed.

male-exposure and female-exposure experiments have been considered, since there is no evidence for a qualitative difference between them at present.

CONSIDERATIONS OF MECHANISMS OF MULTIGENERATIONAL CARCINOGENESIS

Mutation and Epimutation as Possible Mechanisms

There are at least 2 possible general categories of mechanism, which are not mutually exclusive. One is structural genotoxic damage, such as a mutation in an oncogene or a tumor suppressor gene or some other gene regulating neoplasia, or a deletion, amplification, or rearrangement affecting expression of some of these genes. Alternatively or in addition, preconception carcinogenesis might result from derangement of epigenetic gametic modulation of gene expression, i.e., epimutation of the imprinted status of genes (Holliday, 1991).

Mutation

Genotoxic carcinogens are generally mutagens, and some of these, especially the alkylating agents, are exceptionally potent in this regard, both in mammalian somatic cells *in vivo* (Russell and Montgomery, 1982) and in mammalian germ cells at specific stages of development. In males, susceptibility is commonly greatest in postspermatogonial (early spermatid) stages (Russell *et al.*, 1992). It stands to reason that if mutational mechanisms exist for preconception carcinogenesis, or for other genetic disease that may result from preconception exposure of male parents to mutagens, these mechanisms should be revealed by experiments with "supermutagenic" carcinogens such as N-nitrosoethylurea. In fact, this has been accomplished. Mouse models of various inborn errors of metabolism such as sarcosinemia have been created by phenotypic screening of offspring of males that were exposed to a large dose of N-nitrosoethylurea and then allowed to recover from the resulting transient infertility prior to mating (Harding *et al.*, 1992). Affected offspring transmit the trait to their descendants in classical Mendelian pattern. This has proved to be a generally useful approach for developing mouse models for diseases where the phenotype is known, but the specific gene that is responsible is not.

Inborn errors of metabolism, however, commonly involve a single gene that, when rendered nonfunctional, causes a metabolic dysfunction that is inherited as a recessive trait because heterozygous individuals are phenotypically normal. Some cancers--notably retinoblastoma--have a comparable genetic basis (although in the case of retinoblastoma the pattern of inheritance is classically described as autosomal **dominance** with incomplete penetrance), but most cancers appear to involve several genes. For this reason it is much less likely that neoplasms would appear as a consequence of protocols such as that described above. The one exception to date is that of the MIN (Multiple Intestinal Neoplasia) mouse, that develops numerous adenomas throughout its intestinal tract. This mutation was generated by the same kind of protocol as described for the sarcosinemia model (Moser *et al.*, 1990); the resulting predisposition to tumor development is inherited as an autosomal dominant trait, and is due to a nonsense mutation in the APC gene that is also mutated in the germ line of humans with familial adenomatous polyposis (Su *et al.*, 1992). The inheritance pattern is, as expected, classically Mendelian, and demonstrates the possibility that a tumor-related mutation can in fact arise in the male germ line as a result of preconception exposure to a potent mutagen.

Epimutation

The other category of mechanism of action of preconception carcinogens relates to the phenomenon of imprinting, based on an increasing body of evidence from both mouse models and human disease patterns. Imprinting in this sense is currently defined as the parental sex-specific epigenetic regulatory processing of gamete genomes, resulting in a permanent effect on the expression of some genes in the individual arising from the gamete; these could include genes influencing cancer development (Hall, 1990; Sapienza, 1991).

The natural functions of the imprinting phenomenon have variously been proposed to be a role in placentation and early development, a contribution to establishment and maintenance of sexual reproduction, or a device used by fathers, on behalf of their offspring to insure maximum nutritional demands on mothers (Hall, 1990; Chandra and Nanjundiah, 1990; Moore and Haig, 1991). Endogenous imprinted genes are just beginning to be uncovered and include growth factor and growth factor receptor genes expressed during early development (Barlow *et al.*, 1991; Bartolomei *et al.*, 1991; DeChiara *et al.*, 1991).

During the early growth of this new field, little attention has been given to the possibility of modulation of imprinting by factors other than or in addition to parental sex, including possible toxic impact. However, there is a large body of literature, most recently critically reviewed by Campbell and Perkins in 1988, showing persistent functional effects of treatment of parents on their unexposed descendants. The greatest volume of work was done in the neuroendocrine context, and ranged from rather generalized treatments such as crowding, handling, and stress, to some specific treatments that resulted in transgenerational effects on directly related parameters, such as thyroxin and thyroidectomy on thyroid stimulating hormone levels, diabetes and pancreatectomy on beta cells and glucose tolerance, morphine on pain response, psychotropic drugs on vaginal opening, body temperature and drug responses, and parathyroidectomy on calcium handling. Effects have also been described in the immune system, and of nutritional deprivation, and of assorted xenobiotics. References may be found in the review by Campbell and Perkins; see also, Auroux, 1990; and Gandley and Silbergeld, this volume.

Some of these investigators have interpreted these findings as evidence for directed adaptation of the physiology of the descendants based on the experience of the ancestor. Although an actual increase in fitness by such a response has yet to be demonstrated experimentally, the idea is only superficially Lamarckian, and differs from the more recently emphasized imprinting phenomenon only in that gene expression is heritably modulated by the parent's physiological or toxicological internal milieu, rather than (or in addition to) parental sex.

This broadening of the concept of genomic imprinting, to include exposure-responsive events and consequences beyond early development, renders it a potential mechanism to account for multigenerational carcinogenesis. For purposes of analysis we have compared this epimutation mechanism with the classical, genotoxic scenario, asking what predictions each of these two mechanisms might suggest with regard to outcome parameters in a multigenerational carcinogenesis situation, and how the actual data agree with the alternative predictions. Four sets of these prediction-evidence dialogues are offered below.

Frequency of Occurrence. When multigenerational carcinogenesis is caused by mutation, the rate might be similar to that for a known gene in the species. The potential rate for epimutation of imprinting *in vivo* is unknown, but, *in vitro*, imprinted silenced genes are awakened with high frequency after application of demethylating agents (Holliday, 1987). The only experiments with enough quantitative data to make an estimate of the incidence of a multigenerational effect as related to known gene mutations are those with mouse lung tumors by Nomura; the increase was 10-100 fold greater than that caused by mutagens at known mouse loci (Nomura, 1984). For structural change to account for this there would have to be either a highly hypermutable locus or many alternative loci whose mutation would give the same result. The data are perhaps more consistent with a change in imprinting, leading to altered gene regulation and increased expression of spontaneous tumors, or perhaps increased rate of somatic mutation in the lung.

Inheritance Patterns and Permanence. A structural gene change should be permanent over an indefinite number of generations, passed on predictably by tumor bearers, with reduced incidence if mating is random, since only a small proportion of the individuals is affected, and independent of the sex of the transmitter. An imprinting effect would be expected to be nonpermanent and might exhibit rather more variable than predictable inheritance if it is impacting on the rate of spontaneous tumors in all or many descendants. It might well be affected by the sex of the transmitter.

On this point there is evidence supporting both models. Again the most definitive data are from Nomura's work, which showed that lung tumor-bearing mice, but not non-bearers, passed the trait on for 3 generations (Nomura, 1978, 1982, 1986). Similarly Tomatis and Goodall (1969) found that the lung tumor effect was gone in the F_3 generation with random mating.

On the other hand, much of the evidence from multigenerational carcinogenesis experiments is more difficult to reconcile with Mendelian predictions. With the mouse lung tumor model, after transplacental benzpyrene there was increased multiplicity that did not diminish at all in 5 generations in spite of random mating (Turusov et al., 1990). Also in mice, lymphoid tumors, resulting from preconception x-ray or transplacental DMBA, maintained a 5- to 7-fold increase over controls, with no diminution after random mating for 3 or 4 generations (Tomatis and Goodall, 1969; Nomura, 1986). In rats, again with random mating, after transplacental methylnitrosourea, mammary tumors were increased to about the same extent in F_2 and F_3 descendants, and a significant effect in pituitary was not seen until the F_3 (Tomatis, 1975). With transplacental ethylnitrosourea significant effects on mammary tumors occurred only in the F_3 animals, but on pituitary tumors in both F_3 and F_4 (Tomatis *et al.*, 1977).

Available data from these studies do not permit analysis of parental sex dependence of transmission of the effects. This will be an important point for future study.

Target Organs and Oncogene Activation Patterns. For structural change, the target organs might well be the same as those responding after direct treatment, and with similar oncogene activation patterns. These could include organs with a low spontaneous tumor incidence. The imprinting model predicts that the increase will be mainly in tumors with high spontaneous occurrence, and that in fact sometimes there will be a decrease. The evidence on this point is more consistent with the imprinting mechanism. At least 3 models with low spontaneous rate and high responsiveness to direct treatment showed a minimal or absent preconception effect, ethylnitrosourea and rat schwannomas (Turusov and Cardis, 1989) or rat liver preneoplastic foci (Ogawa *et al* ., 1983), and N-nitrosodiethylamine and hamster respiratory tract tumors (Ernst *et al.*, 1987). Further, the skin tumors in F_2 mice after transplacental dimethylbenzanthracene did not show the H-*ras* oncogene activation pattern characteristic of directly-induced tumors, or mutations in the P53 tumor suppressor gene (Loktionov *et al.*, 1992; Yamasaki, H., personal communication). Thus the predictions of the mutation model are not confirmed, and in fact the most prominent effects are seen in organs with high spontaneous tumor incidence, mouse lung, mammary and lymphoid, and rat mammary and pituitary, in accordance with the imprinting model. Recent results in C3H mice are confirmatory. In this strain, which has a high spontaneous liver tumor occurrence, there was a pronounced effect of preconception radiation exposure of fathers on incidence of liver tumors in their male progeny (43% with tumor compared with 3% of controls) (Takahashi *et al.*, 1992), in direct contrast to the absence of effect in rat liver where there is little spontaneous incidence. Finally, there was an indication of a decrease in thyroid tumors after preconception ethylnitrosourea, although this lost statistical significance when correction was made for litter effects (Turusov and Cardis, 1989).

Effective Chemicals and Dose Response. Structural change should be caused by genotoxicants only, with some chemical specificity, and there should be a positive dose response relationship as in typical mutagenesis situations. Imprinting derangement would most likely result from protein interactions, with less specificity for chemical structure and a sigmoid dose response typical of needing to saturate a target and then going quickly to maximum effect. Evidence on this point is scanty. No definitively nongenotoxicants have been shown to be effective so far, but only 2 have been tried, 1,1'-(2,2,2-trichloroethylidene)bis[4-chlorobenzene](DDT) (Turusov *et al.*, 1973), and methylthiouracil (Napalkov, 1969). Diethylstilbestrol, which may act through a nongenotoxic pathway, has given positive results in 2 multigeneration tumorigenesis studies (Walker, 1984; Turusov *et al.*, 1992). For the responsive mouse lung tumor model, no genotoxicant has been reported as negative, and all of them, radiation and a wide variety of types of chemicals, all given at high, just subtoxic doses, give about the same effect, 2.5- to 5-fold increases. This may indicate lack of specificity and a sigmoid dose-response curve. More data can be obtained relatively easily on this point and should prove informative.

CONCLUSIONS

In the above considerations, we have analyzed the imprinting hypothesis in the light of data from the rodent literature. The problem may also be viewed from the other perspective: is this hypothesis compatible with what is known about imprinting? For example, is the extent of imprinting sufficient to account for the observed multiple tumor effects? Parental sex-specific imprinting in the mouse involves only a small percentage of the genome as indicated by gross effects in uniparental disomy experiments (Cattanach and Beechey, 1990). However, homologous regions in the human genome contain a remarkably large assortment of oncogenes, tumor suppressor genes, and growth factor genes (Hall, 1990). Thus, even with only about 10% of the chromosomes imprinted, the key genes may be affected. A somewhat higher percentage, up to 20%, of transgenes are imprinted (Reik *et al.*, 1990). More subtle and graded differences are starting to be revealed, suggesting that the phenomenon may be much broader than originally suspected (Sapienza, 1989). If nonparental sex-specific imprinting is in fact a phenomenon, then many genes would be expected to be implicated.

Another consideration is whether the imprint is fully wiped out with each generation, so the observed effects in the F_3 and beyond could not be explained. For parental sex-specific imprinting this would be expected and usually is the case, but there are exceptions among certain genes and transgenes of mice (Hadchouel *et al.*, 1987; Allen *et al.*, 1990; Reik *et al.*, 1990) and possibly in some human conditions such as Huntington's disease (Ridley *et al*, 1988). These exceptions allow the possibility of persistent inheritance of an altered imprint (Chandra and Nanjundiah, 1990; Surani *et al.*, 1990; Surani, 1991). Whether this may occur as a toxic consequence has not yet been explored but obviously is a key question. For the putative more general non-parental sex specific imprinting a wipe-out is not necessary.

It has been argued that gene methylation, which is thought to be part of the mechanism of imprinting, is largely lost during gametogenesis so there is no way to preserve the epimutation to the next generation. However, control of gene expression must persist in these as in all cells or chaos would result; these control mechanisms could contribute to the imprinting effect. Most investigators in this field opine that methylation is not the sole component of the imprinting mechanism (Hall, 1990; Surani, 1991).

In sum, while structural gene mutation probably contributes to the multigenerational carcinogenesis phenomenon in rodents and humans, analysis of the data from the rodent literature compels consideration of epimutation as a component, and nothing in our current knowledge of imprinting forbids this proposal. This epimutation may consist of derangement of parental sex-specific imprinting, and/or resemble the trans-generational functional effects of parental manipulation. The latter still await molecular confirmation and delineation, but may well provide the key to understanding the influence of chemical or radiation exposure of parents on cancer risk in their descendants. Experimental and epidemiological examination of this hypothesis seems rather urgent in the public health context.

ACKNOWLEDGMENTS

The authors appreciate histopathological diagnosis and consultation by Dr. Miriam Anver and Dr. Sabine Rehm and statistical analysis by Charles W. Riggs.

REFERENCES

Allen, N.D., Norris, M.L., and Surani, M.A., 1990, Epigenetic control of transgene expression and imprinting by genotype-specific modifiers, *Cell* 61:853.

Anderson, L.M., Jones, A.B., Riggs, C.W., and Kovatch, R.M., 1989, Modification of transplacental tumorigenesis by 3-methylcholanthrene in mice by genotype at the *Ah* locus and pretreatment with β-naphthoflavone, *Cancer Res.* 49:1676.

Auroux, M., Dulioust, E., Selva, J., and Rince, P., 1990, Cyclophosphamide in the F_0 male rat: physical and behavioral changes in three successive adult generations. *Mutat. Res.* 229:189.

Barlow, D.P., Stoger, R., Herrmann, B.G., Saito, K., and Schweifer, N., 1991, The mouse insulin-like growth factor type-2 receptor is imprinted and closely linked to the *Tme* locus, *Nature* 349:84.

Bartolomei, M.S., Zemel, S., and Tilghman, S.M., 1991, Parental imprinting of the mouse H19 gene, *Nature* 351:153.

Behari, J., Chandra, S.V., and Tandon, S.K., 1978, Comparative toxicity of trivalent and hexavalent chromium to rabbits. III. Biochemical and histological changes in the testicular tissue, *Acta Biol. Med. Ger.* 37:463.

Bunin, G.R., Noller, K., Rose, P., and Smith, E., 1992, Carcinogenesis, *in:* "Occupational and Environmental Reproductive Hazards: A Guide for Clinicians," M. Paul, ed., Williams and Wilkins, Baltimore.

Campbell, J.H., and Perkins, P., 1988, Transgenerational effects of drugs and hormonal treatments in mammals: a review of observations and ideas. *Prog. Brain Res.* 73:535.

Cattanach, B.M., and Beechey, C.V., 1990, Autosomal and X-chromosome imprinting, *Develop.* (Suppl.):63.

Chandra, H.S., and Nanjundiah, V., 1990, The evolution of genomic imprinting, *Develop.* (Suppl.):47.

Danielsson, B.R.G., Dencker, L., Lindgren, A., and Tjalve, H., 1984, Accumulation of toxic metals in male reproductive organs, *Archiv. Toxicol.* Suppl 7:177.

DeChiara, T.M., Robertson, E.J., and Efstratiadis, A., 1991, Parental imprinting of the mouse insulin-like growth factor II gene, *Cell* 64:849.

Diwan, B.A., Kasprzak, K.S., and Rice, J.M., 1992, Transplacental carcinogenic effects of nickel(II) acetate in the renal cortex, renal pelvis and adenohypophysis in F344/NCr rats, *Carcinogenesis* 13:1351.

Ernst, H., Emura, M., Bellmann, B., Seinsch, D., and Mohr, U., 1987, Failure to transmit diethylnitrosamine tumorigenicity from transplacentally exposed F_1 generation Syrian hamsters to the respiratory tract of F_2 and F_3 generations, *Cancer Res.* 47:5112.

Francis, A.J., Anderson, D., Evans, J.G., Jenkinson, P.C., and Godbert, P., 1990, Tumours and malformations in the adult offspring of cyclophosphamide-treated and control male rats-Preliminary communication, *Mutat. Res.* 229:239.

Hadchouel, M., Farza, H, Simon, D., Tiollais, P., and Pourcel, C., 1987, Maternal inhibition of hepatitis B surface antigen gene expression in transgenic mice correlates wtih *de novo* methylation, *Nature* 329:454.

Hall, J.G., 1990, Genomic imprinting: review and relevance to human disease, *Am. J. Hum. Genet.* 46:857.

Harding, C.O., Williams, P., Pflanzer, D.M., Colwell, R.E., Lyne, P.W., and Wolff, J.A., 1992, *sar*: A genetic mouse model for human sarcosinemia generated by ethylnitrosourea mutagenesis, *Proc. Natl. Acad. Sci.* 89:2644.

Haseman, J.K., and Kupper, L.L., 1979, Analysis of dichotomous response data from certain toxicological situations, *Biometrics* 35:281.
Hoey, M.J., 1966, The effects of metallic salts on the histology and functioning of the rat testis, *J. Reprod. Fert.* 12:461.
Holliday, R., 1987, The inheritance of epigenetic defects, *Science* 238:163.
Holliday, R., 1991, Mutations and epimutations in mammalian cells, *Mutat. Res.* 250:351.
IARC Monographs on the Evaluation of Carcinogenic Risks to Humans, 1990, Vol. 49, *Chromium, Nickel, and Welding*, IARC, Lyon, France.
Jacquet, P., and Mayence, A., 1982, Application of the in vitro embryo culture to the study of mutagenic effects of nickel in male germ cells, *Toxicol. Lett.* 11:193.
Loktionov, A., Popovich, I., Zabezhinski, M., Martel, N., Yamasaki, H., and Tomatis, L., 1992, Transplacental and transgeneration carcinogenic effect of 7, 12-dimethylbenz[*a*]anthracene: relationship with *ras* oncogene activation, *Carcinogenesis* 13:19.
Magos, L., 1991, Epidemiological and experimental aspects of metal carcinogenesis: physicochemical properties, kinetics, and the active species, *Environ. Health Perspect.* 95:157.
Moore, T, and Haig, D., 1991, Genomic imprinting in mammalian development: a parental tug-of-war, *Trends Genet.* 7:45.
Moser, A.R., Pitot, H.C., and Dove, W.F., 1990, A dominant mutation that predisposes to multiple intestinal neoplasia in the mouse, *Science* 247:322.
Napalkov, N.P., 1969, Thyroid tumorigenesis in rats treated with 6-methylthiouracil for several successive generations, *in*: "Thyroid Cancer", C.E. Heidinger, ed., Springer, Berlin.
Nomura, T., 1975, Transmission of tumors and malformations to the next generation of mice subsequent to urethan treatment, *Cancer Res.* 35:264.
Nomura, T., 1978, Changed urethan and radiation response of the mouse germ cell to tumor induction, *in*: "Tumours in Early Life in Man and Animals", L. Severi, ed., Perugia Quadrennial Int. Conf. Cancer, Perugia.
Nomura, T., 1982, Parental exposure to X rays and chemicals induces heritable tumours and anomalies in mice, *Nature* 296:575.
Nomura, T., 1984, Quantitative studies on mutagenesis, teratogenesis, and carcinogenesis in mice, *in*: "Problems of Threshold in Chemical Mutagenesis", Y. Tazima, S. Kondo, and Y. Kuroda, eds., Environmental Mutagen Society of Japan, Shizuoka.
Nomura, T., 1986, Further studies on X-ray and chemically induced germ-line alterations causing tumors and malformations in mice, *in*: "Genetic Toxicology of Environmental Chemicals, Part B: Genetic Factors and Applied Mutagenesis," C. Ramel, B. Lambert, and J. Magnusson, eds., Alan R. Liss, New York.
Nomura, T., 1989, Role of radiation-induced mutations in multigeneration carcinogenesis, *IARC Sci. Publ.* 96:375.
Ogawa, K., Yokokawa, K., Tomoyori, T., and Narasaki, M., 1983, Initiation of focal hyperplastic hepatic lesions by transplacental administration of ethylnitrosourea in rats of F_1 generation, and no transmission of the effect to F_2 and F_3 generations, *Int. J. Cancer* 31:775.
Olsen, J.H., Brown, P.D., Schulgen, G., and Jensen, O.M., 1991, Parental employment at time of conception and risk of cancer in offspring, *Eur. J. Cancer* 27:958.
Olshan, A.F., Breslow, N.E., Daling, J.R., Falletta, J.M., Grufferman, S., Robison, L.L., Waskerwitz, M., and Hammond, G.D., 1990, Wilms' tumor and paternal occupation, *Cancer Res.* 50:3212.
Reik, W., Howlett, S.K., and Surani, M.A., 1990, Imprinting by DNA methylation: from transgene to endogenous gene sequences, *Develop.* (Suppl.):99.
Ridley, R.M., Frith, C.D., Crow, T.J., and Conneally, P.M., 1988, Anticipation in Huntington's disease is inherited through the male line but may originate in the female, *J. med. Genet.* 25:589.
Russell, L.G., Hunsicker, P.R., Cacheiro, N.L.A., and Rinchik, E.M., 1992, Genetic, cytogenetic, and molecular analyses of mutations induced by melphalan demonstrate high frequencies of heritable deletions and other rearrangements from exposure of postspermatogonial stages of the mouse, *Proc. Natl. Acad. Sci.* 89:6182.
Russell, L.B., and Montgomery, C.S., 1982, Supermutagenicity of ethylnitrosourea in the mouse spot test. Comparison with methylnitrosourea and ethylnitrosourethane, *Mut. Res.* 92:193.
Sapienza, C., 1989, Genome imprinting and dominance modification, *Ann. N.Y. Acad. Sci.* 564:24.
Sapienza, C., 1991, Genome imprinting and carcinogenesis, *Biochim. Biophys. Acta* 1072:51.
Su, L., Kinzler, K.W., Vogelstein, B., Preisinger, A.C., Moser, A.R., Luongo, C., Gould, K.A., and Dove, W.F., 1992, Multiple intestinal neoplasia caused by a mutation in the murine homolog of the APC gene, *Science* 256:668.
Sunderman, F.W., 1986, Carcinogenicity and mutagenicity of some metals and their compounds, *IARC Sci. Publ.* 71:17.

Surani, M.A., Allen, N.D., Barton, S.C., Fundele, R., Howlett, S.K., Norris, M.L., and Reik, W., 1990, Developmental consequences of imprinting of parental chromosomes by DNA methylation, *Phil. Trans. R. Soc. Lond.* B 326:313.
Surani, M.A., 1991, Influence of genome imprinting on gene expression, phenotypic variations and development, *Human Reprod.* 6:45.
Takahashi, T., Watanabe, H., Dohi, K., and Ito, A., 1992, ^{252}Cf relative biological effectiveness and inheritable effect of fission neutrons in mouse liver tumorigenesis, *Cancer Res.* 52:1948.
Tomatis, L., and Goodall, C.M., 1969, The occurrence of tumours in F_1, F_2, and F_3 descendants of pregnant mice injected with 7,12-dimethylbenz(*a*)anthracene, *Int. J. Cancer* 4:219.
Tomatis, L., Hilfrich, J., and Turusov, V., 1975, The occurrence of tumours in F_1, F_2, and F_3 descendants of BD rats exposed to N-nitrosomethylurea during pregnancy, *Int. J. Cancer* 15:385.
Tomatis, L., Ponomarkov, V, and Turusov, V., 1977, Effects of ethylnitrosourea administration during pregnancy on three subsequent generations of BDVI rats, *Int. J. Cancer* 19:240.
Tomatis, L., Narod, S., and Yamasaki, H., 1992, Transgeneration transmission of carcinogenic risk, *Carcinogenesis* 13:145.
Turusov, V.S., Day, N.E., Tomatis, L., Gati, E., and Charles, R.T., 1973, Tumors in CF-1 mice exposed for six generations to DDT, *J Natl. Cancer Inst.* 51:983.
Turusov, V.S., and Cardis, E., 1989, Review of experiments on multigeneration carcinogenicity of alkylating agents: discussion of design, experimental models and analysis, *IARC Sci. Publ.* 96:105.
Turusov, V.S., Nikonova, T.V., and Parfenov, Y.D., 1990, Increased multiplicity of lung adenomas in five generations of mice treated with benz(*a*)pyrene when pregnant, *Cancer Lett.* 55:227.
Turusov, V.S., Trukhanova, L.S., Parfenov, Y.D., and Tomatis, L., 1992, Occurrence of tumours in the descendants of CBA male mice prenatally treated with diethylstilbestrol, *Int. J. Cancer* 50:131.
Walker, B.E., 1984, Tumors of female offspring of mice exposed prenatally to diethylstilbestrol, *JNCI* 73:133.

MALE-MEDIATED REPRODUCTIVE TOXICITY: EFFECTS ON THE NERVOUS SYSTEM OF OFFSPRING

Robin E. Gandley[1,3] and Ellen K. Silbergeld[1,2]

[1]Program in Toxicology
[2]Department of Epidemiology & Preventative Medicine
University of Maryland at Baltimore
660 West Redwood Street, Room 466
Baltimore, MD 21201-1596
[3]Magee-Women's Research Institute (current address)
204 Craft Avenue, Room 630
Pittsburgh, PA 15213

INTRODUCTION

Male reproductive toxicity deals primarily with infertility measures. The possibility of toxic effects subtly manifested in the offspring of exposed males is rarely considered. This is possibly due to the lack of research and testing dealing with this type of toxicity. Paternally-mediated effects on the development of offspring are not commonly assessed in toxicity studies. This oversight is partly due to the standard reproductive toxicity testing parameters of EPA and other regulatory agencies. Also, it is due to the lack of attention in epidemiological studies to exposure or co-exposure of fathers to reproductive and developmental toxicants. In both cases, developmental toxicity studies concentrate on maternal exposures both prior to pregnancy and during gestation. Developmental toxicity due to paternal exposure is rarely considered, even in the cases of recognized reproductive toxicants.

Lead is an example of a known reproductive and developmental toxicant with very little research to determine developmental effects due to paternal exposure. Numerous epidemiological studies have monitored lead levels in mothers and offspring and attempted to assess harmful effects. The majority of these studies have not at any time measured paternal exposure. A common assumption is that paternal exposures to lead result in infertility or reduced fertility because only normal sperm are capable of fertilizing oocytes. Lead can thus serve as an example of a commonly studied toxicant with very little data relating to male-mediated effects on development. It will be utilized to illustrate how conventional reproductive and developmental toxicology test designs can be adapted to assess male-mediated effects on reproductive outcome.

Male-Mediated Developmental Toxicity, Edited by D.R. Mattison
and A.F. Olshan, Plenum Press, New York, 1994

SPECTRUM OF REPRODUCTIVE TOXICITY TESTING FOR MALES

The possibility of male-mediated effects on neurodevelopment in offspring from paternal exposure has rarely been considered during studies of reproductive or developmental toxicity. The spectrum of male reproductive toxicity testing generally deals solely with direct effects on the exposed male (Gray, *et al.*, 1988). Pregnancy outcome studies are only recommended in the assessment of **female** reproductive toxicity and teratology. Studies similar in design to those suggested for females should also be utilized to detect male-mediated effects.

Table 1. EPA Proposed Testing Parameters for Reproductive and Developmental Toxicity

Parameters tested in male rats	
body and organ weight and histopathology	endocrinology
sexual maturation milestones	pituitary hormones
sperm counts, motility, morphology	
Parameters tested in female rats	
body and organ weight and histopathology	fertility
sexual maturation milestones	litter size
intrauterine growth and survival	offspring malformations

More sensitive testing parameters are vital to answering questions dealing with male reproductive outcome. Until adequate testing for male-mediated effects are conducted, consistent with those assessed in females, it will be difficult to even detect the possibility of male-mediated neurodevelopmental effects on offspring.

The list of chemicals that have been associated with adverse reproductive outcomes for males is based primarily on infertility.

Table 2. Some Reported Male Reproductive Toxicants

lead	mercury
cadmium	alcohol and other solvents
glycol ethers	pesticides (picloram, 2,4- D)
2, 3, 7, 8- TCDD	cocaine
morphine	cytotoxic antineoplastic drugs
vinyl chloride	combustion byproducts (firefighting)
dibromochlorophropane	
n-methylformamide	radiation

To assess these toxicants, the dose range was selected at levels high enough to affect normal spermatogenesis. More sensitive endpoints of toxicity at lower doses are rarely tested. Even with the testing paradigms now recommended by the EPA, the outcomes of subsequent pregnancies fathered by exposed males are not assessed.

EPIDEMIOLOGICAL EVIDENCE FOR NEURODEVELOPMENTAL EFFECTS

Epidemiological associations have been reported between paternal exposures and central nervous system malformations. Olshan *et al.* (1991) reviewed reports of anencephaly, spina bifida, and hydrocephaly related to paternal occupational exposures. While these studies have identified effects common to specific professions, they do not identify the actual doses or the agents causing adverse neurological outcomes. Unlike animal models, epidemiological studies cannot clearly define exposure, duration or dose. Occupational exposures are rarely limited to one specific agent. A need exists to develop model systems to test the observations reported for occupational risks to determine causative agents and mechanisms.

No studies of neurobehavioral function have been undertaken to assess associations between paternal exposures and developmental outcome in children. To assess the affect of male exposure to toxicants on the growth and development of the offspring, several non-conventional aspects of study design must be considered. First, the dosing must not be so high as to eliminate male fertility since failure to produce offspring precludes the detection of developmental defects. Next, the dosing must be limited to only paternal exposure. If both maternal and paternal dosing is conducted, the offspring exposure will become very difficult to assess. Third, much more sensitive endpoints are needed to determine risks for paternally mediated effects of toxicity. Model systems can take advantage of the ability to isolate paternal versus maternal effects of exposure by isolating exposures. In this way, the sensitivity levels of male versus female effects can be assessed. Presently, the female is assumed to be more sensitive due to the present testing schematic where more sensitive female outcomes are assessed.

A case-control study which assessed occupational lead exposure of parents of offspring suffering from strabismus was conducted by examining potential for exposure by job classification (Hakim *et al.*, 1991). The study suggested a possibility of a weak association between strabismus in offspring and paternal lead exposures. However, the definition of the period of potential lead exposure in this study was the period from conception through 9 months. This exposure period may not thoroughly cover the potential for paternal effects. This definition does not take into account occupational exposure of the father prior to conception. The only means of paternal effects assessed with this definition is due to direct contamination of the living area by the occupationally exposed father. To assess male-mediated effects on development from direct damage at conception, the window of exposure must consider paternal exposure during the period of spermatogenesis prior to conception.

EXPERIMENTAL EVIDENCE

Only two experimental studies exist to suggest that lead dosing of males prior to breeding can result in deficits in neurobehavioral learning tasks and neuronal development (Brady, *et al.*, 1975; Silbergeld, *et al.*, 1991). The lack of information is due to the failure to use models that separate maternal and paternal exposures. Studies isolating maternal lead exposure or combining parental exposures are commonly conducted. Rarely are breeding studies focused on paternal exposures. Other evidence for male-mediated effects was reported in a two-generational study which examined the reproductive outcomes of the F1 males from lead treated fathers. These animals were not directly exposed to lead at any point after conception. Decreased fertility of this F1 generation was found in breeding studies with control females. Both reduced litter size and pregnancy rate were reported (Stowe and Goyer, 1971).

MODEL SYSTEM

The following model was utilized to study relatively low levels of paternal lead exposure (0, 25, and 250 ppm lead in drinking water) on embryo-fetal development. Male rats (35 days old) were exposed to lead acetate for 35 days prior to breeding. This treatment period should result in paternal exposure approximately from the last spermatogonial mitotic division through the production of mature spermatozoa in the epididymis. Only those sperm that complete the transit time in the epididymis are capable of fertilization.

Table 3. Spermatogenic Cycles of Different Species

Species	Seminiferous Cycle	Transit Time[1]
Man	64 days	2-5 days
Rat	48-53 days	6-10 days
Rabbit	48-51 days	4-10 days

[1] Transit time is the time required for mature spermatozoa to be produced in the epididymis.
(de Kretser and Kerr, 1988; Robaire and Hermo, 1988)

In this model, the female and the conceptus are never directly exposed to lead treated drinking water. The only opportunity for direct exposure to lead is during insemination (either in seminal fluid or incorporated into sperm).

Breeding studies with lead-treated males were carried out with naive control females (70 days old) using superovulation and natural breeding techniques. Superovulation was necessary to facilitate timed breeding and retrieval of 2-cell embryos in numbers large enough to enable electrophoresis. Fertilization capacity of treated males could be assessed by comparing the number of available oocytes to the number of 2-cell embryos produced. Natural breeding was necessary to assess the viability of the pre-implantation embryos produced. Evidence of miscarriage at gestational day 17 was assessed during the collection of embryos for hippocampal culture production. Litter size and viability were determined in litters at term. By combining data from natural breeding and superovulation, questions of fertility and miscarriage can be addressed.

The 2-cell embryos were retrieved by flushing the oviducts of superovulated females with a Modified Krebs Ringer Bicarbonate media with 1 mg/mL hyaluronidase to aid removal of cumulus (Whittingham, 1971). The techniques for producing and culturing 2-cell mouse embryos are well documented (Howlett, 1987; Pratt, 1987). Rats are less desirable due to culturing difficulty and irregular cycling of the adult female. Ideally, a mouse model is utilized if *in vitro* culturing is necessary. Normally when rats are used in reproductive studies requiring superovulation, immature female rats would be utilized to increase the efficiency of superovulation response. However, because we have been using the Sprague-Dawley rat in many studies of lead toxicity, we have utilized rats in the work to date.

In vivo fertilization by treated males allowed the embryos to remain *in vivo* until the 2-cell stage. *In vivo* fertilization also permitted examination of only those embryos which were products of sperm naturally capable of fertilizing an oocyte. The embryos were kept in culture for a labeling period of 3 hours. Since ideal culturing conditions for rat 2-cell embryos have not been defined, the time an embryo was in culture was kept to a minimum. The results of the breeding studies are shown in Table 4.

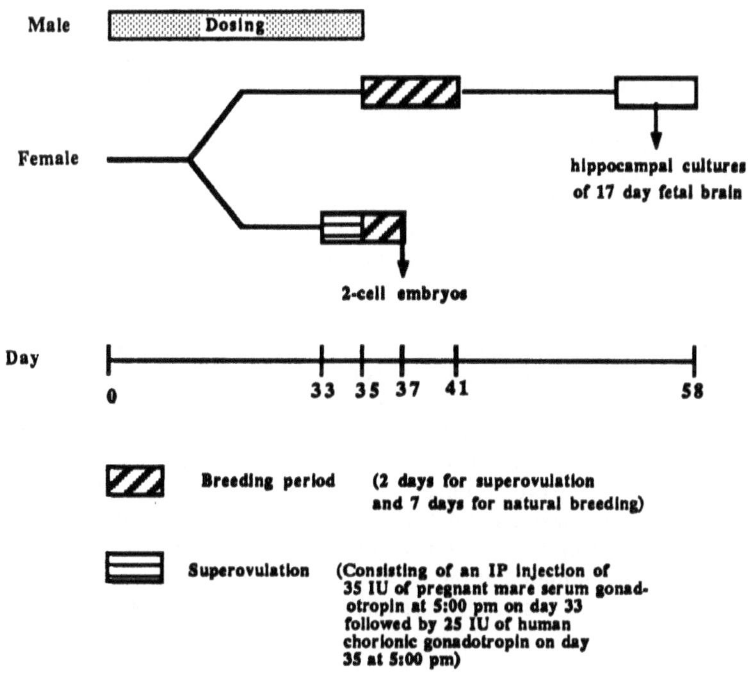

Figure 1. **Experimental Design.** Male animals were dosed for 35 days via their drinking water. Doses were 0, 25, 250 pm lead acetate. The males were then bred to control females. During the breeding period, the animals were provided with control drinking water. Two days following breeding of superovulated females, two-cell embryos were collected. Hippocampal cultures were prepared from 17 day fetal brain.

Table 4. Effects of Lead on Male Fertility

Lead Exposure	Blood Lead (mcg/dL)	Free Breeding[a]	Superovulation[b]
Control	0 - 2	83 % (30)[c]	84 % (31)[d]
25 pm	15 - 23	85 % (20)	81 % (43)
250 pm	29 - 60	65 % (20)	77 % (46)

[a] Fertility using free breeding was measured as the percentage of females producing litters from the given breeding pairs.
[b] Fertility using superovulated control females was measured as 2-cell embryos produced versus the total number of ooctyes available.
[c] Number in parentheses is the number of free breeding pairs
[d] Number in parentheses is the number of superovulated females.

The percentage of breeding pairs producing litters was reduced at the high paternal dose. No reduction in litter size or pup weight was seen with paternal dose. The superovulation data could be used to determine if the reduction in the number of litters was due to a fertilization problem, or due to early embryo loss. The result of superovulation found no significant difference in 2-cell embryo production with paternal dose. Pregnant females sacrificed at 17 days of gestation showed no sign of miscarriage. Therefore, the reduction in litters produced at the high dose must result from late pre-implantation loss or early post-implantation loss.

Effects of lead were detected using two techniques which are more sensitive than those recommended by the EPA. Neurodevelopment was assessed by techniques to detect alterations in brain growth and protein synthesis. The gels were loaded on counts per minute incorporated into protein. Examination of fetal hippocampal brain cultures showed a decrease in dendritic connections between otherwise normal appearing neurons.

Protein synthesis in these cultures was also monitored using ^{35}S-methionine incorporation (Silbergeld, *et al.*, 1991). The total incorporation and two-dimensional electrophoretic patterns of the proteins produced were altered with paternal exposure (O'Farrell, 1975 ; Aoki, *et al.*, 1990). However, lead affects the synthesis of many proteins in the hippocampus. To narrow the search for the mechanism of toxicity, a more simplistic model was necessary.

Assuming that the effects observed in the fetal and neonatal hippocampus were paternally transmitted, some indication of response should be observed in the early embryo. With this assumption, the two-cell embryo was examined. Expression of the embryonic genome is initiated at the two-cell stage in mice and rats (Flach, *et al.*, 1982; Bolton, *et al.*, 1984). Thus, this is the initial expression of paternally derived DNA.

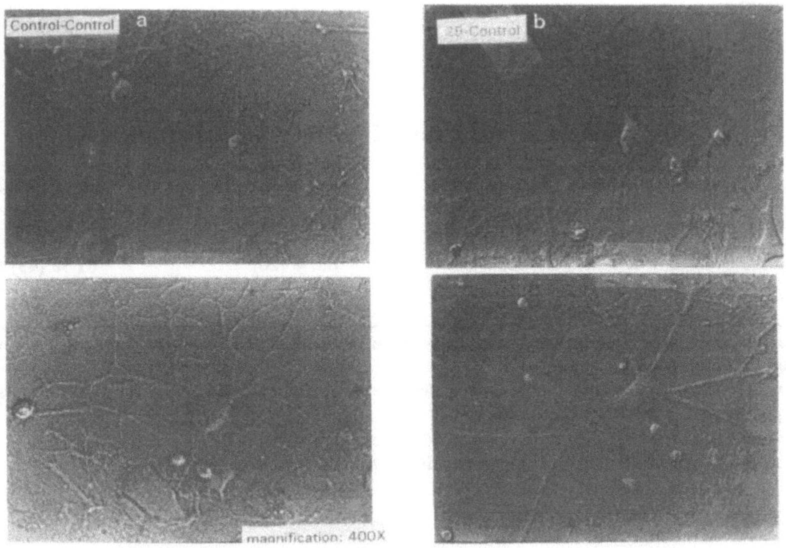

Figure 2. Fetal hippocampal cultures. 28 day cultures of 17 day fetal hippocampal cells, of **2a** control and **2b** low (25ppm) paternally dosed fetuses. The cultures demonstrated abnormal dendritic connections between otherwise normal appearing neurons.

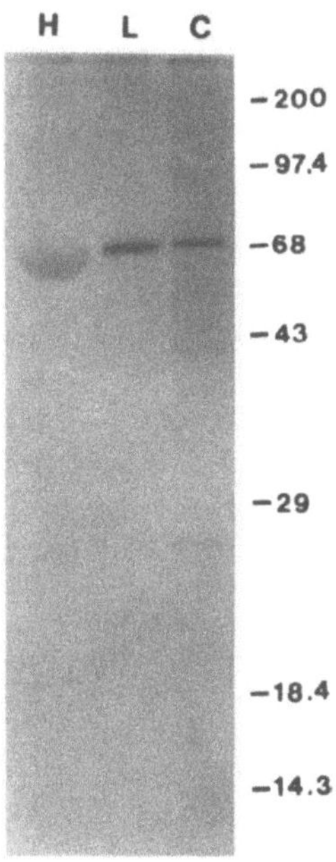

Figure 3. 1-Dimensional Gel Electrophoresis. A representative autoradiograph of a one-dimensional SDS slab gel of 2-cell embryos with ^{35}S-methionine incorporation (3 hrs. in Modified Krebs Ringer Bicarbonate) is pictured. The 2-cell embryos were obtained from superovulated control females mated with control (C), low (L), and high (H) lead dosed males. The gel was loaded with a pool of embryos with 10,000 cpm of ^{35}S-methionine incorporated. This gel was standardized by equal amounts of incorporated ^{35}S-methionine (TCA precipitation). Silver stained gels standardized to embryo number (not shown) also demonstrated a dose response.

Gene expression at the two-cell stage is much more limited. The major product of the two-cell mouse embryo is a group of closely related proteins which are referred to as "transcription requiring complex" (Poueymirou, *et al.*, 1987). The major protein product at the 2-cell stage demonstrated a change related to paternal lead treatment (Fig. 3). To further characterize the protein band observed, 2-dimensional SDS gels were produced.

Analysis of the patterns of protein synthesis in 2-dimensional SDS gel electrophoresis showed the presence of a newly synthesized set of proteins at approximately 70 kDa and pI 6.2-6.8 (Fig. 4). The synthesis of these proteins was increased in a dose-dependent manner in embryos according to paternal lead exposure (Gandley, *et al.*, 1992).

The identity and function of these proteins are unknown. However, an electrophoretically similar group of proteins has been identified and partially characterized in mouse 2-cell embryos. These proteins are α-amanitin sensitive (which indicates they are products of embryonic genome and not associated with parental transcripts) and appear for a limited period during early pre-implantation development of the embryo (Flach, *et al.*, 1982; Bensaude, *et al.*, 1983; Conover, *et al.*, 1991).

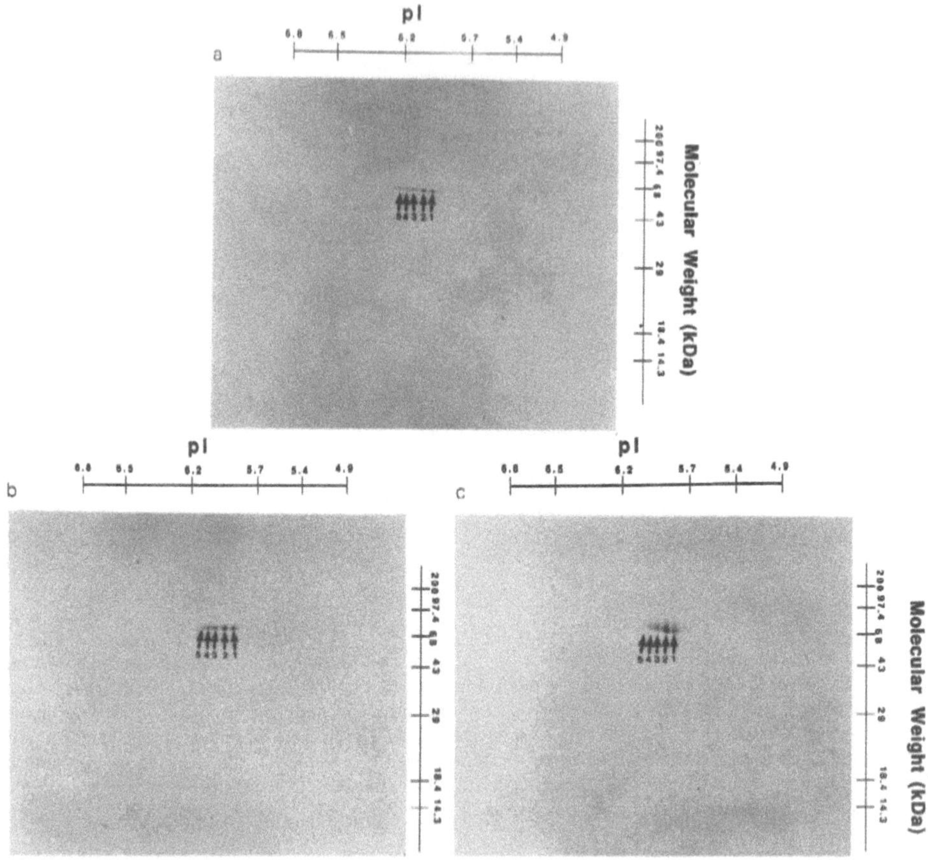

Figure 4. 2-Dimensional SDS Gel Electrophoresis of radiolabeled 2-cell embryos. **4a** Control 2-cell embryos, **4b** 25 ppm paternally treated 2-cell embryos, **4c** 250 pm maternally treated 2-cell embryos were radiolabeled with ^{35}S-methionine and 9000 incorporated cpms were loaded per gel. Autoradiography was a three week exposure. The isoforms of proteins 1,2,3,4,5 are identified in each gel. A dose response in the volume of the five proteins is demonstrated from the control to the low dose and from the low dose to the high dose. These 2-cell embryos were not directly exposed to lead prior to conception.

PROPOSED MECHANISMS OF MALE-MEDIATED REPRODUCTIVE AND DEVELOPMENTAL LEAD TOXICITY

Presently identified reproductive targets of lead toxicity include the hypothalamus, pituitary, gonads, and hepatic metabolism of steroids. Within the gonads there is some evidence for effects of lead on androgen receptors, Leydig and Sertoli cells, and gonadal hormone processing (Thomas and Brogan, 1983; Nathan, *et al.*, 1992). *In vivo* exposure of lead (with blood lead levels >40 mcg/ dL) in both humans and rodents can result in changes in sperm shape and function, as well as in number and viability (Lancranjan, *et al.*, 1975; Hall, 1981; Hilderbrand, *et al.*, 1973;

Overstreet, *et al.*, 1980). These effects, which are seen at high dose, may be associated with reductions in male fertility but are unlikely to explain effects of lead on the embryo.

Three general mechanisms of male-mediated developmental lead toxicity are proposed, based on the experimental results observed, our current knowledge of developmental biology, and previous reports dealing with mechanisms of lead toxicity. First, lead may be delivered to the oocyte as a microinjection at conception. A microinjection of lead at conception assumes that high enough levels of lead exist in/on the sperm or in the ejaculate to enter the conceptus and cause adverse effects. The actual amount of lead present in the sperm or ejaculate of treated males has not been measured. The actual level of Pb present at conception would be significantly less than the amount of Pb in the ejaculate due to dilution with vaginal and uterine fluids. A more probable source of Pb levels sufficiently high to cause adverse effects would be Pb either in/on the sperm. No measures of genetic material or lead levels in treated sperm have been made. While lead is known to form inclusion bodies (dense intracellular deposits of high lead concentration) in the germinal epithelium of birds (Gullvag, *et al.*, 1975), no inclusion bodies have been detected in sperm heads using electron microscopy with a micro Z x-ray detector (Johansson and Pellicciari, 1988). This technique should detect inclusion bodies, but it does not rule out overall increases in lead in/on the sperm.

A second mechanism by which lead may exert effects in embryos may or may not require direct exposure to lead. For lead to directly effect the developing fetus, it must act prior to implantation to affect a change that can be carried through later stages of development without actually being present in each cell. Direct genotoxicity would enable an effect of Pb to persist from the 2-cell stage to an adult organism. Lead could affect the genetic material of sperm prior to fertilization. A possible genetic mechanism is supported by the report of decreased fertility on the F_1 generation (Stowe and Goyer 1971).

Very little is known of gene-level events associated with lead toxicity. Lead has been found to be a renal carcinogen (Boyland, *et al.*, 1962; Van Esch, *et al.*, 1962). Lead exposure produces clastogenic changes in lymphocytes of exposed men (Sahu, *et al.*, 1989; Borella and Giardino, 1991; Winder, 1988; Al-Hakkak, *et al.*, 1986). Lead can also cause DNA breakage at mM concentrations (Muro and Goyer, 1969; Zelikoff, *et al.*, 1988). If genotoxicity is the effect being observed at the 2-cell stage and in the 17 day hippocampal cultures, the effect must be subtle enough to allow an F_1 generation to be produced. Moreover, the genetic affect must not only be subtle enough not to cause increased fetal loss, but also specific in its point of action to cause reproducible effects in the hippocampus of many offspring.

Alternatively, lead may cause epigenetic effects that alter gene transcription in the embryo. The mechanisms involved in the regulation and transcription processes of the pre-implantation embryo are not clearly defined. The role of the sperm in the onset and maintenance of cell division is vital but poorly defined. Epigenetic factors such as regulatory proteins, RNA or activating enzymes in the sperm may be necessary for normal early embryonic development. Lead could act to alter one or more epigenetic factors necessary in early development and as a consequence cause a cascading effect in later development, for example, through effects on zinc-binding fingerloop proteins or other transcription factors. Alterations in charge (i.e., methylation) of the sperm content may also cause an effect in the conceptus. If the effects of lead in the F_1 generation are epigenetic, they would not be expressed in subsequent generations.

CONCLUSION

We have utilized a model system to assess both reproductive and developmental toxicity of paternal exposures to lead, which provides novel information related to male-mediated effects on development. If such effects of toxicants occur below the levels inducing reproductive toxicity, more sensitive measures of both sperm and offspring are necessary to determine possible toxic mechanisms. Concern for affected offspring should be foremost in the assessment of paternal reproductive toxicity studies. The experimental design described in the example assesses both fertility and viability of offspring by combining 2-cell embryos production and natural breeding and delivery. Breeding the F_1 generation may also be very useful for determination of possible genetic or reproductive effects from paternal exposure.

Male-mediated effects specifically directed at neurodevelopment may be difficult to determine. However, with the long period of sensitivity from fertilization to postnatal maturity, neurodevelopmental effects of paternal exposure may be an excellent system for identifying possible toxicants.

REFERENCES

Al-Hakkak, Z.S., Hamamy, H.A., Murad, A.M.B., and Hussain, A.F., 1986, Chromosome aberrations in workers at a storage battery plant in Iraq. *Mutat. Res.* 171:53-60.

Aoki, Y., Lipsky, M., and Fowler, B.A., 1990, Alteration of protein synthesis in primary cultured rat proximal tubular cells by exposure to gallium, indium and arsenite, *Toxicol. Appl. Pharm.* 106: 462-468.

Bensaude, O., Babinet, C., Morange, M. and Jacob, F., 1983, Heat shock proteins, first major products of zygotic gene activity in mouse embryos, *Nature* 305: 331-3.

Bolton, V.N., Oades, P.J. and Johnson, M.H., 1984, The relationship between cleavage, DNA replication and gene expression in the mouse 2-cell embryo. *J. Embryol. Exp. Morph.* 79: 139-163.

Borella, P., and Giardino, A., 1991, Lead and cadmium at very low doses affect in vitro immune response of human lymphocytes, *Environ. Res.* 55:165-77.

Boyland, E., Dukes, C.E. Grover, P.L., and Mitchley, B.C.V., 1962, The induction of renal tumors by feeding lead acetate to rats. *Br. J. Cancer* 16:283-288.

Brady, K., Herrera, Y., and Zenick, H., 1975, Influence of paternal lead exposure on subsequent learning ability of offspring, *Pharmacol. Biochem. Behav.* 3: 561-5.

Conover, J.C., Temeles, G.L., Zimmermann, J.W., Burke, B., and Schultz, R.M., 1991, Stage-specific expression of a family of proteins that are major products of zygotic gene activation in the mouse embryo, *Devel. Biol.* 144: 392-404.

de Kretser, D.M., and Kerr, J.B., 1988, The cytology of the testis, in: "The Physiology of Reproduction," E. Knobil, and J.D. Neill, eds., Raven Press, New York.

Flach, G., Johnson, M.H., Braude, P.R., Taylor, R.A.S. and Bolton, V.N., 1982, The transition from maternal to embryonic control in the 2-cell mouse embryo. *EMBO J.* 1: 681-6.

Gandley, R.E., Couture, M., Silbergeld, E.K., Anderson, L.D., and Fowler, B.A., 1992, Paternal exposure to lead (Pb) alters initial genomic expression in offspring, *Toxicologist* 12:213.

Gray, L.E., Ostby, J., Sigmon, R., Ferrell, J., Rehnberg, G., Linder, R., Cooper, R., Goldman, J., and Laskey, J., 1988, The development of a protocol to assess reproductive effects of toxicants in the Rat. *Reprod. Toxicol.* 2: 281-7.

Gullvag, B.M., Ophus, E.M., and Eskeland, B., 1975, Lead poisoning of Japanese quail: An analysis of different body tissues using atomic absorption spectrophotometry and transmission electron microscopy, *ACTA Zoolog.* 56: 163-176.

Hakim, R.B., Stewart, W.F., Canner, J.K., and Tielsch, J.M., 1991, Occupational lead exposure and strabismus in offspring: a case-control study, *Am. J. Epidem.* 133: 351-6.

Hall, J.L., 1981, Relationship between semen quality and human sperm penetration of zona-free hamster ova, *Fertil. Steril.* 35: 457-63.

Hilderbrand, D.C., Olds, M., Der, R. and Hafim, M.S., 1973, Effects of lead acetate on reproduction, *Amer. J. Obst. Gynecol.* 115: 1058-1065.

Howlett, S.K., 1987, Qualitative analysis of protein changes in early mouse development, in "Mammalian Development," M. Monk, ed., IRL Press, Washington D.C.

Johansson, L., and Pellicciari, C.E., 1988, Lead-induced changes in the stabilization of the mouse sperm chromatin, *Toxicol.* 51: 11-24.
Lancranjan, I, Posecu, H., Galvenescu, O., 1975, Reproductive ability of workmen occupationally exposed to lead. *Arch. Environ. Health* 30: 396-401.
Muro, L.A. and Goyer, R.A., 1969, Chromosomal damage in experimental lead poisoning, *Arch. Path.* 87:660-663.
Nathan, E., Huang, H.F.S., Pogach, L., Giglio, W., Bogden, J.D., and Seebode, J., 1992, Lead acetate does not impair secretion of sertoli cell function marker proteins in the adult Sprague-Dawley rat, *Arch. Environ. Health* 47: 370-75.
O'Farrell, P.H., 1975, High resolution two-dimensional electrophoresis of proteins, *J. Biol. Chem.* 250: 4005-4021.
Olshan, A.F., Teschke, K., and Baird, P.A., 1991, Paternal occupation and congenital anomalies in offspring, *Am. J. Indust. Med.* 20:447-75.
Overstreet, J.W., Yamagimachi, R., Katz, D.F., and Hanson, F.W.,1980, Penetration of human spermatozoa into the human zona pellucida and the zona-free hamster egg.: A study of fertile donors and infertile patients, *Fertil. Steril.* 33:534-542.
Poueymiro, W.T., and Schultz, R.M., 1987, Differential effects of activators of cAMP-dependent protein kinase and protein kinase C on cleavage of one-cell mouse embryos and protein synthesis and phosphorylation in one- and two-cell embryos, *Devel. Biol.* 121: 489-498.
Pratt, H.P.M., 1987, Isolation, culture, and manipulation of pre-implantation mouse embryos, in "Mammalian Development," M. Monk, ed.,IRL Press, Washington D.C.
Robaire, B. and Hermo, L., 1988, Efferent ducts, epididymis, and vas deferens: structure, function and their regulation, in:"The Physiology of Reproduction," E. Knobil, and J.D. Neill, eds., Raven Press, New York.
Sahu, R.K., Katisifis, S.P., Kinney, P.L., and Christie, N.T.,1989, Effects of nickel sulfate, lead sulfate, and sodium arsenite alone and with UV light on sister chromatid exchanges in cultures human lymphocytes, *Molecul. Tox.* 2:129-39.
Silbergeld, E.K., Akkerman, M., Fowler, B.A., Albuquerque, E.X., and Alkondon, M., 1991, Lead: male-mediated effects on reproduction and in utero neurodevelopment, *Toxicologist* 11: 235.
Stowe, H.D., and Goyer, R.A., 1971, The reproductive ability and progeny of F_1 lead-toxic rats. *Fertil. Steril.* 22: 755-760.
Thomas, J.A., and Brogan, W.C., 1983, Some actions of lead on the sperm and on the male reproductive system, *Prog. Clin. Biol.* Res. 117: 127-34.
Van Esch, G.J., Van Genderen, H., and Vink, H.H., 1962, The induction of renal tumors by feeding of basic lead acetate to rats. *Br. J. Cancer* 16:289-297.
Winder, C., 1988, Reproductive and chromosomal effects of occupational exposure to lead in the male, *Reprod. Toxicol.* 3: 221-233.
Whittingham, D.G., 1971, Culture of mouse ova, *J. Repro. Fert. Suppl.* 14: 7-21.
Zelikoff, J.T., Li, J.H., Hartwig, A., Wang, X.W., Costa, M., and Rossman, T.G., 1988, Genetic toxicology of lead compounds, *Carcinog.* 9:1727-1732.

PATERNAL OCCUPATION AND BIRTH DEFECTS

Andrew F. Olshan and Patricia G. Schnitzer

Department of Epidemiology
School of Public Health
University of North Carolina
Chapel Hill, NC 27599-7400

INTRODUCTION

The etiology of many birth defects among humans is not well understood, with most of the epidemiologic and laboratory research focusing on maternal factors. It has been stated that approximately 60% of all human birth defects have no known cause (Kalter and Warkany, 1983). The potential role of paternal exposures has not been extensively investigated. This has been due, in part, to the prevailing view that male-mediated developmental effects are unlikely (Brown, 1985). However, recent laboratory and epidemiologic investigations have reinforced earlier animal data suggesting that paternal exposures may be more important than previously suspected. The published literature on paternal occupational exposures and the risk of birth defects in offspring will be reviewed in this chapter.

EPIDEMIOLOGIC FINDINGS

The relationship between paternal occupational exposures and birth defects has only recently been explored in several large studies. With the exception of malformations such as neural tube defects, there are a limited number of multiple studies available to evaluate consistency among findings. Further, direct comparison with previous studies is complicated by the fact that different classifications and groupings of birth defects and occupations were used in each study.

Table 1 provides a description of the key characteristics of each study including the number of cases, source of the comparison group, source of exposure information, time period of exposure and potential confounding factors that were considered. The table is divided into three parts based upon the basic study design; the first a description of the case-control studies; the second part the cohort studies; and third a cross-sectional study. There have been a total of 22 published studies that have examined paternal occupation and birth defects. The earliest paper was published in 1976 although most were published since 1980. The majority, 16 of 22, were case-control studies, 5 were cohort studies and 1 was a cross-sectional study.

Male-Mediated Developmental Toxicity, Edited by D.R. Mattison
and A.F. Olshan, Plenum Press, New York, 1994

Table 1. Paternal Occupation and Birth Defects Study Characteristics

Case-Control Studies/ Reference	Study Location, Time Period	Case Group & Source	No. of Cases	Comparison Group & Source	Source of Exposure Information	Time Period of Exposure	Potential Confounders Considered
Fedrick (1976)	Oxfordshire & Berkshire, England 1965-1972	Anencephalus cases from 5 sources (hospital records, registry, death certs, etc)	177	Infants born to area residents from vital records	Paternal occupation from birth certificate	At birth	None stated
Edmonds et al. (1978)	Kanawha County, WV 1970-1974	CNS defect cases from 2 local hospitals, and birth, fetal death, and death certificates	41	Live born infants of city residents from vital records	Paternal occupation history from family member interview	5 yrs. prior to birth	Education, maternal age, SES, previous fetal death, other children with anomalies
Erickson et al. (1979)	Atlanta, GA	Infants with birth defects registered with the Metropolitan Atlanta Congenital Defects Program (MACDP)	989	Malformed infants registered with MACDP	Paternal occupation from maternal interview	Time of conception	None stated
Hemminki et al. (1980)	Finland 1967-1977	CNS, oral cleft, musculoskeletal malformations from the Finnish Register of Congenital Malformations	3,300	Previous birth in the district identified from the maternity welfare records	Paternal occupation from Maternity Welfare Center records	During pregnancy	None stated
OPCS Monitor (1982)	England & Wales 1974-1976, 1977-1979	All birth defects reported to the Congenital Malformation Programme	48,510	All births from vital records	Paternal occupation from malformation notification form	At birth	None

Table 1. (Continued)

Case-Control Studies/ Reference	Study Location, Time Period	Case Group & Source	No. of Cases	Comparison Group & Source	Source of Exposure Information	Time Period of Exposure	Potential Confounders Considered
Olsen (1983)	Funen Cty, Denmark 1972-1976	CNS, GI, extremity defects from the Funen County Register of Congenital Malformations	702	Children with medical conditions other than congenital defects	Paternal occupation from birth certificate	At birth	Residence, birth order, paternal age, maternal age
Polednak & Janerich (1983)	Upstate NY 1968-1974	Neural tube defects, from birth certificates	171	Live births identified from vital records	Paternal occupation from birth certificate	At birth	Maternal age, race
Erickson *et al.* (1984)	Atlanta, GA 1968-1980	"Major" birth defects registered with the Metropolitan Atlanta Congenital Defects Program (MACDP)	4,815	Atlanta area live births from vital records	Paternal interview; exposure score estimated by a specialist panel		Maternal age, education, alcohol consumption, birth defect in relative
McDowall (1985)	England & Wales 1980-1982	All birth defects reported to the Congenital Malformation Monitoring Programme	24,922	All births from vital records	Paternal occupation from malformation notification form	At birth	Maternal age
Papier (1985)	Rehovot, Israel 1966-1976	All malformed babies born at Kaplan Hospital	1,481	All live births at Kaplan Hospital	Paternal occupation from maternal interview	During pregnancy	None stated
Sever *et al.* (1988)	Benton & Franklin Counties, WA 1957-1980	All congenital anomalies identified from hospital records, birth and fetal death certificates	672	Live births without defects from delivery room records	Employment records at Hanford Nuclear Site	Before conception	Hospital, maternal age, birth year, race, sex of offspring

Table 1. (Continued)

Case-Control Studies/ Reference	Study Location, Time Period	Case Group & Source	No. of Cases	Comparison Group & Source	Source of Exposure Information	Time Period of Exposure	Potential Confounders Considered
McDonald *et al.* (1989)	Montreal, Canada 1982-1984	Chromosomal and selected developmental defects from hospital records and maternal questionnaire	605	Infants without defects from hospital records	Paternal occupation from maternal interview	At conception	Maternal age, gravidity, previous SAB, ethnicity, education, smoking, alcohol consumption
Olshan *et al.* (1989)	British Columbia, Canada 1952-1973	Down Syndrome identified from the British Columbia Health Surveillance Registry	1,008	Live births from vital records	Paternal occupation from birth certificate	At birth	Maternal age
Brender & Suarez (1990)	Texas 1981-1986	Anencephaly from live birth and infant death records	585	Live births without defects from vital records	Paternal occupation from birth certificate	At birth	Ethnicity, race
Olshan *et al.* (1990)	British Columbia, Canada 1952-1973	All defects from the British Columbia Health Surveillance Registry	89	Live births from vital records	Paternal occupation from birth certificate	At birth	Parental age, race, Previous pregnancy outcomes, sex, birth weight, prematurity
Olshan *et al.* (1991)	British Columbia, Canada 1952-1973	20 birth defect categories from the British Columbia Health Surveillance Registry	14,415	Live births from vital records	Paternal occupation from birth certificate	At birth	Parental age, race, previous pregnancy outcomes

Table 1. (Continued)

Retrospective Cohort Studies/ Reference	Study Location, Time Period	Exposure Group & Source	No. Exposed	Comparison Group & Source	Source of Birth Defect Information	Time Period of Exposure	Paternal Confounders Considered
Cohen *et al.* (1980)	United States	Male dentists who use inhalation anesthetics; American Dental Association survey	Not stated	Male dentists not using inhalation anesthetics; Q'aire	Mail questionnaire (paternal)	Previous 10 yrs	Maternal age & smoking, previous SAB or birth defect
Smith *et al.* (1982)	New Zealand 1969-1980	Chemical (2,4,5-T) applicators; New Zealand Agricultural Chemicals Board survey	548	Agricultural contractors without chemical use; Q'aire	Mail questionnaire (paternal)	2 yr prior to birth	Maternal age, race, smoking
Townsend *et al.* (1982)	Midland, MI 1939-1975	Wives of male Dow Chemical Co. employees working in dioxin exposed areas; company records	370	Dow Chemical Co. employees not working in exposed areas; interview	Maternal interview	Anytime prior to conception	Age, smoking, gravidity, meds during pregnancy, plus many more
Roan *et al.* (1984)	United States 1978	Agricultural pilots; National Agricultural Aviation Association membership lists	314	Brother or brother-in-law of ag pilots; Q'aire	Mail questionnaire (paternal)		Age, education, tobacco, alcohol, & prescription drug use
Savitz *et al.* 1984	United States 1979	Oil, Chemical, & Atomic Workers (OCAW) Union members engaged in halogenated hydrocarbon use or manufacturer	804	OCAW union members without halogenated hycrocarbon exposure; Q'aire	Mail Q'aire, wives could assist with reproductive history	From hire date	Age, race, duration of employment
Cross-Sectional Study/Reference							
Schwartz *et al.* (1986)	Brawley, CA 1975-1978	Agricultural occupations from maternal hospital delivery records	990	Nonagricultural occupations from hospital records	Hospital records	At birth	Age, prenatal care, Spanish surname, rural residence, multiparity

Most studies were of limited sample size to evaluate specific birth defects. Even those studies with a larger number of cases were limited, for example the study of 3,300 cases in Finland included only 10 very broad occupation groups and 3 malformation groups. The cohort studies also had generally small samples. In most studies, cases of birth defects were ascertained from various sources, most from either a registry system or interview with one or both parents. An important consideration is how the investigators categorized or grouped birth defect cases for analysis. Six studies combined all defects into a single group, five analyzed a single defect, and twelve studies analyzed more than one defect. However, many of the studies evaluating more than one defect included combined categories such as all heart defects. Father's occupation was derived from a variety of sources, mostly vital records (e.g., birth certificate) or interview. Occupation was usually self-reported and recorded primarily as a job title.

In this review, study findings were considered to show an association if the authors reported either a statistically significant risk ratio estimate or in some cases, if the risk ratio was greater than 1.5, regardless of statistical significance. The review of findings has been organized with a summary of the studies with mostly negative results presented first, followed by a description of the positive findings, according to occupation or industry group.

Several general studies including broad occupation and/or birth defect groups have, for the most part, reported a lack of association. These studies include those Hemminki *et al.* (1980), Papier (1985), and McDonald *et al.* (1989). Hemminki *et al.* (1980) evaluated 10 broad occupational groups in relation to 3,300 cases of central nervous system (CNS), oral cleft and musculoskeletal malformations. No associations were reported by the authors. Papier (1985) reported no association with 10 paternal occupation categories and 1,481 cases of birth defects diagnosed at a single hospital in Israel. A survey of paternal occupation and pregnancy outcome among 47,882 women (605 birth defect cases) in Montreal (McDonald *et al.*, 1989) did not report any association with chromosomal defects and 24 occupation groups. Employment in food and beverage processing was related to an increased risk of what they termed "developmental defects." Several studies have focused on specific paternal occupations or exposures in relation to a variety of birth defects including pesticide applicators (Smith *et al.*, 1982), agricultural pilots (Roan *et al.*, 1984), dioxin workers (Townsend *et al.*, 1982), and Vietnam veterans (Erickson *et al.*, 1984). These studies have, in general, not yielded significant associations.

Two large studies that have examined paternal occupation and birth defects include the analyses of vital statistics data from England and Wales (OPCS, 1982; McDowall, 1985) and British Columbia (Olshan *et al.*, 1989; 1990; 1991). Due to the size and potential importance of these studies, they will described in some detail. In the OPCS studies, birth defects and father's occupation information was derived from the congenital malformation notification system in England and Wales for the periods 1974-1979 (OPCS, 1982) and 1980-1982 (McDowall, 1985). Standardized congenital malformation ratios were calculated using national birth data. The earlier analysis (1974-1979) did not adjust for mother's age and analyzed all defects as a group. The 1980-1982 analysis included 24,922 birth defect cases categorized into 15 defect groups and controlled for maternal age (McDowall, 1985).

The British Columbia (B.C.) case-control study used a population based registry system to identify 14,415 cases of birth defects (20 defect groups) for the period 1952-1973 (Olshan *et al.*, 1991). Two matched controls per case were identified from the birth files of British Columbia. Paternal occupation was derived from the linked birth certificate for both cases and controls. Adjustment for potential confounding factors was carried out using variables available on the birth certificate. It is often the case that the British Columbia and the OPCS studies were the only ones with a sufficient number of subjects to analyze individual birth defects, thus they constitute the majority of the published findings and discussion.

Table 2 presents selected results from published studies organized according to occupation or industry group. Some of the key exposures that are encountered in the occupation or industry are noted below.

Table 2. Paternal Occupation and Birth Defects Selected Findings

Occupation/Exposure	Birth Defect+	No. Exposed Cases	OR/RR (95% CI)'	Reference
PRINTERS:				
Printer	Anencephalus	5	6.72 (2.82, 15.98)	Fedrick (1976)
Printer	Cleft palate	2	undefined (1.80∞)	Erickson (1979)
Printer	Club foot	22	2.18 (1.17, 4.05)	Olshan (1991)
Printer	Polydactyly	31	176	McDowall (1985)
Printer	Rectal atresia	5	354	McDowall (1985)
Printer	Atresia of the urethra	7	4.50 (0.97, 20.81)	Olshan (1991)
Engraver, etcher	Polydactyly	2	936	McDowall (1985)
PAINTS & PIGMENTS:				
Painter/decorator	Anencephalus	9	2.52 (1.34, 4.75)	Fedrick (1976)
Painter	ASB	8	2.98 (0.94, 9.42)	Olshan (1991)
Painter	Anencephalus	22	3.43 (1.83, 6.43)	Brender (1990)
Painter	Spina bifida	7	3.21 (0.91, 11.36)	Olshan (1991)
Painter	CNS Defects	3	4.9 (1.4, 17.1)	Olsen (1983)
Painter	ASB	8	275	McDowall (1985)
Painter	PDA	11	2.34 (1.00, 5.45)	Olshan (1991)
Painter	Cleft palate	3	5.9 (0.89, 30.62)	Erickson (1979)
Painter	Cleft palate	9	3.36 (1.19, 9.46)	Olshan (1991)
Painter	Polydactyly	8	392	McDowall (1985)
Painter	Syndactyly	8	459	McDowall (1985)
Painter	Talipes	20	246	McDowall (1985)
Painter	Reduction def.	4	399	McDowall (1985)
TRANSPORTATION-RELATED:				
Transport/communication	Anencephalus	22	1.69 (1.5, 2.49)	Fedrick (1976)
Auto mechanic/service station	Neural tube defects	6	2.50	Polednak (1983)
Vehicle mechanic	Cleft palate	19	2.12 (1.13, 3.99)	Olshan (1991)
Other mechanics	Down syndrome	26	3.27 (1.57, 6.80)	Olshan (1989)
Service station attn.	Hypospadias	9	3.18 (1.06, 9.53)	Olshan (1991)
Mechanic, nec	Hypospadias	21	1.64 (0.89, 3.00)	Olshan (1991)
Service station attn.	Spina Bifida	5	8.30 (0.92, 74.66)	Olshan (1991)
Railway/other transport	PDA	10	1.90 (0.81, 4.48)	Olshan (1991)
Railway/other transport	Atresia of the urethra	11	3.14 (1.22, 8.12)	Olshan (1991)
Railway porter	All	*	550	OPCS (1982)
Railway station men	ASB	6	312	McDowall (1985)
Railway station men	Cleft lip & palate	7	347	McDowall (1985)
Driver of road goods vehicles	ASB	75	145	McDowall (1985)
Driver of road goods vehicles	Circ. system defect	70	129	McDowall (1985)
Driver of road goods vehicles	Cleft lip & palate	74	134	McDowall (1985)
Driver of road goods vehicles	Down syndrome	45	142	McDowall (1985)
Driver of road goods vehicles	Hypospadias	87	135	McDowall (1985)
Driver of road goods vehicles	All	*	112	OPCS (1982)
Bus & coach driver	Circ. system defect	20	171	McDowall (1985)
Bus/coach driver	Talipes	43	139	McDowall (1985)

Table 2. (Continued)

Occupation/Exposure	Birth Defect+	No. Exposed Cases	OR/RR (95% CI)§	Reference
Motor vehicle oper.	Cleft palate	49	1.86 (1.25, 2.78)	Olshan (1991)
Motor vehicle oper.	Cleft lip	34	1.66 (1.04, 2.67)	Olshan (1991)
Motor vehicle oper.	Cleft lip & palate	58	1.49 (1.04, 2.11)	Olshan (1991)
Motor vehicle oper.	Syndactyly	46	2.92 (1.75, 4.88)	Olshan (1991)
Crane/driver/oper.	Cleft lip & palate	10	241	McDowall (1985)
Material moving equip. oper.	Down syndrome	19	1.88 (0.93, 3.82)	Olshan (1989)
AGRICULTURAL:				
Farmer	Neural tube defects	5	5.00	Polednak (1983)
Farm & ranch worker	Anencephalus	13	1.73 (0.84, 3.55)	Brender (1990)
Farm manager/worker	Down syndrome	59	2.03 (1.25, 3.03)	Olshan (1989)
Agricultural worker	Limb reduction defects	50	2.3 (1.47, 3.57)	Schwartz (1986)
Farming, nec	Hypospadias	5	405	McDowall (1985)
Gardener/groundsmen	Hypospadias	19	184	McDowall (1985)
SALES-RELATED:				
Other clerk/cashier (not retail)	ASB	75	133	McDowall (1985)
Other clerk/cashier (not retail)	Circ. system defect	80	130	McDowall (1985)
Other clerk/cashier (not retail)	Cleft lip & palate	97	155	McDowall (1985)
Other clerk/cashier (not retail)	Polydactyly	54	134	McDowall (1985)
Other clerk/cashier (not retail)	Talipes	222	137	McDowall (1985)
Other clerk/cashier (not retail)	Hypospadias	96	133	McDowall (1985)
Sales rep.	Rectal atresia	10	230	McDowall (1985)
Shop salesmen & asst.	ASB	27	210	McDowall (1985)
Sales, shop asst.	Hypospadias	26	163	McDowall (1985)
Shop asst.	All	*	137	OPCS (1982)
Clerk	Atresia of the urethra	19	1.60 (0.86, 2.98)	Olshan (1991)
Clerk	All	*	141	OPCS (1982)
Store & dispatch clerk	Down syndrome	3	563	McDowall (1985)
OTHER CHEMICALS:				
Chemical, gas, petroleum plant oper.	ASB	20	186	McDowall (1985)
Chemical, gas, petroleum plant oper.	Talipes	49	161	McDowall (1985)
Chemical, gas, petroleum plant oper.	Hypospadias	24	177	McDowall (1985)
Chemical, gas, petroleum plant oper.	Tracheoesophageal fistula	6	446	McDowall (1985)
Rubber process worker	ASB	6	418	McDowall (1985)
METALS WORKERS:				
Sheetmetal, iron & other metal worker	VSD	46	1.49 (1.01, 2.18)	Olshan (1991)
Sheetmetal, iron & other metal worker	ASD	32	1.55 (0.96, 2.50)	Olshan (1991)
Sheetmetal, iron & other metal worker	Down syndrome	34	1.57 (0.92, 2.69)	Olshan (1991)
Sheetmetal, iron & other metal worker	Syndactyly	19	1.62 (0.87, 3.03)	Olshan (1991)
Furnace, forge, foundry worker	Anencephalus	2	3.78 (1.04, 13.83)	Fedrick (1976)
Foundry, smelter worker	Cleft palate	14	1.77 (0.87, 3.63)	Olshan (1991)
Metal plate worker	Spina bifida	8	249	McDowall (1985)
Welder	Spina bifida	23	171	McDowall (1985)

Table 2. (Continued)

Occupation/Exposure	Birth Defect+	No. Exposed Cases	OR/RR (95% CI)'	Reference
Welder & cutter	Cleft lip	6	4.75 (0.91, 24.83)	Olshan (1991)
Metallurgist	Talipes	7	266	McDowall (1985)
MINING:				
Miner, not coal	ASB	5	321	McDowall (1985)
Metal miner	Other heart defects	10	1.94 (0.80, 4.71)	Olshan (1991)
Metal miner	Cleft palate	10	2.41 (0.96, 6.05)	Olshan (1991)
Underground coal miner	All	*	161	OPCS (1982)
Unskilled laborer --coal mining	ASB	32	265	McDowall (1985)
Unskilled laborer --coal mining	Circ. system defect	43	356	McDowall (1985)
Unskilled laborer --coal mining	Talipes	87	261	McDowall (1985)
Unskilled laborer --coal mining	Hypospadias	33	223	McDowall (1985)
WOOD PRODUCTS:				
Forestry & logging	Anencephalus	57	1.57 (1.09, 2.27)	Olshan (1991)
Forestry & logging	Spina bifida	46	1.90 (1.24, 2.90)	Olshan (1991)
Forestry & logging	ASD	54	2.03 (1.35, 3.05)	Olshan (1991)
Forestry & logging	Other heart defects	103	1.50 (1.14, 1.98)	Olshan (1991)
Forestry & logging	Cleft lip	25	1.50 (0.84, 2.67)	Olshan (1991)
Forestry & logging	Syndactyly	24	2.03 (1.05, 3.89)	Olshan (1991)
Forestry & logging	Pyloric stenosis	61	1.64 (1.14, 2.35)	Olshan (1991)
Carpenter/woodworker	Anencephalus	28	1.53 (0.93, 2.52)	Olshan (1991)
Carpenter	Circ. system defect	17	143	McDowall (1985)
Carpenter	Cleft lip & palate	25	1.58 (0.93, 2.68)	Olshan (1991)
Carpenter/woodworker	Hypospadias	47	1.84 (1.21, 2.79)	Olshan (1991)
Paper worker	Spina bifida	14	1.77 (0.83, 3.76)	Olshan (1991)
Paper worker	Pyloric stenosis	20	1.53 (0.85, 2.75)	Olshan (1991)
Paper worker	Undescended testicle	20	1.54 (0.82, 2.89)	Olshan (1991)
Plywood mill worker	PDA	12	2.52 (1.08, 5.87)	Olshan (1991)
Plywood mill worker	Dislocation of hip	13	2.71 (1.08, 6.81)	Olshan (1991)
Plywood mill worker	Pyloric stenosis	10	4.12 (1.41, 12.07)	Olshan (1991)
Sawmill worker	PDA	38	1.54 (1.00, 2.37)	Olshan (1991)
Sawmill worker	other heart defects	74	1.49 (1.09, 2.05)	Olshan (1991)
Sawmill worker	Down syndrome	43	1.43 (0.90, 2.66)	Olshan (1989)
Sawmill worker	Dislocation of hip	55	1.50 (1.04, 2.17)	Olshan (1991)
Sawmill worker	Obstructive renal defects	20	1.50 (0.82, 2.75)	Olshan (1991)
MACHINIST:				
Machine tool operator	ASB	77	189	McDowall (1985)
Machine tool operator	Circ. system defect	77	182	McDowall (1985)
Machine tool operator	Cleft lip & palate	87	202	McDowall (1985)
Machine tool operator	Down syndrome	39	162	McDowall (1985)
Machine tool operator	Polydactyly	51	179	McDowall (1985)
Machine tool operator	Syndactyly	54	219	McDowall (1985)
Machine tool operator	Talipes	222	196	McDowall (1985)
Machine tool operator	Reduction def.	26	185	McDowall (1985)
Machine tool operator	Tracheoesophageal fistula	14	279	McDowall (1985)
Machine tool operator	Hypospadias	94	186	McDowall (1985)
Machinist/production fitter	Talipes	183	121	McDowall (1985)
Office machine operator	All	*	144	OPCS (1982)
LABORERS:				
Handler, Laborer, nec	Anencephalus	25	1.72 (0.99, 2.97)	Olshan (1991)
Handler, Laborer, nec	Spina bifida	21	1.62 (0.89, 2.94)	Olshan (1991)
Builder	Circ. system defect	28	163	McDowall (1985)
Builder	Syndactyly	19	186	McDowall (1985)

Table 2. (Continued)

Occupation/Exposure	Birth Defect+	No. Exposed Cases	OR/RR (95% CI)$	Reference
Builder	Talipes	66	143	McDowall (1985)
Roofer, glazer	Tracheoesophageal fistula	4	407	McDowall (1985)
Construction worker	Syndactyly	23	2.60 (1.31, 5.17)	Olshan (1991)
Construction worker	Obstructive renal defects	15	2.36 (1.10, 5.05)	Olshan (1991)
Handler/Laborer, nec	Syndactyly	12	1.88 (0.83, 4.23)	Olshan (1991)
Laborer & unskilled worker				
--textiles	ASB	14	408	McDowall (1985)
--textiles	Polydactyly	7	293	McDowall (1985)
--textiles	Talipes	24	256	McDowall (1985)
--other	ASB	73	150	McDowall (1985)
--other	Polydactyly	56	166	McDowall (1985)
--other	Syndactyly	42	157	McDowall (1985)
--other	Talipes	206	157	McDowall (1985)
--other	Reduction def.	26	160	McDowall (1985)
--other	Rectal atresia	21	216	McDowall (1985)
--other	Hypospadias	85	145	McDowall (1985)
--nec	Circ. system defect	77	159	McDowall (1985)
ELECTRICAL WORKERS:				
Electrical engineer	Cleft lip & palate	5	334	McDowall (1985)
Electrical engineer/ technician	Neural tube defects	4	4.00	Polednak (1983)
Electrician	Syndactyly	33	175	McDowall (1985)
Electrician, electrical worker	ASD	28	1.57 (0.94, 2.60)	Olshan (1991)
Cablejoiner, lineman	Spina bifida	7	458	McDowall (1985)
Radio/TV mechanic	Tracheoesophageal fistula	3	567	McDowall (1985)
NUCLEAR WORKERS:				
Exposure to ionizing radiation (10 mSv)	Neural tube defects	7	1.46 (0.98, 4.5)	Sever (1988)
Exposure to ionizing radiation (100 mSv)	Neural tube defects	2	5.6 (0.81, 36)	Sever (1988)
FOOD-RELATED:				
Brewery, vinery process worker	ASB	4	393	McDowall (1985)
Food service worker	VSD	21	1.67 (0.89, 3.16)	Olshan (1991)
Kitchen porter	Cleft lip & palate	4	396	McDowall (1985)
Waiter/waitress	Down syndrome	6	282	McDowall (1985)
Food processor	Cleft palate	14	1.72 (0.83, 3.55)	Olshan (1991)
Food processor	Cleft lip & palate	16	2.88 (1.36, 6.09)	Olshan (1991)
Food processor	Down syndrome	26	1.79 (0.96, 3.31)	Olshan (1991)
Food processor	Hypospadias	22	1.55 (0.85, 2.81)	Olshan (1991)
TEXTILE WORKERS:				
Textile worker	PDA	4	7.60 (0.84, 68.58)	Olshan (1991)
Weaver	Down syndrome	5	511	McDowall (1985)
Spinner, doubler, twister	Tracheoesophageal fistula	3	705	McDowall (1985)
Knitter	All	*	163	OPCS (1982)
OTHER OCCUPATIONS:				
Pharmacist	Anencephalus	3	673	McDowall (1985)
Physician/surgeon	Cleft lip	6	3.28 (0.82, 13.19)	Olshan (1991)
Underwriter/broker	Talipes	29	181	McDowall (1985)
Architect	Syndactyly	7	259	McDowall (1985)
Physical, geological scientist	Rectal atresia	4	398	McDowall (1985)

Table 2. (Continued)

Occupation/Exposure	Birth Defect+	No. Exposed Cases	OR/RR (95% CI)§	Reference
Medical technician	Circ. system defect	3	497	McDowall (1985)
Hospital porter	ASB	6	319	McDowall (1985)
Manager, nec	ASB	33	246	McDowall (1985)
Manager, nec	Circ. system defect	40	264	McDowall (1985)
Manager, nec	Cleft lip & palate	29	189	McDowall (1985)
Manager, nec	Down syndrome	25	215	McDowall (1985)
Manager, nec	Polydactyly	17	176	McDowall (1985)
Manager, nec	Syndactyly	22	244	McDowall (1985)
Manager, nec	Talipes	92	240	McDowall (1985)
Manager, nec	Reduction def.	15	314	McDowall (1985)
Manager, nec	Rectal atresia	11	376	McDowall (1985)
Manager, nec	Hypospadias	43	252	McDowall (1985)
Fireman	VSD	10	2.70 (1.02, 7.18)	Olshan (1990)
Fireman	VSD	10	2.37 (0.98, 5.70)	Olshan (1991)
Fireman	ASD	10	5.91 (1.60, 21.83)	Olshan (1990)
Fireman	ASD	10	4.75 (1.36, 16.55)	Olshan (1991)
Teacher	Circ. system defect	47	138	McDowall (1985)
Teacher	Down syndrome	47	189	McDowall (1985)
Teacher/Librarian	Obstructive renal defects	19	3.09 (1.44, 6.65)	Olshan (1991)
Teacher	All	*	111	OPCS (1982)
Personal service worker	Anencephalus	9	2.75 (1.02, 7.41)	Olshan (1991)
Personal service worker	Spina bifida	7	3.02 (0.96, 9.48)	Olshan (1991)
Janitor	Hydrocephalus	7	5.04 (1.23, 20.64)	Olshan (1991)
Janitor	VSD	13	2.45 (1.10, 5.45)	Olshan (1991)
Janitor	Other heart defects	14	2.35 (1.07, 5.13)	Olshan (1991)
Janitor	Down syndrome	12	3.26 (1.02, 10.44)	Olshan (1989)
Janitor	Atresia of the urethra	6	3.28 (0.82, 13.19)	Olshan (1991)
Fisherman/hunter	ASD	12	2.05 (0.87, 4.83)	Olshan (1991)
Fisherman/hunter	Dislocation of hip	15	1.83 (0.87, 3.87)	Olshan (1991)
Student	Cleft lip	6	10.00 (1.17, 85.8)	Olshan (1991)
Police officer	Cleft lip & palate	29	158	McDowall (1985)
Policeman, guard	Polydactyly	8	3.38 (1.08, 10.62)	Olshan (1991)
Armed forces	Polydactyly	9	2.44 (0.93, 6.40)	Olshan (1991)
Literary, artistic, or sports worker	Cleft lip & palate	4	995	McDowall (1985)
Author, journalist	Hypospadias	9	238	McDowall (1985)
Butcher	All	*	116	OPCS (1982)

+ Abbreviations used: ASB=anencephalus and spina bifida; PDA=patent ductus arteriosus; VSD=ventricular septal defect; ASD=atrial septal defect.

§ Standardized Malformation Ratio (SMR) reported from McDowall (1985) and OPCS (1982). SMR of 100 equals relative risk of 1.0. We were unable to calculate confidence intervals for the SMRs, but all findings presented are statistically significant at $p < 0.05$.

* Number of exposed cases not stated or ascertainable from the paper.

Printers

Printers are exposed to metals and metal oxides, carbon black, ink mists, heat, and solvents. Printers have been associated with an increased risk of several defects including CNS defects (Fredrick, 1976), clubfoot, atresia of the urethra (Olshan *et al.*, 1991), and cleft palate (Erickson, *et al.*, 1979).

Painters

Painters are exposed to paints, solvents, metals, epoxy resins and other agents. They have been associated with CNS defects as well as cleft palate. The association with CNS defects has been reported in five independent studies with relative risk estimates ranging from 2.5 to 4.9 (Fedrick, 1976; Olshan *et al.*, 1991;Brender and Suarez, 1990; Olsen, 1982; McDowall, 1985). Two studies (Erickson *et al.*, 1979; Olshan *et al.*, 1991) found an association between painters and cleft palate in offspring.

Mechanics

Mechanics are exposed to solvents, oils, and metal dusts. Two studies reported an elevated risk for neural tube defects (Poldenak and Janerich, 1983; McDowall, 1985). Increased odds ratios have also been reported for Down syndrome and hypospadias (Olshan *et al.*, 1989; 1991).

Transportation-Related Occupations

This group includes motor vehicle operators or drivers with exposure to combustion products. Various studies have indicated that men in these occupations were at increased risk of having a child with a variety of birth defects including neural tube defects (Poldenak, 1983; McDowall, 1985; Olshan *et al.*, 1991), oral clefts (Olshan *et al.*, 1991; McDowall, 1985) and Down syndrome (Olshan *et al.*, 1989; McDowall, 1985).

Agricultural Occupations

This group includes farmers and farm workers with potential exposure to pesticides, fertilizers, and plant and animal organisms. An excess risk for neural tube defects was found in two studies (Poldenak and Janerich, 1983; Brender and Suarez, 1990).

Chemical Workers

Chemical workers, depending upon their particular industry of employment, are exposed to a variety of substances including oils, fuels, paints, pesticides, solvents, and acids. This group showed some association with central nervous system and gastrointestinal system defects (McDowall, 1985).

Metal Workers

Metal workers including welders, sheetmetal, iron and foundry workers are exposed to metal oxides and dusts, solvents, polycyclic aromatic hydrocarbons, and electromagnetic fields . Associations with several types of defects in offspring including central nervous system defects (Fedrick, 1976; McDowall, 1985), cardiovascular system defects (Olshan *et al.*, 1991), oral clefts (Olshan *et al.*, 1991), and Down syndrome (Olshan *et al.*,1989) have been reported.

Wood-Related Occupations

These occupations include forestry and logging workers, sawmill workers, carpenters, and paper workers. They have exposure to wood dust, glues, and paints. Elevated relative risks have been reported for a number of individual birth defects including central nervous system defects, cardiovascular defects, oral clefts, and defects of the musculoskeletal, gastrointestinal, and genitourinary systems (Olshan *et al.*, 1991).

Machinists

Machinists are exposed to metal dusts, solvents, lubricating oils, and cutting fluids. Elevated relative risk estimates for defects of the central nervous system, cardiovascular system, gastrointestinal system, as well as, Down syndrome, and oral clefts have been reported for machinists (McDowall, 1985).

Electronics-Related Occupations

This group includes occupations such as electricians, electronics workers and utility workers with potential exposure to electromagnetic fields and solvents. Two studies reported an association with central nervous system defects (Polednak and Janerich, 1983; McDowall, 1985).

Other Occupations

Other specific occupations that were found to be at an increased risk of having a child with a birth defect included textile workers, firemen, and janitors (Olshan *et al.*, 1990; 1991). A series of special analyses of the British Columbia data indicated that compared with either policemen or men in other occupations, firemen had an increased risk of ventricular septal and atrial septal defects in offspring (Olshan *et al.*, 1990). Firemen, who are exposed to a wide range of known mutagens and carcinogens, warrant further investigation.

SUMMARY OF FINDINGS

The limited number of published studies of paternal occupation and birth defects have suggested several occupations and related exposures that warrant further study. These include janitors, forestry and logging workers, sawmill workers, carpenters and wood workers, motor vehicle operators, painters, electricians, electrical and electronics workers, firemen, welders, and textile workers. Suggestive associations with related occupational exposures include wood and wood products, metals, solvents, and pesticides.

METHODOLOGIC LIMITATIONS OF PREVIOUS STUDIES

The studies reviewed here are subject to the methodologic limitations that many epidemiologic studies suffer from (Shaw and Gold, 1988). One source of potential bias in this context is the misclassification of exposure. A major source of misclassification is the use of birth certificates to ascertain paternal job title and industry at the time of birth (Shaw *et al.*, 1990). This error is nondifferential with respect to case or control status and would result in an relative risk estimate of the effect biased towards the value of 1.0 (no association) (Rothman, 1986). An analogous problem occurs when birth defects with different etiologies are combined into categories. This grouping of defects results in etiologic heterogeneity, which again leads to an underestimate of the true relative risk.

An important problem in epidemiologic studies of human teratogens is differential prenatal survival. Many studies, including those reviewed here, are based on the detection of birth defects in live births. If the exposure of interest increases the risk of fetal loss, as well as birth defects in liveborns, the relative risk estimate may be distorted. Generally, this differential prenatal survival will lead to the dilution of the relative risk towards the null. (Khoury *et al.*, 1989).

Confounding due to factors associated with birth defects and the exposures of interest may also introduce bias. At present, there are very few definitive risk factors for many birth defects so the role of confounding cannot be fully evaluated. Most of the epidemiologic studies of birth defects reviewed here had limited power to detect associations for individual defects and specific occupations. In an effort to increase power many investigators created combined categories of heterogeneous birth defects and occupations. Unfortunately, these groupings may contribute to the misclassification problems noted above.

The most important limitation, in general, is that most of the epidemiologic studies suffer to some degree from the misclassification of paternal occupational exposures. It is important to note that this would tend to bias the relative risk estimates towards the null, that is, these studies may have yielded **underestimates** of the true relative risk.

CONCLUSIONS

The research on paternal occupation and birth defects is still in the hypothesis generating phase. There have not been an adequate number of both informative positive and negative epidemiologic studies conducted to date to be able to begin to definitively characterize the risk associated with paternal occupation. However, some leads have been provided and only further epidemiologic study and guidance from our laboratory colleagues will ultimately resolve the issue.

REFERENCES

Brender, J.D., Suarez, L., 1990, Paternal occupation and anencephaly, *Am. J. Epidemiol.* 131:517-21.

Brown, N.A., 1985, Are offspring at risk from their father's exposure to toxins? *Nature* 316:110.

Cohen, E.N., Gift, H.C., Brown, B.W., Greenfield, W., Wu, M.L., Jones, T.W., Whicher, C.E., Driscoll, E.J., Brodsky, J.B., 1980, Occupational disease in dentistry and chronic exposure to trace anesthetic gases, *JADA* 101:21-31.

Edmonds, L.C., Anderson, C.E., Flynt, J.W. Jr., James, L.M., 1978, Congenital central nervous system malformations and vinyl chloride monomer exposure: A community study. *Teratol.* 17:137-142.

Erickson, J.D., Cochran, W.M., Andersen, C.E., 1979, Parental occupation and birth defects, a preliminary report. *Contrib. Epidemiol. Biostat.* 1:107-17.

Erickson, J.D., Mulinare, J., McClain, P.W., Fitch, T.G., James, L.M., McClearn, A.B., Adams Jr., M.J., 1984, Vietnam veterans; risks for fathering babies with birth defects. *J. Am. Med. Assoc.* 252:903-12.

Fedrick, J., 1976, Anencephalus in the Oxford Record Linkage Study Area. *Develop. Med. Child Neurol.* 18:643-56.

Hemminki, K., Mutanen, P., Luomak, K., Saloniemi, I., 1980, Congenital Malformations by the parental occupation in Finland. *Int. Arch. Occup. Environ. Health* 46:93-8.

Kalter, H., Warkany, J., 1983, Congenital malformations: Etiologic factors and their role in prevention. Part One. *New Eng. J. Med.* 308:424-31.

Khoury, M.J., Flanders, W.D., James, L.M., Erickson, J.D., 1989b, Human teratogens, prenatal mortality and selection bias. *Am. J. Epidemiol.* 130:361-370.

McDonald, A.D., McDonald, J.C., Armstrong, B., Cherry, N.M., Nolin, A.D., 1989, Fathers' occupation and pregnancy outcome. *Br. J. Ind. Med.* 46:329-33.

McDowall, M.E., 1985, Occupational reproductive epidemiology. Series SMPS 50. London: Her Majesty's Stationery Office.
Olsen, J., 1982, Risk of exposure to teratogens amongst laboratory staff and painters. *Dan. Med. Bull.* 30:24-8.
Olshan, A.F., Baird, P.A., Teschke, K., 1989, Paternal occupational exposures and the risk of Down Syndrome. *Am. J. Hum Genet.* 44:646-51.
Olshan, A.F., Teschke, K., Baird, P.A., 1990, Birth defects among offspring of firemen. *Am. J. Epidemiol.* 131:312-21.
Olshan, A.F., Teschke, K., Baird, P.A., 1991, Paternal occupation and congenital anomalies in offspring. *Am. J. Indust. Med.* 20:447-75.
OPCS Monitor., 1982, Congenital malformations and parents' occupation. MB 3 82/1. Office of Population Censuses and Surveys, London.
Papier, C.M., 1985, Parental occupation and congenital malformations in a series of 35,000 births in Israel. In: "Prevention of physical and mental congenital defects, Part B: Epidemiology, early detection and therapy, and environmental factors." ed. M. Marois, pp 291-4, Alan Liss, Inc, New York, NY.
Polednak, A.P., Janerich, D.T., 1983, Uses of available record systems in epidemiologic studies of reproductive toxicology. *Am. J. Ind. Med.* 4:329-48.
Roan, C.C., Matanoski, G.E., McIlnay, C.Q., Olds, K.L., Pylant, F., Trout, J.R., Wheeler, P., Morgan, D.P., 1984, Spontaneous abortions, stillbirths and birth defects in families of agricultural pilots. *Arch. Environ. Health* 39:56-60.
Rothman, K.J., 1986, *Modern Epidemiology*. Boston: Little, Brown and Company. 358 pp. Savitz, D.A., Harley, B., Krekel, S., Marshall, J., Bondy, J., Orleans, M., 1984, Survey of reproductive hazards among oil, chemical, and atomic workers exposed to halogenated hydrocarbons. *Am. J. Ind. Med.* 6:253-254.
Schwartz, D.A., Newsum, L.A., Markowitz, Heifetz, R., 1986, Parental occupation and birth outcome in a agricultural community. *Scand. J. Work Environ. Health* 12:51-54.
Sever, L.E., Gilbert, E.S., Hessol, N.A., McIntyre, J.M., 1988, A case-control study of congenital malformations and occupational exposure to low-level ionizing radiation. *Am. J. Epidemiol.* 127:226-42.
Shaw, G. M., Gold, E.B., 1988, Methodologic considerations in the study of parental occupational exposures and congenital malformations in offspring. *Scan. J. Work Environ. Health* 14:344-355.
Shaw, G.M., Malcoe, L.H., Croen, L.A., Smith, D.F., 1990, An assessment of error in parental occupation from the birth certificate. *Am. J. Epidemiol.* 131:1072-1079.
Smith, A.H., Fisher, D.O., Pearce, N., Chapman, C.J., 1982, Congenital defects and miscarriages among New Zealand 2,4,5-T sprayers. *Arch. Environ. Health* 37:197-200.
Townsend, J.C., Bodner, K.M., Van Peenan, P.F.D., Olson, R.D., Cook, R.R., 1982, Survey of reproductive events of wives of employees exposed to chlorinated dioxins. *Am. J. Epidemiol.* 115:695-713.

MALE-MEDIATED DEVELOPMENTAL TOXICITY: PATERNAL EXPOSURES AND CHILDHOOD CANCER

Jonathan Buckley

Department of Preventive Medicine
University of Southern California
1420 San Pablo Street
Los Angeles, CA 90033

INTRODUCTION

Investigation of the epidemiology of cancers of childhood provides some unique challenges, and offers the prospect of deeper understanding of the process of carcinogenesis as it applies to both adults and children. It can also be a frustrating and unrewarding pursuit. The question of the effect of paternal exposure on risk of cancer in an offspring illustrates both the fascination of this field, and its limitations.

A key issue, of relevance to this meeting, is the relative contributions of environmental and genetic factors in the pathogenesis of cancers of childhood. In adults, it is widely accepted that the environment we inhabit is much more important than the genes we inherit, although this perception may be biased by the fact that we have, in the past, been better able to observe the environment than the genetic inheritance of an individual. However, adult and childhood cancers are very different--for example, in their histological spectrum--and the roles of environmental and genetic factors for most childhood cancers has yet to be clearly established. Some cancers, most notably retinoblastoma, have an unequivocal heritable component, but the evidence from studies of twins, siblings and offspring of affected children indicates that overall the genetic contribution is modest at best.

On the other hand, there is little compelling evidence for environmental risk factors. Geographic variation in incidence is substantially less than for adult cancers, suggesting that environmental factors may be correspondingly less important. Furthermore, the short average interval between conception to cancer diagnosis should greatly assist those looking for an environmental cause: the fact that so few reproducible findings have come from a fairly substantial research effort in this field can't help but call into question the role of the child's environment.

That is not to say that case-control studies of childhood cancer have been completely unproductive. While attempts to implicate environmental exposures to the

Male-Mediated Developmental Toxicity, Edited by D.R. Mattison
and A.F. Olshan, Plenum Press, New York, 1994

child have not yielded many useful leads, most studies have also included questions on **parental** exposures and there have been numerous reports of significant associations between parental occupation and cancer risk. Parental exposures could be transferred to the child on work clothes or skin, or passed on from mother to a child transplacentally, or via breast milk. Alternatively, and more controversially, the exposure could damage sperm or ova leading to a constitutional genetic defect that manifested as a malignancy. In this paper, I will present some of the data linking paternal exposures to risk of cancer in an offspring, consider the relevant methodological questions and caveats, touch on the issues of imprinting and preferential allele loss and consider possible mechanisms.

DATA

The literature concerning parental occupation and childhood cancer risk has been comprehensively reviewed in two recent publications (Leary *et al.*, 1991; Savitz and Chen, 1990). There is so much between-study variation with respect to case population, control group, data collection method and analysis and interpretation of the data that it is very difficult to compare study findings directly. Some broad generalizations are possible however:

1) Most studies that look for increased risks associated with parental occupational exposures report some statistically significant associations;

2) Although the nature of these associations vary widely, some exposures appear to recur. Specifically, occupational exposures to solvents, petroleum products, paints/pigments, pesticides, radiation and metals have been repeatedly linked to cancers in children; and

3) Paternal occupational exposures are more commonly reported to be significant than are maternal.

Assessment of job-related exposures is problematic. Three different approaches are possible. The first is to classify jobs into broad categories with, presumably, a similar spectrum of exposures: an example can be seen in the study by Fabia and Thuy which showed an association of cancer of all types and hydrocarbon-related occupations. The second approach is to use a job-exposure matrix to infer specific exposures from the reported occupational titles. This method has the advantage of minimizing recall biases (since the respondent need only remember the job title), and can make use of information that has been routinely recorded, such as on birth or death certificates. The disadvantage is that the quality of the exposure measure is directly related to the accuracy and appropriateness of the job-exposure matrix used, and most would agree that currently available matrices have many limitations. Even a perfect matrix can only provide an estimate of the probability that a person in a given job had a given exposure, not whether or not he did. To learn that, there is no alternative to asking the parent (or perhaps his employer) about the specific chemical and physical agents that he was exposed to at work. This might seem to be the ideal approach, provided the parent is available for interview, but is susceptible to recall bias and presumes that the person knows what he is exposed to. The studies by Lowengart *et al.* and Buckley *et al.* relied on self-reported exposures.

Tables 1-3 provide a summary of the data for some of the more commonly reported associations. Table 1 concerns a loosely defined group of chemicals--solvents, petroleum products and paints. There is obviously overlap amongst these categories in that solvents may be petroleum products, and paints or pesticides may contain chemical solvents, etc. As can be seen, associations with this group of chemicals have been reported on many occasions for a range of malignancies.

Table 1. Reported Associations of Childhood Cancer Risk and Parental Exposures to Solvents, Petroleum Products or Paints

Author	Site	N	Source	Exposure	O.R.
		Paternal			
Fabia and Thuy, 1974	All cancer	386	Birth cert.	Hydrocarbon-related occ.	2.1
Sanders *et al.*, 1981	All cancer	6920	Death Cert.	Gas, coke, chem. occ.	PMR =143
Peters *et al.*, 1981	CNS	92	Interview	Solvent exposure	2.8
				Paint exposure	7.0
Hicks *et al.*, 1984	All cancer	298	Interview	Petroleum refining	inf.
Vianna *et al.*, 1984	AL (<1 yr)	60	Interview	Motor vehicle exhaust	2.5
Spitz and Johnson, 1985	Neuroblastoma	157	Birth cert.	Hydrocarbons	3.2
Johnson *et al.*, 1987	CNS	499	Birth cert.	Chem. & Petroleum refining	3.0
Lowengart *et al.*, 1987	AL	123	Interview	Chlorinated solvents	3.5
				Cutting oil	1.7
				Dyes and pigments	4.5
Buckley *et al.*, 1989	AML	204	Interview	Painter	7.0
				Solvents	2.0
				Coal & oil products	1.9
				Petroleum products	2.7
		Maternal			
Shu *et al.*, 1987	AML	94	Interview	Benzene	4.0
				Gasoline	1.7
Buckley *et al.*, 1989	AML	204	Interview	Gasoline	1.7
Buckley *et al.*, 1989	Hepatoblastoma	75	Interview	Oil & coal products	3.7
Wilkins and Sinks, 1990	CNS	110	Interview	Hydrocarbons Paints/pigments	3.7

Table 2 summarizes studies that showed associations with metal exposures. As can be seen, most reports are non-specific, and relate to welding, soldering, exposure to molten metals and/or metal processing. Again, there is little consistency in regards to the tumor types.

Table 2. Reported Associations of Childhood Cancer Risk and Parental Exposures to Metals

Author	Site	N	Source	Metal	O.R.
		Paternal			
Kantor, *et al.*, 1979	Wilms' tumor	149	Birth cert.	Lead occupation	3.7
Wilkins and Sinks, 1984	Wilms' tumor	62	Birth cert.	Boron occupation	3.5
Buckley *et al.*, 1989	AML	204	Interview	Lead exposure	inf.
Wilkins and Koutras, 1988	CNS	491	Birth cert.	Metal Industry	5.3
McKinney *et al.*, 1987	Leuk./ Lymphoma	234	Interview	Furnace/forge/ foundry	4.8
		Maternal			
Buckley *et al.*, 1989	Hepatoblastoma	75	Interview	Metal exposure	7.0
Buckley *et al.*, 1989	AML	204	Interview	Metal dust exposure	6.0
				Metal exposure	4.5
Shu *et al.*, 1987	AML	94	Interview	Metal refining	4.6

Table 3 provides data for pesticide exposures. The term "pesticide" includes herbicides, fungicides and insecticides, and covers a diverse group of chemicals.

Table 3. Reported Associations of Childhood Cancer Risk and Parental Exposures to Pesticides

Author	Site	N	Source	Exposure	O.R.
		Paternal			
Hemminki *et al.*, 1981	All cancer	1600	Maternity rec.	Farmer	7.1
Wilkins and Koutras, 1988	CNS	491	Birth cert.	Farmer	2.4
Buckley *et al.*, 1989	AML	204	Interview	Pesticide exposure	2.7
		Maternal			
Shu *et al.*, 1987	ALL	172	Interview	Pesticide exposure	3.5
Buckley *et al.*, 1989	AML	204	Interview	Pesticide exposure	inf.
		Home Exposure (Either Parent)			
Gold *et al.*, 1979	CNS	73	Interview	Living on farm	4.0
Lowengart, *et al.*, 1987	AL	123	Interview	Household pesticides	3.8
				Garden pesticides	6.5

If nothing else, it is obvious from these data that there is a pressing need for a new generation of studies that focus on specific exposures and attempt to define much more precisely the chemicals involved.

These observations are interesting, but must not be viewed uncritically. Usually the studies were exploratory, with no clear prior hypotheses. Many cancer-exposure or cancer-occupation combinations were tested and some significant associations can be expected by chance. The predominance of solvents, metals, etc. in the findings requires explanation, but this could simply reflect the fact that these exposures are very common in the workplace and/or are more prone to recall bias. The greater number of paternal associations is easily explained by the fact that some studies only considered father's occupation, and men will generally have had more opportunities for exposure (with a larger average number of jobs held, and generally dirtier jobs) which would increase statistical significance for a fixed degree of association. Most investigators attempted to control for possible demographic confounders such as parental age, race, education, and SES or income. However, choice of occupation is so fundamentally linked to every aspect of one's life, that our crude attempts to control for confounding are probably quite inadequate.

The flip side of the coin is that there is obviously a huge amount of non-differential misclassification in all these studies, which would reduce the apparent strength of any association. Furthermore, the broad groupings of disease (for example: all cancers or all CNS tumors),combining all age groups together, and grouping of exposure types (for example: all solvents),would also greatly dilute any association between a single cancer and a specific chemical. There is no consensus on the critical period of exposure: should we be concerned with the job held just prior to conception, during the pregnancy, during the nursing period, or during the first years of the child's life, or should we go back to include all jobs at any age, or perhaps some combinations of these. Thus if any of these findings are real, the actual disease-exposure odds ratio, with no misclassification or dilution, must be considerably larger than the estimates from the data.

We should also consider that these same case-control studies have generally failed to identify, with any consistency, any other environmental risk factors. Does the plethora of significant findings for occupational exposures indicate that this is indeed where we should be looking for the causes of childhood cancers, or are recall biases, confounding, or other methodological problems leading us astray?

PREFERENTIAL ALLELE LOSS

My own inclination, reflecting perhaps a distrusting and skeptical nature, is to believe that the occupational associations are largely spurious. However, better understanding of the genetic mechanisms that lead to childhood cancers has provided a new twist to the story, that lends some credence to the hypothesis that paternal exposures are important.

The unexpected twist was an observation, made initially for Wilms' tumors but later found to apply more generally, that loss of heterozygosity in regions of known or suspected tumor suppressor genes did not affect the paternally- and maternally-derived genes with equal frequency. There was a dramatic bias towards retention of the paternal alleles which could not be ascribed to chance.

Similar results have been reported for a number of other tumors including retinoblastoma, osteosarcoma, and rhabdomyosarcoma. These data have been generally interpreted to show that the initial mutation in a tumor suppressor gene occurred on the paternally-derived chromosome, and that a subsequent deletion of the corresponding segment of the other chromosome, leaves the cell without a functional copy of this gene.

If this phenomenon was limited to the heritable form of the disease, the simplest explanation would be that the father sustained damage to his germinal DNA and passed this to his offspring: the paternal excess could reflect the greater opportunities for exposure in male-dominated jobs, or differences in replication fidelity in spermatogenesis and oogenesis. However the same effect is present in cases that are thought to be sporadic. That is, a phenomenon in a somatic cell is affecting the paternal and maternal chromosomes differently.

IMPRINTING

We are accustomed to thinking of the two chromosomes in a pair as being entirely equivalent, and perhaps this is true for most genes throughout most of our adult life. It is clearly not the case during embryogenesis. For reasons that are only partly understood, segments of one or other chromosome in a pair are "imprinted" during gametogenesis, as a result of which only the maternal or paternal allele is expressed. The cell is thought to imprint a stretch of DNA by heavily methylating cytosines in CpG dinucleotides. Methylation enzymes capable of recognizing a hemimethylated CpG dinucleotide methylate the opposite strand after DNA replication, thus preserving the methylation pattern through multiple cell divisions. In theory, selective suppression of a DNA segment could be maintained throughout life, only being erased and reimprinted in the process of gametogenesis. There is some evidence for imprinting that persists to adult life, but it seems that the effect is most pronounced in early life. One hypothesis is that imprinting has evolved as a means of preventing parthenogenesis: the egg and sperm may possess a full DNA complement, but each expresses a different set of genes and only the combination can make a baby. Whatever the reason, it is clear that imprinting is relevant to many childhood cancers.

How may imprinting affect cancer risk? The most obvious is that a cell could lose the maternal allele (of a tumor suppressor gene, say) while the paternal allele is still suppressed. This suppression could be quite normal, or might have persisted long after it should have been removed: in the latter case, the defect would lie in the mechanisms that regulate the selective demethylation of imprinted regions of the genome. In either case, the retained allele would be found on sequencing to be of wild-type.

Me-C DEAMINATION

An alternative possibility is that imprinted DNA is more mutation-prone than other DNA. In fact methylated cytosine undergo spontaneous deamination to thymine. The unmethylated form will also spontaneously deaminate, but this converts the cytosine to a uracil which is very readily detected and repaired. The C to T transition, creating a GT mismatch, presents more of a dilemma to the cell since either the G or the T could be in error. In fact, repair enzymes greatly favor the correct replacement of the T by C, but in perhaps 5-10% of instances the mutation becomes permanent when the G is replaced by A. It has been found that approximately 40% of point mutations in man are C to T transitions at CpG dinucleotide, from which one can determine that the mutation rate at these hot-spots is approximately 25 times the average at other sites. In is worth noting that the sperm is much more heavily methylated than the ovum.

I have referred to the deamination as a spontaneous chemical process, which indeed it can be. If a substantial fraction of childhood cancers arise from such mutations, the apparent lack of strong genetic or environmental associations could be explained. Of course this is the ultimate nightmare for the epidemiologist who has spent his or her life searching for risk factors for cancers in children. However, this is perhaps being overly pessimistic. There are factors that can significantly influence the rate of deamination--such as nitric oxide--and it may be that the reason we have been so unsuccessful our search for the causes of childhood cancer is that we have been looking in the wrong place. The archetypal carcinogens are chemicals or radiation that cause direct DNA damage; perhaps the main culprits here are chemicals which affect the rate of deamination, the efficiency of repair, or impair the enzymes responsible for maintenance or removal of imprinting.

It is even possible that the answer is, in fact, staring us in the face and the most important causes of childhood cancer is parental, and particularly paternal, occupational exposure. Only time and additional research will tell.

REFERENCES

Buckley, J.D., Sather, H., Ruccione, K., Rogers, P.C.J., Haas, J.E., Henderson, B.E., Hammond, G.D. A case-control study of risk factors for hepatoblastoma. A report from the Childrens Cancer Study Group. *Cancer* 1989; 64:1169-1176.

Buckley, J.D., Robison, L.L., Swotinsky, R., Garabrant, D., LeBeau, M., Nesbit, M.E., Odom, L., Peters, J., Woods, W.G., Hammond, G.D. Occupational exposures of parents of children with acute nonlymphocytic leukemia. *Cancer Res.* 1989; 49:4030-4037.

Fabia, J., Thuy, T.D. Occupation of father at time of birth of children dying of malignant diseases. *Br. J. Prev. Soc. Med.* 1974; 28 :98-100.

Gold, E., Gordis, L., Tonascia, J., Szklo, M. Risk factors for brain tumours in children. *Am J of Epidemiol.* 1979; 109:309-319.

Hemminki, K., Saloniemi, I., Salonen, T., Partanen, T., Vainio, H. Childhood cancer and parental occupation in Finland. *J. Epidemiol. Community Health* 1981; 35:11-15.

Hicks, N., Zack, M., Caldwell, G.G., Fernbach, D.J., Falletta, J.M. Childhood cancer and occupational radiation exposure in parents. *Cancer* 1984; 53:1637-1643.

Johnson, C.C., Annegers, J.F., Frankowski, R.F., Spitz, M.R., Buffler, P.A. Childhood nervous system tumors--an evaluation of the association with paternal occupational exposure to hydrocarbons. *Am. J. Epidemiol.* 1987; 126:605-613.

Kantor, A.F., McCrea-Curnen, M.G., Meigs, J.W., Flannery, J.T. Occupations of fathers of patients with Wilms' tumour. *J. Epidemiol. Community Health* 1979; 33:253-256.

Leary, L.M., Hicks, A.M., Peters, J.M., London, S. Parental occupational exposures and risk of childhood cancer: a review. *Am. J. Ind. Med.* 1991; 20:17-35.

Lowengart, R.A., Peters, J.M., Cicioni, C., Buckley, J., Bernstein, L., Preston-Martin, S., Rappaport, E. Childhood leukemia and parent's occupational and home exposures. *JNCI* 1987; 79:39-46.

McKinney, P.A., Cartwright, R.A., Saiu, J.M.T., Mann, J.R., Stiller, C.A., Draper, G.J., Hartley, A.L., Hopton, P.A., Birch, J.M., Waterhouse, J.A.H., and Johnston, H.E. The inter-regional epidemiological study of childhood cancer (IRESCC): a case control study of aetiological factors in leukaemia and lymphoma. *Arch. Dis. Child.* 1987; 62 :279-287.

Peters, J.M., Preston-Martin, S. Brain tumours in children and occupational exposure of parents. *Science* 1981; 213:235-237.

Sanders, B.M., White, G.C., Draper, G.J. Occupations of fathers of children dying from neoplasms. *J. Epidemiol. Comm. Health* 1981; 35 :245-250.

Savitz, D.A., Chen, J. Parental occupation and childhood cancer: review of epidemiologic studies. *Environ. Health Perspect.* 1990; 88:325-337.

Shu, X.O., Gao, Y.T., Brinton, L.A., Linet, M.S., Tu, J.T., Zheng, W., Fraumeni, J.F. A population-based case-control study of childhood leukemia in Shanghai. Not published, 1987.

Spitz, M.R., Johnson C.C. Neuroblastoma and paternal occupation. *Am. J. Epidemiol.* 1985; 121:924-929.

Vianna, N.J., Kovasznay B, Polan A, Ju C. Infant leukemia and paternal exposure to motor vehicle exhaust fumes. *J. Occup. Med.* 1984; 26:679-682.

Wilkins, J.R., Sinks TH. Paternal occupation and Wilms' tumour in offspring. *J. Epidemiol. Community Health* 1984; 38:7-11.

Wilkins, J.R., Sinks, T.H. Parental occupation and intracranial neoplasms of childhood: results of a case-control interview study. *Am. J. Epidemiol.* 1990; 132:275-292.

Wilkins, J.R., Koutras, R.A. Paternal occupation and brain cancer in offspring: a mortality-based case-control study. *Am. J. Ind. Med.* 1988; 14:299-318.

PATERNAL EXPOSURES AND PREGNANCY OUTCOME: MISCARRIAGE, STILLBIRTH, LOW BIRTH WEIGHT, PRETERM DELIVERY

David A. Savitz

Department of Epidemiology
Campus Box #7400
School of Public Health
University of North Carolina
Chapel Hill, NC 27599-7400

INTRODUCTION

In this overview, the epidemiologic evidence regarding the influence of paternal exposures in the etiology of pregnancy outcomes will be presented. These outcomes, covering the spectrum of fetal survival and development exclusive of birth defects, have been the subject of a number of studies that include paternal factors. Most of the research has focused on miscarriage as the endpoint, and among those, most have been concerned with occupational exposures. Public concern with such issues as Agent Orange exposure among male veterans who served in Vietnam has lead to a concentration on occupational exposures rather than lifestyle factors such as tobacco or alcohol use.

Until very recently, the relationship between father's exposures and pregnancy outcome has been evaluated through the exploitation of existing data or by analyzing data from studies with a focus on some other exposure (typically maternal), with the predictable limitations in the level of detail available regarding paternal factors. However, in the last few years, studies designed explicitly to examine the paternal contribution to miscarriage have appeared (Alcser *et al.*, 1989; Taskinen *et al.*, 1989; Cordier *et al.*, 1991) and yielded some rather interesting results.

In addition to characterizing the state of evidence, an attempt will be made to highlight some of the methodologic issues specific to paternal factors and the pregnancy outcomes of interest. Finally, the needs to advance research in this area will be discussed.

PATERNAL EXPOSURES AND MISCARRIAGE

Research on paternal exposures and miscarriage was initiated in the mid- 1970s with surveys of male dentists exposed to anesthetic gases (Cohen *et al.*, 1974) and vinyl

Male-Mediated Developmental Toxicity, Edited by D.R. Mattison
and A.F. Olshan, Plenum Press, New York, 1994

chloride workers (Infante *et al.*, 1976). In each instance, suggestive evidence was provided that the male's occupational exposures around the time of conception, independent of maternal risk factors, were related to an increased risk of miscarriage. Though the evidence remains inconclusive, in retrospect, these initial efforts opened up an important avenue of research. A total of at least 38 papers have now been published examining the association between father's occupational exposure and miscarriage (Savitz *et al.*, in press), with a modest number of additional papers concerned with paternal tobacco and alcohol use.

As noted above, most of these studies considered paternal exposure and miscarriage as part of surveys with a broader array of interests. As a result, the level of detail on either the father's exposure, the pregnancy outcome, or both is limited. Father's occupational exposures have largely been based on a job title rather than more thorough methods such as detailed workplace descriptions, biological markers, or environmental measurements. Miscarriages have usually been based on maternal report, with paternal reports sometimes used, and in Scandinavia, medical records. Interest has been concentrated on exposure to heavy metals, pesticides, solvents, and anesthetic gases, with limited data on other potentially harmful agents. Selected findings are summarized below, including the most promising or extensively researched topics.

Table 1 summarizes results of three studies addressing paternal mercury and miscarriage, with null results in one (Brodsky *et al.*, 1985) but notably positive associations in two studies that considered exposure in greater detail (Alcser *et al.*, 1989; Cordier *et al.*, 1991). In the latter two studies, a dose-response gradient was observed based on estimated mercury exposure. In addition to the direct information conveyed about paternal mercury and miscarriage, these latter two studies illustrate the potential value of studies that incorporate measures of dose.

Table 1. Summary of Results of Studies of Paternal Exposure to Mercury and Miscarriage

Reference	Exposure	RR	95% CI
Brodsky *et al.*, 1985	Exposure in dentistry	0.9	0.7-1.1
Alcser *et al.*, 1989[a]	Urine Hg 2,000-3,999 ug/l	1.1	0.8-1.6
	Urine Hg 4,000-9,000 ug/l	1.7	1.1-2.5
Cordier *et al.*, 1991[b]	Urine Hg 1-19 ug/l	1.3	1.0-1.7
	Urine Hg 20-49 ug/l	1.7	1.0-3.0
	Urine Hg 50+ ug/l	2.3	1.0-5.2

[a] Cumulative average quarterly urine levels over lifetime.
[b] Average mercury level in period before conception.

Studies of paternal exposure to anesthetic gases and miscarriage occurred in a rather brief period in the mid- to late-1970s. Summarizing results for only the most highly or clearly exposed group in each study (Table 2) provides a consistent indication of increased risk, with relative risk measures of 1.5 to 1.8. Mitigating the value of this replication somewhat is the similarity in methods of all the Cohen *et al.* studies (1974, 1975, 1980), with paternal self-report elicited in mail questionnaires. Limited response proportions, in particular, call the results into question. Although the results are sufficiently impressive to warrant more sophisticated study, the increasingly widespread use of scavenging systems that markedly lower workplace anesthetic gas exposure limits the likelihood of any continuing risk and thus the opportunity for further study.

Solvents constitute a broad set of agents that are of concern as occupational hazards, with an expectation that there would be diversity in their effects, if any, on male reproduction. Taskinen *et al.*'s (1989) study, the results of which are summarized in Table 3, illustrates that diversity in findings across specific types of solvents. Interestingly, the broadest class ("organic solvents") provided the largest relative risk estimate.

Table 2. Summary of Studies of Paternal Exposure to Anesthetic Gases and Miscarriage: Results for Most Exposed Groups

Reference	Exposure	RR	95% CI
Cohen *et al.*, 1974	Operating room technicians	1.8	0.8-4.1
Cohen *et al.*, 1975	Dentists exposed > 3 hours/wk	1.8	1.4-2.2
Tomlin, 1979	Anesthesiologists exposed > 20 hours/wk	1.6	0.9-3.0
Cohen *et al.*, 1980	Dentists exposed 8+ hours/wk	1.5	1/3-1.8

Table 3. Summary of Taskinen *et al.*'s (1989) Study of Paternal Exposure to Solvents and Miscarriage

Exposure	RR	95% CI
Organic solvents	2.3	1.1-5.0
Aromatic hydrocarbons	1.6	1.0-2.4
Styrene	1.3	0.8-2.1
Toluene	1.5	0.9-2.5
Halogenated hydrocarbons	1.1	0.6-1.8
Trichloroethylene	1.0	0.6-2.0
1,1,1-trichloroethane	0.9	0.3-2.3
Aliphatic hydrocarbons	1.5	0.9-2.5
Acetone	1.0	0.6-1.7

A primary inspiration for the entire line of research concerns exposure to Agent Orange, or more specifically the dioxin contaminants of Agent Orange and related herbicides. A number of studies have considered the association with miscarriage (reviewed by Savitz *et al.*, in press), with results for industrial workers and pilots who sprayed herbicides consistently negative. Studies of United States servicemen who served in Vietnam have been inconclusive, in part because of the virtual impossibility of establishing exposure with any certainty. Studies relying on self-reported exposure (Centers for Disease Control, 1988) show increasing risks with increasing perceived levels of exposure, but that is most plausibly due to a reporting or response bias. One study (Stellman *et al.*, 1988) did provide evidence that inferred paternal exposure was associated with increased risk of miscarriage. At present, and very possibly into the future, the relation between paternal dioxin and miscarriage is uncertain.

Ionizing radiation, as an established mutagen, is of particular interest in regard to paternal exposure and miscarriage. However, it has received very little attention, with only three reports (Table 4). Two studies yielded evidence of modest elevations in risk (Boue *et al.*, 1975; Lindbohm *et al.*, 1991) and one had negative results

(McDonald *et al.*, 1989). The evidence is presently insufficient for any conclusions regarding ionizing radiation.

Table 4. Summary of Studies of Paternal Exposure to Ionizing Radiation and Miscarriage

Reference	Exposure	RR	95% CI
Boue *et al.*, 1975	Ionizing radiation	1.3	1.1-1.5
McDonald *et al.*, 1989	Ionizing radiation	0.9	0.6-1.3
Lindbohm *et al.*, 1991	X-rays	1.5	0.6-3.6

Among exposures incurred by men, cigarette smoking is perhaps the most intense and provides the most diverse array of agents. Furthermore, compared to occupational exposures, it is relatively easily ascertained. Thus, it is surprising that it has received such limited attention. Table 5 summarizes results from seven studies, with only the most recent examining smoking as anything other than a confounder of other agents of interest. The results are completely negative except for studies by Lindbohm *et al.* (1991) and Rupa *et al.* (1991), but the issue remains worthy of study on theoretical grounds in spite of the limited epidemiologic support.

Table 5. Summary of Studies of Paternal Smoking and Miscarriage

Reference	RR	CI
Beckman & Nordstrom, 1992	1.0	0.7-1.4
Nordstrom *et al.*, 1983	1.0	0.6-1.6
Halmesmaki *et al.*, 1989	0.9	0.5-1.7
Taskinen *et al.*, 1989	0.6	0.4-1.0
Lindbohm *et al.*, 1991	1.3	0.9-1.9
Rupa *et al.*, 1991	1.4	1.2-1.5
Windham *et al.*, 1992	1.1	0.9-1.4

Several methodological issues are specific to miscarriage as opposed to paternal influences on reproduction more generally. Ascertainment of miscarriage is notoriously difficult, with clear evidence that clinically recognized pregnancies constitute only a subset of all conceptions and implantations (Wilcox *et al.*, 1988). Recall of miscarriage is also fallible, particularly for those losses in the remote past or of short gestation (Wilcox & Horney, 1984). Paternal recall of miscarriages is likely to be far more limited than maternal recall (Selevan, 1985). The key issue for researchers is to be sure that for the spectrum of miscarriages that are of concern (defined by medical treatment, gestational age, or other methods), the completeness of reporting is not related to paternal exposure. Valid studies of medically treated miscarriage are possible and would be informative, for example, in spite of a restricted case definition.

The endpoint of miscarriage is known to be heterogeneous, most obviously in regard to the chromosomal status of the conceptus. The studies at Columbia University (Kline *et al.*, 1989) have shown that risk factors are often specific to chromosomally normal rather than chromosomally abnormal fetuses or even specific to the type of aneuploidy. When we examine paternal exposures and miscarriage in the aggregate, any specific associations are going to be diluted by inclusion of unrelated subsets of cases. However, the logistical difficulties in subdividing miscarriages based on karyotype should be recognized, pinpointing a need for other more tractable approaches to meaningful classification.

Confounding is always a potential concern, but the present state of knowledge regarding risk factors for miscarriage limits specific candidates (Kline *et al.*, 1989). Only maternal age and prior miscarriage are clearly and persuasively related to risk of miscarriage, with maternal tobacco and alcohol use potential risk factors as well. In conclusion, the studies of paternal exposures and miscarriage are numerous but tend to be rather weak in design. Among the specific agents which have been examined, the evidence seems to be strongest for mercury, based on two recent, more rigorous studies. Also, anesthetic gases, in spite of no publications for over a decade, show consistent increased risks.

Solvents, pesticides, ionizing radiation, and hydrocarbons have also been implicated in one or more studies and are worthy of future attention as well. The challenge to extending this avenue of research is to identify opportunities to study large numbers of men with substantial and well-documented exposures who can be linked to carefully measured pregnancy outcomes. Unfortunately, elegance in exposure assessment or pregnancy outcome measurement alone does not compensate for deficiencies in the other.

PATERNAL EXPOSURES AND STILLBIRTH

Limited attention has been devoted to a potential paternal role in stillbirth, perhaps because of the perception that maternal factors rather than fetal development are responsible. Nonetheless, a few studies have provided some suggestive associations linking stillbirth to paternal employment in copper smelting (Beckman & Nordstrom, 1992), cotton production (Rupa *et al.*, 1991), textile employment (Hartz & Jacques, 1978; Savitz *et al.*, 1989), paper and wood, art, and chemicals (Savitz *et al.*, 1989). As noted with respect to miscarriage, the etiologic heterogeneity of stillbirth also poses a major challenge to identifying any paternal contribution.

PATERNAL EXPOSURES AND BIRTH WEIGHT, GESTATIONAL AGE, AND FETAL GROWTH

The literature on paternal exposures and indices of infant growth and maturity are limited to analyses of three surveys that contained data on paternal occupation: the Washington State Birth Certificates (Daniell & Vaughan, 1988); the 1980 National Natality Survey, a national probability sample of US live births (Savitz *et al.*, 1989); and the survey of births in Scotland from 1981 to 1984 (Sanjose *et al.*, 1991). The birth outcome information in these sources should be of reasonably high quality, but in each instance paternal exposures were limited to inferences based on a job title alone.

Table 6 summarizes associations observed in those three studies. It is difficult to pinpoint those industries and agents most deserving of additional research given the limitation in the occupational exposure data. Nonetheless, textiles emerge as an industry of particular concern as well as solvents related to work in the rubber and plastics industries, ceramics, and glass, clay, and stone industries.

Table 6. Suggested Occupational Associations of Paternal Exposure and Infant Health

Birth Weight	Gestational Age	Fetal Growth
Auto body shop painting	Ceramics	Plastics
Construction painting	Textiles*	Textiles*
Ceramics	Glass, clay, stone	Art
Rubber	Mining	Benzene
Textiles	Rubber, plastics	Pesticides
	X-rays	
	Polyvinyl alcohol	

*Observed in more than one study.

Several methodological issues are specific to this set of endpoints. In contrast to the situation with miscarriage, where the basis for subdividing the endpoint is theoretically or logistically limited, suggestions can be made for subdividing indicators of infant size and development. At a minimum, preterm births should be separated from small size, with the latter indicated by such measures as "small for gestational age." Even among preterm births, the pathways of preterm labor, preterm premature rupture of the membranes, and medically indicated preterm delivery due to complications should be distinguished (Savitz *et al.*, 1991). Among preterm deliveries, the fetus (and hence, the father) is believed to primarily influence preterm labor, so that any paternal influence should be concentrated on that pathway as well. Maternal factors presumably determine the integrity of the chorioamnionic membranes and the likelihood of developing complications.

The obvious need is for combining sophisticated exposure assessment with detailed indicators of pregnancy outcome. Medical records contain sufficient data to classify the birth outcome in some detail so that data on cohorts of men with information on exposure could, in principle, be linked to the medical records concerning the birth. Conversely, starting with good birth outcome data, detailed interviews with the fathers could provide information on the lifestyle and occupational factors of interest. These avenues have not yet been pursued.

FUTURE DIRECTIONS FOR STUDIES OF PATERNAL EXPOSURE AND PREGNANCY OUTCOME

Based in part on the empirical observations reviewed above and on personal judgments, some directions for future studies in this area will be offered. Underlying these suggestions is the belief that the research generically is worthy of pursuit. The possibility of a paternal influence on fetal viability and growth is clear based on laboratory research and suggestive epidemiologic findings, but more sophisticated studies are needed to assess whether the phenomenon is actually operative in humans.

The emphasis on occupational exposures over other "lifestyle" factors is perhaps excessive. More sophisticated studies of such intense lifestyle exposures as tobacco, alcohol, medications, illicit drugs, and diet are needed. Relative to typical occupational exposures, the intensity of these behavioral exposure sources is markedly greater and potentially more influential, yet very little research has been done to address them. In addition, intense occupational exposures could be studied. Such conditions rarely exist in the United States, but developing countries and parts of Eastern Europe may have sizable workforces exposed to more extreme levels of toxic agents. If well-designed studies of such men yielded no evidence of adverse pregnancy outcomes, then the markedly lower levels experienced elsewhere would be exonerated as well.

Finally, there is a real opportunity to work backwards from data sources with detailed outcome data to obtain paternal exposures in some detail. Numerous examples of linkage to a job title can be found in the literature, but through detailed assessment of self-reported data, occupational and lifestyle factors can be evaluated in much greater detail than they have in past studies.

Pursuit of more rigorous studies is guaranteed to be informative and advance our understanding of whether or when the hypothesized phenomenon occurs. Reducing misclassification and subdividing exposures and outcomes into more homogeneous groups will either indicate an absence of associations with more certainty or identify specific and presumably more pronounced associations if etiologic relations are truly present. The justification and methods for pursuing such studies are apparent.

REFERENCES

Alcser, K.H., Brix, K.A., Fine, L.J., Kallenbach, L.R., and Wolfe, R.A., 1989, Occupational mercury exposure and male reproductive health. *Am. J. Ind. Med.* 15:517-29.

Beckman, L., and Nordstrom, S., 1982, Occupational and environmental risks in and around a smelter in northern Sweden. *Hereditas* 97:1-7.

Boue, J., Boue, A., and Lazar, P., 1975, Retrospective and prospective epidemiological studies of 1500 karyotyped spontaneous human abortions. *Teratology* 12:11-26.

Brodsky, J.B., Cohen, E.N., Whitcher, C., Brown, B.W., and Wu, M.L., 1985, Occupational exposure to mercury in dentistry and pregnancy outcome. *JADA* 111:779-780.

The Centers for Disease Control Vietnam Experience Study, 1988, Health status of Vietnam veterans, III. Reproductive outcomes and child health. *JAMA* 259:2715-2719.

Cohen, E.N., for the Ad Hoc Committee on the Effect of Trace Anesthetics on the Health of Operating Room Personnel, American Society of Anesthesiologists, 1974, Occupational disease among operating room personnel: A National Study. *Anesthesiology* 41:321-340.

Cohen, E.N., Brown, B.W., Bruce, D.L., Cascorbi, H.F., Corbett, T.H., Jones, T.W., and Whitcher, C.E., 1975, A survey of anesthetic health hazards among dentists. *JADA* 90:1291-1296.

Cohen, E.N., Brown, B.W., Wu, M.L., Whitcher, C.E., Brodsky, J.B., Gift, H.C., Greenfield, W., Jones, T.W., and Driscoll, E.J., 1980, Occupational disease in dentistry and chronic exposure to trace anesthetic gases. *JADA* 101:21-31.

Cordier, S., Deplan, F., Mandereau, L., and Hemon, D., 1991, Paternal exposure to mercury and spontaneous abortions. *Br. J. Ind. Med.* 48:375-381.

Daniell, W., and Vaughan, T.L., 1988, Paternal employment in solvent related occupations and adverse pregnancy outcomes. *Br. J. Ind. Med.* 45:193-197.

Halmesmaki, E., Valimaki, M., Roine, R., Ylikahri, R., and Ylikorkala, O., 1989, Maternal and paternal alcohol consumption and miscarriage. *Br. J. Obstet. Gynecol.* 96:188-191.

Hartz, S.C., and Jacques, P.F., 1978, "Occupational Health Surveillance of Reproductive Outcomes: A Case-Control Study," NIOSH Contract No. 210-78-0057, National Institute for Occupational Safety and Health, Cincinnati, Ohio.

Infante, P.F., Wagoner, J.K., McMichael, A.J., Waxweiler, R.J., and Falk, H., 1976, Genetic risks of vinyl chloride. *Lancet* I:734-736.

Kline, J., Stein, Z., and Susser, M., 1989, "Conception to Birth. Epidemiology of Prenatal Development," Oxford University Press, New York.

Lindbohm, M.-L., Sallmen, M., Anttila, A., Taskinen, H., and Hemminki, K., 1991, Paternal occupational lead exposure and spontaneous abortion. *Scand. J. Work Environ. Health* 17:95-103.

McDonald, A.D., McDonald, J.C., Armstrong, B., Cherry, N.M., Nolin, A.D., and Robert, D., 1989, Father's occupation and pregnancy outcome. *Br. J. Ind. Med.* 46:329-333.

Rupa, D.S., Reddy, P.P., and Reddi, O.S., 1991, Reproductive performance in population exposed to pesticides in cotton fields in India. *Environ. Res.* 55:123-128.

Savitz, D.A., Blackmore, C.A., and Thorp, J.M., 1991, Epidemiology of preterm delivery: etiologic heterogeneity. *Am. J. Obstet. Gynecol.* 164:467-471.

Savitz, D.A., Sonnenfeld, N.L., and Olshan, A.F., 1992, Review of epidemiologic studies of paternal occupational exposure and spontaneous abortion. *Am. J. Ind. Med.* (in press).

Selevan, S.G., 1985, Design of pregnancy outcome studies of industrial exposures. In K. Hemminki, M. Sorsa, and H. Vainio (editors), "Occupational Hazards and Reproduction," Hemisphere Publishing Company, Washington, D.C., pp. 219-229.

Stellman, S.D., Stellman, J.M., and Sommer, J.F., 1988, Health and reproductive outcomes among American Legionnaires in relation to combat and herbicide exposure in Vietnam. *Environ. Res.* 47:150-174.

Taskinen, H., Anttila, A., Lindbohm, M.-L., Sallmen, M., and Hemminki, K.,1989, Spontaneous abortions and congenital malformations among the wives of men occupationally exposed to organic solvents. *Scand. J. Work Environ. Health* 15:345-352.

Tomlin, P.J., 1979, Health problems of anesthetists and their families in the West Midlands. *Br. Med. J.* 1:779-784.

Wilcox, A.J., Weinberg, C.R., O'Connor, J.F., Baird, D.D., Schlatterer, J.P., Canfield, R.E., Armstrong, E.G., and Nisula, B.C., 1988, Incidence of early loss of pregnancy. *N. Engl. J. Med.* 319:189-194.

Wilcox, A.J., and Horney, L.F., 1984, Accuracy of spontaneous abortion recall. *Am. J. Epidemiol.* 120:727-733.

Windham, G.C., Swan, S.H., and Fenster, L., 1992, Parental cigarette smoking and the risk of spontaneous abortion. *Am. J. Epidemiol.* 135:1394-1403.

PATERNAL EXPOSURES AND EMBRYONIC OR FETAL LOSS: THE TOXICOLOGIC AND EPIDEMIOLOGIC EVIDENCE

Jennifer M. Ratcliffe

Department of Epidemiology
The School of Public Health
McGavern-Greenberg Hall, CB# 7400
University of North Carolina
Chapel Hill, NC 27599-7400

INTRODUCTION

The purpose of this review is to critically examine the toxicologic and epidemiologic evidence for a paternal contribution to embryonic and fetal loss (hereafter mainly referred to simply as "fetal loss") due to preconceptual exposure of the male to chemical and physical agents. The starting point for this review is the fertilized ovum; effects of agents on semen quality or the capacity for fertilization will not be reviewed. The putative contribution of paternal factors to adverse reproductive outcomes such as fetal loss in humans has, until recently, received little attention, and few attempts have been made to correlate toxicologic data from experimental animals with epidemiologic data. Further, the mechanisms by which preconceptual paternal exposures may affect development of the embryo have not been clearly elucidated to date, in part because, although techniques for characterizing damage to paternal chromosomes, genes and DNA are at a relatively advanced stage of development (National Research Council 1989), other e.g. epigenetic or non-genetic mechanisms have yet to be established.

Hitherto, much of the epidemiologic data on factors affecting loss of the embryo or fetus has focused almost exclusively on maternal risk factors. This review hopes to demonstrate that paternal exposures also have a known role in this outcome, at least with respect to genetically-mediated effects. This knowledge may significantly alter our approach both to the conduct of epidemiologic studies, so that both male and female risk factors are taken into account, and to our interpretation of the role of maternal risk factors.

In the review that follows, toxicologic evidence for a relationship between paternal exposures and fetal loss is discussed first, starting with a background section briefly describing and discussing methodologic issues and ending with a discussion of animal evidence on possible mechanisms of effect. The agents given as examples of experimentally-induced fetal loss have been selected primarily on the basis of their

Male-Mediated Developmental Toxicity, Edited by D.R. Mattison
and A.F. Olshan, Plenum Press, New York, 1994

potential for significant **human** exposure either as pharmacologic agents (medical and recreational), or as environmental or occupational hazards; the list is not intended to be comprehensive. The epidemiologic data are described next, starting with a brief introduction to what is known generally about the etiology of fetal loss in humans and a discussion of methodologic issues. The epidemiologic literature is relatively sparse; in the majority of studies, occupations are the object of study rather than specifically defined exposures. Finally, a discussion of the gaps in the toxicologic and epidemiologic literature is presented, and some of the biologic and methodologic issues which need to be addressed in future studies are identified.

TOXICOLOGIC STUDIES OF PATERNAL EXPOSURES AND FETAL LOSS

Virtually all the available animal data on paternally-mediated fetal loss come from the study of **dominant lethal mutations** in rodents. Rieger *et al.* (1991) defined a dominant lethal mutation as "any mutation (mainly structural or numerical chromosome aberrations) which kills an individual heterozygous for it." The dominant lethal assay was developed approximately 20 years ago as a method for determining the mutagenicity of chemical agents and not primarily as a method for studying paternal exposures and fetal loss [see, for example, Ehling *et al.* 1968, Epstein and Rohrborn (1971) and Bateman (1973)]. A number of parameters are measured in the assay (Table 1). The usual approach to the assay involves acute or subacute treatment of the male (often administering a single dose), followed by mating the treated male with a series of untreated females either for an entire spermatogenic cycle (e.g. for 8 weeks post-exposure in mice or 10 weeks in rats), or on specific days after exposure (Green *et al.* 1987). This allows the stage(s) of spermatogenesis which have been affected to be determined (Table 2). For example, matings in 1-2, 3-4, 5-7 and 5-9 after treatment represent respectively post-meiotic, meiotic and pre-meiotic stages of spermatogenesis in the mouse; corresponding periods in the rat are 1-4, 5-6 and 7-9 weeks. Pre- and/or post-implantation losses are determined after sacrificing the mated, untreated females usually at 13 to 17 days of the three week pregnancy. The difference between the number of corpora lutea and implantation sites can be scored to calculate pre-implantation losses/fertilization failures. [Using the *in vivo* assay as described above, pre-implantation loss due to a dominant lethal effect in the male cannot be distinguished from reduced fertilization or other, non-genetic causes of early embryonic death (Rohrborn 1968; Bateman 1973). This can be done with a less widely used in vitro technique which measures the ability of a two-cell stage embryo to develop to the trophectoderm stage (Goldstein and Spindle, 1976), or by sacrificing groups of females on the second day of gestation, recovering ova from the oviducts and determining whether cleavage has occurred (Kratochvilova 1978). Resorption sites and live and dead fetuses are scored to obtain post-implantation losses. A check on mating frequency and numbers of pregnant females is kept to avoid biasing results because of sterility or changes in mating behavior.

The principal characteristics of dominant lethal mutations in the rodent are, first, that they result in embryonic death either at an early cleavage stage so that implantation does not occur, or soon after implantation; for most of the chemicals thus far investigated, fetal losses do not appear to occur at a later stage (M. Shelby, personal communication). Secondly, different germ cell stages usually show different sensitivities to the agent the male is treated with (see paper by L. Russell, this

Proceedings). For example, many cancer chemotherapy drugs selectively affect post-meiotic stages; the chemotherapy drug 6-mercaptopurine, however, primarily affects differentiating spermatogonia (Generoso *et al.* 1975). The majority of environmental agents appear to selectively affect post-meiotic sperm stages in mice and rats, according to Shelby *et al.* (in press). Finally, there may be species and strain differences in response to a given agent; for example, rats exhibit a positive dominant lethal response to dibromochloropropane but mice exhibit a negative response (Rao *et al.* 1983, Generoso *et al.* 1985).

The main advantages of the dominant lethal assay are that it is comparatively easy to conduct and simple and unambiguous to score (at least for post-implantation losses); in the rodent, the assay is of relatively short duration; if attention is paid to study design issues such as sample size (Generoso and Piergorsch, in press), it appears to be a relatively sensitive test of germ cell embryolethal mutagenicity without requiring large numbers of animals; it permits the stage(s) of spermatogenesis affected to be identified; and it permits dose-response relationships to be established. The largest disadvantage of the assay is, arguably, that since the assay was developed as a mutagenicity test and not primarily as a method for investigating paternally-mediated fetal loss, the question of whether the rodent is an adequate model for fetal loss in humans and paternal exposures has not, thus far, been addressed. For example, although the stages of spermatogenesis in humans are the same as those in the rodent, it cannot be assumed that they will demonstrate the same stage-specific sensitivity as rodent sperm. Of equal importance are the considerable differences between rodent and human **female** reproductive physiology which may critically affect fertilization and the embryolethal response to a given male germ cell mutagen.

Table 1. Principal Parameters Measured in the Dominant Lethal Assay

Mating frequency
Proportion of pregnant females
Number of corpora lutea/female
Total implants/female
Proportion of pregnant females + dead implants
Number of "resorptions" or deciduomata/female
Number of dead embryos/female
Number of live embryos/female

Finally, it should be noted that, theoretically at least, a sperm-mediated effect may not be distinguishable from an effect due to transmission of an agent to the egg in seminal fluid or attached to the sperm. While many substances do pass into seminal fluid (Mann and Lutwak-Mann 1982), it is not known what if any reaches the egg and whether an embryolethal effect would ensue.

b) Indirect effects on chromosomes or genes (e.g. via effects on sperm protamine)
c) Epigenetic effects affecting gene expression.

Direct Effects on Chromosomes or Genes

Embryonic or fetal loss following male treatment is usually considered to be the result of induced whole chromosome loss, chromosome translocation, or breakage (clastogenicity) at one or more stages of spermatogenesis. In animals, evidence for the cytogenetic basis of embryolethality rests largely on the work of authors such as Brewen *et al.* (1975), who treated male mice with the experimental mutagen methyl methanesulfonate followed by serial mating. Cytogenetic analyses of first cleavage metaphases showed chromosome aberrations, including shattered chromosomes, double fragments, chromatid interchanges and chromatid deletions. The sensitivity of a given spermatogenic stage to chromosomal damage correlated closely with the proportion of pre- and post-implantation deaths reported by Ehling at different times of mating after dosing with the same agent (Ehling 1977).

One of the mechanisms by which numerical chromosome errors can occur is by interference with spindle formation during meiosis. For example, the drugs vincristine, vinblastine, and colchicine, which, in addition to their ability to directly damage DNA in sperm (Zhang and Sun 1992), can bind to tubulin, which is the protein in microtubules that result in the formation of the spindle during cell division, and thereby interfere with normal chromosome and chromatid segregation (Singer and Himes 1992). Only 10 compounds have been identified as known or suspect spindle poisons to date, but this may be due to, until very recently, the lack of an adequate test system (Miller and Adler 1992).

At the molecular level, one of the principal mechanisms by which chromosome aberrations may be produced is considered to be by alkylation of certain target sites on the DNA. Generoso *et al.* (1984) have postulated that, in post-meiotic germ cells, dominant lethal aberrations can be produced in several ways. First, adducts may be formed primarily of the N-7 position of guanine, and to a lesser extent the N-3 position of adenine. These adducts are subsequently lost by hydrolysis, leading to the formation of intermediate lesions. At fertilization these are converted to chromosome exchanges or deletions before pronuclear DNA synthesis and the first chromosome replication; these may be lethal or produce heritable translocations. Methylnitrosurea and ethylene oxide are examples of chemicals which produce such adducts. Other DNA binding sites associated with dominant lethal mutations are the oxygen of the phosphate backbone, (which forms alkylphosphotriesters with alkylating agents) and the oxygen in bases such as the O-6 position in guanine (which forms e.g. alkylguanine products). If these alkylation products are not repaired, they persist up to the first post-fertilization pronuclear DNA synthesis and chromosome replication, whereupon they result in chromatid and isochromatid deletions which are lethal. Ethylnitrosurea and benzo(a)pyrene are examples of chemicals which primarily produce dominant lethals by this mechanism (Generoso *et al.* 1984).

Although the evidence that dominant lethal effects resulting from chromosome aberrations due to alkylation is extensive, alkylation may not be the only direct mechanism involved in producing aberrations nor is chromosomal damage necessarily the only direct mechanism. The possibility that a dominant lethal effect might be caused by a direct non-clastogenic effect on sperm DNA, such as point mutations or microdeletions, cannot be ruled out on theoretical grounds. It will probably not be possible to resolve this question until sequencing of sperm DNA becomes more technically feasible.

Table 2. Serial Mating in the Dominant Lethal Assay and Stages of Spermatogenesis Tested in the Mouse

Week of Mating	Germ Cell Stage Tested	Comment
1	Epididymal sperm	
2	Late spermatids	
		No more DNA repair
3	Early spermatids	
4	Secondary spermatocytes	
		Meiosis
		Last DNA synthesis
5	Primary spermatocytes	
6	Differentiating spermatogonia	
7	Stem cell spermatogonia	

Mouse spermatogenesis = 45 - 50 days
Rat spermatogenesis = 48 - 63 days
Human spermatogenesis = 64 - 72 days

In addition to the measurement of fetal loss itself following exposure of the male, animal and, if available, human data on the pharmacokinetics and metabolism of a given agent are important in determining the potential for toxicity, the relevant route(s) of exposure and the relevant time period of exposure in relation to the time of fertilization. Such data must also be taken into account in determining whether extrapolation across species is appropriate. Consideration of the bioavailability of the agent to the germ cells (e.g. does it readily cross the blood-testis barrier?), the half life of the agent in the body, the disposition of the agent in different body compartments and whether the parent compound or a metabolite is the active agent can be used to estimate, for example, what the dose to the gonad is over a given time period. For agents that are quickly metabolized, such as cocaine or methadone, this may be a few hours following a single exposure. For agents which may be accumulated in the body, the gonad may continue to be exposed after the cessation of exposure, depending on the half life and bioavailability of the agent. (Polychlorinated biphenyls, for example, are stored in body fat but are not released from this storage depot, in contrast to heavy metals such as lead which are, albeit slowly, released from storage depots such as bone after cessation of exposure).

Examples of pharmaceutical and recreational drugs, and occupational and environmental agents that have been shown to produce fetal loss after paternal exposure in at least one rodent species are given in Table 3.

PATERNAL EXPOSURES AND FETAL LOSS: POSSIBLE MECHANISMS

Theoretically, there are a number of ways in which fetal loss might be induced by xenobiotic agents:

a) Direct effects on chromosomes (aneuploidy or clastogenicity) or genes;

Table 3. Dominant Lethal Studies of Paternal Exposures and Fetal Loss: Examples of Agents

CANCER CHEMOTHERAPY DRUGS	
	e.g. cyclophosphamide, busulfan, vincristine, adriamycin, 6-mercaptopurine
OTHER PHARMACEUTICAL DRUGS	
	e.g. chloramphenicol, ergotamine
RECREATIONAL DRUGS	
	e.g. cannabinoids, ethanol
PESTICIDES	
	e.g. dibromochloropropane, captan
INDUSTRIAL CHEMICALS	
	e.g. acrylamide, trimethylphosphate, ethylene oxide, chloroprene
PHYSICAL AGENTS	
	e.g. ionizing radiation

Indirect Effects on Chromosomes or Genes

Alteration of chromatin condensation. Condensation of the sperm chromatin during spermiogenesis is the function of the conversion of histone proteins into sperm-specific protamines, which, by means of the sulfhydryl groups they possess, form disulfide bonds to condense the DNA. For at least certain agents, such as ethylene oxide and acrylamide, alkylation of cysteine sulfhydryl groups in sperm protamine was correlated with the germ cell stage known to be most sensitive to the induction of dominant lethals (late spermatids - early spermatozoa) (Sega and Owens 1978, Sega 1991). These authors postulated that blocking of the normal chromatin condensation in the nucleus, thereby rendering the chromosomes vulnerable to breakage, was a primary mechanism of causing dominant lethality.

Alteration of DNA repair capacity. A variety of repair enzymes exist which can recognize and repair damage to double stranded DNA. Sega (1980) has described how unscheduled DNA synthesis (UDS), indicating DNA repair, occurs in certain sperm stages after treatment with alkylating agents. Some DNA repair is, however, error prone, and the repair enzymes may in some cases be inhibited or saturated. Zhang and Sun (1992) for example, found that the amount of UDS produced by administration of vincristine to male mice decreased at higher doses, suggesting some inhibition of DNA repair capacity. This may result in the persistence of dominant lethal gene mutations. Further, certain commonly used compounds, such as caffeine and alcohol, have been found to act as inhibitors of DNA repair enzymes at least in rodents (Garro *et al.* 1986, Matsuda and Tobari 1989) and may exert a dominant lethal effect via this mechanism.

Epigenetic Effects on Gene Expression

While there is no direct evidence at present that certain alkylating agents might produce their embryolethal effects epigenetically by alteration of gene expression, as well as by direct mutational activity, this is theoretically possible and awaits investigation. In this context, the recent discovery of imprinting may be relevant. The parental genomes inherited from the sperm and egg are not functionally equivalent during development, due to a group of genes whose expression differs according to whether they are of paternal or maternal origin; both genomes are required for normal embryonic development (Sapienza 1990). The process by which certain genes are "marked" differently in the maternal and paternal genomes is referred to as imprinting, and is thought to be due to an epigenetic mechanism such as DNA methylation, which alters gene expression (Sanford *et al.* 1987, Sapienza 1990). Such findings suggest that a xenobiotic agent with alkylating properties might cause embryonic death by interfering with the normal DNA methylation pattern involved in the imprinting process and in subsequent paternal gene expression.

EPIDEMIOLOGIC STUDIES OF PATERNAL EXPOSURES AND FETAL LOSS

In humans, at least 50% of all conceptions are estimated to result in fetal loss, of which approximately 50% of early losses (before 16 weeks) and 25% of losses after 16 weeks are estimated to be chromosomally abnormal, although no data are available for very early losses before about 6 weeks (Warburton *et al.* 1980). Among fetal losses that can be karyotyped (which excludes fetal tissue that cannot be successfully cultured as well as early losses which are unrecognized or which do not occur in a setting where fetal tissue can be collected), most of the observed chromosomal abnormalities are numerical; less than 5% were structural in several series (see review by Warburton *et al.* 1980). Not all aneuploidies are fatal; some of the autosomal trisomies and the sex chromosome aneuploidies are usually viable. The relative proportions of numerical and structural chromosome aberrations appear to differ from that observed in rodents. This difference may, however, be artifactual. In the cytogenetic analyses of first cleavage metaphases in rodent studies, numerical anomalies are not scored as they frequently are a result of the preparation process (J.B. Bishop, personal communication). Further, as noted above, successful techniques for scoring aneuploidies have only recently been developed (Miller and Adler 1992). In humans, the difficulty of recovering and culturing early fetal losses may mean that structural aberrations, along with monosomies, are lost too early to detect. Although in cases of chromosomally abnormal fetal loss of genetic origin the relative maternal v. paternal contribution can be established, the proportion of fetal losses due to de novo chromosome aberrations in sperm is presently unknown. Further, 50% of chromosomally normal fetal losses are morphologically abnormal, but again, the potential contribution of a sperm-mediated, as opposed to a teratogenic, effect is unknown.

Methodologic Issues

Measurement of embryonic or fetal loss. It has not been established whether paternally-mediated embryonic or fetal loss in humans would show the same characteristics as described above for rodents, but there would be clear biological and methodological implications if this were the case. If most embryolethal agents acting through the male produce early loss before or soon after implantation, as in the rodent, the study of clinically recognized fetal loss, which accounts for only 10 to 15% of all conceptions (Bloom 1981), may considerably underestimate or miss a potential effect. Early loss of the conceptus can be directly detected only by prospectively measuring human chorionic gonadotrophin (Wilcox *et al.* 1988), and this almost certainly misses pre-implantation losses (Baird *et al.* 1991). Time to recognized conception can be measured prospectively or retrospectively, but clearly cannot distinguish between impaired fertility and early, unrecognized embryonic or fetal loss. To date, all the cohort studies of paternal exposures and fetal loss have been retrospective (and case-control studies are by definition retrospective), so that there are no direct data on the role of paternal factors and early fetal loss to date.

Measurement of exposure. Determining the relevant time period of exposure to measure is critical to the investigation of an exposure-effect relationship. If only post-meiotic sperm stages were affected in humans by certain agents, as observed in rodents, this would mean that only the spermatogenic cycle exposed before conception would be relevant for study. Further, investigating a history of fetal loss after the cessation of exposure (with the exception of cumulative toxins which also continue to be bioavailable to the testis) would be inappropriate and may yield false negative findings. If, however, stem cell spermatogonia were genetically damaged, leading to a viable germ cell line carrying dominant lethal mutations (which is theoretically possible, although it has not been clearly demonstrated in an animal model to date) this would affect the remainder of the individual's reproductive lifetime. In such cases, the study of fetal loss at any time after cessation of exposure would be appropriate. (It should also be noted, however, that selection against such a germ cell population might be expected to occur resulting in a very low frequency in mature sperm).

In retrospective studies, it is frequently more difficult to reconstruct exposure histories than in prospective studies, and, unless actual industrial hygiene data are available, it may not be possible to investigate dose-response relationships, except perhaps by crude exposure categories. Further, multiple exposures are considerably more common than single exposures, particularly in the occupational setting. This not only makes the etiology of any observed association between paternal occupation and fetal loss more difficult to establish, but also means that direct comparisons with toxicologic data cannot be made (see below). A further exposure-related problem is that, particularly in the case of certain industrial exposures, the father may expose his partner via contact with contaminated skin or clothing. It may then not be possible to clearly distinguish between a sperm-mediated (or seminally transmitted) effect on the conceptus from an effect due to secondary exposure of the mother.

Sample size considerations. Large numbers of person years (of women of reproductive age) are required in cohort studies to detect an effect on fetal loss with adequate statistical power. [For example, 100 pregnancies per comparison group are required to detect a 2.4 fold increase in the frequency of clinically recognized fetal loss with 95% power, assuming a one-sided alpha of 0.05 and a frequency of 15% in the control group (Bloom 1981). The current birth rate for all U.S. women of childbearing age (15 to 44 years) is 67/1000 women (U.S. Bureau of the Census 1991); again assuming a 15% fetal loss rate, approximately 1298 person years must be included in each group to yield 100 pregnancies.] Sample sizes may be limited by the size of the available cohort and the participation rate, in which case the study design may need to be reconsidered. A similar problem in case-control or registry-based studies is that, unless a common exposure is under investigation (such as alcohol use), or the study has been specifically designed to include potential exposure to a given agent in both cases and controls (e.g. performing a nested case-control study of fetal loss within a cohort study) it may not be possible to find enough subjects with a given exposure to yield adequate statistical power.

Other, general methodologic issues that need to be considered in epidemiologic studies include adequate consideration of potentially confounding factors, (including for example maternal risk factors), participation bias, the possibility of recall error and bias in remembering past events such as exposures and risk factors, and the fact that referent groups may be exposed to agents that could also affect reproduction.

Table 4. Fetal Loss in Epidemiologic Studies by Occupation*

Floriculture workers	+
Petroleum foundries (+ solvents)	+
Mfr of rubber products (+solvents)	+
"Solvent-exposed jobs"	+
Waste water treatment	+
Motor vehicle mechanics	+
Dentists	+
Metal plate workers	+
Auto fume exposure	+
Professors, lawyers	+
Stainless steel welders	+
Elemental mercury workers	+/-
Auto body shop workers	+/-
Agricultural pilots	-
Painters (construction)	-
Fiberglass workers	-
Mfr of halogenated hydrocarbons	-
Operating room staff	-
Dry cleaning workers	-
Printers	-

*A complete list of references for the above studies is available from the author on request.
+ or - : paper's authors report positive (or negative) association; +/- =both positive and negative studies found.

Table 5. Toxicologic and Epidemiologic Studies of Paternal Exposures and Fetal Loss*

	Species		
Agent	Rat	Mouse	Human
Chloroprene	+	+	+
Lead compounds	-	+/-	+
Vinyl chloride	n	-	+/-
DBCP	+	-	+/-
2,4-D;2,4,5-T;TCDD#	-	-	+/-
Inorganic mercury compounds	n	-	+/-
Ethanol	+/-	+/-	+/-
Halogenated anesthetic gases	n	-	-
Cannabinoids	n	+/-	-

2,4-D = 2,4-dichorophenoxyacetic acid; 2,4,5-T = 2,4,5-trichlorophenoxyacetic acid; TCDD = 2,3,7,8-tetrachlorodibenzo-p-dioxin.
*A complete list of references to the above studies is available from the author on request.
+ or - : paper's authors report positive (or negative) association; +/- =both positive and negative studies found; n = no study found.

Examples of Studies

To date, most epidemiologic studies have investigated paternally-mediated fetal loss by occupation rather than by specific exposure (Table 4). Despite limitations, a number of studies suggest associations between paternal occupations and fetal loss. Few investigators have, however, attempted to determine which specific exposure(s) might be responsible for observed associations (see also below).

There are a relatively few epidemiologic studies of specific exposures, and, at least in some cases, it cannot be assumed that there are no other exposures present. Further, there are even fewer human studies for which animal data also exist, so that it is difficult and probably misleading to make interspecies comparisons based on a few examples. Examples of agents for which both toxicologic and epidemiologic data exist are shown in Table 5, which does not claim be comprehensive, but does show some differences in findings across species. Whether such differences are methodologic or biological remains to be established.

DISCUSSION

It is clear from the toxicologic evidence reviewed above that paternal exposure to a number of chemical and physical agents can result in an increased rate of fetal loss, at least in rodents. Animal studies have also demonstrated a number of mechanisms by which such effects could be mediated. Further, animal studies have permitted the study of dose-response relationships and the pharmacokinetics and metabolism of chemical compounds, which are required to determine the potential for gonadal toxicity, relevant routes and timing of exposure, and interspecies differences. There are also some suggestive (and in some cases replicated) epidemiologic findings of associations between paternal exposures or occupations and fetal loss which warrant further investigation. There are, however, significant gaps in both the toxicologic and epidemiologic literature, and there are also a number of outstanding biologic and methodologic questions which need to be addressed in future studies. With respect to gaps in the literature, there are a number of toxicologic studies which have demonstrated a marked effect of a paternal exposure on fetal loss and to which there is significant human exposure, but few or no epidemiologic studies have yet been conducted. Examples of such agents are cancer chemotherapy drugs, X-rays, ethylene oxide and acrylamide. Conversely, there are several epidemiologic studies which have suggested that certain paternal exposures may be associated with increased fetal loss, but for which no or insufficient toxicologic data exist. Examples are nitrous oxide, inorganic mercury, rubber chemicals, welding fumes, gasoline, and certain common solvents. Finally, there are a number of agents to which a large proportion of the male population may be exposed (often voluntarily) and which are thus of considerable public health significance, but for which neither adequate toxicologic nor epidemiologic data currently exist. Examples are alcohol, caffeine, cannabinoids, cocaine, methadone, tobacco smoke and lead. One of many reasons for such gaps may be a lack of familiarity with literature in fields other than one's own, and prompts the question of whether more interdisciplinary exchanges, meetings, and conferences should be instituted (see Multidisciplinary Issues workshop discussion, this Proceedings).

Of the issues that need to be addressed in future studies, some of the more critical include the following (in no particular order of importance): What are relevant exposures to study (based on e.g. mutagenicity, extent of human exposure etc.)? What are relevant routes of human exposure (inhalation, dermal or ingestion) and can these be modeled in animal studies? What is the relevant pre-conceptual period of exposure (based on pharmacokinetics and sperm stage(s) affected)? Is the rodent dominant lethal assay an adequate model for human paternally-mediated fetal loss (and how can we go about answering this question)? Which epidemiologic study designs are methodologically adequate (e.g. what are adequate sample sizes etc.)? Is the study of paternal exposures and early fetal loss technically feasible in humans? In epidemiological studies of occupations and fetal loss which have shown a positive association, which exposure(s) are the most likely etiologic agents (based on the toxicologic literature)? Can sperm-mediated effects be distinguished from effects due to transmission of an agent by seminal fluid or sperm, or, in the case of human exposures, due to contamination of the mother?

In conclusion, it is clear that the study of paternal exposures and fetal loss in humans is an emerging field with many, as yet, unanswered questions, that we have not yet determined the applicability of available animal models of paternally-mediated fetal loss to humans, and that there are a large number of agents to which the male population is exposed which have yet to be studied. The resolution of such questions, together with further toxicologic and epidemiologic studies of exposures, may also have a significant impact on women's reproductive health research if it emerges that certain risk factors for fetal loss that have hitherto been ascribed to the female have been done so erroneously.

REFERENCES

Baird, D.D., Weinberg, C.R., Wilcox, A.J., McConnaughey, D.R., Musey, P.I., and Collins, D.C., 1991, Hormone profiles of natural conception cycles ending in early, unrecognized pregnancy loss, *J. Clin. Endocrinol. Metab.* 72:793.

Bateman, A.J., 1973, The dominant lethal assay in the mouse, *Agents Actions* 3:73.

Bloom, A.D. (Ed.), 1981, Guidelines for studies of human populations exposed to mutagenic and reproductive hazards, March of Dimes Birth Defects Foundation., White Plains, N.Y.

Brewen, J.G., Payne, H.S., Jones, K.P., and Preston, R.J., 1975, Studies on chemically induced dominant lethality. I. The cytogenetic basis of MMS -induced dominant lethality in post-meiotic male germ cells, *Mutat. Res.* 33:239.

Ehling, U.H., 1977, Dominant lethal mutations in male mice, *Arch. Toxicol.* 38:1.

Ehling, U.H., Cumming,R.B., and Malling, H.V., 1968, Induction of dominant lethal mutations by alkylating agents in male mice, *Mutat. Res.* 5:417.

Epstein, S.S., and Rohrborn, G., 1971, Recommended procedures for testing genetic hazards from chemicals, based on the induction of dominant lethal mutations in mammals. *Nature* 230:45.

Garro, A.J., Espina, N., Farinati, F., and Lieber, C.S., 1986, Ethanol and the repair of deoxyribonucleic acid, *Alcohol Health Res. World*, 10:26.

Generoso, W.M., Cain, K.T., Cornett, C.C., and Cacheiro, N.L.A., 1984, DNA target sites associated with chemical induction of dominant lethal mutations and heritable translocations in mice. *Genet.* new Fr. Ont. Proc. Int. Congr. 15th 1:347.

Generoso, W.M., Cain, K.T., and Hughes, L.A., 1985, Tests for dominant lethal effects of 1,2-dibromo-3-chloropropane (DBCP) in male and female mice, *Mutat. Res.* 156:103.

Generoso, W.M., and Piegorsch, W.W., (in press), Dominant lethal tests in male and female mice, in: Methods in Reproductive Toxicology, Chapin, R.E., Heindel, J.J., eds., Academic Press, New York.

Generoso, W.M., Preston, R.J., and Brewen, J.G., 1975, 6-Mercaptopurine, an inducer of cytogenetic and dominant lethal effects in pre-meiotic and early meiotic stages of male mice, *Mutat. Res.* 28:437.

Goldstein, L.S., and Spindle, A.I., 1976, Detection of X-ray induced dominant lethal mutations in mice: An in vitro approach, *Mutat. Res.* 41:289.

Green, S., Lavappa, K.S., Manandhar, M., Sheu, C-J., Whorton, E., and Springer, J.A., 1987, A guide for mutagenicity testing using the dominant lethal assay, *Mutat. Res.* 189:167.
Kratochvilova, J., 1978, Evaluation of pre-implantation loss in dominant lethal assay in the mouse, *Mutat. Res.* 54:47.
Mann, T. and Lutwak-Mann,C., 1982, Passage of chemicals into human and animal semen: mechanisms and significance, *CRC Critical Rev. Toxicol.* 11:1.
Matsuda, Y. and Tobari, I., 1989, Repair capacity of fertilized mouse eggs for X-ray damage induced in sperm and mature oocytes, *Mutat. Res.* 210:35.
Miller, B.M. and Adler, I.-D., 1992, Aneuploidy induction in mouse spermatocytes, Mutagenesis 7:69.
National Research Council, 1989, Biologic Markers in Reproductive Toxicology, National Academy of Sciences, Washington, D.C.
Rao, K.S., Burek, J.D., John, J.A., Schwetz, B.A., Bell, T.J., Potts, W.J., and Parker, C.M., 1983, Toxicologic and reproductive effects of inhaled 1,2-dibromo-3-3 chloropropane in rats, *Fundam. Appl. Toxicol.* 3:104.
Rieger, R., Michaelis, A., and Green, M.M., 1991, Glossary of Genetics, Springer-Verlag, Berlin.
Rohrborn, G., 1968, Mutagenicity studies in mice. I. The dominant lethal method and the control problem, *Humangenetik* 6:345.
Sanford, J.P., Clark, H.J., Chapman, V.M., and Rossant, J., 1987, Differences in DNA methylation during oogenesis and spermatogenesis and their persistence during early embryogenesis in the mouse, *Genes Dev.* 1:1039.
Sapienza, C., 1990, Parental imprinting of genes. *Sci. Amer.* 263:52.
Sega, G., 1980, Relationship between unscheduled DNA synthesis and mutation induction in male mice, *Basic Life Sci.* 15:373.
Sega, G.A., 1991, Adducts in sperm protamine and DNA vs. mutation frequency, *Prog. Clin. Biol. Res.* 372:521.
Sega, G.A., and Owens, J.G., 1978, Ethylation of DNA and protamine by ethyl methanesulfonate in the germ cells of male mice and the relevancy of these molecular targets to the induction of dominant lethals, *Mutat. Res.* 52:87.
Shelby, M.D., Bishop, J.B., Mason, J.M., and Tindall, K.R., (in press), Fertility, reproduction and genetic disease: studies on the mutagenic effects of environmental agents on mammalian germ cells, *Environ. Health Perspect.*
Singer, W.D., and Himes, R.H., 1992, Cellular uptake and tubulin binding properties of four vinca alkaloids, *Biochem. Pharmacol.* 43:545.
Warburton, D., Stein, Z., Kline, J., and Susser, M., 1980, Chromosome abnormalities in spontaneous abortion: data from the New York City study, in: Human Embryonic and Fetal Death, Porter,I.H., Hook, E.B., eds., Academic Press, New York.
Wilcox, A.J., Weinberg, C.R., O'Connor, J.F., Baird, D.D., Schlatterer, J.P., Canfield, R.E., Armstrong, E.G., and Nisula, B.C., 1988, Incidence of early loss of pregnancy. *N. Engl. J. Med.* 319:189.
U.S. Bureau of the Census., 1991, Fertility of American Women: June 1991, U.S. Government Printing Office, Washington, D.C.
Zhang, Y., and Sun, K., 1992, Unscheduled DNA synthesis induced by the antitumor drug vincristine in germ cells of male mice, *Mutat. Res.* 281:25.

REPRODUCTIVE OUTCOMES AMONG MEN TREATED FOR CANCER

John J. Mulvihill

Department of Human Genetics
Graduate School of Public Health
University of Pittsburgh
Pittsburgh, Pennsylvania 15261

ILLUSTRATIVE CASE

In 1983, at the National Naval Medical Center (Bethesda, Maryland), I consulted on a patient who had metastatic colon cancer. He had received chemotherapy with 5-fluorouracil and methotrexate two and a half months and one month before conception and again around the time of conception of a pregnancy. In the fourth month of pregnancy, the obstetricians asked for genetic advice about this couple. The man's history was also unusual because his father had died of lymphoma, and ureteral transitional cell carcinoma had occurred in a brother and two paternal cousins. Ultrasound examination of the fetus and amniocentesis were recommended to monitor the pregnancy. The fetus was found to have a deletion in the long arm of chromosome 1. Was this chromosomal defect due to exposure of the man's germ cells to chemotherapy? The unusual circumstances might support such a conclusion but the mother had the same minor chromosomal deletion and was clinically normal; hence, the deletion was called a normal variant in this family. A normal baby was born, six weeks after his father's death.

WHY STUDY CANCER SURVIVORS?

The situation was a complicated and rare one, but one likely to increase in frequency in the United States. It can be calculated that there are some 42,000 men who are in reproductive years in the United States and are survivors of Hodgkin's disease, testicular cancer or childhood cancer, a renal transplantation or an autoimmune disorder that required cytotoxic therapy, like that given for cancer (Table 1). The therapies for men with cancer comprise several groups: physical agents (namely ionizing radiation), antimetabolites, alkylating agents, antibiotics, and alkaloids. The agents are potent mutagens since they are intended to interfere with DNA metabolism. They are often given not as a single agent, but rather as combinations. So, in contrast to the pure exposures of experimental investigations of

Male-Mediated Developmental Toxicity, Edited by D.R. Mattison
and A.F. Olshan, Plenum Press, New York, 1994

male reproductive toxicity, the human exposure to cancer treatment is complex. But, unlike other human exposures, the timing and dosage of exposure to cancer therapy can be precisely known. Moreover, dosage can be independently corroborated by medical records and occasionally by direct *in vivo* measurements, for example, with blood levels of DNA adducts.

Table 1. Males of Reproductive Age after Cancer Therapy

DISORDER	ANNUAL U.S. FREQUENCY
ALL CANCERS (<45 YEARS)	32,800
HODGKIN'S DISEASE	3,200
TESTICULAR CANCER	2,800
CHILDHOOD CANCER	1,600
TRANSPLANT RECIPIENTS	?1,300
IMMUNE DISORDERS	?

ENDPOINTS

The endpoints that can be studied are those mentioned throughout this volume. Some are clinical endpoints that do not require laboratory collaboration, such as spontaneous abortion, infertility, and birth defects, especially the so-called sentinel phenotypes. A sentinel phenotype is a clinical disorder or syndrome that occurs sporadically as a consequence of a single highly penetrant mutant gene that is a dominant trait of some frequency and low fitness and that is uniformly expressed and accurately diagnosable, with minimal effort at or near birth (Mulvihill and Cziezel, 1983). In *Mendelian Inheritance in Man* (McKusick, 1992), there are about 40 traits that meet this definition. Together they have a combined frequency around one in a thousand. So sentinel phenotypes, like achondroplasia or aniridia, are powerful endpoints, clinically relevant, almost certainly representing new mutation, and possibly attributed to environmental exposure. But, they are rare events and large numbers of persons have to be counted to observe them.

Other endpoints require laboratory collaboration: sperm counts, chromosomal abnormalities, and variations in proteins or nucleic acids in offspring. With many tests on single subjects, fewer exposed people are needed to achieve statistical power. But, the relevance of molecular abnormalities to the health of clinically normal persons is problematic.

There are many barriers to reproduction by men (and women) after cancer and its treatment. The focus of this chapter is the narrow issue of testicular impairment and germ cell mutation. However, reproductive potential of cancer survivors could be limited by the many concomitant associations of the experience of cancer in childhood and adolescence: impaired education, growth, and development, limited job opportunities and income, sterility per se, dismemberment, or fear and uncertainty about the future.

Many variables influence even a single measure of reproduction or reproductive capacity, even one as discrete as gonadotrophin levels. For example, young men were studied who had received cyclophosphomide, not for cancer but for immunologic renal disease in childhood (Rivkees and Crawford, 1988). In general, the frequency of gonadal dysfunction, as reflected by gonadotrophin levels in blood, increased with the dose of cyclophosphamide; but, it also varied with age at exposure or, more precisely, with the pubertal status. The post-pubertal testis is much more sensitive to sterilization by cyclophosphamide than the pre-pubertal testis, that is slightly impaired. Comparable data are not available for females, to my knowledge.

By contrast, there is a large male-female difference in reproductive outcomes among long-time survivors of Wilms' tumor of the kidney. In a network of seven pediatric hospitals (Li *et al.*, 1987), 30% of 114 pregnancies of women who survived Wilms' tumor and who underwent abdominal radiation had adverse outcomes, defined as fetal or neonatal death or birth weight under 2.5 kg. The rates of such outcomes among female survivors without abdominal radiation and among the wives of male survivors were 0 and 3%, respectively. A recent postal survey of the British general practitioners who treated survivors of cancer diagnosed in childhood between 1946 and 1977, identified 20 Wilms' tumor survivors (Hawkins and Smith, 1989). The birth weight of offspring of women who received abdominal radiation was 2584 g; among the wives of male survivors and women who did not undergo radiation, birthweight was 3146 g, a highly significant difference. In short, the offspring of men with Wilms' tumor did not suffer these consequences.

THE FIVE-CENTER STUDY

Design

To clarify several issues concerning late reproductive and other effects among cancer survivors, a team at the National Cancer Institute struck collaborations with five cancer registries around the nation in the late 1970s (Mulvihill *et al.*, 1987; Byrne *et al.*, 1987). The first goal was to assemble a cohort of long-term survivors of childhood and adolescent cancer. Cases had either histologically confirmed cancer or clinically diagnosed brain tumors. Diagnosis took place from 1945 through 1974, under the age of 20 years. Finally, survival of 5 years and attainment of 21 years were required by an arbitrary date of study cutoff.

These criteria produced a peculiar set of 2498 eligible cancer survivors. Because three decades of cancer treatment were included, half of the cases had surgery only; these represent a control group within the study itself that could be used to distinguish the effects of therapy from any determinants of cancer risk. One-third had radiotherapy only, and another third had some chemotherapy, often far less than the multi-agent chemotherapy that pediatric oncologists presently give. Twenty-one percent of study subjects were pre-pubertal boys and girls, so we were able to do some comparisons between pre-pubertal and post-pubertal exposures. The collected data included an interview about interval medical and social history, infertility, any ill health of the offspring and, importantly, consent for records so that certain medical events, such as cancer, birth defects, and infertility, could be documented by actual medical records or death certificates. Cases' permission was also gained to talk to their brothers and sisters, who were selected as controls in a ratio of two controls per one case.

Results

One early analysis of the large database addressed Wilms' tumor (Byrne *et al.*, 1988), specifically the 26 female and 21 male survivors and 77 of their siblings (as controls). The apparent excess of fetal deaths in females was due to five miscarriages in one woman with a bicornuate uterus. Again, female, but not male survivors had a significant excess (four-fold) of adverse, liveborn outcomes. One out of 37 offspring had Wilms' tumor: a female with unilateral Wilms' tumor had a male with bilateral tumors and also a ventricular septal defect (Mulvihill *et al.*, 1987). These findings were consistent with other studies and, in a sense, validated the study design.

A major analysis addressed infertility, measured as the time from first marriage to first pregnancy (Byrne *et al.*, 1987). There were big differences, by type of therapy, among the 595 male and 637 female cases. By the crude and the adjusted fertility rate, men had a greater loss of fertility than women, especially those men exposed to alkylating agents. The mechanism of infertility seemed to be germ cell aplasia, worse in men than women.

If fertility was maintained--and it was in many--was there any increase in genetic disease among the offspring? One measure of genetic disease is cancer. Of the 2308 offspring, seven cancers occurred (Mulvihill *et al.*, 1987). This number was no greater than expected either in the controls (cousins of the offspring) or in population estimates, in a study that had about an 80% power to detect the tripling. By inspection, most of the cases' offspring who had cancer had a known hereditary or familial type of cancer: two with retinoblastoma, one with multiple endocrine neoplasia, and one with Wilms' tumor. So, there seemed to be no excess of cancer apart from the known hereditary and familial syndromes.

As another measure of genetic disease, we arbitrarily defined "genetic disease" as a known or probable cytogenetic syndrome, a single gene trait, or one of 15 simple malformations tracked by the U.S. Centers for Disease Control. To detect what could represent a mutation (in contrast to possible genetic disorders already in the family), we established that "sporadic" occurrences of genetic disease had to have no affected relatives with a similar genetic disease and that the other group, "familial," had a similarly affected relative. We scored any recessive disorder as "familial," because we assumed both parents were carriers and neither had a new mutation from germ cell exposure to the cancer treatments. Offspring achieved an average of 11.5 years, and 75 (3.4%) of them had a genetic disease by our definition. In comparison, the frequency in cousins, the offspring of sibling controls, was 2.8%, not a statistically significant difference.

A few so-called sentinel phenotypes occurred, but almost all of them were already in the family. One control offspring had a sporadic genetic disease: a congenital cataract, but we could not be sure it was a new dominant mutation. One child of a case survivor had albinism; by a stretch of reasoning, he could represent a new mutation, under the assumption that the disorder was X-linked and the mother had been well examined by an ophthalmologist and found to have normal retinas. These assumptions are unlikely; more credible is the interpretation that the child's disorder is autosomal recessive and that both parents inherited and passed on the mutant allele. In short, there was no clear instance of a sentinel phenotype in the offspring of our cancer survivors. As a nested case-control study, the parents of individuals with sporadic genetic disease had "potentially mutagenic therapy" no more often than the parents of the larger number of offspring without genetic disease. By chi square, there is no significant difference in a study that had an 87% power to detect a doubling.

The Male Experience

The preceding preliminary analyses on both sexes have been presented only as an abstract (Mulvihill *et al.*, 1986), but additional tabulations have been done that allow inspection of the data on males only. The frequency of genetic disease in the offspring of males was 2.6%, lower than the frequency in offspring of females (4.0%). The frequency of fetal deaths was 10.7% among pregnancies by male survivors and 11.0% among pregnancies of female survivors. The sex ratio in offspring of males was 1.04 (male:female), compared to 1.01 in offspring of females (and 1.00 and 1.04 respectively, in offspring of controls). The sex ratio is used to examine the possibility of X-linked lethal mutuations in human beings.

Table 2. Sex Ratio (M:F) of Offspring (Hawkins 1991)

Offspring of:		Number	Survivors	Controls
Males	first born	46	0.53*	1.02
	-later	38	1.24	1.01
Females	first born	33	0.94	1.13
	-later	58	0.82	1.28

*Born 10 years after therapy.

OTHER STUDIES

A similar study in the U.K. had the same strategy of examining many issues in the long-term survivors of childhood cancer, except that general practitioners were the informants and not the survivors themselves (Hawkins and Smith, 1989; Hawkins, Draper and Smith, 1989). Offspring of male survivors had a large deficit of male offspring, but only in firstborn offspring (Table 2; Hawkins 1991). All were born 10 years after all therapy stopped, and sex ratios were otherwise normal.

At the Roswell Park Cancer Institute, Buffalo (Green *et al.*, 1991), 102 liveborn or stillborn pregnancies occurred among 60 survivors who had had chemotherapy. There was no difference in the rate of congenital anomalies in the offspring of males versus females. The one curious finding was that, out of the 20 individuals treated with dactinomycin, two children had children with congenital heart disease, both involving septal defects. In the five center study above, (Byrne *et al.*, 1992), none of the 52 offspring born to 36 survivors (11 men and 25 women) who had been treated with dactinomycin had a major birth defect.

FUTURE OPPORTUNITIES

Another retrospective cohort study is just beginning. The Childhood Cancer Survivor Study, spearheaded by Dr. Leslie Robison, University of Minnesota, will interview 25,000 long-term survivors of childhood and adolescent cancer for information on many chronic health and genetic consequences. Again cancer and birth defects in offspring will be enumerated, and the issue of infertility will be explored especially the possibility of early menopause that might be attributed to exposure of the ovary to chemotherapy. For economy, no collection of biological specimens was proposed in the original design.

I wonder if it is apt to revisit the theoretic DNA-based methods for detecting germ cell mutageneses that were proposed at a meeting co-sponsored by the U.S. Department of Energy and the International Commission for Protection Against Environmental Agents, Mutagens and Carcinogens (Delehanty *et al.*, 1988). One of the possible speculative strategies was to collect DNA from the cancer survivor, all children and their other parent. Perhaps some method, such as subtraction hybridization with artificially constructed oligonucleotides could be applied. Simplistically, could one subtract out all the child's DNA from one parent, and anneal the residual unbound fragments to the DNA of the other parent? In short, all the child's DNA should be accounted for by the mother's and father's. Any child's DNA that remains would be prima facie evidence of new mutation and would be a small collection of fragments that could be characterized.

Regardless of the exact laboratory strategy, some protein or nucleic acid-based search seems necessary to gain sufficient power to rule out, e.g., a relative risk for germ cell mutation, in the order of 1.5. Otherwise, additional cohorts of 10,000 or 25,000 long-term survivors will be needed.

COUNSELLING

For now, clinical advice has to be given to patients (Mulvihill, 1993). I conclude that there are enormous theoretical concerns of exposure to ionizing radiation and chemotherapy of the male cancer survivor patient. Cancer therapy is supposed to interfere with DNA. There are limited empirical data on their offspring, but they give room for clinical reassurance about the lack of an excess of birth defects or general diseases by the above definitions.

Germ cell mutation surely does occur in human beings and I think there are compelling reasons to press very hard on the issue in cancer survivors, in a sense, a worst case exposure. If genetic epidemiologists cannot find an environmental germ cell mutagen in cancer survivors by a very sensitive means, strong reassurance can be made to the larger population about the probable lack of genetic diseases due to lower environmental exposure.

REFERENCES

Bryne, J., Mulvihill, J.J., Myers, M.H., *et al.*, 1987, Effects of treatment on fertility in long-term survivors of childhood or adolescent cancer. *N Engl J Med.* 317:1315-1321.

Byrne, J., Mulvihill, J.J., Connelly, R.R., *et al.*, 1988, Reproductive problems and birth defects in survivors of Wilms' tumor and their relatives. *Med Pediatr Oncol.* 16:233-240.

Byrne, J., Nicholson, H.S., Mulvihill, J.J., 1992, Absence of birth defects in offspring of women treated with dactinomycin. (Letter to the Editor) *N Engl J Med.* 326:137.

Delehanty, J. White, R.L. and Mendelsohn, M.L., 1986, Approaches to determining mutation rates in human DNA. *Mutation Res.* 167:215-232.

Green, D.M., Zevon, M.A., Lowrie G., *et al.*, 1991, Congenital anomalies in children of patients who received chemotherapy for cancer in childhood and adolescence. *N Engl J Med.* 325:141-146.

Hawkins, M.M., 1991, Is there evidence of a therapy-related increase in germ cell mutation among childhood cancer survivors? *J Natl Cancer Inst.* 83:1643-1650.

Hawkins, M.M. and Smith, R.A., 1989, Pregnancy outcomes in childhood cancer survivors: Probable effects of abdominal irradiation. *Int J Cancer.* 43:399-402.

Hawkins, M.M., Draper, G.J., and Smith, R.A., 1989, Cancer among 1348 offspring of survivors of childhood cancer. *Int J Cancer.* 43:975-978.

Li, F.P., Gimbrere, K., Gelber, R.D., *et al.*, 1987, Outcome of pregnancy in survivors of Wilms' tumor. *JAMA.* 257:216-219.

McKusick, V.A., 1992, Mendelian Inheritance in Man. Tenth edition. The Johns Hopkins University Press, Baltimore.

Mulvihill, J.J., Byrne, J., Steinhorn, S.C., *et al.*, 1986, Genetic disease in offspring of survivors of cancer in the young. *Am J Hum Genet.* 39:A72.

Mulvihill, J.J., 1993, Genetic counseling of the cancer patient, In Cancer: Principles and Practice in Oncology, DeVita, V.T., Jr., Hellman, S., Rosenberg, S.A., Eds. J. B. Lippincott, Philadelphia. 2529-2537.

Mulvihill, J.J., Myers, M.H., Connelly, R.R., *et al.*, 1987, Cancer in offspring of long-term survivors of childhood and adolescent cancer. *Lancet.* 2:813-817.

Mulvihill, J.J. and Czeizel, A., 1983, Perspectives in mutation epidemiology: A 1983 view of sentinel phenotypes. *Mutation Res.* 123:345-361.

Rivkees, S.A. and Crawford, J.D., 1988, The relationship of gonadal activity and chemotheropy-induced gonadal damage. *JAMA.* 259:2123-2125.

GENETIC EFFECTS OF ATOMIC-BOMB EXPOSURE

Robert W. Miller

Chief, Clinical Epidemiology Branch
National Cancer Institute, EPN-400
Bethesda, MD 20892

INTRODUCTION

In 1948 James V. Neel, Jr., M.D. initiated a huge epidemiologic study of the genetics effects of exposure to the atomic bomb in Hiroshima or Nagasaki three years earlier. He was well prepared for this endeavor, having trained as a physician after receiving his doctoral degree in Drosophila genetics. His work in Japan began when he was assigned by the U.S. Army as Acting Director of the Atomic Bomb Casualty Commission (ABCC), an agency of the National Academy of Sciences-National Research Council. Neel combined his scholarship with outstanding administrative ability in originating and conducting a series of genetic studies in Hiroshima and Nagasaki for 44 years thus far.

He learned that pregnant women in Japan were given an extra ration of rice beginning in the fifth month of pregnancy. When they registered for this supplement, they were entered in into the Genetics Program of the ABCC. A genetics questionnaire originated then, with information about parental exposure and the pregnancy to date. Almost all infants at that time were delivered by midwives. Neel enlisted their cooperation, and ABCC was notified of the delivery of each child, at which time the questionnaire was completed, with information concerning the birth and the health of the infant. An ABCC physician checked all pregnancy terminations reported as abnormal and a ten percent sample of infants reported as normal. Autopsies were performed when possible on stillborn infants and those who died in the first 6 days of life. A random sample of children was examined and anthropometric measurements obtained at nine months of age in Hiroshima (N = 9845) and Nagasaki (N = 8653).

The sex ratio (males/females) at birth was also evaluated because the effects would be different, depending on which parent was exposed to the atomic bomb. A radiation-induced **recessive** lethal mutation on a maternal X chromosome transmitted to daughters would have no effect, but would have a lethal effect on sons, thus lowering the sex ratio. A **dominant** lethal mutation on a paternal X chromosome would be transmitted only to daughters and would raise the sex ratio.

Male-Mediated Developmental Toxicity, Edited by D.R. Mattison
and A.F. Olshan, Plenum Press, New York, 1994

UNTOWARD PREGNANCY OUTCOMES

No genetic effects of radiation were detected from six measures made in the first nine months of life: the frequencies of stillbirths, deaths under 7 days of age, congenital malformations, birthweight, anthropometrics at nine months of age or sex ratio.

F_1 MORTALITY

With the passage of time further measures of genetic effects were made. No effect was found on the F_1 mortality among 31,159 children of parents one or both of whom were exposed to the bomb, 1946-85. The 115 deaths from cancer were unrelated to radiation dose.

CYTOGENETICS

In 1955 there were calls for termination of the ABCC in the belief that further effects, genetic or otherwise, were unlikely to be found. In 1956 Tjio and Levan reported that the human had 46 chromosomes. Cytogenetic studies flourished thereafter. When study was made of 8,322 children whose parents were exposed to the A-bomb and 7,976 controls, no excess was found of balanced or unbalanced autosomal structural rearrangements, sex chromosome abnormalities or autosomal trisomy.

MUTATION ALTERING CHARGE OR FUNCTION

In 1974, with the advent of electrophoresis for protein analysis, Neel proposed that a search be made for mutations that alter protein charge and/or function in the F_1 generation of A-bomb survivors. A battery of 30 serum and erythrocyte proteins were studied from 13,052 children whose parent(s) was (were) exposed within 2000 meters of the bomb and 10,609 whose parents were beyond 2000 meters. No excess was found in the group whose parents had been under 2000 meters. Three mutations that altered electrophoretic motility were found in the equivalent of 667,404 locus products of the proximal group, and three were found in 466,881 tests of the control group.

GAMETIC DOUBLING DOSE

Recent reconsideration of all the data on the genetics of radiation available from mice provided an estimate of the doubling dose that is in "satisfactory agreement with the higher estimate based on humans." The minimal gametic doubling doses for humans were estimated to be 1.7-2.2 Sv for acute radiation exposure, and 3.4-4.5 Sv for chronic exposure (extrapolated from experiments on mice).

OVERVIEW

Since 1948, the progression of genetic studies, from relatively crude clinical observations to increasingly sophisticated laboratory studies, has revealed no genetic effects of ionizing radiation in the survivors of the atomic bombs. There is no doubt that genetic effects occurred as they do in every species studied experimentally, but as large as the numbers of persons studied in Japan were, they were not large enough to reveal genetic damage through the measures used to date.

The substantial data from the genetics studies of A-bomb survivors provides a basis for evaluating claims such as that by Gardner *et al.* (1990) of a relationship between preconception radiation of the father and the occurrence of a cluster of childhood leukemia. Neel and Schull argue that the study by Gardner *et al.* (1990) is based on very low, poorly defined exposures, with a leukemogenic effect that was not found in the Japanese studies. Also, against a causal relationship is the lack of an increase in birth defects or a cancer such as retinoblastoma, with a well defined germ cell mutation, as contrasted with leukemia which has none.

THE FUTURE

Studies of cancer incidence and mortality of the F_1 generation will be extended. With the development of new genetic laboratory techniques attempts are being made to study mutations at the level of DNA. Since 1985 "immortalized" lymphocytes (by Epstein-Barr infection) have been obtained from family trios: the child, mother, and father. The goal is to store lymphocytes from 500 trios. New mutations can be identified by comparing findings in the child with those in each parent. At the DNA level, the use of homologous probes in parallel studies of mice and humans may enhance comparison of the genetic effects of radiation doses in the range of those received by A bomb survivors in Hiroshima and Nagasaki.

REFERENCES

Awa, A.A., Honda, T., Neriishi, S., Sufuni, T., Shimba, H., Ohtaki, K, Nakano, M., Kodama, Y., Itoh, M., Hamilton, H.B., 1987: Cytogenetic study of the offspring of atomic bomb survivors, Hiroshima and Nagasaki. In "Cytogenetics," (Eds. Obe G, Basler A) Berlin, Springer-Verlag, pp. 166-83.*

Gardner, M.J., Snee, M.P., Hall, A.J., Powell, C.A., Downes, S., Terrell, J.D., 1990: Results of case-control study of leukaemia and lymphoma among young people near Sellafield nuclear plant in West Cumbria. *Br Med J* 300:423-9.

Neel, J.V., Lewis, S.E., 1990: The comparative radiation genetics of human and mice. *Annu Rev Genet* 24:328-62.

Neel, J.V., Mohrenweiser, H., Satoh, C., Hamilton, H.B., 1977: A consideration of two biochemical approaches to monitoring human populations for a change in germ cell mutation rates. RERF TR4-77, Hiroshima, Japan, pp 19.

Neel, J.V., Satoh, C., Goriki, K., Asakawa, J., Fujita, M., Takahashi, N., Kageoka, T., Hazama, R., 1988: Search for mutations altering protein charge and/or function in children of atomic bomb survivors: final report. *Am J Hum Genet* 42:663-76.*

Neel, J.V., Schull, W.J., 1956: "The Effect of Exposure to the Atomic Bombs on Pregnancy Termination in Hiroshima and Nagasaki," Washington, D.C., National Academy of Sciences-National Research Council, Pub. no. 461, pp. 241.*

Neel, J.V., Schull, W.J., eds, 1991: The Children of Atomic Bomb Survivors: A Genetic Study. National Academy Press, pp. 491.*

Neel, J.V., Schull, W.J., Awa, A.A., Satoh, C., Kato, H., Otake, M., Yoshimoto, Y., 1990: The children of parents exposed to atomic bombs: Estimate of the genetic doubling dose of radiation for humans. *Am J Hum Genet* 46:1053-72.*

Schull, W.J., Neel, J.V., Hashizume, A., 1966: Some further observations on the sex ratio among infants born to survivors of the atomic bombings of Hiroshima and Nagasaki. *Am J Hum Genet* 18:328-38.

Tjio, J.H., Levan, A., 1956: The chromosome number of man. *Hereditas* 42:1-6.
Yoshimoto, Y., Schull, W.J., Kato, H., Neel, J.V., 1946-85: Mortality among the offspring (F_1) of atomic bomb survivors, Technical Report Series, RERF TR 1-91, Hiroshima, Japan.

*These articles were reprinted in Neel, J.V., and Schull, W.J., eds, 1991: The Children of Atomic Bomb Survivors: A Genetic Study, National Academy Press, pp. 491 (as shown above).

BIOLOGICAL FACTORS RELATED TO MALE MEDIATED REPRODUCTIVE AND DEVELOPMENTAL TOXICITY

Robert L. Brent

Department of Pediatrics
Jefferson Medical College
Alfred I. duPont Institute
1600 Rockland Road, Box 269
Wilmington, DE 19899

INTRODUCTION

Over the past 30 years the area of reproductive problems and reproductive failure has assumed a greater proportion of the interest of clinicians, epidemiologists, teratologists, geneticists and toxicologists. Much has been learned about the reproductive risks of exposing pregnant women to drugs and chemicals and physical environmental agents. Both epidemiological and animal studies have assisted in understanding the mechanism of some of these reproductive toxicants and the actual risk of being exposed to these agents. There also has been great advances in the field of clinical and basic genetics so that we have a much better understanding of the impact of genetic disease and our ability to diagnose these disorders. We are also aware of the fact that each generation has a small proportion of genetic disease not due to inherited genetic traits, but due to new genetic disease resulting from mutations that occurred during the development of germ cell mutations that produce mature eggs and sperm. We are still in the early stages of understanding the process of mutation and are presently investigating the impact of environmental agents on the induction of mutations as well as the alteration of reproductive performance.

This paper is concerned with the impact of environmental agents on the reproductive capacity of the male. While a superficial view of this problem may make it appear that this subject is only concerned with the mutagenic effects of environmental agents, a more in-depth analysis reveals that reproductive problems encompasses more than the matter of mutagenesis (Table 1).

Therefore male mediated developmental toxicity can result from mutagenic exposures to the male gonadocytes resulting in nucleotide sequencing abnormalities or chromosomal abnormalities as well as exposing the maternal organism to toxic levels of reproductive toxicants contained in the ejaculate. While concern about the potential risks of agents in the ejaculate is new and based predominantly on animal studies, concern about the magnitude of the mutagenic problem has persisted for years (Table 2).

Male-Mediated Developmental Toxicity, Edited by D.R. Mattison
and A.F. Olshan, Plenum Press, New York, 1994

Table 1. Reproductive Risks in Human Populations

Reproductive Risk	Frequency
Immunologically plus clinically diagnosed spontaneous abortions/10^6 conceptions	**350,000**
Clinically recognized spontaneous abortion/10^6 pregnancies	**150,000**
Genetic Diseases in 10^6 births	**110,000**
Multifactorial or polygenic (genetic-environmental interactions)	90,000
Dominantly inherited disease	10,000
Autosomal and sex-linked genetic disease	1200
Cytogenetic (chromosomal abnormalities)	5000
New mutations	3000
Major congenital malformations/10^6 births (many are genetic in etiology)	**30,000**
Prematurity/10^6 births	**40,000**
Fetal growth retardation/10^6 births	**30,000**
Stillbirths/10^6 pregnancies (>20 weeks)	**20,900**
Infertility	**15% of couples**

Table 2. Controversy Over the Magnitude of the Genetic Risk from Environmental Agents

"The threat of genetic damage is our number one health problem" (Legator 1968).

"Chemical mutagenesis is certainly a very small problem as we see it at present. We view it as 1% of the very broad problem of human genetics. We do not propose setting up whatever Dr. Crow means by a monitoring system" (Shannon 1968).

"If the public flatly refuses to take decisive action on the basis of the massive volumes of data linking cigarette smoking to lung cancer, how can you expect people to act vigorously on the more hazy and abstruse things like a chemical that may or may not be mutagenic in man" (Crow 1969).

"There is no realistic way to predict a safe human level of a substance shown to be teratogenic, carcinogenic or mutagenic in animals" (Epstein 1971).

"I think it is absolutely essential that we do not delude ourselves about the magnitude and complexity of our task. The general public is easily scared and when they are scared, they may form pressure groups to push governmental agencies into action. These agencies are scientifically naive and have to rely on our advice. We should be very careful not to give advice that is itself naive; that is advice based on oversimplified tests and facile interpretations" (Auerbach 1971).

"It would be foolish to advise anything other than extreme caution over the exposure of fathers to chemical mutagens when our understanding of the quantitative risks to future generations is so rudimentary" (Brown 1985).

"The world is full of mutagens, carcinogens and reproductive toxins, and it always has been. The important issue is the human exposure dose. Fortunately the exposure is usually minuscule" (Ames 1989).

"Nature's toxic chemicals are the major carcinogens and mutagens ingested by humans. These natural chemicals in plants are present at a level 10,000 times the prevalence of man-made pesticide residues. There is a tendency for laymen to think of chemicals as being only man-made and to characterize them as toxic" (Ames 1989).

Let us examine the various categories of reproductive failure and risks and their known causes in order to determine whether environmental exposures to the male population contribute measurably to any of these categories of reproductive failure. While an understanding of reproductive risks can be obtained from epidemiological studies, animal studies and basic research, the foundation for determining the cause and importance of clinical diseases is derived from studying human populations with epidemiological techniques.

THE EPIDEMIOLOGY OF REPRODUCTIVE FAILURES AND REPRODUCTIVE RISKS

One of the most difficult and complicated areas of epidemiological research is the evaluation of reproductive risks. The reason for the complexity of this area of research is:

1. Reproductive failures of various types are common, affect a substantial portion of the population and have occurrence patterns that vary and are therefore are not always predictable in various populations (Table 1, Brent 1985).

2. Reproductive problems include a wide variety of pathological conditions that may or may not be mechanistically related to each other.

3. The known etiologies for reproductive problems include both genetic and environmental problems. Furthermore, the environmental factors include a host of external chemical, infectious and physical agents as well as intrinsic disease processes

that may occur before or during pregnancy in women and before fertilization in men (Brent 1985). Even the frequency of spontaneous abortions, chromosomal aberrations and mutations at the molecular level vary with the age of the parents. As an example Lian *et al.* (1986) studied the association between paternal age and the occurrence of birth defects using data collected in Metropolitan Atlanta. Older fathers had a somewhat higher risk for having babies with defects, when all types of defects were combined; an equivalent association for older mothers was not found. Yet we are all aware of the fact that older mothers have an increased risk for having a child with Downs syndrome with increasing age. Logistic regression analyses also indicated modestly higher risks for older fathers for having babies with ventricular septal defects and atrial septal defects and substantially higher risks for having babies with defects classified in the category chondrodystrophy and babies with situs inversus.

MALE MEDIATED REPRODUCTIVE TOXICITY

How do investigators determine whether a particular drug, chemical or physical agent presents a reproductive risks to the male population? The study of human populations not only necessitates a knowledge of epidemiology but a broad understanding of the field of reproductive failures. Since all types of reproductive problems have multiple etiologies, the epidemiologist has to recognize that differences between the incidence of reproductive problems in the exposed and non-exposed groups may be due to differences in the study groups. Therefore, positive associations may be unrelated to the environmental exposure to the male population.

Reproductive failures for which the male may be responsible can be categorized as follows:

1. **Infertility** may be due to hereditary disease, congenital malformations of the reproductive tract, infection and venereal disease, mechanical problems, psychological illness, medication, environmental chemical and physical agents, chronic illness, and ignorance.

2. **Spontaneous abortion** can result from inherited or acquired chromosomal abnormalities or mutations from the father; and hypothetically from toxic substances in the ejaculate. It has been estimated that up to 50% of all fertilized ova in the human are lost within the first three weeks of development (Hertig 1967, Tables 1 and 3). The World Health Organization (1970) estimated that 15% of all clinically recognizable pregnancies end in spontaneous abortion, while 50-60% of the spontaneously aborted fetuses have chromosomal abnormalities (Boue *et al.*, 1975; Simpson 1980). This means that, as a conservative estimate, 1173 clinically recognized pregnancies will result in approximately 176 miscarriages. The true incidence of pregnancy loss is much higher, but undocumented pregnancies are not included in this risk estimate. While we know little about the mechanisms which result in the in utero death of defective embryos (Warkany 1978), it is perhaps more important to understand the circumstances which permit abnormal embryos to survive to term.

3. **Stillbirth** can result from inherited or induced mutations in the male genome resulting in congenital malformations or chromosomal abnormalities;

4. **Prematurity** can result from inherited or acquired chromosomal abnormalities or mutations from the father resulting in congenital malformations;

5. **Fetal Growth Retardation** can result from inherited or acquired chromosomal abnormalities or mutations from the father resulting in congenital malformations or genetic disease;

6. **Hereditary Diseases** are predominantly inherited but may be due to spontaneous mutations or prenatally acquired chromosomal abnormalities. While there is no question that many environmental drugs and chemicals have mutagenic

potential there is disagreement as to whether environmental exposure of males to mutagenic agents contributes very much to the incidence of genetic disease. The offspring of the Atomic Bomb survivors had no measurable increase in induced mutations following exposure to an agent which has proven mutagenic potential (Neel *et al.*, 1990; Miller, this volume). By contrast, there is substantial data to indicate that environmental drugs, chemicals and ionizing radiation can induce cancer in human populations (Herbst *et al.*, 1971; BEIR VI 1990). The greater potential for mutagenic agents to induce clinical cancer rather than hereditary disease is very likely due to the loss of mutations and mutated cells during the process of meiosis, fertilization, implantation and organogenesis rather than to the fact that environmental mutations are not produced. (Table 4)

Table 3. Estimated Outcome of 100 Pregnancies Versus Time from Conception*

Time from Conception	Percent Survival to Term	Percent Death During Interval
Preimplantation 0-6 days	25	54.55
Postimplantation 7-13 days	55	24.66
14-20 days	73	8.18
3-5 weeks	79.5	7.56
6-9 wk	90	6.52
10-13 wk	92	4.42
14-17 wk	96.26	1.33
18-21 wk	97.56	0.85
22-25 wk	98.39	0.31
26.29 wk	98.69	0.30
30-33 wk	98.98	0.30
34-37 wk	99.26	0.34
38 + wk	99.32	0.68

* Data from Kline and Stein (1985). An estimated 50 to 70 percent of all human conceptions are lost in the first 30 weeks of gestation (Hertig 1967) and 78 percent are lost before term (Robert and Lowe 1975).

Table 4. Biologic Filtration

Many chemical agents have both mutagenic and carcinogenic potential. Yet there are numerous instances in which epidemiological studies have demonstrated a causal relationship between exposures to chemicals and drugs and an increase in cancer in the exposed population. Populations exposed to the same "mutagenic" agents rarely demonstrate an increase in mutations. Why?

Loss of affected germ cells during gametogenesis.
Decrease capacity of mature eggs or sperm to fertilize or be fertilized.
Embryonic loss during preimplantation or very early organogenesis.
Lapse in time from exposure of gonadocytes to fertilization resulting in the loss of damaged gonadocytes.
Low or no "exposures" in the targeted population.

Congenital Malformations can result from inherited or acquired chromosomal abnormalities or mutations from the father. The etiology of congenital malformations can be divided into three broad categories: unknown, genetic, and environmental factors (Table 5).

Table 5. Etiology of Human Congenital Malformations Observed During the First year of Life*

Suspected Cause	**Percent of Total**
Unknown	**65-75**
Polygenic	
Multifactorial (gene-environment interactions)	
Spontaneous errors of development	
Synergistic interactions of teratogens	
Genetic	**10-25**
Autosomal and sex-linked genetic disease	
New mutations	
Cytogenetic (chromosomal abnormalities)	
Environmental	**10**
Maternal conditions: Alcoholism; diabetes; endocrinopathies; phenylketonuria; smoking and nicotine; starvation; nutritional	4
Infectious agents: Rubella, toxoplasmosis, syphilis, herpes, cytomegalic inclusion disease, varicella, Venezuelan equine encephalitis, parovirus B 19	3
Mechanical problems (deformations): Amniotic band constrictions; umbilical cord constraint; disparity in uterine size and uterine contents	1-2
Chemicals, prescription drugs, high-dose ionizing radiation, hyperthermia	<1

*Adapted from Brent (1976, 1985) and Brent and Holmes (1988).

The etiology of the majority of human malformations, approximately 65-75%, is unknown (Wilson 1973; Brent 1976, 1985; Heinonen *et al.*, 1977); however, a significant proportion of congenital malformations of unknown etiology is likely to be polygenic, that is, due to two or more genetic loci (Carter 1976; McLaughlin 1977) or at least have an important genetic component. Thus inheritance from the father or mutations in the father's genome could have as much impact on the incidence of polygenic disease as does the mother. Malformations with an increased recurrent risk, such as cleft lip and palate, anencephaly, spina bifida, certain congenital heart diseases, pyloric stenosis, hypospadias, inguinal hernia, talipes equinovarus, and congenital dislocation of the hip, can fit the category of multifactorial disease, as well as the category of polygenic inherited disease (Carter 1976; Fraser 1976). The multifactorial/threshold hypothesis (Fraser 1976) involves the modulation of a continuum of genetic characteristics by intrinsic and extrinsic (environmental) factors.

Spontaneous errors of development may account for some of the malformations that occur without apparent abnormalities of the genome or without environmental influence. It has been postulated that there is some probability for error during embryonic development based on the fact that embryonic development is a complicated process, similar to the concept of spontaneous mutations (Brent 1964, 1976, 1985).

ENVIRONMENTAL RISK PARAMETERS OR MODIFIERS

There are several stages of the reproductive life-cycle of the male exposed to reproductive toxicants that can have an effect on the next generation. Environmental toxicants can have a direct effect on spermatogenesis by altering the genome in spermatogonia, spermatocytes or mature sperm. The resulting sperm can be killed, altered to diminish the possibility of fertilization or have changes in the genome that may or may not be manifested in the next generation. Genomic changes in the spermatogonia may persist and produce abnormal sperm for multiple cycles of spermatogenesis. Genomic alterations in spermatocytes and sperm have a narrow window of opportunity to transmit these genetic changes to the next generation. These reproductive effects are primarily, but not exclusively, stochastic phenomena (Table 6). The exposure of cells involved in spermatogenesis can result in:

Mutations
Karyotype abnormalities
Cytotoxicity
Imprinting pattern alterations
Epigenetic effects

The factors which determine the hazard of the exposure are the 1) stage of the spermatogenesis cycle exposure for the fertilizing sperm, 2) the potency and characteristics of the reproductive toxin, 3) the dose or exposure of the reproductive toxin experienced by the developing gonadal cells, 4) the nature of the reproductive effect i.e. threshold versus stochastic response, 5) the length of elapsed time from exposure to fertilization.

The direct effect of environmental agents on the genome of gonadal cells is the predominant risk that concerns investigators interested in male mediated reproductive toxicity, but there are theoretical concerns and animal studies that have examined the possibility of transmitting toxic agents to the embryo via the sperm or ejaculate. The importance of sperm cells or the ejaculate as vectors of toxic agents to the embryo in the human is presently a matter for conjecture. These are primarily threshold phenomena (Table 6).

Ejaculate containing toxic substances
Sperm containing toxic substances

Table 6. Comparison of Stochastic Phenomena and Threshold Phenomena in the Etiology of Diseases Produced by Environmental Agents and the Risk of Occurrence

Phenomena	Pathology	Site	Diseases	Risk	Definition
Threshold phenomena	Multiple sites of subcellular and cellular injury	Great variation in etiology, affecting many cell and organ processes	Malformation, death, growth retardation, etc.	Completely disappears below a certain threshold dose	The incidence and severity of disease increases with exposure.
Stochastic phenomena	Damage to a single cell may result in disease	DNA	Cancer, mutation	Exists at all exposures, although at low exposure, the risk is below the spontaneous risk	The incidence of disease increases with exposure but not the severity.

It would appear to be very unlikely that sperm can carry toxic agents in quantities that could damage the developing embryo, since the sperm volume is so much smaller than the egg. For the first two weeks of human development the embryo is very sensitive to the lethal effects of embryotoxic agents and quite resistant to their teratogenic effects (Wilson and Brent 1953; Brent and Bolden 1968; Brent 1980; Russell and Russell 1950, 1954; Generoso *et al.*, 1988; Pampfer and Streffer 1988). For these reasons it appears to be very unlikely that sperm could transmit teratogenic doses of toxicants to the embryo. Alternatively, it is not unreasonable to suggest that multiple exposures to an ejaculate containing embryotoxic agents can affect the developing embryo.

IDENTIFICATION OF REPRODUCTIVE TOXICANTS AND HUMAN TERATOGENS

The discovery of human teratogens and reproductive toxicants have come primarily from human epidemiological studies. Animal studies and *in vitro* studies can be very helpful in determining the mechanism of teratogenesis and the pharmacokinetics related to teratogenesis (Brent 1988). Without decisive evidence of human reproductive toxicity it is difficult to provide estimates of the hazard of male exposures to specific agents such as drugs, chemicals and physical agents.

Several considerations are helpful in establishing that an environmental exposure causes reproductive toxicity in the human: (1) Human epidemiologic studies should consistently report that exposure to an agent is associated with an increased incidence of reproductive effects. In the case of mutations, it would be necessary to demonstrate an increase in the incidence of new mutations, i.e., genetic disease in the exposed population. Historically, this has been a very difficult task. With regard to exposures to the developing embryo from chemicals or drugs contained in the ejaculate, it would be necessary to demonstrate an increase in the incidence of congenital malformations in the exposed population. This type of exposure, if teratogenic, would most likely result in a reproducible constellation of specific malformations and developmental problems. If the agent is mutagenic the exposed population can demonstrate an increase in genetic disease, including genetically determined malformations. (2) For common exposures, secular trend data should support the suggestion that there has been an change in the mutation rate or in the incidence of malformations. (3) An animal model mimics the reproductive effect suspected in humans at exposures which are experienced by men. (4) The reproductive effect or teratogenic effect should increase with dose. 5) The suggested reproductive effect should be biologically plausible and not contradict established scientific principles, although there may be, and indeed, there have been exceptions to this rule (Brent 1978, 1983, 1986; Shepard 1986). This approach is of greatest value when utilized for the evaluation of environmental agents that have been in use for some time or for evaluating new agents that have a similar mechanism of action, function, chemical structure, pharmacology or physical effect of other agents which have been extensively studied.

As an example, the complexity of determining the etiology of reproductive problems is represented by the attempt to study spontaneous abortion. Epidemiological investigations of the causes of spontaneous abortions must deal with formidable problems:

(1) A majority of spontaneous abortions are due to chromosomal abnormalities that are unrelated to environmental exposures that may have occurred during pregnancy.

(2) The risk of abortion changes with each day of pregnancy (Table 3), so that matching controls is essential in order to eliminate the selection of two populations with different spontaneous abortion rates.

(3) Attempts to control for the hidden incidence of therapeutic abortions have only limited success (Susser 1983; Olsen 1984). "The existence of high rates of induced abortion in the population may distort currently employed values of the rate of spontaneous abortion" (Susser 1983).

The evaluation of new drugs, physical agents or chemicals must depend more on *in vivo* and *in vitro* animal testing and the application of the principles of biologic plausibility, since these data are usually not available from human populations. Even if the agent is closely related to a previously used compound, closely related compounds may have markedly different reproductive risks. It is well known that very minor changes in the thalidomide molecule can eliminate its teratogenic effect (Wuest 1968) and similar results have been reported with mutagenic agents.

The purpose of *in vitro* and *in vivo* testing for reproductive effects in the male is to determine whether a new environmental agent presents a reproductive risk; namely, a mutagenic effect or some other evidence of reproductive toxicity. Once there is a suggestion that an agent has a reproductive effect, then *in vivo* and *in vitro* systems can be used to study the pharmacokinetics of a drug effect or the mechanisms of action of the agent. In most instances the actual human reproductive risk cannot be obtained from *in vivo* animal studies or *in vitro* studies (Brent 1964, 1980 1988; Schardein 1988; Lyon *et al.*, 1985). Each *in vitro* test system has some unique features that have been attractive to some investigators. With the proliferation of these systems it is obvious that the cost of performing a combination of these techniques could be more than a whole animal reproductive study. More important than these unique features or the possible reduced cost, is an important principle, "the nature of these tests indicate that they can NEVER be more predictive of teratogenicity or embryotoxicity than *in vivo* systems" (Schardein 1988). Schardein (1988) stated it quite eloquently, "the real dilemma in their use is eliminating procedures in animals and at the same time making tests more predictive; an incongruity to say the least."

In vitro tests offer the reproductive toxicologist an opportunity to study various facets of their field. They can be used to study 1) mutagenesis at the nucleotide or chromosomal level, 2) normal embryonic development and differentiation, 3) some mechanisms of teratogenesis, mutagenesis and embryotoxicity, 4) to study pharmacokinetics and the effect of isolated or combined metabolic products, and 5) to screen for cytotoxicity and interference with differentiation.

It is also obvious that *in vitro* testing, using a single system, is going to fail to delineate important reproductive toxicity effects, including: 1) the magnitude of the human genetic risk, 2) the risk of infertility, fetal growth problems and other perinatal problems referred to in Table 1, 3) the risk of late central nervous system effects (brain histogenesis, behavior), 4) unique unpredictable embryotoxic specific effects such as aspermatogenesis, cardiovascular and hemodynamic changes, vascular disruption, yolk sac dysfunction or chorioplacental effects); 5) failure to differentiate between recuperable versus non-recuperable growth retardation; 6) the risk of transplacental carcinogenesis; 7) the lethal effects occurring during the preimplantation period; and 8) effects not previously reported.

A REVIEW OF PUBLICATIONS DEALING WITH MALE MEDIATED DEVELOPMENTAL TOXICITY

1. Epidemiological Studies

The background incidence of reproductive pathology is high, affecting a large segment of the population (Table 1). Small increases in the incidence of reproductive pathology may be difficult to document. While most human epidemiological studies have not found an increase in translocations or mutations following drug, chemical or physical agent exposures, various degrees of infertility or embryonic loss have been reported in some of these studies.

a. Reproduction in survivors following chemotherapy and radiation

The largest group of patients that have been studied are groups that have been treated for cancer and have received chemotherapeutic agents, immunosuppressants or radiation. Survivors of either cancer in childhood or as adults have been studied in order to determine whether the exposure to reproductive toxicants has diminished the survivors' reproductive performance.

Mulvihill and Byrne (1989) reviewed the literature pertaining to the normalcy of the offspring of cancer survivors. The results of their study is summarized in Table 7. The incidence of malformations in the offspring of over 1000 cancer survivors was 3.85% and therefore within the range that one would expect in a untreated or unexposed population. It is of interest to note that women survivors were predominant in these study populations.

Table 7. Epidemiological Studies Of The Offspring Of Cancer Survivors*

Exposed Parents	Fetal Loss	Elective Abortion	Normal newborn	Total major Defects
825	336	127	1194	46
		Percent malformations in 1194 offspring		
		3.85%		

* Data from Mulvihill and Byrne (1989).

Nygaard *et al.* (1991) studied the reproductive performance of childhood leukemia survivors. There were 981 male and female leukemia survivors. In this large series of patients, 299 had reached the age of 18 without a relapse. From these adult survivors, there were 48 offspring, one of whom was malformed. There were no myeloid leukemia survivors who became parents. Of 131 male survivors with acute lymphatic leukemia, only 4 became parents. Of 149 female survivors with acute lymphatic leukemia, 23 became the parents of 41 children. This study also reflected a suggestion that male reproductive capacity was more adversely affected, possibly because of a difference in therapy between the sexes or a difference in the response to the therapy. The study reported that 18-24 Gy to the CNS may lower the rate of first births. Reduced testicular size and impaired sperm production have been associated with cranial and testicular radiation. 4.0 Gy to the ovaries may result in permanent sterility in young females. 1.4-3.0 Gy to the testes may cause permanent aspermia in males. Although the population size of this study was small, the authors reported no childhood malignancies or genetic diseases diagnosed in the offspring and no increase in congenital malformations in the offspring. On the other hand, while these same authors (Nygaard *et al.*, 1991) did not find an increase in the frequency of malignancies in the offspring of the cancer survivors, this was not the case in the survivors themselves. Among a cohort of 981 children who were followed 4.3-26.5 years after cessation of anti-leukemic therapy, eight patients who were in remission of acute lymphoblastic leukemia (ALL) developed a distinctively new malignant disease. The second malignant neoplasms (SMN) included brain tumors, basal cell carcinomas, thyroid cancer, leiomyosarcoma and finally rhabdomyosarcoma in a patient who also had suffered from Hodgkins disease while still on anti-leukemic treatment. Individuals treated for childhood ALL are at increased risk of a new malignancy, and this seems mainly to be associated with previous irradiation. This study reflects the concept of biologic filtration described in Table 4, which indicates that the same exposure to mutagenic or cytotoxic agents present a much greater risk of inducing clinically recognized malignant disease in the exposed rather than genetic disease in their offspring.

Byrne *et al.* (1988) studied the reproductive problems and birth defects in survivors of Wilms' Tumor and their relatives. In this retrospective cohort study of 47 Wilms' tumor survivors and their 77 sibling controls, female survivors had a fourfold excess risk (risk ratio, 4.1; 95% confidence interval, 1.7-10.1) for any adverse live birth outcome, including birth defects, compared with their sibling controls. Wives of male survivors had no apparent excess risk for problem pregnancies. The implication suggested by this study is that radiation to the female child's abdomen and genital organs resulted in deleterious reproductive outcomes (prematurity, increase neonatal mortality and increased neonatal mortality and IUGR), but that these effects were unrelated to any alteration of the genome of the ova.

Green and Hall (1988) studied the reproductive histories of 93 patients who had been treated for Hodgkin's disease, were 18 or older and were at least five years post treatment. Forty-eight pregnancies were reported by 22 patients or their spouses. Fifteen female patients had 33 pregnancies and 7 male patients reported 14 pregnancies in their spouses. While the authors reported no increase in birth defects or childhood malignancies in the offspring, the number of cases was quite small. There was no attempt to determine whether infertility was increased or whether there was a greater effect in the male or female.

Ionizing radiation has been demonstrated to be mutagenic in both *in vivo* animal studies and *in vitro* systems. Furthermore, radiation can readily produce sterility in both males and females if the dose is high enough. The largest study ever undertaken to examine the genetic effects of ionizing radiation occurred after the Atomic Bomb detonation in Hiroshima and Nagasaki. The offspring of the Atomic Bomb survivors who were exposed to a single dose of radiation had no measurable increase in induced mutations following analysis of the incidence of chromosome abnormalities in children, neonates and abortuses born to the irradiated parents (Okada *et al.* 1975). On the other hand, Martin *et al.* (1986) reported induced chromosomal aberrations were present in human sperm years after therapeutic radiation. But the authors did not perform a controlled study to determine whether these findings were associated with a decrease in reproductive performance.

b. Reproductive performance of various occupational groups

While the fertile survivors of cancer treatment, in most studies, have not had a measurable increase in the frequency of malformations or other fetal effects in their offspring, this has not been the case with epidemiological studies dealing with the association of certain occupations with reproductive problems and abnormal offspring. Hoglund *et al.* (1992) studied the weight and length of children of male spray painters.

The course and outcome of the pregnancies of the wives of 80 spray painters and 80 electronics workers were recorded from birth registers, hospital records, and a questionnaire. The mean length and weight of the children of spray painters at birth were slightly lower than those of the children of electronics workers. No differences were recorded for serious complications of pregnancy, malformations, or the clinical course after birth.

Gardner *et al.* (1990) reported an increase in leukemia in the offspring of fathers exposed occupationally in nuclear power plants. This particular research report has raised a storm of controversy and discussion. Since the exposure to ionizing radiation was very low and there were no other increases in "genetic" disease in the population, critics of this research claim that the results are not biologically plausible. Obviously, further studies of this and similar populations are warranted.

Bonde *et al.* (1990) studied the fertility of male welders. There were 3702 welders with a follow-up of 47,674 person-years. The investigators reported a slight decrease in fertility in the exposed workers. This was consistent with other reports which have reported delayed conception, and decreased semen quality (Mortensen 1988; Rachootin and Olsen 1983; Jelnes 1988; Bonde 1990).

Olshan *et al.* (1991) identified 14,415 live born children with birth defects from a population based registry in British Columbia. They then determined the paternal occupations of the children in order to determine whether there was an increased association of birth defects with particular occupations. Certain occupations had an increased odds ratio for some birth defects. The authors explained that this study was exploratory in nature, "had several limitations," and "the results must be viewed with caution." Some of the associations can be discounted on the basis of biological plausibility, since many birth defects have such a small genetic component, that mutagenic agents in the environment would be unlikely to increase the incidence of these particular malformations. But the authors point out that another mechanism for occupationally related birth defects is that the chemical associated with the job may be present in the ejaculate and therefore act as a teratogen during embryonic development.

Lindbohm *et al.* (1991) studied the effects of paternal occupational exposure on the frequency of spontaneous abortion. The frequency of medically diagnosed spontaneous abortion was determined in a group of 99,196 pregnancies in Finland from 1975 to 1980. A job exposure classification was developed to classify men and women according to their occupation, job title and industry. In 10% of the pregnancies, the husbands were exposed to one or more mutagens and the rate of spontaneous abortions was unaffected. Four of the 25 "mutagenic" chemicals were associated with an increased relative risk for spontaneous abortion. The investigators were reluctant to draw any casual relationship until the findings were confirmed and exposures could be measured.

While there are many other articles in the literature dealing with the reproductive risks of occupational exposures, the results are inconclusive with regard to both the existence and the magnitude of the risk. This risk of spontaneous abortion and congenital malformations from occupational exposures was extensively summarized by Tasksinen (1990). She reviewed 149 publications dealing with this subject. Although there were studies with positive findings, she said, "It has been difficult to distinguish the occupational causes of spontaneous abortion and congenital malformations from other factors related to the parent's characteristics and their living environment," and, "There is a lack of conclusive epidemiologic studies on this topic."

2. Analysis of Secular Trend Epidemiological Data for the Purpose of Evaluating Male-Mediated Developmental Toxicity

While the analysis of secular trends of various diseases has been useful in some fields, they have proven to be less valuable in evaluating genetic disease. The key characteristics of disease categories that lend themselves to secular analysis are:

1) There should be substantial numbers of the population that can be identified as exposed and non-exposed. 2) The diseases being studied should occur in a low frequency and be rather unique. 3) There should be accurate epidemiological data that have been collected over a period of years. 4) The diseases or abnormality being studied should be readily recognized.

Male-mediated reproductive disease, as a separate entity, does not readily lend itself to secular trend analysis. Genetic diseases are quite common and are manifested in approximately 11% of births. We do not have as yet satisfactory epidemiologic data on the incidence of genetic disease. Most mutagens do not have a propensity to produce one type of genetic disease or a particular syndrome, but affect the genome in a random manner. Therefore, small increases in genetic disease would be very difficult to ascertain. Furthermore, we already can approximate the "spontaneous mutation rate," and it appears that environmentally induced mutations would be an extremely small part of the inherited and spontaneous group of genetic disease.

Chromosome aberrations might lend themselves to secular trend analysis since we have better data on the frequency of these diseases. But in many instances the abnormal karyotype is derived from the mother which would diminish the sensitivity of this approach. Infertility or sterility would also lend itself to secular trend analysis, if we had better data on the frequency of infertility along with better data on its etiology. Finally, clinical problems such as prematurity, fetal growth retardation and stillbirth have excellent epidemiological data, but also have multiple etiologies and major contributions from maternal health problems.

Therefore, secular trend analysis is as yet not a fruitful area for the evaluation of the importance of male-mediated developmental toxicity.

3. The Use of *In Vivo* Animal Studies to Study Male-Mediated Developmental Toxicity

There has been extensive animal and cellular research in the field of gene and chromosomal alterations from environmental mutagenic and cytotoxic agents. In contradistinction to the human epidemiological studies, nucleotide sequence changes and chromosomal abnormalities have been readily produced in the gonadocytes of experimental animals and *in vitro* cellular systems. Even the offspring of exposed males have demonstrated genomic changes, although to a lesser extent than the females. This is expected because of the loss of some of these genomic changes during the process of development (Table 4).

Animals whose testes or sperm were exposed to ionizing radiation or chemical mutagens have fathered offspring with an excessive *in utero* mortality, translocations and point mutations or who themselves have various degrees of infertility. These effects in animals are related to the dose, dose-rate, spermatogenesis stage, the interval between exposure and insemination and the nature of the toxicant.

One of the most frequently studied mutagens has been ionizing radiation. Grahn *et al.* (1984) evaluated the extent of genetic injury in hybrid male mice exposed to low doses of ^{60}Co-γ-rays or fission neutrons. Dose-response functions were linear or linear-quadratic for testicular weight loss and the presence of abnormal sperm over the full dose range. The RBEs (relative biological effectiveness) for neutrons were between 5 and 6 (Table 8). They were between 7 and 9 at lower doses (< 10 rad) for translocations (Table 9). RBEs for postimplantation and total dominant lethal rates were 5-6 when the dose was above 10 rad and 10-14 when the dose was below 10 rad.

Table 8. Percentage of Abnormal Epididymal Sperm Listed According to Dose and Number of Weeks After Irradiation*

Dose	Percentage of Abnormal Epididymal Sperm After 4-6 Weeks Irradiation			
(rad)	4 weeks	5 weeks	6 weeks	5+6 weeks
0	2.40	2.58	2.14	2.36
Neutrons				
1	2.23	2.97	2.40	2.68
2.5	3.20	4.50	2.13	3.32
10	3.40	8.63	5.70	7.17
40	8.89	18.44	30.88	24.64
γ-Rays				
22.5	2.73	5.13	2.33	3.73
145	8.71	11.17	22.63	16.90

* Data from Grahn *et al.* (1984).

The RBEs for preimplantation losses were between 15 and 25 when the dose was above 10 rad and possibly higher when the dose was below 10 rad, although the data are statistically "noisy." While these studies clearly indicated that radiation of the testes produced abnormal sperm and karyotype abnormalities, the authors did not study the genetic changes in surviving offspring of the irradiated males.

Table 9. Percentage of Cells with Reciprocal Chromosome Translocations and their Distribution by Number Per Cell*

Dose (rad)	Number of Cells with 0, 1, 2 or 3 translocations				
	n	0	1	2	% ± S.E.
Colony control	29	6042	2	0	0.033±0.023
Neutrons					
1	11	3158	2	0	0.063±0.033
2.5	11	3591	9	0	0.25 ± 0.10
5	7	1804	6	0	0.33 ±0.19
10	7	1881	19	0	1.00±0.26
20	4	1077	8	0	0.74±0.28
40	4	1175	25	0	2.08±0.26
γ−Rays 22.5	7	2046	4	0	0.195±0.092
45	7	1593	7	0	0.44±0.05
45	5	1593	7	0	1.52±0.31
145	5	1556	24	0	1.52±0.31

* Data from Grahn *et al.* (1984).

Nomura *et al.* (1988) did examine the offspring of irradiated animals and animals treated with other mutagenic agents. These authors observed a large and significant increase of phenotypic anomalies in the progeny of ICR parent mice treated before mating with X-rays (Table 11), urethane, 7,12-dimethylbenz[a]anthracene, ethylnitrosourea (ENU), and 4-nitroquinoline 1-oxide, but the increase was not significant with furylfuramide. Major types of induced anomalies were cleft palate, dwarfs, open eyelid, tail anomalies, and exencephalus. Dwarfs, open eyelid and tail anomalies were predominant types of viable anomalies and were inherited as if they were dominant mutations with varying expressivity or penetrance. These authors also observed that the incidence of prenatal anomalies increased with the dose. Spermatogonia were less sensitive to X-rays and urethane than spermatozoa, while ENU induced a very high incidence of prenatal anomalies following spermatogonial treatment. In contrast to the previous publications in which X-rays were utilized, there was a clear, almost linear increase of anomalies in the dose range from 0-216 rad after spermatogonial exposure. There is a significant increase in the frequency of malformations following exposure to x-ray during spermatogenesis (Table 11), but the increase is numerically small, when compared to the frequency of induced malformations following the exposure of developing embryos during organogenesis to similar doses of radiation. Furthermore, this increase in the frequency of malformations was observed in the fetuses, but was not observed in the live-born mouse fetuses (Table 10).

Table 10. Comparison of Anomaly Rate in the F_1 Offspring after Spermatogonial Treatment with X-rays *

Age at Treatment	F_1 fetuses	Live-born mice
Day-8 embryo	0/60 (0.0)	0/255 (0.0)
Day-14 fetus	1/72 (1.4)	0/175 (0.0)
21 days	5/297 (1.7)	0/62 (0.0)
63-65 days	9/496 (1.8)	0/239 (0.0)

* Data from Nomura (1988).

A number of laboratories have attempted to perform a risk assessment of the mutagenic effect of various environmental agents. This is not an easy task because the actual impact of the mutagenic effects is related to dose, time of gonadal development and time after exposure that conception occurs. Furthermore the best methodology for determining human risks is to utilize human epidemiological data. The problem of using human data is that you have to accept the exposures to which the population have been exposed, even when the exposures are very low. When dealing with low risk phenomena, one needs large populations to demonstrate an effect. In many of the human epidemiological studies the populations are so small that even if there were a small reproductive effect, it would not be discerned. Animal investigators have capitalized on this deficiency, since they can use multiple doses that are high and they can permit conception at times post-exposure that will maximize any genetic or cytogenetic effects.

Van Buul (1984) studied the enhancement of radiosensitivity for the induction of translocations in mouse stem cell spermatogonia following treatment with chemotherapeutic agents. The effects of pretreatment of mouse spermatogonial stem cells with cyclophosphamide (100 and 200 mg/kg) and adriamycin (2.5, 5 and 7 mg/kg) on the induction of chromosomal translocations by high doses (800 or 900 rad) of X-rays applied 24 hours later are summarized in Table 11. Both compounds were able to alter the chromosomal radiosensitivity of surviving stem cells. It is concluded that depletion of differentiation and differentiated spermatogonia is sufficient for triggering stem cells into a more radiosensitive phase.

The mutagenic potential of ^{125}I was studied by Lavu *et al.* (1984) using the dominant lethal (DL) test (Table 12). Dominant lethality represents embryonic death resulting from the chromosomal breakage in the gametes of parents. Significant pre-implantation losses were observed. Dead implantations per pregnant female in the isotope treated groups showed a significant increase from controls indicating induced levels of post-implantation losses. All the stages of spermatogenesis, i.e., spermatozoa, spermatid, spermatocyte and spermatogonia were found to be sensitive to the induction of post-implantation losses, the spermatid stage being the most sensitive.

4. Estimating human genetic risks from *in vivo* animal experiments

The following investigators have designed experiments in an effort to convert the results of exposure of animals to mutagenic agents into actual human clinical genetic risks. Benova *et al.* (1985, Table 13) utilized ionizing radiation as the mutagen in mice. Gamma rays, deuterons and neutrons were utilized in these studies. Using conditions which would maximize the genetic manifestations in the F_1 generation the

Table 11. Induction of Chromosomal Aberrations in Stem Cell Spermatogonia and Testis Weights Following Combined Treatments with Adriamycin (A) or Cyclophosphamide (CP) and X-Rays With A 24 Hour Interval Between Exposure to the Chemical and Irradiation*

Treatment	Number of mice	Number of cells analyzed	% Translocations ± SEM	Testis weights in mg ± SEM
0	11	2200	0.1 ± 0.1	133 ± 5
CP 200 mg/kg...	5	1000	0.2 ± 0.1	135 ± 9
A 5 mg/kg...	5	1000	0.7 ± 0.2	107 ± 8
800 rad	5	930	10. 5 ± 2.6	79 ± 10
CP 200 mg/kg... + 800 rad	5	950	14.9 ± 1.2	74 ± 8
A 7 mg/kg...+ 800 rad	5	900	21.5 ± 5.8	57 ± 7

CP = cyclophosphamide; A = adriamycin
* Data from van Buul (1984).

Table 12. Number and Percentage Distribution of Dead Implantation Pregnant Female Mice Mated with Males Treated with Different Doses of ^{125} Iodine*

Group	Pregnant females	Number of dead implan-tations	Number of pregnant females with dead implantations					
			0	1	2	3	4	5
Control	393	139	213 54.2%	101 25.7%	31 7.9%	7	0	0
5mCi	357	183	174 48.7%	115 32.2%	38 10.6%	18 5.0%	6 1.7%	6 1.7%
10mCi	354	205	150 42.3%	119 33.6%	56 15.8%	15 4.2%	8 2.3%	7 2.0%
15mCi	355	208	147 41.4%	101 28.5%	70 19.7%	20 5.6%	4 1.1%	13 3.7%

* Data from Lavu *et al.* (1984).

investigators estimated the genetic results of exposing a population to 0.01 Gy (1 rad, Table 13).

Even if these results were transformed into genetic diseases with clinical manifestations it would be impossible to identify the increase in genetic disease unless there was a very large number of individuals in the exposed population. For example, Benova *et al.* predicted that 0.01 Gy would result in 39 additional translocations (Table 13). Since there are approximately 5000 chromosomal abnormalities per million live births, it is obvious that even if the exposed and control populations were each one million, statistical analysis would not differentiate between 5000 and 5039 karyotype abnormalities. Similarly, 25 induced clinical abortions would be unidentifiable in one million births containing 150,000 spontaneous abortions (Table 1, 13).

Table 13. Genetic Radiation Risk Assessment Based on Experimental Mutagenesis in Laboratory Animal (Mice)*

1) Estimated risk of 0.01 Gy gamma irradiation 39 translocations / million conceptions. 5 cases of multiple congenital malformations 25 additional clinical abortions / million conceptions 49 abortions in undiagnosed pregnancies.
2) Estimated risk of chronic gamma irradiation of 1.3 mrad / minute, 10 mrad / minute and 17 mrad / minute was three to ten times less effective.
3) 4.2 GeV deuterons proved inferior in effectiveness to gamma irradiation.
4) Chronic exposure to 4.1 MeV neutrons, 80 mrad / minute, was 7 times as effective as chronic gamma irradiation.

*Data from Benova *et al.* (1985), Med. Acad. Res. Inst. Radiobiol. Sofia.

The International Commission for Protection Against Environmental Mutagens and Carcinogens (ICPEMC) has issued a number of reports dealing with the basic science of mutagenesis and the impact of environmental mutagenesis on human disease. The Committee 4 Final Report dealt with the estimation of genetic risks from environmental mutagens (Lyon *et al.* 1985, Table 15 and 16). The committee concluded that in estimating the genetic hazards of environmental mutagens there are major problems in extrapolating from experimental data to human situations. While the committee suggested methods of improving our ability to extrapolate from animal data, they also believed that we were far from achieving that goal.

The Committee stated, "At present extrapolation can rarely be justified except where there are data on mutagenicity in the germ cells of animals." Thus the committee had little confidence in utilizing mutation data from *in vitro* cellular systems for determining mutagenic risks in humans. They also suggested that the genetic dose of a mutagen may be significantly different than the exposure dose. Utilizing *in vivo* animal data the committee estimated the Relative Mutagenic Effect, which is the ratio of the spontaneous plus induced mutations divided by the spontaneous mutation rate (Table 14, 15). The estimates indicate that the risks for most of the mutagenic drugs that were studied were greater in the first three months after treatment. The difficulty with this data is that we do not have any human epidemiological data that agrees with the estimates obtained from animal research. First generation genetic disease predicted from animal data indicates increases in genetic disease from 2% to 70%

Table 14. Cases of Genetic Disease Expected to be Induced Per 10^6 Births by About Three Months Treatment With Some Cytotoxic Drugs, and Relative Risk*

	First three months (post meiotic stages)					
Drug	Generation 1		Generation 2		Total all Gens.	
	Cases	R.R.	Cases	R.R.	Cases	R.R.
Adriamycin	9,000	1.09	7,600	1.07	60,000	1.6
CP	180,000	2.7	150,000	2.4	1,200,000	12
MMC	2,100	1.02	1,800	1.02	14,000	1.1
Myleran	-	-	-	-	-	-
Natulan	31,000	1.3	26,000	1.25	210,000	3.0
TEM	1,000,000	10	840,000	9.0	6,600,000	64
Thio TEPA	76,000	1.7	63,000	1.6	510,000	5.9

R.R. = relative risk, assuming normal incidence of 105,000 cases of genetic disease per 10^6 live births
Results have been rounded to two significant figures
CP = cyclophosphamide: MMC = mitomycin C
* Data from Lyon *et al.* (1985).

during the first three post-meiotic months, depending on the drug (Table 14). The total number of patients with induced genetic disease over many generations would be substantial, provided that there would be continuous or protracted exposure to the mutagenic agent. If conception occurs after the first three postmeiotic months from the discontinuation of treatment, then the incidence of induced genetic disease is diminished (Table 15). There is no human epidemiological data to support these estimates and in fact the data that is available would indicate that the risks are lower.

Table 15. Cases of Genetic Disease Expected to be Induced Per 10^6 Births by About Three Months Treatment With Some Cytotoxic Drugs, and Relative Risk to be Expected in the Exposed Population*

	After three-month interval (spermatogonial stages)					
Drug	Generation 1		Generation 2		Total all Gens.	
	Cases	R.R.	Cases	R.R.	Cases	R.R.
Adriamycin	36,000	1.3	30,000	1.3	240,000	3.3
CP	4,800	1.05	4,200	1.04	42,000	1.4
MMC	6,400	1.06	5,600	1.05	56,000	1.5
Myleran	2,500	1.02	2,200	1.02	22,000	1.2
Natulan	18,000	1.2	160,000	1.2	160,000	2.5
TEM	25,000	1.2	22,000	1.2	220,000	3.1
Thio TEPA	20,000	1.2	17,000	1.2	130,000	2.2

R.R. = relative risk, assuming normal incidence of 105,000 cases of genetic disease per 10^6 live births
Results have been rounded to two significant figures
CP = cyclophosphamide: MMC = mitomycin C
* Data from Lyon *et al.* (1985).

5. Biological Plausibility Of Published Estimated Risks Of Male Mediated Developmental Toxicity

a. Possible mechanisms of male mediated reproductive toxicity

Almost all of the theoretical reproductive risks of exposing males to chemicals, drugs and physical agents have been demonstrated in experimental animals or *in vitro* systems:

Mutations
Karyotype abnormalities
Gonadocyte cytotoxicity
Ejaculates containing toxic substances

There are even animal experiments that suggest that pregnant females exposed to ejaculates containing toxic doses of drugs or chemicals could induce embryonic losses or congenital malformations (Hales *et al.*, 1986).

There is very little possibility that Sperm containing drugs or chemicals can deliver toxic doses of these agents to the ovum or the embryo that would result in an increase in congenital malformations although it is remotely possible that the early zygote might be killed. This was discussed in an earlier section of this paper.

b. Super mutagens and site-directed mutagenesis

There have been suggestions that a small percentage of chemicals or drugs that are already in existence or are yet to be synthesized, may have much greater mutagenic potential than the array of mutagenic agents that have been studied. While with the aid of complicated molecular biology techniques we can synthesize agents that have the capacity for site directed mutagenesis, it is very unlikely that such compounds would be created by chance in the laboratory. Furthermore since mutagens damage the DNA or the chromosomes they are also cytotoxic at higher doses. In a sense this is "natures way" of protecting the genome from so-called "super mutagens."

c. Imprinting pattern alterations and epigenetic factors

Genomic imprinting in mammals may provide an explanation for a remarkably diverse set of observations or conditions whose genetic transmission and expression does not conform to the predictions of single gene inheritance - an explanation for traits that do not Mendelize (Hall 1990). The concept of imprinting has broad implications with regard to the mechanisms of gene expression. With regard to male mediated reproductive toxicity, imprinting may have implications pertaining to special mutagenic susceptibility of some genes on the male genome. It would appear that the marking or imprinting of DNA normally occurs during gametogenesis in some areas of the mammalian genome and is reversible; that is it is not a mutation or permanent change, but rather a modification which can normally be "wiped off" or reestablished when germ cells are produced in the next generation. The transgenic mouse work suggests that in those cases where the transgene is imprinted, differential expression of the transgene is associated with methylation. While the concept of imprinting explains the differential expression of genes on the male and female genome, the molecular mechanisms involved are still being investigated. Whether imprinting makes some sites on the male genome more susceptible to mutation is yet to be determined (Hall 1990).

d. Discrepancies among the human epidemiological data and between the human and animal data

An extensive and thorough study of the reproductive risks of paternal occupation found both positive and negative odds ratios for the frequency of birth defects (Olshan 1991). Taskinen (1990) also reported negative and positive relative risks for abortion and congenital malformations in paternal occupation studies. Both these articles should be reviewed for their bibliographic citations and analysis. Table 17 refers to some of these studies.

Epidemiology studies concerned with teratogenic agents can expect to observe an increase in a cluster of malformations that may eventually be identifiable as a syndrome. Furthermore, repeat studies will tend to report these findings if the agent being investigated is truly teratogenic. Paternal occupation studies would be expected to duplicate teratology epidemiological studies only if the ejaculate contained teratogenic agents from paternal occupational exposure. While this is theoretically possible, it is unlikely that teratogens contained in the male ejaculate will account for even a small percentage of the positive studies reported in Table 16.

On the other hand if investigators are primarily concerned with the possibility that the father's exposure is modifying the genome in the male gonadocytes (spermatogonia, spermatocytes and sperm), then the epidemiologists have a very difficult task. One would not expect to find any specificity in the increase in the mutation rate or the incidence of malformations, since mutations at the molecular level and chromosome level will primarily be randomly induced. Furthermore, only a small percentage of malformations have a simple genetic basis, so that genetically induced malformations will be overshadowed by the large number of congenital malformations that occur and do not have a primarily genetic basis. Probably, the one reproductive effect that may lend itself to epidemiological investigation is the frequency of spontaneous abortion. In fact, many of the studies dealing with the reproductive impact of the paternal occupation have dealt with the frequency of abortion.

Table 16. Studies of the Relative Risk of Birth Defects and Reproductive Failure Among Paternal Occupation Groups

Bonde *et al.*, 1990	3702 Danish metal workers accounting for 47674 person-years	Reduction in fertility with welding of mild steel, but not stainless steel
Brender and Suarez 1990	Vital records case control study of 727 cases with anencephaly	Positive association for painters and solvent exposure
Erickson *et al.*, 1970	233 cases of CNS abnormalities	Positive association for managers in business and repair services
Erickson 1984	Vietnam war veterans (Agent orange)	No increased risk for birth defects
Heminski *et al.*, 1980	Studied 3,300 birth defect cases	No assoc. in occupational groups
Lindbohm *et al.*, 1991	Spontaneous abortion frequency was evaluated in 99,196 pregnancies in Finland	In four of 25 job classifications, there was an increased relative risk for abortion. Overall the OR = 1.0
McDonald 1989	47,882 women with 605 birth defects	No. assoc. with chromosomal defects
MacDowall 1985	24,922 birth defect cases	Positive and negative associations. The positive associations were for a number of blue-collar jobs. Negative associations were for accountants and sales.
Olshan, 1991	22,192 birth defects in 39 categories. Final analysis included 14,415 cases of 5 defects or more per occupation. A monumental study.	More positive and negative associations.
Olsen 1983	CNS malformations	Positive association with painters.
Papier 1985	1,481 cases of birth defects	No assoc. with paternal occupation
Poldenak & Janerich 1983	171 cases of anencephaly and/or spina bifida	Positive association
Roan *et al.*, 1984	Agriculture pilots	No increased risk for birth defects
Sever 1988	Nuclear plant employees	No increased risk for birth defects
Smith *et al.*, 1982	Pesticide applications	No increased risk for birth defects
Taskinen 1990	an extensive review article dealing with parental occupation exposures	Six studies of spontaneous abortion in exposed fathers indicated that 3 were positive and 3 were negative.
Townsend 1982	Dioxin workers	No increased risk for birth defects

There are two other difficulties that have to be explained. If patient populations that have survived large doses of chemotherapeutic agents and radiation do not clinically demonstrate an increase in mutations or birth defects in their offspring, why should we expect low exposures of chemicals in the work place to induce readily discernable genetic disease, congenital malformations and/or abortion? One hypothesis is that the mutagenic effect of some mutagens may be diminished at higher exposures because many of the cells exposed to the mutagen do not survive. Thus one could theorize that chronic low occupational exposures may have a measurable reproductive effect when acute high exposures do not. Another possible explanation for the differences in reproductive problems between the occupational and cancer populations is that the abnormal gonadocytes are depleted with the passage of time in the cancer survivors. Thus, protracted occupational exposures to mutagens may be more effective because the abnormal gonadocytes are constantly being replenished, while cancer survivors may have years to deplete their population of abnormal gonadocytes before procreation is initiated. This is clearly demonstrated in the animal investigations of Alavanti and Searle in which even a short-time interval following exposure can decrease the mutagenic effect (1985, Table 17).

Table 17. Male Mice Administered 10 Gy X-Irradiation at Ages 12-17 Weeks of Age (5 Gy + 5 Gy, 24 hours apart)*

Radiation day	Number of mice	Testis wt (g)	Spermatocytes examined	Translocations per cell	Cells with translocations
16-19	7	84.2	680	0.57	45.7
39-42	5	106.1	500	0.35	30.4
64-66	6	112.3	211	0.35	28.8

* Data from Alavanti and Searle (1985).

The second difficulty pertains to the significance of the positive results in the paternal occupations studies. Each investigator or reviewer indicates the possibility that the positive results may be due to some type of bias in the collection of the data or the selection of the subjects. This matter must be resolved since in some of the studies, as much as 30% of the significant findings would indicate that some jobs provide protection if the data were interpreted to have a causal relationship.

One important result of discussing the topic of environmental mutagenesis is that it forces scientists to put environmental problems into perspective. It should be apparent to all scientists with any interest in this field that although there is a legitimate controversy regarding the magnitude of the problem of environmental mutagenesis, there is no such controversy regarding environmental carcinogenesis. Human studies demonstrate quite clearly that high doses of mutagenic chemicals and ionizing radiation have produced an increase in cancer in exposed populations. Since there is a significant overlap in the mechanisms involved in mutagenesis and carcinogenesis, this observation should be a significant lesson to biologists, oncologists geneticists and epidemiologists. Cancer may be a more sensitive indicator of the presence of environmental mutagens than changes in the incidence of genetic disease. Table 18 summarizes the relative sensitivity of three targets of ionizing radiation. Epidemiological studies and animal studies have indicated that a increases in genetic disease is the most difficult to document because the manifestations of new mutations and clinical genetic disease are so difficult to delineate. Therefore it would appear that monitoring the incidence of cancer would be the most efficient method of recognizing new environmental mutagens.

Table 18. Relative Sensitivity to the Reproductive Effects of Ionizing Radiation

Acute, High Dose	Protracted, Low Dose
Embryotoxicity Cancer Induction	Cancer Induction
Mutagenesis	Mutagenesis Embryotoxicity

SUMMARY

There has been renewed interest in the impact of environmental toxins on the male reproductive capacity. Although the interest has been long-standing, society's interest in environmental toxicity, improved work safety and the proliferation of new chemicals has more recently focused on this area. There has been a long-standing polarization of feelings among scientists with regard to the risk of male mediated reproductive effects (Table 2). There are three categories of scientists who have demonstrated an interest in male reproductive toxicity: 1) Scientists who are concerned about occupational male mediated risks and believe that the risks are significant. 2) Scientists who believe that male mediated reproductive risks exist as a problem, but that the actual risks are very small, when compared to other environmental risks. 3) Scientists who are more concerned with the notoriety seeking potential of exaggerating the mutagenic risks from environmental chemicals.

CONCLUSIONS

1) There are substantial data from animal experiments and epidemiological studies that indicate that drugs and chemical can decrease male fertility by decreasing the number of sperm or by causing the production of abnormal sperm. These results have been observed in occupational epidemiological studies and in patient populations who have survived cancer chemotherapy or radiation therapy. In the population of male cancer survivors there is a suggestion that the males are at greater risk for infertility than the females. Thus, subfertility or infertility in men are proven risks from certain exposures to some chemicals and chemotherapeutic agents.

2) There is extensive mammalian animal investigations to indicate that males exposed to radiation and some chemicals before inseminating a female have offspring with an increase in dominant lethals, point mutations and chromosome abnormalities. It is true that the maximum manifestation of some of these effects has a narrow window of time for expression, because of the differential sensitivity of spermatogonia, spermatocytes and sperm and because the altered spermatocytes and sperm have a limited time for transmitting the genetic alteration. Even in animal studies where the experimental conditions have been established to maximize the mutagenic effect, very low exposures to mutagenic agents result in offspring that cannot be differentiated from the controls.

3) Patients who have survived therapeutic doses of radiation and chemotherapy as children or young adults may have problems with infertility, but the fertile population does not appear to have a measurably increased risk of spontaneous abortion or offspring with congenital malformations. It is true that even in the larger studies the populations are small. It is therefore possible that if there were a small risk for the induction of chromosomal aberrations and genetically caused congenital malformations that these studies would not be able to recognize them. Furthermore,

patients frequently wait years following cancer treatment before having children and of course there may be two decades between childhood cancer treatment and procreation. The very low risk of transmitting induced mutations to the offspring of treated cancer survivors, is a reflection of the very low risk of mutagenesis from environmental mutagens in general.

4) Occupational studies that have examined the question of male mediated reproductive toxicity have reported that males exposed to some occupational tasks have a greater risk for spontaneous abortions and offspring with birth defects. Of course there are also many negative studies that have been published. In fact, in some reports indicating that some occupational tasks are associated with odds ratios that are increased for reproductive effects, there are other job classifications that report significantly decreased odds ratios. Overall, it does appear the reproductive risks are greater in the blue collar population. Extensive discussions in these reports deal with whether the results are due to actual toxic exposures or to case selection bias or data collection bias. Further research in this area is obviously needed. The field could benefit from the combined efforts of reproductive specialists, epidemiologists, geneticists and toxicologists. It would have been helpful if the epidemiology studies had attempted to measure exposures in the populations being studied.

5) Although the matter of toxic chemicals in the ejaculate was mentioned, there are only animal data to support this concept as a legitimate line of investigation. It does not appear that any of the epidemiological studies determined whether toxic chemicals were present in the ejaculate of the male subjects. In fact there would have to be significant animal investigation to lay down the basis for conducting an appropriate human study.

6) "Mutation monitoring" is not an efficient method of determining the mutagenic potential of environmental agents. Monitoring the population for an alteration in cancer incidence would be a far more efficient process for discovering many environmental mutagens.

REFERENCES

Adler, I.D., Ashby, J., and Wurgler, F.E.: Screening for Possible Human Carcinogens and Mutagens: A Symposium Report. *Mutation Res.*, 213:27-39, 1989.

Al-Shawaf, T., Nolan, A., Harper, J., Serhal, P., and Craft, I.: Case Report - Pregnancy Following Gamete Intra-Fallopian Transfer (GIFT) with Cryopreserved Semen From Infertile Men Following Therapy to Lymphomas or Testicular Tumour: Report of Three Cases. *Human Reproduction*, 6(3):365-366, 1991.

Alavanti, D. and Searle, A.G.: Effects of Post-Irradiation Interval on Translocation Frequency in Male Mice. *Mutation Res.*, 142:65-68, 1985.

Albanese, R.: Review - Mammalian Male Germ Cell Cytogenetics. *Mutagenesis*, 2(2):79-81, 1987.

Albertini, R.J.: Somatic Gene Mutations In Vivo as Indicated by the 6-Thio-Guanine-Resistant T-Lymphocytes in Human Blood. *Mutat. Res.*, 150:411-422, 1985.

Ames, B.N.: Mutagenesis and Carcinogenesis: Endogenous and Exogenous Factors. *Environ. and Molec. Mutagenesis*, 14(16):66-77, 1989.

Anderson, D., and Styles, J.A.: The Bacterial Mutation Test. *Br. J. Cancer*, 37:924-930, 1978.

Asby, J.: Series: Current Issues in Mutagenesis and Carcinogenesis', No. 26. Genotoxicity Data Supporting the Proposed Metabolic Activation of Ethyl Carbamate Urethane) to a Carcinogen: The Problem Now Posed by Methyl Carbamate. *Mutation Res.*, 260:307-308, 1991.

Auletta, A.E., Martz, A.G., and Parmar, A.S.: Mutagenicity of Nitrosourea Compounds for Salmonella Typhimurium. *J. Natl. Cancer Inst.*, 60:1495-1497, 1978.

Auroux, M.R., Dulioust, E., Selva, J., and Rince, P.: Cyclophos-phamide in the F_0 Male Rt: Physical and Behavioral Changes in Three Successive Adult Generations *Mutation Res.*, 229:189-200, 1990.

Auroux, M.R., Dulioust, E.J.B., Nawar, N.N.Y., Yacoub, S.G., Mayaux, M.J., Schwartz, D., and David, G.: Antimitotic Drugs in the Male Rat Behavioral Abnormalities in the Second Generation. *J. Androl.*, 9:153-159, 1988.

Baird, P.A.: Measuring Birth Defects and Handicapping Disorders in the Population: The British Columbia Health Surveillance Registry. *Can Med. Assoc. J.* 36:109-111, 1987.
Baverstock, K.F: DNA Instability, Paternal Irradiation and Leukaemia in Children Around Sellafield. *Int. J. Radiat. Biol.*, 60(4):581-595, 1991.
Benedict, W.F., Baker, M.S., Haroun, L., Choi, E. and Ames, B.N.: Mutagenicity of Cancer Chemotherapeutic Agents in the Salmonella Microsome Test. *Cancer Res.*, 37:2209-2213, 1977.
Bennett, J. and Pedersen, R.A.: Early Mouse Embryos Exhibit Strain Variation in Radiation-Induced Sister-Chromatid Exchange: Relationship with DNA Repair. *Mutation Res.*, 126:153-157, 1984.
Bever, M.B.: Ionizing Radiation: Genetic Effects in Humans. In Bever, M.B. (Ed) Encyclopedia of Materials Science and Engineering. Pergamon Press, pp. 2421-2422, 1986.
Boerjan, M.L. and Saris, L.A.: The Effects of Spermatozoal Irradiation with X-Rays on Chromosome Abnormalities and on Development of Mouse Zygotes After Delayed Fertilization. *Mutation Res.*, 256:49-57, 1991.
Bond, D.J., and Chandley,A.C.: "Aneuploidy" Oxford Monographs on Medical Genetics, Vol. 11, New York: Oxford University Press, 1983.
Bonde, J.P., Hansen, K.S., and Levine, R.J.: Fertility Among Danish Male Welders. *Scand. J. Work Environ. Health*, 16:315-322, 1990.
Brandriff, B.F., Gordon, L.A., Segraves, R., and Pinkel, D.: The Male-Derived Genome After Sperm-Egg Fusion: Spatial Distribution of Chromosomal DNA and Paternal-maternal Genomic Association. *Chromosoma*, 100:262-266, 1991.
Brender, J.D., and Suarez, L.: Paternal Occupation and Anencephaly. *Am. J. Epidemiol.* 131: 571-21, 1990.
Brent, R.L.: Another Possible Cause of Spontaneous Mutations. U.S.A.E.C.D. U.R.-313, 1954.
Brent, R.L.: Protecting the Public From Teratogenic and Mutagenic Hazards. *J. Clin. Pharmacol.* 12:61-70, 1972.
Brown, N.A.: Are Offspring at Risk From Their Father's Exposure to Toxins? *Nature*, 316:110, 1985.
Burgess, W.A.: "Recognition of Health Hazards in Industry. New York: John Wiley & Sons, 1981.
Byrne, J., Mulvihill, J.J., Connelly, R.R., *et al.*: Reproductive Problems and Birth Defects in Survivors of Wilms' Tumor and Their Relatives. *Med. Pediatr. Oncol.*, 16:233-240, 1988.
Cassidy, S.B., Gainey, A.J., and Butler, M.G.: Occupational Hydrocarbon Exposure Among Fathers of Prader-Willi Syndrome Patients With and Without Deletions of 15q. *Am. J. Hum. Genet.*, 44:806-810, 1989.
Cattanach, B.M. and Rasberry, C.: Genetic Effects of Combined Chemical-X-Ray Treatments in Male Mouse Germ Cells. *Int. J. Radiat. Biol.*, 51(6):985-996, 1987.
Chellman, G.J., Hurtt, M.E., Bus, J.S., and Working, P.K.: Role of Testicular Versus Epididymal Toxicity in the Induction of Cytotoxic Damage in Fischer-344 Rat Sperm by Methyl Chloride. *Reproductive Toxicology*, 1(1):25-35, 1987.
Cordier, S., Deplan, F., Mandereau, L., and Hemon, D.: Paternal Exposure to Mercury and Spontaneous Abortions. *Br. J. Ind. Med.*, 48:375-381, 1991.
Cralley, L.V., and Cralley, L.J.: "Industrial Hygiene Aspects of Plant Operations." Vol. 1-3, New York: MacMillan Publishing Co., 1982-1985.
Crow, J.F.: Population Perspective. In Hilton, B., Callahan, D., Harris, M., Condlife, P., and Berkley, B. (Eds.), "Ethical Issues in Human Genetics," Plenum Publishing Corp., New York, pp. 73.
Czeizel, A. and Kis-Varga, A.: Mutation Surveillance of Sentinel Anomalies in Hungary, 1980-1984. *Mutation Res.*, 186:73-79, 1987.
Czeizel, A., Sankaranarayanan, K., Losonci, A., Rudas, T., and Keresztes, M.: The Load of Genetic and Partially Genetic Diseases in Man. II. Some Selected Common Multifactorial Diseases: Estimates of Population Prevalence and of Detriment in Terms of Years of Lost and Impaired Life. *Mutation Res.*, 196:259-292, 1988.
Davis, D.L.: Paternal Smoking and Fetal Health. *The Lancet*, 337:123, Jan. 12, 1991.
Davis, D.L., Friedler, G., Mattison, D., and Morris, R.: Male-Mediated Teratogenesis and Other Reproductive Effects: Biologic and Epidemiologic Findings and A Plea for Clinical Research. *Reproductive Toxicology*, 6:289-292, 1992.
Dearfield, K.L., Auletta, A.E., Cimino, M.C., and Moore, M.M.: Considerations in the U.S. Environmental Protection Agency's Testing Approach for Mutagenicity. *Mutation Res.*, 258:259-283, 1991.
Delehanty, J., White, R.L., and Mendelsohn, M.L. Approaches to Determining Mutation Rates in Human DNA. *Mutation Res.*, 167:215-232, 1986.

Dexeus, F.H., Logothetis, C.J., Chong, C., Sella, A., and Ogden, S.: Genetic Abnormalities in Men with Germ Cell Tumors. *J. Urol.*, 140:80-84, 1988.

Drake, J.W.: Mechanisms of Mutagenesis. *Environ. & Molec. Mutagenesis*, 14, Supplement 16:11-15, 1989.

Dulioust, E.J.B., Nawar, N.Y., Yacoub, S.G., Ebel, A.B., Kempf, E.H., and Auroux, M.R.: Cyclophosphamide in the Male Rat: New Pattern of Anomalies in the Third Generation. *J. Androl.*, 10:296-303, 1989.

Edwards, A.A., Lloyd, D.C., and Purrott, R.J.: Radiation Induced Chromosome Aberrations and the Poisson Distribution. *Rad. and Envirornm. Biophys.*, 16:89-100, 1979.

Ehling, U.H. and Neuhauser-Klaus, A.: Induction of Specific-Locus and Dominant Lethal Mutations in Male Mice by 1-Methyl-1-Nitrosourea MNU). *Mutation Res.*, 250:447-456, 1991.

Ehling, U.H. and Neuhauser-Klaus, A.: Induction of Specific-Locus and Dominant Lethal Mutations in Male Mice by Chlormethine. *Mutation Res.*, 227:81-89, 1989.

Ehling, U.H. and Neuhauser-Klaus, A.: Induction of Specific-Locus Mutations in Male Mice by Ethyl Methanesulfonate (EMS). *Mutation Res.*, 227:91-95, 1989.

Ehling, U.H.: Germ-Cell Mutations in Mice: Standards for Protecting the Human Genome. *Mutation Res.*, 212:43-53, 1989.

Ehling, U.H.: Induction of Specific-Locus Mutations in Male Mice by Diethyl Sulfate (DES). *Mutation Res.*, 214:329, 1989.

Engel, E. and DeLozier-Blanchet, C.D.: Uniparental Disomy, Isodisomy, and Imprinting: Probable Effects in Man and Strategies for Their Detection. *Am. J. Med. Genet.*, 40:432-439, 1991.

Erickson, J.D., Mulinare, J., McClain,P.W., Fitch, T.G., James, L.M., McClearn, A.B., and Adams, Jr., M.J.: Vietnam Veterans; Risks for Fathering Babies with Birth Defects. *J. Am. Med. Assoc.* 252:903-912, 1984.

Erickson, J.D., Cochran, W.M., and Andersen, C.E.: Parental Occupation and Birth Defects, A Preliminary Report. *Contrib. Epidemiol. Biostat.* 1:107-117, 1979.

Farrer, L.A., Cupples, L.A., Kiely, D.K., Conneally, P.M., and Myers, R.H.: Inverse Relationship Between Age at Onset of Huntington Disease and Paternal Age Suggests Involvement of Genetic Imprinting. *Am. J. Hum. Genet.* (In press).

Favor, J., Neuhauser-Klaus, A., and Ehling, U.H.: Radiation-Induced Forward and Reverse Specific Locus Mutations and Dominant Cataract Mutations in Treated Strain BALB/c and DBA/2 Male Mice. *Mutation Res.*, 177:161-169, 1987.

Favor, J., Neuhauser-Klaus, A., Kratochvilova, J., and Pretsch, W.: Towards an Understanding of the Nature and Fitness of Induced Mutations in Germ Cells of Mice: Homozygous Viability and Heterozygous Fitness Effects of Induced Specific-Locus, Dominant Cataract and Enzyme-Activity Mutations. *Mutation Res.*, 212:67-75, 1989.

Favor, J., Strauss, P.G., and Erfle, V.: Molecular Characterization of a Radiation-Induced Reverse Mutation at the Dilute Locus in the Mouse. *Genet. Res. Camb.*, 50:219-223, 1987.

Favor, J.: Mammalian Germ Cell Mutagenesis Data and Human Genetic Risk: *Biol. Zent. Bl.*, 108:309-321, 1989.

Fedrick, J.: Anencephalus in the Oxford Record Linkage Study Area. Develop. Med. Child. Neurol. 18:643-656, 1976.

Ficsor, G., Oldford, G.M., Loughlin, K.R., Pands, B.B., Dubien, J.L., and Ginsberg, L.C.: Comparison of Methods for Detecting Mitomycin C-Ethyl Nitrosourea-Induced Germ Cell Damage in Mice: Sperm Enzyme Activities, Sperm Motility, and Testis Weight. *Environmental Mutagenesis*, 6:287-298, 1984.

Fleiss, J.L.: The Mantel-Haenszel Estimator in Case-Control Studies with Varying Numbers of Controls Matched to Each Case. *Am. J. Epidemiol.* 120:1-3, 1984.

Fossa, S.D., Abyholm, T., Normann, N., and Jetne, V.: Post-Treatment Fertility in Patients with Testicular Cancer: III. Influence of Radiotherapy in Seminoma Patients. *Br. J. Urol.*, 58:315-319, 1986.

Franza, B.R. Jr., Oeschger, N.S., Oeschger, M.P. and Schein, P.S.: Mutagenic Activity of Nitrosourea Antitumor Agents. *J. Natl. Cancer Inst.*, 65:149-154, 1980.

Freud, A., Canfi, A., Sod-Moriah, U.A., and Chayoth, R.: Neonatal Low-Dose Gamma Irradiation-Induced Impaired Fertility in Mature Rats. *Isr. J. Med. Sci.*, 6:611-615, 1990.

Friedler, G.: Effects of Limited Paternal Exposure to Xenobiotic Agents on the Development of Progeny. *Neurobehav. Toxicol. Teratol.*, 76:739-743, 1985.

Friedler, G., and Cicero, T.J.: Paternal Pregestational Opiate Exposure in Male Mice: Neuroendocrine Deficits in Offspring. *Res. Comm. Substance Abuse*, 8:109-116, 1987.

Garber, J.E.: Long-Term Follow-Up of Children Exposed In Utero to Antineoplastic Agents. Seminars in Oncology, 16(5):437-444, 1989.

Generoso, W.M., Cain, K.T., Cornett, C.V., and Frome, E.L.: Comparison of Two Stocks of Mice in Spermatogonial Response to Different Conditions of Radiation Exposure. *Mutation Res.*, 249:301-310, 1991.

Genesca, A., Barrios, L., Miro, R., Caballin, M.R., Benet, J., Fuster, C., Bonfill, X., and Egozcue, J.: Lymphocyte and Sperm Chromosome Studies in Cancer-Treated Men. *Hum. Genet.*, 84:353-355, 1990.

Grahn, D., Carnes, B.A., and Farrington, B.H.: Genetic Injury in Hybrid Male Mice Exposed to Low Doses of ^{60}Co γ-Rays or Fission Neutrons. II. Dominant Lethal Mutation Response to Long-Term Weekly Exposures. *Mutation Res.*, 162:81-89, 1986.

Grahn, D., Carnes, B.A., Farrington, B.H., and Lee, C.H.: Genetic Injury in Hybrid Male Mice Exposed to Low Doses of ^{60}Co γ-Rays or Fission Neutrons. I. Response to Single Doses. *Mutation Res.*, 129:215-229, 1984.

Green, D.M., Fine,W.E., and Li, F.P.: Offspring of Patients Treated for Unilateral Wilms' Tumor in Childhood. *Cancer*, 49:2285-2288, 1982.

Green, D.M., Hall, B., and Zevon, M.A.: Pregnancy Outcome After Treatment for Acute Lymphoblastic Leukemia During Childhood or Adolescence. *Cancer*, 64:2335-2339, 1989.

Green, D.M., Zevon, M.A., and Hall, B.: Achievement of Life Goals by Adult Survivors of Modern Treatment for Childhood Cancer. *Cancer*, 67:206-213, 1991.

Green, D.M., Zevon, M.A., Lowrie, G., Seigelstein, N., and Hall, B.: Congenital Anomalies in Children of Patients Who Received Chemotherapy for Cancer in Childhood and Adolescence. *N. Engl. J. Med.*, 325:141-146, 1991.

Green, D.M., Zevon, M.A., Lowrie, G., Seigelstein, N., and Hall, B.: Congenital Anomalies in Children of Patients who Received Chemotherapy for Cancer in Childhood and Adolescence. *N. Engl. J. Med.*, 325:141-146, 1991.

Green, S., Auletta, A., Fabricant, J., Kapp, Robert, Manandhar, M., Sheu, C., Springer, J., and Whitfield, B.: Current Status of Bioassays in Genetic Toxicology - The Dominant Lethal Assay. *Mutation Res.*, 154:49-67, 1985.

Gridley, T.: Insertional Versus Targeted Mutagenesis in Mice. *The New Biologist*, 3(11):1025-1034, 1991.

Grosovsky, A.J. and Little, J.B.: Evidence for Linear Response for the Induction of Mutation in Human Cells by X-Ray Exposures Below 10 Rads. Proc. *Natl. Acad. Sci.*, 82:2092-2095, 1985.

Hakoda, M., Akiyama, M., Kyoizu i, S., Awal, A.A., Yamakido, M., and Otake, M.: Increased Somatic Cell Mutant Frequency in Atomic Bomb Survivors. *Mutat. Res.*, 210:39-48, 1988.

Hales, B. F. and Robaire, B.: Reversibility of the Effects of Chronic Paternal Exposure to Cyclophosphamide on Pregnancy Outcome in Rats. *Mutation Res.*, 229:129-134, 1990.

Hales, B. F., Smith, S., and Robaire, B.: Cyclophosphamide in the Seminal Fluid of Treated Males: Transmission to Females by Mating and Effect on Pregnancy Outcome. *Toxicology and Applied Pharmacology*, 27:602-611, 1986.

Hales, B.F., Smith, S., and Robaire, B.: Cyclophosphamide in the Seminal Fluid of Treated Males: Transmission to Females by Mating and Effect on Progeny Outcome. *Toxicology and Applied Pharmacology*, 84:423-430, 1986.

Hall, J.G.: Genomic Imprinting: Review and Relevance to Human Diseases. *Am. J. Hum. Genet.*, 48:857-873, 1990.

Hansen, S.W., Berthelsen, J.G., and von der Moaase, H.: Long-Term Fertility and Leydig Cell Function in Patients Treated for Germ Cell Cancer with Cisplatin, Vinblastine, and Bleomycin Versus Surveillance. *J. Clin. Oncol.* 8:1695-1698, 1990.

Hawkins, M.M., Smith, R.A., and Curtice, L.J.: Childhood Cancer Survivors and Their Offspring Studied Through A Postal Survey of General Practitioners: Preliminary Results. *J. R. Coll. Gen. Pract.*, 38:102-105, 1988.

Hemminki, K., Mutanen, P., Luoma, K., and Saloniemi, I.: Congenital Malformations by the Parental Occupation in Finland. *Int. Arch. Occup. Environ. Health*, 46:93-98, 1980.

Hens, L., Bonduelle, M., Liebaers, I., Devroey, P., and Van Steirteghem, A.C.: Chromosome Aberrations in 500 Couples Referred for In-Vitro Fertilization or Related Fertility Treatment. *Human Reproduction*, 3(4):451-457, 1988.

Hoglund, G.V., Iselius, E.L., and Knave, B.G.: Children of Male Spray Painters: Weight and Length at Birth. *Br. J. Ind. Med.*, 49:249-253, 1992.

Jenkinson, P.C., Anderson,D., and Gangolli, S.D.: Increased Incidence of Abnormal Fetuses in the Offspring of Cyclophosphamide-Treated Male Mice. *Mutation Research*, 188:57-62, 1987.

Kalen, B.: Epidemiology of Human Reproduction. Boca Raton, Florida, CRC Press, 1991.

Kalter, H., and Warkany J.: Congenital Malformations: Etiologic Factors and Their Role in Prevention. Part One. *New Engl. J. Med.*, 308:424-431, 1983.

Kelly, S.M., Robaire, B., and Hales, B.F.: Paternal Cyclophosphamide Treatment Causes Postimplantation Loss Via Inner Cell Mass-Specific Cell Death. *Teratology*, 45:313-318, 1992.
Kimball, R.F.: The Development of Ideas About the Effect of DNA Repair on the Induction of Gene Mutations and Chromosomal Aberrations by Radiation and by Chemicals. *Mutation Res.*, 196:1-34, 1987.
Knishkowy, B., and Baker, E.L.: Transmission of Occupational Disease to Family Contacts. *Am. J. Ind. Med.*, 9:543-550, 1986.
Kratochvilova, J., Favor, J., and Neuhauser-Klaus, A.: Dominant Cataract and Recessive Specific-Locus Mutations Detected in Offspring of Procarbazine-Treated Male Mice. *Mutation Res.*, 198:296-301, 1988.
Kucerova, M., Gregor, V., Horacek, J., Dolanska, M., and Matejckova, S.: Influence of Different Occupations with Possible Mutagenic Effects of Reproduction and Level of Induced Chromosomal Aberrations in Peripheral Blood. *Mutation Res.* 278:9-22, 1992.
Kyoizumi, S., Nakamura, N., Hakoda, M., *et al.*: Detection of Somatic Mutations at the Glycophorin A Locus in Erythrocytes of Atomic Bomb Survivors Using a Single Beam Flow Sorter. *Cancer Res.*, 49:581-588, 1989.
Langlois, R.G., Bigbee, W.L., Kyoizumi, S., Nakamura, N., Bean, M.A., Akiyama, N., and Jensen, R.H.: Evidence for Increased Somatic Cell Mutations at the Glycophorin A Locus in Atomic Bomb Survivors. *Science*, 236:445-448, 1987.
Lavu, S., Reddy, P.P., and Reddi, O.S.: Dominant Lethal Induction and Testicular Uptake of Iodine-125 in Mice. *Int. J. Radiat. Biol.*, 45(4):331-343, 1984.
Lewis, S.E.: The Biochemical Specific-Locus Test and a New Multiple-Endpoint Mutation Detection System: Considerations for Genetic Risk Assessment. *Environmental and Molecular Mutagenesis*, 18:303-306, 1991.
Li, F.P., and Jaffe, N.: Progeny of Childhood-Cancer Survivors. *Lancet*, 2:707-709, 1974.
Li, F.P., Fine,W., Jaffe, N., Holmes, G.C. and Holmes, F.F.: Offspring of Patients Treated for Cancer in Childhood. *J. Natl. Cancer Inst.*, 62:1193-1197, 1979.
Lian, Z.H., Zack, M.M., and Erickson, J.D.: Paternal Age and the Occurrence of Birth Defects. *Am. J. Hum. Genet.* 39:648-660, 1986.
Lindbohm, M.L., Hemminki, K., Bonhomme, M.G., Anttila, A., Rantala, K., Heikkila, P., and Rosenberg, M.J.: Effects of Paternal Occupational Exposure on Spontaneous Abortions. *Am. J. Public Health*, 81:1029-1033, 1991.
Lindbohm, M.L., Sallmen Markku, A.A., Taskinen, H., and Hemminki, K.: Paternal Occupational Lead Exposure and Spontaneous Abortion. Scand. J. Work Environ. *Health*, 17:95-103, 1991.
Loughlin, K.R. and Agarwal, A.: Use of Theophylline to Enhance Sperm Function. *Archives of Andrology*, 28:99-103, 1992.
Lowery, M.C., Au, W.W., Adams, P.M., Whorton, Jr., E.B., and Legator, M.S.: Male-Mediated Behavioral Abnormalities. *Mutation Res.*, 229:213-229, 1990.
Lowry, R.B., Miller, J.R., Scott, A.E., and Renwick, D.H.G.: The British Columbia Registry for Handicapped Children and Adults: Evolutionary Changes Over Twenty Years. *Can. J. Public Health*, 66:322-326, 1975.
Lyons, M.F., *et al.*: Committee 4 Final Report - Estimation of Genetic Risks and Increased Incidence of Genetic Disease Due to Environmental Mutagens. International Commission for Protection Against Environmental Mutagens and Carcinogens: ICPEMC document 114-1982-5.3. *Biol. Zbl.*, 104:57-87, 1985.
Lyon, M.F. and Renshaw, R.: Induction of Congenital Malformation in Mice by Parental Irradiation: Transmission to Later Generations. *Mutation Res.*, 198:277-283, 1988.
Mann, T., and Lutwak-Mann, C.: Passage of Chemicals into Human and Animal Semen: Mechanisms and Significance. *Crit. Rev. Toxicol.*, 2:1-14, 1982.
Martin-DeLeon, P.A. and Boice, M.L.: Sperm Aging in the Male After Sexual Rest: Contribution to Chromosome Anomalies. *Gamete Res.* 12:151-163, 1985.
Matheson, D., Brusick, D., and Carrano, R.: Comparison of the Relative Mutagenic Activity for Eight Antincoplastic Drugs in the Ames Salmonella Microsome and TK +/- Mouse Lymphoma Assays. *Drug Chem. Toxicol.*, 1:277-304, 1978.
Matney, T.S., Nguyen, T.V., Connor, T.H., Dana, W.J., and Theiss, J.C.: Genotoxic Classification of Anticancer Drugs. Teratogenesis Carcino. *Mutagen.*, 5:319-328, 1985.
Matsuda, Y., Tobari, I., Yamagiwa, J., Utsugi, T., Kitazume, M., and Nakai, S.: b-Ray-Induced Reciprocal Translocations in Spermatogonia of the Crab-Eating Monkey. *Mutation Res.*, 129:373-380, 1984.
McDonald, A.D., McDonald, J.C., Armstrong,B., Cherry, N.M., and Nolin, A.D.: Fathers' Occupation and Pregnancy Outcome. *Br. J. Ind. Med.* 46:329-333, 1989.

McDowall, M.E.: Occupational Reproductive Epidemiology. Series SMPS 50. Oondon: Her Majesty's Stationery Office, 1985.
Meistrich, M.L.: A Method for Quantitative Assessment of Reproductive Risks to the Human Male. *Fundamental and Applied Toxicology*, 18:479-490, 1992 (I).
Meistrich, M.L., Goldstein, L.S., and Wyrobek, A.J.: Long-Term Infertility and Dominant Lethal Mutations in Male Mice Treated with Adriamycin. *Mutation Res.*, 152:53-65, 1985.
Merke, D.P., and Miller, R.W.: Age Differences in the Effects of Ionizing Radiation. In Guzelian, P.S., Henry, C.J., and Olin, S.S. (Eds), "Similarities and Differences Between Children and Adults: Implications for Risk Assessment." ILSI Press, Washington, DC, 1992.
Mulvihill, J.J., and Czeizel, A.: Perspectives in Mutation Epidemiology, 6 A 1983 View of Sentinel Phenotypes. *Mutat. Res.*, 123:345-361, 1983.
Mulvihill, J.J., Byrne, J., Steinhorn, S.A., *et al.*: Genetic Disease in Offspring of Survivors of Cancer in the Young. *Clin. Genet.*, 39:72, Abstract, 1986.
Mulvihill, J.J., Connelly, R.R., Austin, D.F., Cook, J.W., Holmes, F.F., Krauss, M.R., Meigs, J.W., Steinhorn, S.C., Teta, M.J., Myers, M.H., Byrne, J., Bragg, K., Hassinger, D.D., Holmes, G.F., Latourette, H.W., Naughton, M.D., Strong, L.C. and Weyer, P.J.: Cancer in Offspring of Long-Term Survivors of Childhood and Adolescent Cancer. *The Lancet*, Saturday, October 10, 1987.
Murota, T. and Shibuya, T.: The Induction of Specific-Locus Mutations with N-Proply-N-Nitrosourea in Stem-Cell Spermatogonia of Mice. *Mutation Res.*, 264:235-240, 1991.
Myrianthopoulos, N.C., and Chung, C.S.: Congenital Malformations in Singletons: Epidemiologic Survey. Birth Defects Original Article Series. Vol. 10, No. 11, New York, Stratton Intercontinental, 1974.
Nagae, Y., Miyamoto, H., Suzuki, Y., and Shimizu, H.: Effect of Estrogen on Induction of Micronuclei by Mutagens in Male Mice. *Mutation Res.*, 263:21-26, 1991.
Nagao, T.: Frequency of Congenital Defects and Dominant Lethals in the Offspring of Male Mice Treated with Methylnitrosourea. *Mutation Res.*, 177:171-178, 1987.
Narod, S.A., Douglas, G.R., Nestmann, E.R., and Blakey,D.H.: Human Mutagens: Evidence from Paternal Exposure? *Environ. Molecular Mutagenesis*, 11:401-415, 1988.
National Research Council: Biologic Markers in Reproductive Toxicology. Washington, D.C., National Academy Press, 1989.
Neel, J.V., Satoh, C., Goriki, K., *et al.*: Search for Mutations Altering Protein Charge and/or Function in Children of Atomic Bomb Survivors: Final Report. *Am. J. Hum. Genet.*, 42:663-676, 1988.
Neel, J.V., Satoh, C., Hamilton, H.B., *et al.*: Search for Mutations Affecting Protein Structure in Children of Atomic Bomb Survivors: Preliminary Report. *Proc. Natl. Acad. Sci.*, USA, 77:4221-4225, 1980.
Neel, J.V., Schull, W.J., Awa, A.A., Satoh, C., Kato, H., Otake, M., and Yoshimoto, Y.: The Children of Parents Exposed to Atomic Bombs: Estimates of the Genetic Doubling Dose of Radiation for Humans. *Am. J. Hum. Genet.*, 46:1053-1072, 1990.
Neuhauser-Klaus, A. and Chauhan, P.S.: Studies on Somatic Mutation Induction in the Mouse with Isoniazid and Hydrazine. *Mutation Res.*, 191:111-116, 1987.
Neuhauser-Klaus, A. and Lehmacher, W.: The Mutagenic Effect of Calprolactam in the Spot Test with (T X HT) F_1 Mouse Embryos. *Mutation Res.*, 224:369-371, 1989.
Neutra, R.R., Swan, S.H., Hertz-Picciotto, I., Windham, G.C., Wrensch, M., Shaw, G.M., Fenster, L., and Deane, M.: Potential Sources of Bias and Confounding in Environmental Epidemiologic Studies of Pregnancy Outcomes. *Epidemiology*, 3:134-142, 1992.
Nguyen, T.V., Theiss, J.C. and Matney, T.S.: Exposure of Pharmacy Personnel to Mutagenic Antineoplastic Drugs. *Cancer Res.*, 42:4792-4796, 1982.
Nomura, T.: X-Ray- and Chemically Induced Germ-Line Mutation Causing Phenotypical Anomalies in Mice. *Mutation Res.*, 198:309-320, 1988.
Nomura, T., Gotoh, H., and Namba, T.: An Examination of Respiratory Distress and Chromosomal Abnormality in the Offspring of Male Mice Treated with Ethylnitrosourea. *Mutat. Res.*, 229:115-122, 1990.
Norris, M.L., Barton, S.C., and Surani, M.A.H.: Oxford Review of Reproductive Biology. In Milligan, S.R. (Ed.). Oxford University Press, Vol. 12, pp. 225-244, 1990.
Nygaard, R., Clausen, N., Siimes, M.A., Marky, I., Skjeldestqad, F.E., Kristinsson, J.R., Vuoristo, A., Wegelius, R., and Moe, P.J.: Reproduction Following Treatment for Childhood Leukemia: A Population-Based Prospective Cohort Study of Fertility and Offspring. *Med. and Pediat. Oncol.*, 19:459-466, 1991.

Nygaard, R., Garwicz, S., Haldorsen, T., Hertz, H., Jonmundsson, G.K., Lanning, M., and Moe, P.J.: Second Malignant Neoplasms in Patients Treated for Childhood Leukemia. *Acta Paediatr. Scand.* 80:1220-1228, 1991.
Oates, R.D., Sarazen, A.A., and Krane, R.J.: Preservation of Fertility in Patients with Testicular Carcinoma. *World J. Urol.*, 10:52-58, 1992.
Obasaju, M.F., Wiley, L.M., Oudiz, D.J., Raabe, O., and Overstreet, J.W.: A Chimera Embryo Assay Reveals a Decrease in Embryonic Cellular Proliferation Induced by Sperm from X-Irradiated Male Mice. *Radiat. Res.* 118:246-256, 1989.
O'Leary, L.M., Hicks, A.M. and Peters, J.M., and London S.: Parental Occupational Exposures and Risk of Childhood Cancer: Review. *Am. J. Ind. Med.*, 20:17-35, 1991.
Olsen, J.: Risk of Exposure to Teratogens Amongst Laboratory Staff and Painters. *Dan. Med. Bull.*, 30:24-28, 1983.
Olsen, J.H., Brown, P., Schulgen, G., and Jensen, O.M.: Parental Employment at Time of Conception and Risk of Cancer in Offspring. *Eur. J. Cancer*, 27:958-965, 1991.
Olshan, A.F., Breslow, N.E., Daling, J.R., *et al.*: Wilm's Tumor and Paternal Occupation. *Cancer Res.*, 50:3212-3217, 1990.
Olshan, A.F., Teschke, K., and Baird, P.A.: Birth Defects Among Offspring of Firemen: *Am. J. Epidemiol.*, 131:312-321, 1990.
Olshan, A.F., Teschke, K., and Baird, P.A.: Paternal Occupation and Congenital Anomalies in Offspring. *Am. J. Ind. Med.*, 20:447-475, 1991.
OPCS Monitor: Congenital Malformations and Parents' Occupation. MB3 82/1. London: Office of Population Censuses and Surveys, 1982.
Otake, M., Schull, W.J., and Neel, J.V.: Congenital Malformations, Stillbirths, and Early Mortality Among the Children of Atomic Bomb Survivors: A Reanalysis. *Radiat. Res.*, 122:1-11, 1990.
Papier, C.M.: Parental Occupation and Congenital Malformations in a Series of 35,000 Births in Israel. In Marois, M. (Ed.), "Prevention of Physical and Mental Congenital Defects," Part B: Epidemiology, Early Detection and Therapy, and Environmental Factors. New York, Alan R. Liss, Inc., pp. 291-294, 1985.
Paravatou-Petsota, M., Muleris, M., Preieur, M., and Dutrillaux, B.: Diagrammatic Representation for Chromosomal Mutagenesis Studies. III. Radiation-Induced Rearrangements in Pan Troglodytes (Chimpanzee). *Mutation Res.*, 149:57-66, 1985.
Parmeggiani, L.: Encyclopedia of Occupational Health and Safety. Third Edition. Geneva: International Labor Organization, 1983.
Pastorfide, G.B. and Goldstein, D.P.: Pregnancy After Hydatidiform Mole. *Obstet. Gynecol.*, 42:67-70, 1973.
Peters, J.M., Preston-Martin, S., and Yu, M.C.: Brain Tumors in Children and Occupational Exposure of Parents. *Science*, 213:235-237, 1981.
Polednak, A.P., and Janerich,D.T.: Uses of Available Record Systems in Epidemiologic Studies of Reproductive Toxicology. *Am. J. Ind. Med.*, 4:329-348, 1983.
Pomerantseva, M.D., Goloshchapov, P.V., Vilkina, G.A., and Shevchenko, V.A.: Genetic Effect of Chronic Exposure of Male Mice to g-Rays. *Mutation Res.*, 141:195-200, 1984.
Pomerantseva, M.D., Ramaya, L.K., Shevchenko, V.A., Vilkina, G.A., and Lyaginskaya, A.M.: Evaluation of the Genetic Effects of ^{238}Pu Incorporated into Mice. *Mutation Res.*, 226:93-98, 1989.
Prejean, J.D., and Montgomery, J.A.: Structure Activity Relationships in the Carcinogenicity of Anticancer Agents. *Drug Metab. Rev.*, 15:619-646, 1984.
Pueyo, C.: Natulan Induces Forward Mutations to L-Arabinose-Resistance in Salmonella Typhimurium. *Mutat. Res.*, 67:189-192, 1979.
Pylkkanen, L., Jahnukainen, K., Parvinen, M., and Santti, R.: Testicular Toxicity and Mutagenicity of Steroidal and Non-Steroidal Estrogens in the Male Mouse: *Mutation Res.*, 261:181-191, 1991.
Roan, C.C., Matanoski, G.E., McIlnay, C.Q., Olds, K.L., Pylant, F., Trout, J.R., Wheller, P. and Morgan, D.P.: Spontaneous Abortions, Stillbirths, and Birth Defects in Families of Agricultural Pilots. *Arch. Environ. Health*, 39:56-60, 1984.
Robins, J., Breslow, N., and Greenlands, S.: Estimators of the Mantel-Haenszel Variance Consistent in Both Sparse Data and Large-Strata Limiting Models. *Biometrics*, 42:311-323, 1986.
Ross, G.T.: Congenital Anomalies Among Children Born of Mothers Receiving Chemotherapy for Gestational Trophoblastic Neoplasms. *Cancer*, 37: Suppl. 2:1043-1047, 1976.
Rothman, K.J.: Modern Epidemiology. Boston: Little, Brown and Company, 1986.
Russell, L.B. and Bangham, J.W.: The Paternal Genome in Mouse Zygotes is Less Sensitive to ENU Mutagenesis than the Maternal Genome. *Mutation Res.*, 248:203-209, 1991.

Russell, L.B., Hunsicker, P.R., Cacheiro, N.L.A., and Generoso, W.M.: Induction of Specific-Locus Mutations in Male Germ Cells of the Mouse by Acrylamide Monomer. *Mutation Res.*, 262:101-107, 1991.

Russell, L.B., Hunsicker, P.R., Cacheiro, N.L.A., Bangham,J.W., Russell, W.L., and Shelby, M.D.: Chlorambucil Effectively Induces Deletion Mutations in Mouse Germ Cells. *Proc. Natl. Acad. Sci.*, 86:3704-3708, 1989.

Russell, L.B.: Functional and Structural Analyses of Mouse Genomic Regions Screened by the Morphological Specific-Locus Test. *Mutation Res.* 212:23-32, 1989.

Russell, W.L. and Kelly, E.M. Mutation Frequencies in Male Mice and the Estimation of Genetic Hazards of Radiation in Men. *Proc. Natl. Acad. Sci.*, 79:542-544, 1982.

Russell, W.L. and Kelly, E.M.: Specific-Locus Mutation Frequencies in Mouse Stem-Cell Spermatogonia at Very Low Radiation Dose Rates. *Proc. Natl. Acad. Sci.*, 79:539-541, 1982.

Russell, W.L. Comments on Mutagenesis Risk Estimation. *Genetics* 92:s187-s194, May Supplement, 1979.

Russell, W.L., Carpenter, D.A., and Hitotsumachi, S.: Effect of X-Ray and Ethylnitrosourea Exposures Separated by 24 hours on Specific-Locus Mutation Frequency in Mouse Stem-Cell Spermatogonia. *Mutation Res.*, 198:303-307, 1988.

Russell, W.L., Kelly, E.M., and Phipps, E.L.: Induction of Specific-Locus Mutation in the Mouse by Tritiated Water. Behaviour of Tritium in the Environment, IAEA-SM-232185, International Atomic Energy Agency, Vienna, pp. 489-497, 1979.

Russell, W.L.: Dose Response, Repair, and No-Effect Dose Levels in Mouse Germ-Cell Mutagenesis. Problems of Threshold in Chemical Mutagenesis, pp. 153-160, 1984.

Russell, W.L.: Mutation Frequencies in Female Mice and the Estimation of Genetic Hazards of Radiation in Women. Proc. Natl. Acad. Sci., 74(8):3523-3527, 1977.

Rustin, G.J.S., Booth, M., Dent, J., Salt, S., Rustin, F. and Bagshawe, K.D.: Pregnancy After Cytotoxic Chemotherapy for Gestational Trophoblastic Tumours. *BMJ*, 288:103-106, 1984.

Sanjose, S., Roman,E., and Beral, V.: Low Birthweight and Preterm Delivery, Scotland, 1981-84: Effect of Parents' Occupation. *Lancet*, 338:428-431, 1991.

Sankaranarayanan, K.: Invited Review: Prevalence of Genetic and Partially Genetic Diseases in Man and the Estimation of Genetic Risks of Exposure to Ionizing Radiation. *Am. J. Hum. Genet.*, 42:651-662, 1988.

Sankaranarayanan, K.: Mobile Genetic Elements, Spontaneous Mutations, and the Assessment of Genetic Radiation Hazard in Man. Eukaryotic Transposable Elements as Mutagenic Agents, Banbury Report 30, pp. 319-336, 1988.

Savitz, D.A., Schwingl, P.J., and Keels, M.A.: Influence of Paternal Age, Smoking, and Alcohol Consumption on Congenital Anomalies. *Teratology*, 44:429-40, 1991.

Searle, A.G. and Edward, J.H. The Estimation of Risks From the Induction Recessive Mutations After Exposure to Ionisi Radiation. *J. Med. Genet.*, 23:220-226, 1986.

Seethalakshmi, L., Flores, C., Carboni, A.A., Bala, R., Diamond, D.A., and Menon, M.: Cyclosporine: Its Effects on Testicular Function and Fertility in the Prepubertal Rat. *J. Androl.*, 11:17-24, 1990.

Seino, Y., Nagao, M., Yahagi, T., Hoshi, A., Kawachi, T., and Sugimura, T.: Mutagenicity of Several Classes of Antitumor Agents to Salmonella Typhimurium TA98, TA100 and TA92. *Cancer Res.*, 38:2148-2156, 1978.

Selby, P.B. and Niemann, S.L.: Non-Breeding-Test Methods for Dominant Skeletal Mutations Shown by Ethylnitrosourea to be Easily Applicable to Offspring Examined in Specific-Locus Experiments. *Mutation Res.* 127:93-105, 1984.

Selby, P.B.: Applications in Genetic Risk Estimation of Data on the Induction of Dominant Skeletal Mutations in Mice. In de Serres, F.J. and Sheridan, W. (Eds). "Utilization of Mammalian Specific Locus Studies in Hazard Evaluation and Estimation of Genetic Risk." Plenum Publishing Corp., pp. 191-210, 1983.

Selby, P.B.: Radiation Genetics. In "The Mouse in Biomedical Research," Vol. 1, Academic Press Inc., pp. 264-283, 1981.

Selby, P.B.: Radiation-Induced Dominant Skeletal Mutations in Mice: Mutation Rate, Characteristics, and Usefulness in Estimating Genetic Hazard to Humans from Radiation. In Okada, S., Imamura, M., Terashima, T. and Yamaguchi, H. (Eds). "Radiation Research Proceedings of the 6th Intern. Cong. of Radia. Res." Toppan Printing Co., Tokyo, Japan, pp. 537-544, 1979.

Sever, L.E., Gilbert, E.S., Hessol, N.A., and McIntyre, J.M.: A Case-Control Study of Congenital Malformations and Occupational Exposure to Low-Level Ionizing Radiation. *Am. J.Epidemiol.* 127:226-234, 1988.

Shaw, G.M. and Gold, E.B.: Methodological Considerations in the Study of Parental Occupational Exposures and Congenital Malformations in Offspring. *Scand. J. Work Environ. Health*, 14:344-355, 1988.
Shaw, G.M., Malcoe, L.H., Croen, L.A. and Smith, D.F.: An Assessment of Error in Paternal Occupation From the Birth Certificate. *Am. J.Epidemiol.*, 131:1072-1079, 1990.
Shevchenko, V.A., Ramaya, L.K., Pomerantseva, M.D., Lyaginskaya, A.M., and Dementiev, S.I.: Genetics Effects of ^{131}I in Reproductive Cells of Male Mice. *Mutation Res.*, 226:87-91, 1989.
Skare, J.A. and Schrotel, K.R.: Validation of an In Vivo Alkaline Elution Assay to Detect DNA Damage in Rat Testicular Cells. *Environ. Mutagenesis*, 7:563-576, 1985.
Smith, A.H., Fisher, D.L., Pearce, N., and Chapman, C.J.: Congenital Defects and Miscarriages Among New Zealand 2,4,5-T Sprayers. *Arch. Environ. Health*, 37:197-200.
Sonta, S., Yamada, M., and Tsukasaki, M.: Failure of Chromosomally Abnormal Sperm to Participate in Fertilization in the Chinese Hamster. *Cytogenet. Cell Genet.*, 57:200-203, 1991.
Stiller, C.A., Lennox, E.L., and Wilson, L.M.: Incidence of Cardiac Septal Defects in Children with Wilms' Tumour and Other Malignant Diseases. *Carcinogenesis*, 8:129-132, 1987.
Strigini, P., Pierluigi, M., Forni, G.L., Sansone, R., Carobbi, S., Grasso, M., and Bricarelli, F.D.: Effect of X-Rays on Chromosome 21 Nondisjunction. *Am. J. Med. Genet.*, 7:155-159, 1990.
Surani, M.A.H., Allen, N.D., Barton, S.C., *et al.*: Developmental Consequences of Imprinting of Parental Chromosomes by DNA Methylation. *Phil. Trans R Soc. B.*, 326:13-27, 1990.
Tanaka, T.: Effects of Piperonyl Butoxide on F_1 Generation Mice. *Toxicology Letters*, 60:83-90, 1992.
Taskinen, H.K.: Effects of Parental Occupational Exposures on Spontaneous Abortion and Congenital Malformation: *Scand. J. Work Environ. Health*, 16:297-314, 1990.
Thomas, D.C., Siemiatycki, J., Dewar, R., Robins, J., Goldberg, M., and Armstrong, B.G.: The Problem of Multiple Interference in Studies Designed to Generate Hypotheses. *Am. J. Epidemiol.*, 122:1080-1095, 1985.
Tomatis, L., Cabral, J.P.R., Likhachev, A.J., and Ponomarkov, V.: Increased Cancer Incidence in the Progency of Male Rats Exposed to Ethylnitrosourea Before Mating. *Int. J. Cancer*, 28:475-478, 1981.
Tomatis, L., Narod, S., and Yamasaki, H.: Transgeneration Transmission of Carcinogenic Risk. *Carcinogenesis*, 13:145-151, 1992.
Townsend, J.C., Bodner, K.M., Van Peenan, P.F.D., Olson, R.D. and Cook, R.R.: Survey of Reproductive Events of Wives of Employees Exposed to Chlorinated Dioxins. *Am. J. Epidemiol.*, 115:695-713, 1982.
Trasler, J.M., Hales, B.F., and Robaire, B.: Paternal Cyclophosphamide Treatment of Rats Causes Fetal Loss and Malformations Without Affecting Male Fertility. *Nature*, 316 (6024):144-146, 1985.
Turusov, V.S., Trukhanova, L.S., Parfenov, Y.D., and Tomatis, L.: Occurrence of Tumours in the Descendants of CBA Male Mice Prenatally Treated with Diethylstilbestrol. *Int. J. Cancer*, 50:131-135, 1992.
Ujeno, Y.: Epidemiological Studies on Disturbances of Human Fetal Development in Areas with Various Doses of Natural Background Radiation. I. Relationship Between Incidences of Down's Syndrome or Visible Malformation and Gonad Dose Equivalent Rate of Natural Background Radiation. *Archives of Environmental Health.*, 40(3):177-180, 1985.
van Buul, P.P.W., Richardson, Jr., J.F., and Goudzwaard, J.H.: The Induction of Reciprocal Translocations in Rhesus Monkey Stem-Cell Spermatogonia: Effects of Low Doses and Low Dose Rates: *Radiat. Res.*, 105:1-7, 1986.
van Buul, P.P.W.: Enhanced Radiosensitivity for the Induction of Translocation in Mouse Stem Cell Spermatogonia Following Treatment with Cyclophosphamide or Adriamycin. *Mutation Res.*, 128:207-211, 1984.
van Buul, P.P.W.: X-Ray-Induced Translocations in Marmoset (Callithrix Jacchus) Stem-Cell Spermatogonia. *Mutation Res.*, 129:231-234, 1984.
Vine, M.F., Hulka, B.S., Margolin, B.H., *et al.*: Cotinine Concentrations in Semen, Urine, and Blood of Smokers and Nonsmokers. Manuscript submitted for publication.
Wallace, W.H.B., Shalet, M., Lendon, M., and Morris-Jones, P.H.: Male Fertility in Long-Term Survivors of Childhood Acute Lymphoblastic Leukaemia. *Int. J. Andrology*, 14:312-319, 1991.
Wassom, J.S.: Origins of Genetic Toxicology and the Environmental Mutagen Society. Environmental and Molecular Mutagenesis, 14(16):1-6, 1989.
Wijsman, E.M.: Recurrence Risk of a New Dominant Mutation in Children of Unaffected Parents. *Am. J. Hum. Genet.*, 48:654-661, 1991.
Wilcox, A.J., and Horney., L.F.: Accuracy of Spontaneous Abortion Recall. *Am. J. Epidemiol.*, 120:727-733, 1984.

Wilcox, A.J., Weinberg, C.R., and O'Connor, J.F.: Incidence of Early Loss of Pregnancy. *N. Engl. J. Med.*, 319:189-194, 1988.
Williams, G.M.: Methods for Evaluating Chemical Genotoxicity. *Annu. Rev. Pharmacol. Toxicol.*, 29:189-211, 1989.
Windham, G.C., Shusterman, D., Swan, S.H., Fenster, L., and Eskenazi, B.: Exposure to Organic Solvents and Adverse Pregnancy Outcome. *Am. J. Ind. Med.*, 20:241-259, 1991.
Workshop Report from the Division of Research Grants, National Institutes of Health. Site-Specific Mutagenesis - A Chemical Pathology Study Section Workshop. *Cancer Res.*, 49:758-763, 1989.
Yajima, N., Kondo, K., and Morita, K.: Reverse Mutation Tests in Salmonella Typimurium and Chromosomal Aberration Tests in Mammalian Cells in Culture on Fluorinated Pyrimidine Derivatives. *Mutat. Res.*, 88:241-254, 1981.
Yost, G.S., Horstman, M.G., Walily, A.F.E., Gordon, W.P., and Nelson, S.D.: Procarbazine Spermatogenesis Toxicity: Deuterium Isotope Effects Point to Regioselective Metabolism in Mice. *Tox. and Applied Pharm.*, 80:316-322, 1985.
Young, J.L. Jr., Ries, L.G., Silverberg, E., Horm, J.W. and Miller, R.W.: Cancer Incidence, Survival, and Mortality for Children Younger Than Age 15 Years. *Cancer*, 58:598-602, 1986.
Zimmering, S., Thompson, E., Aquavella, J., and Reeder, B.: Dose-Response Relationship for Ethyl Nitrosourea-Induced Sex-Linked Recessive Lethals in Germ Cells of the Female Drosophila Melanogaster at Relatively Low Doses. *Mutation Res.*, 226:81-85, 1989.

ANTIOXIDANT PREVENTION OF BIRTH DEFECTS AND CANCER

B. N. Ames, P. Motchnik, C.G. Fraga,
M.K. Shigenaga, and T.M. Hagen

Division of Biochemistry and Molecular Biology
University of California
Berkeley, CA 94720

INTRODUCTION

Metabolism, like other aspects of life, involves trade-offs. Damage to DNA and protein by oxidant by-products of normal metabolism is massive. We argue that this damage (the same as that produced by radiation) is a major contributor to aging, to degenerative diseases of aging, such as cancer, heart disease, cataracts, and brain dysfunction, and to DNA damage in sperm. Antioxidant defenses against this damage include vitamins C and E and carotenoids, the main sources of which are dietary fruits and vegetables. Low dietary intake of fruits and vegetables doubles the risk of most types of cancer as compared to high intake (about four or five portions per day); it also markedly increases the risk of heart disease and cataracts. We will argue also that this deficiency is a major cause of childhood cancer and birth defects.

We have shown that oxidative damage to sperm DNA is increased markedly when dietary ascorbate is insufficient to keep seminal fluid ascorbate to an adequate level. A sizeable percentage of the U.S. population, including most smokers, consumes inadequate levels of dietary ascorbate. The 1000 ppm of NO_x in cigarette smoke causes a general depletion in antioxidant levels in smokers. Smokers need to eat two to three times as much ascorbate as non-smokers in order to maintain the same blood levels of ascorbate, but they rarely do. We review the data suggesting that smoking fathers may increase the risk of birth defects and childhood cancer in their offspring.

Since only 9% of the U.S. population eats the recommended five portions of fruits and vegetables per day, widespread changes in diet could have major impact on human health.

AGING AND OXIDATION

Evolutionary biologists have argued that aging is inevitable because of several tradeoffs (Kirkwood, 1984, 1992; Williams, 1957; Williams and Nesse, 1991). One tradeoff is that a considerable proportion of an animal's resources is devoted to

Male-Mediated Developmental Toxicity, Edited by D.R. Mattison
and A.F. Olshan, Plenum Press, New York, 1994

reproduction at a cost to maintenance, which means that the maintenance of somatic tissues is less than that required for indefinite survival Metabolism has costs: oxidant by-products of normal energy metabolism extensively damage DNA, proteins, and other molecules in the cell, and this damage accumulates with age (Ames and Shigenaga, 1992; Fraga *et al.*, 1990; Stadtman, 1992). A second tradeoff is that nature selects for many genes that have immediate survival value, but that may have long term deleterious consequences. The oxidative burst from phagocytic cells, for example, protects against death from bacterial and viral infections, but contributes to DNA damage, mutation, and cancer (Weitzman and Gordon, 1990; Cohen *et al.*, 1991).

Degenerative Diseases of Aging

The degenerative diseases associated with aging include cancer, heart disease, brain dysfunction, and cataracts. The degeneration of somatic cells during aging appears, in good part, to contribute to these diseases. The relationship between cancer and age in various mammalian species illustrates this point. Cancer increases with about the fifth power of age in both short-lived species, such as rats, and in long-lived species, such as humans. One important factor in longevity appears to be basal metabolic rate, which is about seven times higher in a rat than a human and which could markedly affect the level of endogenous oxidants and other mutagens produced as by-products of metabolism. The level of oxidative DNA damage appears to be roughly related to metabolic rate in at least four mammalian species (Adelman *et al.*, 1988).

Oxidation and Damage to DNA, Protein, and Lipids

Oxidative damage to cells has been postulated to be the most significant endogenous damage leading to aging (Harman, 1981). Superoxide, hydrogen peroxide, and hydroxyl radicals, which are the mutagens produced by radiation, are also by-products of normal metabolism (Wagner *et al.*, 1992; Von Sonntag, 1987). Lipid peroxidation gives rise to mutagenic lipid epoxides, lipid hydroperoxides, lipid alkoxy and hydroperoxy radicals, and enals (α, β-unsaturated aldehydes) (Marnett *et al.*, 1985). Singlet oxygen, a high energy and mutagenic form of oxygen, can be produced by transfer of energy from light, the respiratory burst from neutrophils, or lipid peroxidation (Ravanat *et al.*, 1992).

Animals have numerous antioxidant defenses, but since these defenses are not perfect, some DNA is oxidized. Oxidatively damaged DNA is repaired by enzymes that excise the lesions, which are then excreted in the urine. We have developed methods to assay several of these excised damaged bases in the urine of rodents and humans (Adelman *et al.*, 1988; Park *et al.*, 1992). We estimate that the number of oxidative hits per cell per day is about 100,000 in the rat and about 10,000 in the human. DNA repair enzymes efficiently remove most, but not all, the lesions formed (Ames and Shigenaga, 1992). Oxidative lesions in DNA accumulate with age, so that by the time a rat is old (2 years) it has about two million DNA lesions per cell, which is about twice that in a young rat. Mutations also accumulate with age. For example, the somatic mutation frequency in human lymphocytes is about nine times greater in elderly people than in neonates (Grist *et al.*, 1992).

Mitochondrial DNA (mtDNA) from rat liver has more than ten times the level of oxidative DNA damage than does nuclear DNA from the same tissue (Richter *et al.*, 1988). This increase may be due to a lack of mtDNA repair enzymes, lack of histones protecting mtDNA, and the proximity of mtDNA to oxidants generated during oxidative phosphorylation. The cell defends itself against this high rate of damage by a constant turnover of mitochondria, thus presumably removing those damaged mitochondria that produce increased oxidants. Despite this turnover, oxidative lesions appear to accumulate with age in mitochondrial DNA at a higher rate than in nuclear DNA (Hagen and Ames, unpublished). Oxidative damage could also account for the higher level of mutations in mtDNA that accumulate with age (Cortopassi *et al.*, 1992; Wallace, 1992).

Endogenous oxidants also damage proteins. Stadtman and his colleagues (Stadtman, 1992; Oliver *et al.*, 1987; Oliver *et al.*, 1990) have shown that the proteolytic enzymes that hydrolyze oxidized proteins are not sufficient to prevent an age-associated accumulation of oxidized proteins. In two human diseases associated with premature aging, Werner's syndrome and progeria, oxidized proteins accumulate at a much higher rate than is normal (Stadtman, 1992). Fluorescent age pigments, which are thought to be due in part to cross-links between protein and lipid peroxidation products, also accumulate with age (Halliwell and Gutteridge, 1989).

SOURCES AND EFFECTS OF OXIDANTS

Four endogenous sources appear to account for most of the oxidants produced by cells: 1) As a consequence of normal aerobic respiration, mitochondria consume molecular oxygen, reducing it by sequential steps to produce H_20. Inevitable by-products of this process are superoxide, hydrogen peroxide, and hydroxyl radical. About 10^{12} oxygen molecules are processed by each rat cell daily, and the leakage of partially reduced oxygen molecules is about 2%, yielding about $2x10^{10}$ superoxide and hydrogen peroxide molecules per cell per day (Chance *et al.*, 1979). 2) Phagocytic cells destroy bacteria or virus-infected cells with an oxidative burst of NO, $O_2^{\cdot-}$, H_2O_2, and $^{\cdot}OCl$. Chronic infection by viruses, bacteria, or parasites, results in a chronic phagocytic activity and consequent chronic inflammation, which is a major risk factor for cancer. Chronic infections are particularly prevalent in third world countries (see below). 3) Cytochrome P450 enzymes in animals constitute one of the primary defense systems against natural toxic chemicals from plants, the major source of dietary toxins (Ames *et al.*, 1990). The induction of these enzymes, prevent acute toxic effects from foreign chemicals, but also results in oxidant by-products that damage DNA (Park, J.-Y. K. and Ames, B.N., unpublished). 4) Peroxisomes, which are organelles responsible for degrading fatty acids and other molecules, produce as a byproduct hydrogen peroxide, which is then degraded by catalase. Evidence suggests that, under certain conditions, some of the peroxide escapes degradation, resulting in its release into other compartments of the cell and in increased oxidative DNA damage (Kasai *et al.*, 1989).

Three exogenous sources may significantly increase the large endogenous oxidant load. 1) The NO_x in cigarette smoke (about 1000 ppm) causes oxidation of macromolecules (Kiyosawa *et al.*, 1990; Pryor *et al.*, 1986; Frei *et al.*, 1991; Reznick *et al.*, 1992) and depletes antioxidant levels (Schectman *et al.*, 1991; Duthie *et al.*, 1991; Bui *et al.*, 1991). It is therefore likely to contribute significantly to the pathology of smoking. Smoking is a risk factor for heart disease as well as a wide variety of cancers in addition to lung cancer (Shopland *et al.*, 1991; Peto *et al.*, 1992; Kneller *et al.*, 1992a; Honjo *et al.*, 1992). 2) Normal diets contain plant food with large amounts of natural phenolic compounds, such as chlorogenic and caffeic acid, that can generate oxidants by redox cycling (Ames *et al.*, 1990; Gold *et al.*, 1992), although it is not yet clear whether the resulting increment to the oxidant load is significant. 3) Iron (and copper) salts promote the generation of oxidizing radicals from peroxides (Fenton chemistry). It has been argued that too much dietary copper or iron, particularly heme iron (which is high in meat), is a risk factor for heart disease and cancer in normal men (Lauffer, 1992; Salonen *et al.*, 1992; Salonen *et al.*, 1991; Sullivan, 1989). Men who absorb significantly more than normal amounts of dietary iron (hemochromatosis disease) are at high risk for both cancer and heart disease (Lauffer, 1992).

Chronic Infection, Inflammation and Cancer

Leucocytes and other phagocytic cells combat bacteria, parasites, and virus-infected cells by destroying them with powerful oxidants: NO, $O_2^{\cdot-}$, H_2O_2 and $^{\cdot}OCl$. These oxidants protect humans from immediate death from infection, but cause oxidative damage to DNA and also stimulate cell division; therefore they contribute to mutation and cancer. Asbestos pathology also appears to be due to chronic inflammation (Korkina *et al.*, 1992; Marsh and Mossman, 1991.) Antioxidants appear to inhibit some of the pathology of chronic inflammation (Korkina *et al.*, 1992; Marsh and Mossman, 1991; Morrison *et al.*, 1981; Sanders and Mahaffey, 1983; Sobala *et al.*, 1991; Kneller *et al.*, 1992b; Arroyo *et al.*, 1992).

Chronic infections contribute to about one-third of the world's cancer, mainly in poorer countries. Hepatitis B and C viruses infect about 500 million people, mainly in Asia and Africa, and are a major cause of hepatocellular carcinoma (Beasley, 1987; Tabor and Kobayashi, 1992; Yu, *et al.*, 1991). About two million people die prematurely every year from hepatitis. Hepatitis B was first established as a cause of liver cancer in Taiwan, which has now vaccinated all children against the virus. Africa and China have not been able to afford widespread vaccination. Since there are no known cures for these viral infections, agents that inhibit inflammation may be the most effective treatments. It remains to be seen whether antioxidant vitamins, such as vitamin C, and anti-inflammatory drugs, such as aspirin, are indeed successful in inhibiting inflammation and cost-effective as well. *Helicobacter pylori* bacteria, which infect the stomachs of roughly one-third of the world population, appear to be the major cause of stomach cancer, ulcers, and gastritis (Sobala *et al.*, 1991; Kneller *et al.*, 1992b; Parsonnet *et al.*, 1991; Dooley *et al.*, 1989; Cover and Blaser, 1992). In wealthy countries the disease is usually asymptomatic, which indicates that the inflammation is at least partially suppressed, possibly by adequate levels of dietary antioxidants (Howson *et al.*, 1986). Specific antibiotic treatments usually eliminate the bacteria. Another major chronic infection is schistosomiasis, which is caused by a parasitic worm that is widespread in China and Egypt. The Chinese worm lays its eggs in the colon, producing inflammation that often leads to colon cancer (Chen, 1988). The Egyptian worm lays eggs in the bladder, promoting bladder cancer (Chen and Mott, 1989). *Opisthorchis viverrini*, a liver fluke, infects millions of people in Thailand and Malaysia. The flukes lodge in bile ducts and increase the risk of cholangiocarcinoma (Srivantanakul *et al.*, 1988). *Chlonorchis sinensis* infections in millions of Chinese increase their risk for biliary tract cancer (Shanmugaratnam, 1956).

OXIDANT STRESS, BIRTH DEFECTS AND CHILDHOOD CANCER

Oxidation and Sperm Damage

Oxidative lesions in sperm DNA are increased 250% when levels of dietary ascorbate are insufficient to keep seminal fluid ascorbate to an adequate level (Fraga *et al.*, 1991) (Table 1 and Figure 1). A sizable percentage of the U.S. population ingest inadequate levels of dietary ascorbate, particularly single males, the poor, and smokers (Patterson and Block, 1991). Married males have about a 12-year longer life expectancy than do single males, about eight years of which appear to be due to reduced heart disease and cancer (Cohen, 1991). This increase may be due in part to the beneficial influence of wives on promoting good nutrition and minimizing their husbands' smoking and drinking. The oxidants in cigarette smoke deplete the antioxidants in plasma. Smokers must eat two to three times more ascorbate than non-smokers to achieve the same level of ascorbate in blood (Schectman *et al.*, 1991; Duthie *et al.*, 1991; Bui *et al.*, 1991), but they rarely do. In a comparison of sperm from smokers and nonsmokers Viczian found that the number of sperm and the percent of mobile sperm decrease significantly in smokers, and this decrease is dependent on the dose and duration of smoking (Viczian, 1968; Evans *et al.*, 1981). The percentage of pathological sperm increases significantly in smokers and is also dependent on the dose and duration of smoking. Paternal smoking, in particular, appears to increase the risk of birth defects and childhood cancer in offspring (see below), and paternal age exacerbates this risk (Crow, 1993) (see below). Due to the high division rate of sperm compared to eggs, the germ line mutation rate in the father greatly exceeds that of the mother. Thus, nutritionally inadequate diets and smoking of fathers result not only in damage to their own DNA but to the DNA of their sperm, an effect that can reverberate down future generations.

Paternal Smoking and Mutations to the Germline

"The much larger number of cell divisions between zygote and sperm than between zygote and egg, the increased age of fathers of children with new dominant mutations, and the greater evolution rate of pseudogenes on the Y chromosome than on autosomes all point to a much higher [germline] mutation rate in males than in females, as first pointed out by Haldane from the study of X-linked hemophilia" (Crow, 1993). Haldane estimated the male germline mutation rate to be about 10 times that of the female. Table 1 illustrates the relation of seminal fluid ascorbate to oxidative DNA damage in sperm.

Table 1. Dietary and Semen Ascorbic Acid Content and oxo^{8}dG in Sperm DNA

	Baseline	Depletion	Marginal	Repletion
Ascorbate Intake (mg/d)	250	5	10 or 20	60 or 250
Dietary period (d)	7-14	32	28	28
Semen Ascorbate (μM)	399±55	203±72*	115±25*	422±100
% oxo^{8}dG	100	198*	248*	158

*Significantly different from baseline values at $p < 0.05$.

Maternal Smoking and Birth Defects

Maternal smoking causes many problems, such as intrauterine growth retardation, leading to spontaneous abortion, premature birth, and low infant birthweight. But numerous epidemiological studies over several decades in various countries have failed to consistently find a link between maternal cigarette smoking and genetic damage to the fetus. These studies, however, have not examined the following critical factor in determining the link between maternal smoking and birth defects: since a fetus develops from eggs that were formed when the mother was *in utero*, it is the smoking status of the grandmother at the time that she was pregnant with the mother that is probably the most important factor. Khoury *et al.* (Khoury, 1987; Khoury *et al.*, 1989) note a dose-response relation between the daily amount of maternal smoking and the risk of cleft lip/palate in offspring. They do not control for paternal smoking, and it is possible that smoking mothers are likely to have smoking mates; thus, the apparent maternal effect may really be a paternal effect.

Paternal Smoking and Birth Defects

We review below those studies that examined paternal smoking and birth defects and childhood tumors. Most studies on smoking and birth defects or childhood tumors looked at maternal smoking.

Comstock and Lundin (1967) found that neonatal death rate among infants of smoking fathers who smoked was 17.2 per 1000 live births compared to 11.9 (adjusted for sex of child and education of father) among infants of nonsmoking fathers and 26.5 among infants of parents who both smoked. Koo *et al.* (1988) found that wives of smoking fathers have more miscarriages and abortions than those of non-smoking fathers. Windham *et al.* (1992) found paternal smoking (greater than one-half pack/day) showed a greater spontaneous abortion rate (OR = 1.3), but this disappeared on adjustment.

Hearey *et al.* (1984) observed an association between paternal smoking and increased neural tube defects (4/8 cases vs. 1/17 controls) in offspring.

Mau and Netter (1974) studied the smoking habits of fathers of 5200 newborns. The rate of major malformations among nonsmoking fathers was 0.8% and among smoking fathers 2.1%. Maternal smoking had no influence on the rate of birth defects. The most striking increase in birth defects concerned major facial clefts. 0.1% of infants of nonsmoking fathers had facial clefts, compared to 0.5% of fathers who smoked 1-10 cigarettes per day and 0.7% of fathers who smoked more than 10

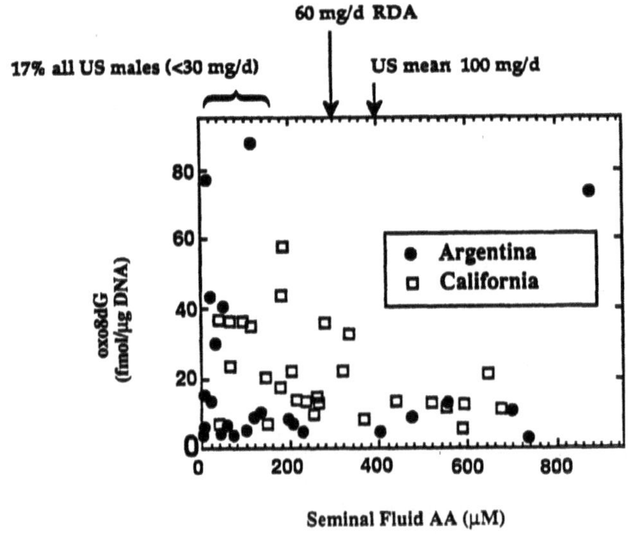

Figure 1. Data from Fraga, et al., Ascorbic acid protects against endogenous oxidative damage in human sperm, *Proc. Natl. Acad. Sci.* U.S.A (1991), as replotted by P. Motchnik.

cigarettes per day. There was also an increase in perinatal mortality (unrelated to major birth defects) if the father smoked more than 10 cigarettes per day, even if the mother did not smoke (4.3% compared to 2.8%). (By facial cleft the author probably meant cleft lip/palate rather than cleft of the total face, which is a very rare malformation.)

Savitz *et al.* (1991) analyzed single live births among 14,685 Kaiser members who participated in the child health studies. Father's cigarette smoking was more common among children with cleft lip (with or without cleft palate), hydrocephalus, ventricular septal defect, and urethral stenosis. However, inverse associations between paternal smoking and birth defects were more common than positive associations. The authors specifically mention genetically altered sperm as a possible cause of birth defects in infants of smoking fathers. They note that several of the anomalies associated with infants of older fathers were also increased among fathers who smoked. The concordance in defects associated with both advanced paternal age and paternal smoking was especially notable for ventricular septal defects and hydrocephalus.

Schmid (1989), in discussing the Mau and Netter study, attributed the effect of paternal smoking on birth defects to passive smoke. However, passive smoking is not a probable explanation for the link between paternal smoking and birth defects because the amount of smoke that reaches the embryo is insignificant compared to the amount that reaches the embryo when the mother herself smokes, and maternal smoking does not appear to significantly increase the risk of birth defects.

Schmidt (1986) made an argument similar to ours concerning smoking and genetic damage: "Tobacco smoke contains numerous mutagenic substances. They reach the male gonads via the blood. They show their mutagenic action here openly much more stronger than on egg-cells because the spermatogenesis continues over the whole male reproductive period whereas the formation of eggs is already completed in the fetal phase."

Seidman *et al.* (1990) provides a table on the incidence of major and minor congenital malformations by paternal and maternal smoking levels. There was a nonsignificant increase in the incidence of major and minor malformations in the offspring of smoking fathers who mate with nonsmoking mothers. This author does not break down the malformations by specific defect.

Von Estel *et al.* (1982) found an increase in cleft lip-palate (5/176 smokers; 1/354 controls).

Zhang *et al.* (1992) examined 1012 birth defects in Shanghai and found that offspring of smokers (vs. non-smokers) had over a doubling of spina bifida, anencephalus, cleft palate, undescended testicle, hernia of diaphragm and anomalies of the eye, the ear, and the foot.

Smoking and Childhood Tumors

Grufferman *et al.* (1982) found that paternal smoking, but not maternal smoking, increased the risk of rhabdomyosarcoma in childhood (relative risk of 3.9). RMS is the sixth most common childhood cancer, with an annual incidence of about 4 cases per million children and a peak incidence at about ages 3-4. The authors suggest that a direct carcinogenic effect of paternal cigarette smoke may be introduced in a prezygotic manner.

Grufferman *et al.* (1983) stated that, "In our recent study of childhood rhabdomyosarcoma (RMS) we found an increase in the risk of RMS among children whose fathers smoked cigarettes. However, there was no association between RMS and mothers' smoking. We hypothesize that differential germ cell damage from cigarette smoking underlies our observations and that this risk of germ cell damage from cigarette smoking and from other environmental exposures is greater for men than for women. The increased susceptibility for male germ cells may be due to the number and timing of meiotic and mitotic cell divisions. In males, germ cells undergo large numbers of meiotic and mitotic divisions throughout the reproductive years. In contrast, in females, generally only one oocyte matures and completes meiosis each month of the reproductive years. Thus, there are very large male-female differences in the number of rapidly dividing germ cells during the reproductive years, and it is rapidly proliferating cells which are most susceptible to genetic damage."

Howe *et al.* (1989) in a study of 31 brain tumors in children found no effect of paternal smoking.

John *et al.* (1991) found associations with paternal smoking during the 12 months prior to conception in the absence of maternal smoking during the first trimester were found for all childhood cancers combined (OR=1.2), acute lymphocytic leukemia (OR=1.4), lymphomas (OR=1.6), and brain cancer (OR=1.6). These correlations, however, appear fairly weak, at best equivocal. These figures are similar to those for maternal smoking alone. After adjustment for paternal education, maternal smoking during the first trimester of pregnancy, in the absence of paternal smoking, was associated with an increased risk for all cancers combined (OR=1.3), acute lymphocytic leukemia (OR=1.9), and lymphomas (OR=2.3).

Johnston *et al.* (1986) found in their study of children with germ cell tumors, the smoking pattern of fathers was similar for cases and controls.

Kramer *et al.* (1987) in a study of 101 children with neuroblastoma found an odds ratio of 1.6 (P = 0.08 one-sided) when offspring of smokers and non-smokers were compared.

Magnani *et al.* (1990) found that both maternal and paternal smoking up to the child's birth were associated with non-Hodgkin lymphoma in childhood. After adjusting for socioeconomic status, the odds ratio for paternal smoking was 6.7 and for maternal 1.7. The author says that the odds ratio for paternal smoking was not correlated with number of cigarettes. The study showed no correlation between acute lymphocytic leukemia and parental smoking. Their results did not reproduce the findings of Grufferman *et al.* (1982).

Neutel and Buch's study (1971) only took into consideration the effect of maternal smoking on childhood cancer. For cancers of all sites, the children of mothers who smoked during pregnancy had a relative risk of 1.3, which does not seem significant.

Preston-Martin *et al.* (1982) found that there was a possibly significant increase in risk of childhood brain tumors (1.5 odds ratio) if during pregnancy the mother lived with a smoker (who was usually the child's father). "Our finding that maternal smoking itself was not related to disease but that living with a smoker (usually the child's father) was may indicate that paternal exposures are important."

Sandler *et al.* (1985) analyzed cancers of all sites, except basal cell cancer of the skin, among people 15-59 years old. Cancer risk was increased 50% among people whose fathers had smoked. (Paternal smoking was defined as the father having smoked before the child reached 10 years of age.) Increased risk associated with paternal smoking was not explained by demographic factors, social class, or individual smoking habits. There was only a slight increase in overall cancer risk associated with maternal smoking (relative risk of 1.1). But maternal and paternal smoking were both associated with risk for hematopoietic cancers, and a dose-response relationship was seen. The relative risk for hematopoietic cancers increased from 1.7 when one parent smoked to 4.6 when both parents smoked. However, the study included a wide range of ages and did not give isolated information on the teenage cases. Therefore, the study does not assess the risk of paternal smoking specifically on childhood cancer.

Stewart *et al.* (1958) found the incidence of paternal smoking was similar among children with cancer and without cancer. However, this study did not distinguish between pipe and cigarette smoking. The smoker of pipes generally doesn't inhale much smoke.

Wilkins and Geidenberger (1992) studied 103 children with brain tumors and found paternal smoking (greater than one pack/day) was associated with a non-significant increase (OR = 1.4) in risk.

We believe the suggestive findings of a large effect of paternal smoking on childhood cancer and birth defects warrant a major effort in this area.

ANTIOXIDANTS PROTECT AGAINST DISEASE

Fruit and Vegetable Consumption Lowers Risk of Degenerative Diseases

Block and her colleagues have recently reviewed the 172 studies in the epidemiological literature that relate the amount of fruit and vegetable consumption to cancer incidence (Block *et al.*, 1992) (Table 2). The data are extremely consistent and show a large protective effect. The quarter of the population with low dietary intake of fruits and vegetables compared to the quarter with high intake has double the cancer rate for most types of cancer (lung, larynx, oral cavity, esophagus, stomach, colon and rectum, bladder, pancreas, cervix, and ovary). As Bjelke (1973) said 20 years ago: "For colorectal cancer, the large majority of the population may be at increased risk compared to the minority with very high vegetable intakes." Data on the types of cancer known to be associated with hormone levels are not as consistent and show less protection: for breast cancer the protective effect was about 30% (Block *et al.*, 1992; Howe *et al.*, 1990). There is also literature on the protective effect of fruit and vegetable consumption on heart disease and stroke (Hennekens *et al.*, in press). Only 9% of Americans eat five servings of fruits and vegetables per day, the intake recommended by the National Cancer Institute and the National Research Council (Block *et al.*, 1992; Patterson *et al.*, 1990). European countries with low fruit and vegetable intake (e.g., Scotland) are generally in poorer health and have higher rates of heart disease and cancer than countries with high intake (e.g., Greece) (James, 1988).

The cost of fruits and vegetables is an important factor in discouraging consumption. Poorer people spend a higher percentage of their income on food, eat less fruits and vegetables (Patterson and Block, 1988), and have shorter life expectancy than wealthier people.

Dietary Antioxidants

The effect of the dietary intake of the antioxidants ascorbate, tocopherol, and carotenoids is difficult to disentangle by epidemiological studies from other important vitamins and ingredients in fruits and vegetables (Block *et al.*, 1992; Block, 1992). Nevertheless, several arguments suggest that the antioxidant content of fruits and vegetables is a major contributor to their protective effect. 1) Biochemical data, discussed above shows that oxidative damage is massive and is likely to be the major endogenous damage to DNA, proteins, and lipids. 2) Studies showing that oxidative damage to sperm DNA is increased when dietary ascorbate is insufficient. 3) Epidemiological studies and intervention trials on prevention of cancer and heart disease in people taking antioxidant supplements are suggestive, though larger studies need to be done (Hennekens *et al.*, in press; Gridley *et al.*, 1992).

Clinical trials using antioxidants will be the critical test for many of the ideas discussed here. 4) Studies on oxidative mechanisms or epidemiology on antioxidant protection for individual degenerative diseases are discussed below. Table 2 is a review of epidemiological studies showing protection by fruits and vegetables against cancer.

Table 2. Fruits and Vegetables Protection Against Cancer*

Cancer Site	Fraction of Studies Showing a Protective Effect (p=0.05)	Relative Risk (Median)
Epithelial		
Lung	24/25	2.2
Oral	9/9	2.0
Larynx	4/4	2.3
Esophagus	15/16	2.0
Stomach	17/19	2.5
Pancreas	9/11	2.8
Cervix	7/8	2.0
Bladder	3/5	2.1
Colorectal	20/35	1.9
Miscellaneous	6/8	—
Hormone-Dependent		
Breast	8/14	1.3
Ovary/Endometrium	3/4	1.8
Prostate	4/14	1.3
	Total 129/172	

*Data from Block, Patterson and Subar, *Nutr. Cancer*, 18:1-29, 1992.

Small molecule dietary antioxidants such as Vitamin C (ascorbate), Vitamin E (tocopherol), and carotenoids have generated particular interest as anticarcinogens and as defenses against degenerative diseases (Frei and Ames, 1992; Bendich and Butterworth, 1991; Gaby *et al.*, 1991; Reznick, 1992). Most carotenoids have antioxidant activity, particularly against singlet oxygen, though only β-carotene can be metabolized to Vitamin A (retinal) (Krinsky, 1989; Rousseau *et al.*, 1992; Palozza and Krinsky, 1992). We have called attention to a number of previously neglected physiological antioxidants including urate, bilirubin, carnosine, and ubiquinol (Ames *et al.*, 1981; Frei *et al.*, 1990; Stocker *et al.*, 1987; Kohen *et al.*, 1988). Ubiquinone (CoQ_{10}), for example, is the critical small molecule for transporting electrons in mitochondria for the generation of energy. Its reduced form, ubiquinol, is an effective antioxidant in membranes and lipoproteins (Frei *et al.*, 1990; Stocker *et al.*, 1991). Optimal levels of dietary ubiquinone/ubiquinol could be of importance in many of the degenerative diseases.

Antioxidants and Cancer

A critical factor in mutagenesis is cell division (Cohen *et al.*, 1991; Ames *et al.*, in press; Ames and Gold, 1990). When the cell divides, an unrepaired DNA lesion has a certain probability of giving rise to a mutation. Thus an important factor in mutagenesis, and therefore carcinogenesis, is the cell division rate in the precursors of tumor cells. Stem cells are important as precursor cells in cancer because they are not on their way to being discarded. Increasing their cell division rate would increase mutation. As expected, there is little cancer in non-dividing cells. Such diverse agents as chronic infection (see below), high levels of particular hormones (Henderson *et al.*, 1982), or chemicals at doses causing cell killing (Cohen *et al.*, 1991; Ames and Gold, 1990; Moalli *et al.*, 1987; Columbano *et al.*, 1990) result in increased cell division and an increased risk for cancer.

Oxidants form one important class of agents that stimulate cell division (Cerutti, 1991; Amstad *et al.*, 1992). This may be related to the stimulation of cell division that occurs during the inflammatory process accompanying wound healing (Cohen *et al.*, 1991). Antioxidants therefore can decrease mutagenesis, and thus carcinogenesis, in two ways: by decreasing oxidative DNA damage and by decreasing cell division. Of great interest is the understanding of mechanisms by which tocopherol and carotenoids can prevent cell division (Boscoboinik *et al.*, 1991a; Boscoboinik *et al.*, 1991b; Bertram *et al.*, 1991; Wolf, 1992).

There is an increasing literature on the protective role of dietary tocopherol, ascorbate, and β-carotene in lowering the incidence of a wide variety of human cancer (Block, 1991; Enstrom *et al.*, 1992; Byers and Perry, 1992). Antioxidants can counteract the induction of cancer in rodents by a variety of carcinogens (Daoud and Griffin, 1980; Kushida *et al.*, 1992; Reddy and Hirota, 1979). Two of the major causes of cancer, cigarette smoke and chronic inflammation, both appear to involve oxidants in their mechanism of action. Almost all of the epidemiological studies that examined the relation between antioxidant levels and cigarette-induced lung cancer showed a statistically significant protective effect of antioxidants (Block *et al.*, 1992; Block 1991; Byers and Perry, 1992). Antioxidants inhibit much of the pathology of cigarette smoke in rodents (Bagnasco *et al.*, 1992; Leuchtenberger and Leuchtenberger 1977). Inflammatory reactions release large amounts of NO, a radical, nitrosating agent, and mutagenic oxidant (Arroyo *et al.*, 1992; Wink *et al.*, 1991; Nathan, 1992). Ascorbate inhibits nitrosation under physiological conditions (Knight *et al.*, 1991). Antioxidants help to protect against the carcinogenic effects of chronic inflammation, as discussed above.

The Optimum Level of Antioxidants

The epidemiological evidence and the guidelines of the National Cancer Institute and the National Research Council/National Academy of Sciences suggest that at least two fruits and three vegetables per day is a desirable intake. Since ascorbate, tocopherol, and β-carotene supplements are inexpensive and high doses are remarkably non-toxic, there is a school that believes that supplements are desirable. There is suggestive, but inadequate epidemiological and biochemical evidence bearing on the question (Gridley *et al.*, 1992; Enstrom *et al.*, 1992). What is clear is that fruits and vegetables contain many necessary micronutrients in addition to antioxidants, some of which also can prevent mutations. Folic acid, for example, is required for the synthesis of the nucleotides in DNA. Inadequate intake has been shown to cause chromosome breaks and increased cancer and birth defects (Bendich and Butterworth 1991; MacGregor *et al.*, 1990). Folate deficiency may be a risk factor for myocardial infarction as well (Stampfer *et al.*, 1991). Niacin is required for making poly (ADP-ribose), a component of DNA repair. Other micronutrients are also likely to be part of our defense systems.

The U.S. Recommended Daily Allowances (RDAs) for ascorbate and tocopherol intake--there is no guideline for β-carotene--are not adequate for several reasons: 1) The amount recommended, e.g., 60 mg/day for ascorbate, is primarily for avoiding an observable deficiency syndrome, e.g., scurvy, and is not necessarily the amount for optimum lifetime health, which is usually not known. 2) A recommended blood level of each antioxidant, e.g., 60 μM ascorbate, would be a more desirable standard. People vary considerably in the intake required to keep their blood level adequate. A smoker, for example, needs to take in several times as much ascorbate as a non-smoker to keep the blood level the same. Infections also cause an oxidative stress by activating phagocytic cells. The observation that antioxidant inadequacy is associated with oxidative damage to DNA of the germ line as well as somatic cells, emphasizes the urgency of determining adequate blood levels (Fraga *et al.*, 1991).

Since only 9% of Americans, and fewer in most other countries, are eating 5 fruits and vegetables per day, there is a great opportunity to improve health by increasing consumption.

ACKNOWLEDGEMENTS

This paper has been adapted in part from Ames, B.N., Shigenaga, M.K. & Hagen, T. "Oxidants, Antioxidants, and the Degenerative Diseases of Aging," (1993). *Proc. Natl. Acad. Sci. USA* 90:7915-7922. We appreciate the criticisms of B. Terracini, P.N. Lee, and the substantial help of M. Profet.

REFERENCES

Adelman, R., Saul, R.L., Ames, B.N. (1988) Oxidative damage to DNA: Relation to species metabolic rate and life span, *Proc. Natl. Acad. Sci. USA.* 85:2706.

Ames, B.N. and Shigenaga, M.K., (1992) Oxidants are a major contributor to aging, *in*: "Ann. N.Y. Acad. Sci.," Franceschi, C., Crepaldi, G., Cristofalo, V.J., Masotti, L., and Vijg, J., ed., The New York Academy of Sciences, New York, 663:85.

Ames, B.N., Profet, M., and Gold, L.S., (1990) Dietary pesticides (99.99% all natural), *Proc. Natl. Acad. Sci. USA.* 87:7777.

Ames, B.N., Cathcart, R., Schwiers, E., and Hochstein, P., (1981) Uric acid provides an antioxidant defense in humans against oxidant- and radical-caused aging and cancer: A hypothesis, *Proc. Natl. Acad. Sci. USA.* 78:6858.

Ames, B.N., Shigenaga, M.K., and Gold, L.S., DNA lesions, inducible DNA repair, and cell division: three key factors in mutagenesis and carcinogenesis, *Environ. Health Perspect.* (In press).

Ames, B.N., and Gold, L.S., (1990) Chemical carcinogenesis: too many rodent carcinogens, *Proc. Natl. Acad. Sci. USA.* 87:7772.

Amstad, P.A., Krupitza, G., and Cerutti, P. A., (1992) Mechanism of c-fos induction by active oxygen, *Cancer Res.* 52:3952.

Arroyo, P.L., Hatch-Pigott, V., Mower, H.F., and Cooney, R.V., (1992) Mutagenicity of nitric oxide and its inhibition by antioxidants, *Mutat. Res.* 281:193.

Bagnasco, M., Bennicelli, C., Camoirano, A., Balansky, R.M. and De Flora, S., (1992) Metabolic alterations produced by cigarette smoke in rat lung and liver, and their modulation by oral N-acetylcysteine, *Mutagenesis.* 7:295.

Beasley, R.P., (1987) Hepatitis B virus, *Cancer.* 61:1942.

Bendich, A., and Butterworth, C.E., Jr., Ed. (1991) Micronutrients in Health and in Disease Prevention. Marcel Dekker, Inc., New York.

Bertram, J.S., Pung, A., Churley, M., Kappock, T.J.I., Wilkins, L.R., and Cooney, R. V., (1991) Diverse carotenoids protect against chemically induced neoplastic transformation, *Carcinogenesis.* 12:671.

Bjelke, E., (1973) "Epidemiologic studies of cancer of the stomach, colon, and rectum; with special emphasis on the role of diet." (University Microfilms International, Ann Arbor, Michigan).

Block, G., Patterson, B., and Subar, A., (1992) Fruit, vegetables and cancer prevention: A review of the epidemiologic evidence, *Nutr. Cancer.* 18:1.

Block, G., (1992) The data support a role for antioxidants in reducing cancer risk, *Nutr. Reviews.* 50:207.

Block, G., (1991) Vitamin C and cancer prevention: the epidemiologic evidence, *Am. J. Clin. Nutr.* 53:270.

Boscoboinik, D., Szewczyk, A., and Azzi, A., (1991a) Alpha-tocopherol (vitamin E) regulates vascular smooth muscle cell proliferation and protein kinase C activity, *Arch. Biochem. Biophys.* 286:264.

Boscoboinik, D., Szewczyk, A., Hensey, C., and Azzi, A., (1991b) Inhibition of cell proliferation by alpha-tocopherol. Role of protein kinase C, *J. Biol. Chem.* 266:6188.

Bui, M. H., Sauty, A., Collet, F., and Leuenberger, P., (1991) Dietary vitamin C intake and concentrations in the body fluids and cells of male smokers and nonsmokers, *J. Nutr.* 122:312.

Byers, T., and Perry, G., (1992) Dietary carotenes, vitamin C, and vitamin E as protective antioxidants in human cancers, *Annu. Rev. Nutr.* 12:139.

Cerutti, P.A., (1991) Oxidant stress and carcinogenesis, *Eur. J. Clin. Invest.* 21:1.

Chance, B., Sies, H., and Boveris, A., (1979) Hydroperoxide metabolism in mammalian organs, *Physiol. Rev.* 59:527.

Chen, M., (1988) Progress in assessment of morbidity due to *Schistosoma japonicum* infection, *Trop. Dis. Bull.* 85:2.

Chen, M., and Mott, K., (1989) Progress in assessment of morbidity due to *Schistosoma haematobium* infection, *Trop. Dis. Bull.* 86:2.

Cohen, B.L., (1991) Catalog of risks extended and updated, *Health Phys.* 61:317.

Cohen, S.M., Purtilo, D.T., and Ellwein, L.B., (1991) Pivotal role of increased cell proliferation in human carcinogenesis, *Mod. Pathol.* 4:371.

Columbano, A., Ledda-Columbano, G.M., Ennas, M.G., Curto, M. Chelo, A., and Pani, P., (1990) Cell proliferation and promotion of liver carcinogenesis: Different effect of hepatic regeneration and mitogen-induced hyperplasia on the development of enzyme-altered foci, *Cell.* 11:771.

Comstock, G.W., and Lundin, F.E.J., (1967) Parental smoking and perinatal mortality, *Am. J. Obst. Gynec.* 98:708.

Cortopassi, G.A., Shibata, D., Soong, N-W., and Arnheim, N., (1992) A pattern of accumulation of a somatic deletion of mitochondrial DNA in aging human tissues, *Proc. Natl. Acad. Sci. USA.* 89:7370.

Cover, T.L., and Blaser, M.J., (1992) *Helicobacter pylori* and gastroduodenal disease, *Ann. Rev. Med.* 43:135.

Crow, J., (1993) How much do we know about spontaneous human mutation rates? *Environ. Mol. Mutagen.* 21:122.

Daoud, A.H., and Griffin, A.C., (1980) Effect of retinoic acid, butylated hydroxytoluene, selenium and sorbic acid on azo-dye hepatocarcinogenesis, *Cancer Lett.* 9:299.

Dooley, C.P., Cohen, H., Fitzgibbons, P.L., Bauer, M., Appleman, M.D., Perez-Perez, G.I., and Blaser, M.J., (1989) Prevalence of Helicobacter pylori infection and histologic gastritis in assymptoriatic persons, *N. Engl. J. Med.* 321:1562.

Duthie, G.G., Arthur, J.R., and James, W.P.T., (1991) Effects of smoking and vitamin E on blood antioxidant status, *Am. J. Clin. Nutr.* 53:1061.

Enstrom, J.E., Kanim, L.E., and Klein, M.A., (1992) Vitamin C intake and mortality among a sample of the United States population, *Epidemiol.* 3:194.

Evans, H.J., Fletcher, J., Torrance, M., and Hargreave, T.B., (1981) Sperm abnormalities and cigarette smoking, *Lancet.* 1:627.

Fraga, C.G., Motchnik, P.A., Shigenaga, M.K., Helbock, H.J., Jacob, R.A., and Ames, B.N., (1991) Ascorbic acid protects against endogenous oxidative damage in human sperm, *Proc. Natl. Acad. Sci. USA.* 88:11003.

Fraga, C.G., Shigenaga, M.K., Park, J.-W., Degan, P., and Ames, B.N., (1990) Oxidative damage to DNA during aging: 8-hydroxy-2'-deoxyguanosine in rat organ DNA and urine, *Proc. Natl. Acad. Sci. USA.* 87:4533.

Frei, B., Forte, T.M., Ames, B.N., and Cross, C.E., (1991) Gas phase oxidants of cigarette smoke induce lipid peroxidation and changes in lipoprotein properties in human blood plasma: Protective effects of ascorbic acid, *Biochem. J.* 277:133..

Frei, B., and Ames, B.N., (1992) Small molecule antioxidant defenses in human extracellular fluids, *in*: "The Molecular Biology of Free Radical Scavenging Systems," J. Scandalios, ed., Cold Spring Harbor Laboratory Press, Cold Spring Harbor, New York.

Frei, B., Kim, M.C., and Ames, B.N., (1990) Ubiquinol-10 is an effective lipid-soluble antioxidant at physiological concentrations, *Proc. Natl. Acad. Sci. USA.* 87:4879.

Gaby, S.K., Bendich, A., Singh, V.N., and Machlin, L.J., (1991) "Vitamin Intake and Health," Marcel Dekker, Inc., New York.

Gold, L.S., Slone, T.H., Stern, B.R., Manley, N.B., and Ames, B.N., (1992) Rodent carcinogens: Setting priorities, *Science.* 258:261.

Gridley, G., McLaughlin, J.K., Block, G., Blot, W.J., Gluch, M., and Fraumeni, J.F.,Jr., (1992) Vitamin supplement use and reduced risk of oral and pharyngeal cancer, *Am. J. Epidemiol.* 135:1083.

Grist, S.A., McCarron, M., Kutlaca, A., Turner, D.R., and Morley, A.A., (1992) *In vivo* human somatic mutation: frequency and spectrum with age, *Mutat. Res.* 266:189.

Grufferman, S., Wang, H.H., DeLong, E.R., Kimm, S.Y.S., Delzell, E.S., and Falletta, J.M., (1982) Environmental factors in the etiology of rhabdomyosarcoma in childhood., *J. Natl. Cancer Inst.* 68:107.

Grufferman, S., Delzell, E.S., Maile, M.C., and G. Michalopoulos, G., (1983) Parents' cigarette smoking and childhood cancer, *Med. Hypotheses.* 12:17.
Haldane, J.B.S., (1947) The rate of mutation of the gene for hemophilia and its segregation ratios in males and females, *Ann. Eugen.* 13:262.
Halliwell, B., and Gutteridge, J.M.C., (1989) "Free Radicals in Biology and Medicine," Clarendon Press, Oxford.
Harman, D., (1981) The aging process, *Proc. Natl. Acad. Sci. USA.* 78:7124.
Hearey, C.D., Harris, J.A., Usatin, M.S., Epstein, D.M., Ury, H.K., and Neutra, R.R., (1984) Investigation of a cluster of anencephaly and spina bifida, *Am. J. Epidemiol.* 120:559.
Henderson, B.E., Ross, R.K., Pike, M.C., and Casagrande, J.T., (1982) Endogenous hormones as a major factor in human cancer, *Cancer Res.* 42:3232.
Hennekens, C.H., Goziano, J.M., Manson, J., and Buring, J., Beta-carotene and cardiovascular disease, *in*: "Ann. N.Y. Acad. Sci.," ed., The New York Academy of Sciences, New York (in press).
Honjo, S., Kono, S., Shinchi, K., Imanishi, K., and Hirohata, T., (1992) Cigarette smoking, alcohol use and adenomatous polyps of the sigmoid colon, *Jpn. J. Cancer Res.* 83:806.
Howe, G.R., Hirohata, T., and Hislop, T.G., (1990) Dietary factors and risk of breast cancer: Combined analysis of 12 case-control studies, *J. Natl. Cancer Inst.* 82:561.
Howe, G.R., Burch, J.D., Chiarelli, A.M., Risch, H.A., and Choi, B.C.K., (1989) An exploratory case-control study of brain tumors in children, *Cancer Res.* 49:4349.
Howson, C., Hiyama, T., and Wynder, E., (1986) The decline in gastric cancer: epidemiology of an unplanned triumph, *Epidemiol. Rev.* 8:1.
James, W.P.T., (1988) "Healthy Nutrition," WHO Regional Publications, Copenhagen.
John, E., M., Savitz, D.A., and Sandler, D.P., (1991) Prenatal exposure to parents' smoking and childhood cancer, *Am. J. Epidemiol.* 133:123.
Johnston, H.E., Mann, J.R., Williams, J., Waterhouse, J.A.H., Birch, J.M., Cartwright, R.A., Draper, G.J., Hartley, A.L., Hopton, P.A., and Stiller, C.A., (1986) The inter-regional, epidemiological study of childhood cancer (IRESCC): Case-control study in children with germ cell tumours, *Carcinogenesis.* 7:717.
Kasai, H., Okada, Y., Nishimura, S., Rao, M.S., and Reddy, J.K., (1989) Formation of 8-hydroxydeoxyguanosine in liver DNA of rats following long-term exposure to a peroxisome proliferator, *Cancer Res.* 49:2603.
Khoury, J.J., Gomez-Farias, M., and Mulinare, J., (1989) Does maternal smoking during pregnancy cause cleft lip and palate in offspring, *A.J.D.C.* 143:333.
Khoury, M.J., (1987) Maternal cigarette smoking and oral clefts: A population-based study, *Am. J. Public Health.* 77:623.
Kirkwood, T.B.L., (1992) Comparative life spans of species: Why do species have the life spans they do?, *Am. J. Clin. Nutr.* 55:1191.
Kirkwood, T.B.L., (1984) Towards a unified theory of cellular aging, *Monogr. Dev. Biol.* 17:9.
Kiyosawa, H., Suko, M., Okudaira, H., Murata, K., Miyamoto, T., Chung, M.-H., Kasai, H., and Nishimura, S., (1990) Cigarette smoking induces formation of 8-hydroxydeoxyguanosine, one of the oxidative DNA damages in human peripheral leukocytes, *Free Rad. Res. Comms.* 11:23.
Kneller, R.W., You, W.-C., Chang, Y.-S., Liu, W.-D., Zhang, L., Zhao, L., Xu, G.-W., Fraumeni, J.F.,Jr., and Blot, W.J., (1992a) Cigarette smoking and other risk factors for progression of precancerous stomach lesions, *J. Natl. Cancer Inst.* 84:1261.
Kneller, R.W., Guo, W.-D., Hsing, A.W., Chen, J.S., Blot, W.J., Li, J.-Y., Forman, D., and Fraumeni, J.F.,Jr., (1992b) Risk factors for stomach cancer in sixty-five Chinese counties, *Cancer Epidem. Biomarkers Prevention.* 1:113.
Knight, T.M., Forman, D., Ohshima, H., and Bartsch, H., (1991) Endogenous nitrosation of L-proline by dietary-derived nitrate, *Nutr. Cancer.* 15:195.
Kohen, R., Yamamoto, Y., Cundy, K., and Ames, B.N., (1988) Antioxidant activity of carnosine, homocarnosine, and anserine present in muscle and brain, *Proc. Natl. Acad. Sci. USA.* 85:3175.
Koo, L.C., Ho, J., and Rylander, R., (1988) Life-history correlates of environmental tobacco smoke: A study on nonsmoking Hong Kong Chinese wives with smoking versus nonsmoking husbands, *Soc. Sci. Med.* 26:751.
Korkina, L.G., Durnev, A.D., Suslova, T.B., Cheremisina, Z.P., Daugel-Dauge, N.O., and Afanas'ev, I.B., (1992) Oxygen radical-mediated mutagenic effect of asbestos on human lymphocytes: Suppression by oxygen radical scavengers, *Mutat. Res.* 265:245.
Kramer, S., Ward, E., Meadows, A.T., and Malone, K.E., (1987) Medical and drug risk factors associated with neuroblastoma: A case-control study, *J. Natl. Cancer Inst.* 78:797.
Krinsky, N.I., (1989) Antioxidant functions of carotenoids, *Free Rad. Biol. Med.* 7:617.

Kushida, H., Wakabayashi, K., Suzuki, M., Takahashi, S., Imaida, K., Sugimura, T., and Nagao, M., (1992) Suppression of spontaneous hepatocellular carcinoma development in C3H/HeNCrj mice by the lipophilic ascorbic acid, 2-O-octadecylascorbic acid (CV-3611), *Carcinogenesis.* 13:913.

Lauffer, R.B., Ed. (1992) Iron and Human Disease. CRC Press, Boca Raton.

Leuchtenberger, C., and Leuchtenberger, R., (1977) Protection of hamster lung cultures by L-cysteine or vitamin C against carcinogenic effects of fresh smoke from tobacco or marihuana cigarettes, *Br. J. Exp. Path.* 58:625.

MacGregor, J.T., Schlegel, R., Wehr, C.M., Alperin, P., and Ames, B.N., (1990) Cytogenetic damage induced by folate deficiency in mice is enhanced by caffeine, *Proc. Natl. Acad. Sci. USA.* 87:9962.

Magnani, C., Pastore, G., Luzzatto, L., and Terracini, B., (1990) Parental occupation and other environmental factors in the etiology of leukemias and non-Hodgkin's lymphomas in childhood: A case-control study, *Tumori.* 76:413.

Marnett, L.J., Hurd, H., Hollstein, M.C., Esterbauer, D.E., and Ames, B.N., (1985) Naturally occurring carbonyl compounds are mutagens in *Salmonella* tester strain TA104, *Mutat. Res.* 148:25.

Marsh, J.P., and Mossman, B.T., (1991) Role of asbestos and active oxygen species in activation and expression of ornithine decarboxylase in hamster tracheal epithelial cells, *Cancer Res.* 51:167.

Mau, G., and Netter, P., (1974) Die auswirkungen des vaeterlichen zigarettenconsums auf die perinatale sterblichkeit und die missbildungshaeufigkeit, *Dtsch. Med. Wochenschr.* 99:1113.

Moalli, P.A., MacDonald, J.L., Goodglick, L.A., and Kane, A.B., (1987) Acute injury and regeneration of the mesothelium in response to asbestos fibers, *Am. J. Pathol.* 128:426.

Morrison, D.G., Daniel, J., Lynd, F.T., Moyer, M.P., Esparza, R.J, Moyer, R.C., and Rogers, W., (1981) Retinyl palmitate and ascorbic acid inhibit pulmonary neoplasms in mice exposed to fiberglass dust, *Nutr. Cancer.* 3:81.

Nathan, C., (1992) Nitric oxide as a secretory product of mammalian cells, FASEB. 6:3051.

Neutel, C.I., and Buch, C., (1971) Effect of smoking during pregnancy on the risk of cancer in children, *J. Natl. Cancer. Inst.* 47:59.

Oliver, C.N., Ahn, B.-W., Moerman, E.J., Goldstein, S., and Stadtman, E.R., (1987) Age-related changes in oxidized proteins, *J. Biol. Chem.* 262:5488.

Oliver, C.N., Starke-Reed, P.E., Stadtman, E.R., Liu, G.J., Carney, J.M., and Floyd, R.A., (1990) Oxidative damage to brain proteins, loss of glutamine synthetase activity, and production of free radicals during ischemia/reperfusion-induced injury to gerbil brain, *Proc. Natl. Acad. Sci. USA.* 87:5144.

Palozza, P., and Krinsky, N.I., (1992) β-carotene and α-tocopherol are synergistic antioxidants, *Arch. Biochem. Biophys.* 297:184.

Park, E.-M., Shigenaga, M.K., Degan, P., Korn, T.S., Kitzler, J.W., Wehr, C.M., P. Kolachana, P., and Ames, B.N., (1992) The assay of excised oxidative DNA lesions: Isolation of 8-oxoguanine and its nucleoside derivatives from biological fluids with a monoclonal antibody column, *Proc. Natl. Acad. Sci. USA.* 89:3375.

Parsonnet, J., Friedman, G.D., Vandersteen, D.P., Chang, Y., Vogelman, J.H., Orentreich, N., and Sibley, R.K., (1991) *Helicobacter pylori* infection and the risk of gastric carcinoma, *N. Engl. J. Med.* 325:1127.

Patterson, B.H., and Block, G., (1988) Food choices and the cancer guidelines, *Am. J. Public Health.* 78:282.

Patterson, B.H., and Block, G., (1991) Fruit and vegetable consumption national survey data, *in*: "Micronutrients in Health and in Disease Prevention," A. Bendich and C.E. Butterworth, ed., Marcel Dekker, New York.

Patterson, B.H., Block, G., Rosenberger, W.F., Pee, D., and Kahle, L.L., (1990) Fruit and vegetables in the American diet: Data from the NHANES II survey, *Am. J. Public Health.* 80:1443.

Peto, R., Lopez, A.D., Boreham, J., Thun, M., and Heath, C., Jr., (1992) Mortality from tobacco in developed countries: Indirect estimation from national vital statistics, *Lancet.* 339:1268.

Preston-Martin, S., Yu, M.C., Benton, B., and Henderson, B.E., (1982) Nitroso compounds and childhood brain tumors: A case-control study, *Cancer Res.* 42:5240.

Pryor, W.A., Dooley, M.M., and Church, D.F., (1986) The mechanisms of the inactivation of human alpha-1-proteinase inhibitor by gas-phase cigarette smoke, *Adv. Free Rad. Biol. Med.* 2:161.

Ravanat, J.-L., Berger, M., Benard, F., Langlois, R., Ouellet, R., van Lier, J.E., and Cadet, J., (1992) Phthalocyanine and naphthalocyanine photosensitized oxidation of 2'-deoxyguanosine: distinct type I and type II products, *Photochem. Photobiol.* 55:809.

Reddy, B.S., and Hirota, N., (1979) Effect of dietary ascorbic acid on 1,2-dimethylhydrazine-induced colon cancer in rats, *Fed. Proc.* 38:714.

Reznick, A.Z., (1992) Vitamin E and the aging process, *in*: "Vitamin E, Biochemistry and Clinical Applications," L. Packer and J. Fuchs, ed., Marcel Dekker,Inc., New York.
Reznick, A.Z., Cross, C.E., Hu, M.-L., Suzuki, Y.J., Khawaja, S., Safadi, A., Motchnik, P.A., Packer, L., and B. Halliwell, B., (1992) Modification of plasma proteins by cigarette smoke as measured by protein carbonyl formation, *Biochem. J.* 286:607.
Richter, C., Park, J.-W., and Ames, B.N., (1988) Normal oxidative damage to mitochondrial and nuclear DNA is extensive, *Proc. Natl. Acad. Sci. USA.* 85:6465.
Rousseau, E.J., Davison, A.J., and Dunn, B., (1992) Protection by β-carotene and related compounds against oxygen-mediated cytotoxicity and genotoxicity: Implications for carcinogenesis and anticarcinogenesis, *Free Rad. Biol. Med.* 13:407.
Salonen, J.T., Nyyssonen, K., Korpela, H., Tuomilehto, J., Seppanen, R., and Salonen, R., (1992) Iron sufficiency is associated with hypertension and excess risk of myocardial infarction: The kuopio ischaemic heart disease risk factor study, *Circulation.* 86:803.
Salonen, J.T., Salonen, R., Korpela, H., Suntioinen, S., and Tuomilehto, J., (1991) Serum copper and the risk of acute myocardial infarction: A prospective population study in men in Eastern Finland, *Am. J. Epidemiol.* 134:268.
Sanders, C.L., and Mahaffey, J.A., (1983) Action of vitamin C on pulmonary carcinogenesis from inhaled 239PuO2, *Health Phys.* 43:794.
Sandler, E.P., Everson, R.B., and Wildox, A.J., (1985) Cancer risk in adulthood from early life exposure to parents' smoking, *Am. J. Public Health.* 75:487.
Savitz, D.A., Schwingl, P., and Keels, M.A., (1991) Influence of paternal age, smoking, and alcohol consumption on congenital anomalies, *Teratology.* 44:429.
Schectman, G., Byrd, J.C., and Hoffmann, R., (1991) Ascorbic acid requirements for smokers: Analysis of a population survey, *Am. J. Clin. Nutr.* 53:1466.
Schmid, J., (1989) Rauchen, pille und schwangerschaft, *Schweiz. Rundschau Med.* 78:100.
Schmidt, F., (1986) Rauchen schaedigt die maennliche Zeugungsfaehigkeit, *Andrologia.* 18:445.
Seidman, D. S., Ever-Hadani, P., and Gale, R. (1990) Effect of maternal smoking and age on congenital anomalies, *Obstet. Gynecol.* 76:1046.
Shanmugaratnam, K., (1956) Primary carcinomas of the liver and biliary tract, *Br. J. Cancer.* 10:232.
Shopland, D.R., Eyre, H.J., and Pechacek, T.F., (1991) Smoking-attributable cancer mortality in 1991: Is lung cancer now the leading cause of death among smokers in the United States?, *J. Natl. Cancer Inst.* 83:1142.
Sobala, G.M., Pignatelli, B., Schorah, C.J., Bartsch, H., Sanderson, M., Dixon, M.F., Shires, S., King, R.F.G., and Axon, A.T.R., (1991) Levels of nitrite, nitrate, N-nitroso compounds, ascorbic acid and total bile acids in gastric juice of patients with and without precancerous conditions of the stomach, *Carcinogenesis.* 12:193.
Srivantanakul, P., Sontipong, S., Chotiwan, P., and Parkin, D.M., (1988) Liver cancer in Thailand: Temporal and geographic variations, *J. Gastroenterol. Hepatol.* 3:413.
Stadtman, E.R., (1992) Protein oxidation and aging, *Science.* 257:1220.
Stampfer, M.J., Malinow, M.R., Willett, W.C., Newcomer, L.M., Upson, B., Ullmann, D., Tishler, P.V., and Hennekens, C.H. (1991) A prospective study of plasma homocyst(e)ine and risk of myocardial infarction in US physicians, *JAMA.* 268:877.
Stewart, A., Webb, J., and Hewitt, D., (1958) A survey of childhood malignancies, *Br. Med. J.* 1:1495.
Stocker, R., Yamamoto, Y., McDonagh, A.F., Glazer, A.N., and Ames, B.N., (1987) Bilirubin is an antioxidant of possible physiological importance, Science. 235:1043.
Stocker, R., Bowry, V., and Frei, B., (1991) Ubiquinol-10 protects human low-density lipoprotein more efficiently against lipid peroxidation than does alpha-tocopherol, *Proc. Natl. Acad. Sci USA.* 88:1646.
Sullivan, J.L., (1989) The iron paradigm of ischemic heart disease, *Am. Heart J.* 117:1177.
Tabor, E., and Kobayashi, K., (1992) Hepatitis C virus, a causative infectious agent of non-A, non-B hepatitis: prevalence and structure-summary of a conference on hepatitis C virus as a cause of hepatocellular carcinoma, *J. Natl. Cancer Inst.* 84:86.
Viczian, M., (1968) Ergebnisse von spermauntersuchungen bei zigarettenrauchern, *Zschr. Haut-Geschi. Krkh.* 4:183.
Von Estel, C., Boettcher, A., and Semman, K., (1982) Ueber den einfluss muetterlicher und vaeterlicher rauchgewohnheiten auf das geburtsgewicht und die missbildungsrate des nuegeborenen, *Abl. Gynaekol.* 104:563.
Von Sonntag, C., (1987) The Chemical Basis of Radiation Biology, Taylor & Francis, London.
Wagner, J.R., Hu, C.-C., and Ames, B.N., (1992) Endogenous oxidative damage of deoxycytidine in DNA, *Proc. Natl. Acad. Sci. USA.* 89:3380.
Wallace, D.C., (1992) Mitochondrial genetics: A paradigm for aging and degenerative diseases? *Science.* 256:628.

Weitzman, S.A., and Gordon, L.I., (1990) Inflammation and cancer: Role of phagocyte-generated oxidants in carcinogenesis, *Blood.* 76:655.
Wilkins, J., and Geidenberger, C., (1992) Perinatal exposure to environmental tobacco smoke and risk of childhood brain tumor, *Am. J. Epidemiol.* 136:1010.
Williams, G.C., (1957) Pleiotropy, natural selection and the evolution of senescence, *Evolution.* 11:398.
Williams, G.C., and Nesse, R.M., (1991) The dawn of Darwinian medicine, *Q. Rev. Biol.* 66:1.
Windham, G.C., Swan, S.H., and Fenster, L., (1992) Parental cigarette smoking and the risk of spontaneous abortion, *Am. J. Epidemiol.* 135:1394.
Wink, D.A., Kasprzak, K.S., Maragos, C.M., Elespuru, R.K., Misra, M., Dunams, T.M., Cebula, T.A., Koch, W.H., Andrews, A.W., Allen, J.S., and Keefer, L.K., (1991) DNA deaminating ability and genotoxicity of nitric oxide and its progenitors, *Science.* 254:1001.
Wolf, G., (1992) Retinoids and carotenoids as inhibitors of carcinogenesis and inducers of cell-cell communication, *Nutr. Rev.* 50:270.
Yu, M.-W, You, S.-L., Chang, A.-S., Lu, S.-N., Liaw, Y.-F., and Chen, C.-J., (1991) Association between hepatitis C virus antibodies and heptocellular carcinoma in Taiwan, *Cancer Res.* 51:5621.
Zhang, J., Savitz, D.A., Schwingl, P.J., and Cai, W.-W., (1992) A case-control study of paternal smoking and birth defects, *Int. J. Epidemiol.* 21:273.

QUANTITATIVE RISK ASSESSMENT FOR PATERNALLY-MEDIATED DEVELOPMENTAL TOXICITY

Donald R. Mattison

111 Parran Hall
Graduate School of Public Health
University of Pittsburgh
130 DeSoto Street
Pittsburgh, PA 15261

INTRODUCTION

This contribution presents a suggested approach for quantitative evaluation of risk to the offspring for developmental toxicity from paternal exposure. This proposed approach is different than the traditional risk assessment method for developmental toxicity which has focused exclusively on maternally mediated toxicity (i.e., toxicity resulting from maternal exposure during pregnancy) to the developing embryo or fetus (Fabro 1985, Fabro et al 1982, Faustman et al 1989, Francis et al 1990, Frankos 1985, Gaylor 1989, Gaylor et al 1988, Hart et al 1988, Hemminki et al 1986, Hemminki and Vineis 1985, Jelovsek et al 1989, Jelovsek et al 1990, Kimmel 1989, Kimmel and Gaylor 1988, Kimmel et al 1989, Koeter 1983, Mattison et al 1989, Schardein et al 1985, Schardein 1993, Shepard 1989,). Over the past decade there has been considerable discussion of a range of approaches to characterize risk to the developing embryo or fetus from maternal exposure, including: (i) statistical approaches for dose response modeling such as the no observed adverse effect level (NOAEL)-safety factor[1] (SF) approach to determine the allowable daily intake (ADI), the benchmark dose (B_mD)-SF approach to determine the reference dose for developmental toxicity (R_fD_{dt}); and (ii) biologically based approaches including classical pharmacokinetics (PK), physiologically based pharmacokinetics (PBPK), and biologically based dose response modeling (BBDR). These approaches will be reviewed briefly to suggest a strategy to develop risk assessment techniques for male mediated developmental toxicity (MMDT). In general, this analysis suggests that statistical approaches which characterize the dose-response relationship are most economically applied to the problem of risk assessment for MMDT. However, as concern for the hazard increases, the development of biologically based methods for MMDT will occur.

---

[1] At a symposium on Quantitative Methods for Developmental Toxicity Risk Assessment held in Ottawa, Canada in May 1992, Dr. Bernard Goldstein (Robert Wood Johnson School of Medicine and Dentistry of New Jersey) made a plea that the nomenclature for this term be given careful consideration. Although variously described as a safety or uncertainty factor, Dr. Goldstein argued that this term actually is a protective factor and should be identified as such.

Male-Mediated Developmental Toxicity, Edited by D.R. Mattison and A.F. Olshan, Plenum Press, New York, 1994

RISK ASSESSMENT

Qualitative or quantitative risk assessment provides a formal methodology to protect the public from adverse health effects resulting from use of chemicals in foods or medicinals, or exposure to chemicals in an occupational setting or the environment. While the methods for characterizing risks to human health has evolved over the past forty years, most attention to the development of quantitative risk assessment methods has occurred over the past ten to fifteen years (National Research Council 1983), and has focused predominately on cancer. Risk assessment is one of four risk sciences which include: risk assessment, risk perception, risk communication and risk management. The nature of the other three risk sciences is beyond the scope of this discussion (An excellent introduction to the Risk Sciences can be found in Readings in Risk, edited by Glickman and Gough 1990). By its nature, risk assessment involves both scientific and quasi-scientific considerations and incorporates both qualitative and quantitative data. As a result, risk assessments are contentious and frequently disputed. It is important to understand that this is likely to always be the case with all aspects of risk assessment (Jasanoff 1990).

Risk assessment involves four steps; hazard identification, hazard characterization, exposure assessment and risk characterization. The nature of each step and the approach utilized to combine the data into the ultimate qualitative or quantitative risk assessment must be based on our best understanding of the biology of the process for which risk assessment is being conducted (Mattison 1991, Figure 1).

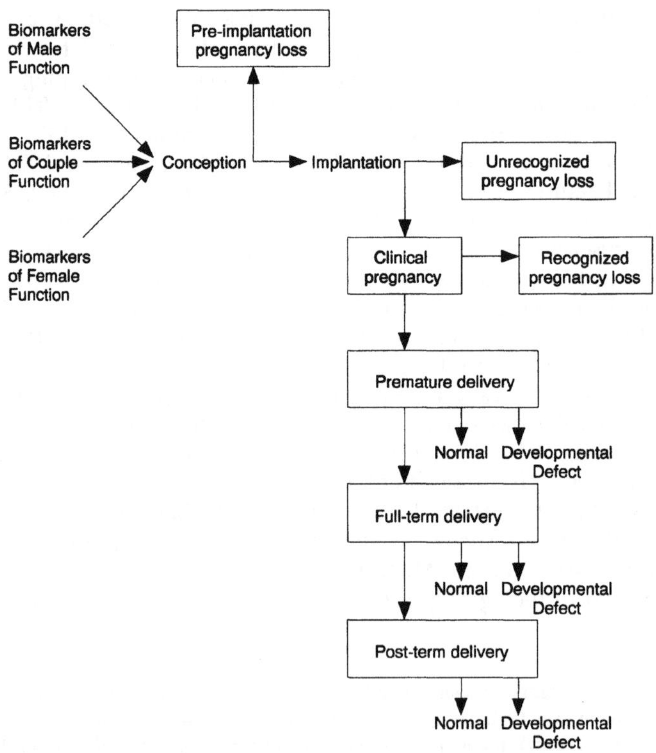

Figure 1. Couple-Based Model for Biomarker-Based Risk Assessment for Male-Mediated-Developmental Toxicity.

In the context of risk assessments conducted for MMDT (Figure 1 and Table 1), the four risk assessment steps have the following general definitions. **Hazard identification** is the identification of chemical, physical or biological agents which act through the male to produce developmental toxicity. It is important at this point to stress differences between MMDT and other forms of toxicity. In most toxicological experiments treatment and response are assessed in the same individual. In MMDT, however, males are treated and the response observed in females or offspring (although male biomarkers may be utilized as surrogates of that response). Therefore hazard identification will require treating the male, breeding and then determining the effect of treatment on reproductive or developmental health in the female or offspring. **Hazard characterization** is the dose-response relationship between male exposure to the chemical, physical or biological agent and developmental toxicity as expressed in the female or offspring. In the process of hazard characterization it is also important to define the site and mechanism of action of the male mediated developmental toxicant. As reproductive and developmental toxicants typically act at specific stages of reproduction or development, it is also important to define the sensitive stages in male reproductive processes (e.g., sensitive stages in the production and maturation of sperm). **Exposure assessment** includes characterization of the amount and duration of exposure. Given the potential for stage specificity in MMDT it is also important to define the timing of exposure with respect to the male reproductive cycle and to conception. **Risk characterization** is the final step in which the information on dose-response relationships, site and mechanism of action, window of sensitivity, and characterization of exposure is combined to determine qualitatively or quantitatively the risk to the individual or population.

Statistical Approaches for Risk Assessment for Developmental Toxicity

NOAEL-SF-ADI Approach: Initially the focus of attention for risk assessment for developmental toxicity was on the identification of the NOAEL and the application of a SF (which ranged from 1 to 1000) to the NOAEL to determine the ADI (Barnes and Dourson 1988, Crump 1984, Dourson and Stara 1983, Dourson et al 1985, Hart et al 1988, Kimmel et al 1989, Mattison et al 1989). This approach assumes that there are thresholds for developmental toxicity and once the threshold is defined it is possible to set a safe level of exposure for humans. Although this approach appears simple to apply, it has definite disadvantages. To determine the NOAEL, some investigators have simply identified the highest dose for which the response is not statistically different from control. Others have used an approach to identify the highest dose that does not have a statistically significant trend for a dose-response relationship. While these approaches are conceptually simple, they present some analytical problems to the risk assessor; the first method does not use all of the data, and in both methods the NOAEL is dependent on the number of animals in the treatment and control groups. As the number gets smaller the NOAEL increases. These characteristics tend to reward small studies with large spacing between the doses.

Applying the NOAEL-SF-ADI approaches for male-mediated developmental toxicity would require treating male experimental animals with multiple doses of a chemical of concern, and then mating these males at appropriate time intervals with females. Several developmental endpoints (Table 1 and Figure 1) are then monitored in the female experimental animals or their offspring. For example, if post-implantation loss occurs and is the endpoint of concern, the number of resorptions or difference between number of implantation sites and live pups is determined. The NOAEL could then be defined as the highest dose at which the number of resorptions or difference between implantation sites and live pups is not significantly different from controls. As previously indicated, as the number of experimental animals at each dose decreases, statistical power decreases and the identified NOAEL increases.

Table 1. Potential Adverse Developmental Effects Following Male Exposure

Reproductive Process	Toxicological Effect of Male Exposure	Endpoint Observed in Epidemiological Studies
Conception	Preimplantation loss	Decreased fecundity Decreased fertility Increased time to pregnancy
Implantation	Preimplantation loss	Decreased fecundity Decreased fertility Increased spontaneous abortion
Embryo Development	Failure or disruption of embryonic development	Increased spontaneous abortion Increased fetal death Increased malformation Increased growth retardation Increased functional deficit Increased premature birth
Fetal growth and development	Failure or disruption of fetal growth and development	Increased fetal death Increased malformation Increased growth retardation Increased premature birth Increased functional deficit

Benchmark Dose-SF-R_fD_{dt} Approach: Another statistical approach which takes advantage of all of the data, uses individual animal responses at all points in the dose-response curve to calculate a B_mD which is in or near the observed treatment range (e.g., a lower confidence interval on the ED_{05} or ED_{10}). The B_mD is then divided by a SF (which can range from 1 to 1000) to determine the R_fD_{dt}. This approach has great appeal because it uses all of the data collected in a developmental toxicity experiment. Some investigators, however, have avoided the approach because of the statistical analysis required. Applying the B_mD-SF-R_fD_{dt} approach for male-mediated developmental toxicity would begin with an experiment as described above. Treated males would be mated to untreated females and the developmental endpoint of concern evaluated (Table 1 and Figure 1). If the endpoint of concern is malformation, the pregnancies are allowed to continue to term and the pups evaluated after delivery. The paternal dose-pup response (malformation) relationship would be fully evaluated and modeled. A given benchmark dose (e.g., the lower 95% confidence interval on ED_{05}) would be determined, and a safety factor applied to determine the R_fD_{dt}. Note that both the NOAEL-SF-ADI and the B_mD-SF-R_fD_{dt} approaches for MMDT assume that there are interactions between the male and female that influence the outcome of the dose-response experiment (Figure 1).

Biological Approaches for Risk Assessment

A range of assumptions are necessary to perform risk assessments for human health, these include the extrapolations necessary from the high doses at which animal experiments are conducted to the low doses typical of human exposure. Other extrapolations include those from animals to humans and extrapolation across route of exposure (e.g., intraperitoneal dosing in animals to pulmonary or dermal exposure in humans). As the risk sciences have developed improved methods for characterizing human health risks, the techniques have increasingly utilized more detailed biological information to decrease the uncertainty about these extrapolations (Antjes et al 1950, Amann 1982, Amann and Howard 1980, Amann and Hamershedt 1980, CECOS 1982, David et al 1979, Ewing and Mattison 1987, Faustman et al 1989, Foote et al 1986 ab, Freireich et al 1966, Guerrero and Rojas 1975, Horning et al 1981, Katoh et al 1989, Katz et al 1981, 1982, Mattison 1990, 1991, 1993, Mattison and Thomford 1989, Mattison et al 1990, 1991). A general consensus has developed which suggests that *to the extent possible risk assessments should be constructed with consideration of the biological principles that govern the endpoint evaluated* (Figure 1 summarizes some of these principles for MMDT). For example, some of the factors which can influence MMDT include: male reproductive status or male fecundity, the dose and duration of exposure, the stages of spermatogenesis effected by the chemical, the distribution of the chemical to the testis and other male reproductive organs, the mechanisms of action of the chemical, and the frequency and timing of intercourse with respect to treatment.

PK Approaches: The classical PK approaches have used species specific information on clearance and volume of distribution to reduce the uncertainty inherent in a dose response relationship based on exposure rather than internal or target organ concentration (National Research Council 1987, 1989). For some chemicals these PK characteristics are known across species. For those that are not, allometric techniques can be used for the cross species extrapolation. In characterizing risks for MMDT the pharmacokinetic information of interest would include delivery of the chemical and its metabolites to the testis, concentrations within the seminiferous tubule and impact on sperm function. Unfortunately, it is not immediately apparent using classical PK approaches which of the compartments represent the testis or seminiferous tubules. While the classical PK approaches are thought to decrease the uncertainty, the compartment volumes and rate constants do not have obvious biological counterparts, for this reason PBPK approaches are being developed.

PBPK Approaches: Successful pregnancy in humans and experimental animals is characterized by change in the mother, placenta and fetus (Mattison and Jelovsek 1987, Mattison et al 1991, Mattison 1990, 1991b). This continuing change complicates the traditional PK approaches as the volumes and rate constants change over the course of pregnancy in ways that are species specific. An approach which is readily adaptable to the changes during pregnancy are the PBPK approaches which specifically include cardiac output, organ volumes and metabolic processing within each individual organ. While PBPK approaches have not been utilized to decrease uncertainty in MMDT risk assessment they should be explored.

BBDR Approaches: As risk assessment for developmental toxicity advances to characterize the amount of the chemical delivered to the target site in the mother, placenta or fetus which is responsible for the developmental effects observed, there is increasing interest in characterizing the response in a more biologically relevant framework. For example, if the mechanism of developmental toxicity involves occupancy of a receptor in the fetal liver, and if the receptor characteristics differ in humans and experimental animals, that difference can be explicitly included in the cross-species extrapolation of the target organ dose-response relationship. This decreases the uncertainty of the extrapolation from experimental animals to humans. This approach allows the potential for differences across species in the mechanisms of developmental toxicity to be explicitly incorporated into risk assessment. While this approach is very data intensive, it offers the potential for some compounds to provide a substantial reduction in the uncertainty of the risk assessment. As the mechanisms for MMDT remain incompletely characterized it is probably premature to consider using this method to characterize male mediated risks to development.

COMPARISON OF RISK ASSESSMENT FOR MMDT WITH TRADITIONAL RISK ASSESSMENT

Like reproductive and developmental toxicity, risk assessment for MMDT is outside the general framework of risk assessment for cancer or end organ toxicity. The traditional risk assessment completes the four steps within the context of an exposure to a specific individual or population. Both exposure and effect are characterized for the same individual. However, risk assessment for MMDT is substantially different. While the exposure is to the male, the response can be measured in the male, pregnant female or offspring. This complicates risk assessment considerably and suggests that the risk assessment process must be considered within a different - couple based framework (Baird et al 1986, Baird and Wilcox 1985, Barrett 1970, 1971ab, 1978, Barrett and Marshall 1969, CECOS 1982, David et al 1979, Generoso et al 1979ab, Generoso 1980, Guerrero and Rojas 1975, Levine et al 1980ab 1981, Mattison and Brewer 1988, Mattison 1991, Meistrich 1983,1984 1988 1989abc, Meistrich and Brown 1983, Menken et al 1987, Mosher and Pratt 1982, 1985, Schwartz et al 1982). In the context of the couple-based approach for characterizing the risk of MMDT, endpoints may include: post-term delivery; developmental defect, including both structural and functional; premature delivery; fetal death; recognized pregnancy loss; unrecognized pregnancy loss and preimplantation pregnancy loss (Figure 1, Table 1). The risk assessment process will be conducted using biological markers of male reproductive function (National Research Council 1987 1989).

BIOLOGICAL MARKERS IN MALE MEDIATED DEVELOPMENTAL TOXICITY RISK ASSESSMENT

The utility of biological markers as tools for characterizing risk of MMDT following male exposure is not well defined. Biologic markers should be utilized to clarify relationships between exposure of the father and developmental disease in the embryo, fetus or offspring (National Research Council 1987,1989). The availability of validated biological markers will allow the development of biomarker-based risk assessment for MMDT. A biomarker-based model of risk assessment should incorporate male factors (MF), female factors (FF) and couple factors (CF) which may influence developmental success.

Biomarkers of Male Function

Biomarkers of male reproductive function characterize the ability of the male to fertilize the female and produce a normal offspring. The biomarkers which *may* be applicable for characterizing the impact of an exposure of the male to developmental success in the female or offspring include: anatomic factors (A_{fm}), ejaculate volume (E_v), ejaculate composition (E_c), sperm number per ejaculate (S_n), sperm motility (S_m), sperm morphology (S_s), and measures of the genetic integrity of the sperm (S_{gi}). These biomarkers contribute to a function which describes the role of male factors in developmental success:

$$\textbf{Male Factors} = MF(A_{fm}, E_v, E_c, S_n, S_m, S_s, S_{gl})$$

Selected biomarkers will be included in risk assessment for MMDT, depending on: availability of data, utility of a given biomarker, and the sensitivity of a specific developmental endpoint to changes in a specific biomarker. Data from the National Toxicology Program (NTP) suggests that sperm concentration, motility and morphology may be reasonable biomarkers in characterizing male fecundity (Table 2), their utility in characterizing MMDT however remains to be defined (Morrissey 1989). The data in this study is limited however by the number of chemicals examined, there were 24 chemicals studied in a testing protocol designed to assess reproductive performance by continuous breeding (Lamb 1989).

Table 2. Utility of selected biomarkers as predictors of male fecundity[a]

Biological Marker	Sensitivity (%)	Specificity (%)	Positive Predictive Value (%)	Negative Predictive Value (%)
Epididymal weight	80	87	80	87
Testis Weight	62	83	80	67
Sperm Motility	69	92	90	73
Sperm Concentration	70	80	70	80
Abnormal Morphology (%)	60	79	67	73
Body Weight	38	58	50	47

[a] Data from Morrissey (1989, Figure 1 page 211)

Biomarkers of Female Function

These biomarkers characterize the ability of the female to be fertilized by the male and produce a normal offspring. The biomarkers which *may* be applicable for characterizing developmental success include: anatomic factors (A_{ff}), ovulatory frequency (O_f), follicular phase characteristics (F_f), luteal phase characteristics (L_f), endometrial function (Ef), tubal function (T_f), DNA repair characteristics of the oocyte (O_{dna}), genetic characteristics of the oocyte (O_{gi}). As described for the male these biomarkers contribute to a function which describes the role of female factors in developmental success:

$$\textbf{Female Factors} = \textbf{FF}(A_{ff}, O_f, F_f, L_f, E_f, T_f, O_{dna}, O_{gi})$$

Biomarkers applicable for a given risk assessment for MMDT will contribute based on their ability to represent the adverse male effect on developmental outcome. The use of an individual biomarker is dependent upon: availability of data and the utility of that biomarker for the developmental endpoints under consideration.

Biomarkers of Couple Specific Function

These biomarkers characterize the ability of a given couple (or breeding pair for experimental animals) to produce normal offspring. The biomarkers which *may* be applicable for characterizing the role of couple specific factors in developmental success include: frequency of intercourse (F_i), male-female interactions (M_f), and female-male interactions (F_m). These biomarkers contribute to a function which describes the role of couple specific factors in developmental success:

$$\textbf{Couple-dependent factors} = \textbf{CF}(F_i, M_f, F_m)$$

As described for MF and FF, the biomarkers which characterize couple function will be included in risk assessment for MMDT depending on: availability of data and utility of a given biomarker to predict the endpoint of concern or sensitivity of the endpoint to the biomarker.

MMDT Risk Assessment

In the context of the couple-specific approach for MMDT, developmental risk is a function of the individual and couple-specific factors and is represented by the equation:

Developmental risk = DR (MF, FF, CF)

Note that the function describing developmental risk is likely to be specific to the endpoint considered. For example, the function describing the male contribution to preimplantation pregnancy loss is likely to be different from the function describing the male contribution to developmental defects observed in the offspring at birth. Animal data is needed to investigate relationships between biomarkers suggested to quantitate the impact of exposures to the male and the risk for developmental toxicity. Methods to properly extrapolate this data from animals to humans, from high to low dose and across other factors are also required. To illustrate this process, we will explore two examples; in the context of the couple-based approach, one using animal data on dominant lethal effects and another hypothetical risk assessment for MMDT.

Table 3. Effect of Nitrobenzene (NB) on MMDT[a]

Endpoint	Control	NB (140 mg/kg)	Percent of Control
	Male Reproductive Biomarkers		
Sperm Production Rate ($x10^6$/day)	24.2	13.1	54%
Epididymal Sperm Number ($x10^6$)	397	137	35%
Abnormal Morphology (%)	0.8%	6.0%	750%
Motile (%)	62%	25%	40%
Velocity (um/sec)	47	43	91%
MNMS/day[b] ($x10^6$)	14.8	3.1	21%
	Breeding Parameters[c]		
Females Pregnant	85%	80%	94%
Implants/Pregnancy	13.7	11.8	86%
Corpora Lutea/Pregnancy	14.6	14.4	99%
Preimplantation Loss	0.9	2.6	289%

[a] Data from Blazak (1989, Table 4 page 167). [b]MNMS/day = number of morphologically normal motile sperm produced per day. [c] Breeding data are from the week prior to sacrifice, during this week each male was caged with two untreated virgin female Sprague Dawley rats.

MMDT: Dominant Lethal Effects

Analysis of NTP data from reproductive effects by continuous breeding (Lamb 1989) suggests that morphologically normal sperm concentration in the ejaculate is a good predictor of male fertility (Table 2 and Morrissey 1989). Using data from an experiment with Nitrobenzene (NB) it is possible to define the effect on early pregnancy loss (dominant lethal effect) quantitatively (Blazak 1989). In this experiment Sprague Dawley rats were treated with NB (140 mg/kg body weight by gavage in corn oil) for five consecutive days. Each male was caged with two virgin female Sprague Dawley rats per week for eight weeks. The parameters explored in this analysis include those suggested to be valid predictors of male fertility (Tables 2 and 3).

Notice that although there were significant alterations in sperm production rate, epididymal sperm number and morphologically abnormal sperm, there was minimal effect on the proportion of females who became pregnant. Preimplantation loss however was increased significantly. Using the authors description of the methodology of the study it is possible to derive estimates of the cycle specific fertility rate and then male, female and couple factors. As the experiment was conducted the treated males were bred to two untreated virgin females each week. The equation describing the relationship between the cumulative pregnancy rate and cycle specific fertility is:

$$F = 1 - (1 - CSFR)^n$$

where F is the cumulative fertility (85% in control and 80% in treated Sprague Dawley rats), CSFR is the cycle specific fertility rate, and n is the number of breeding cycles (n = 2 in this experiment because the data on fertility reported was for the week prior to sacrifice). Using this data it is possible to calculate the appropriate reproductive parameters (Table 4). Estimates of the human parameters are also included for comparison (Mattison 1991).

Table 4. Reproductive Parameters for the Sprague Dawley Rat and Human

Reproductive Parameter	Sprague Dawley Rat		Human	
	Control	NB (140 mg/kg)	Worst Case	Best Case
CSFR	0.613	0.553	0.10	0.30
EPL	0.06	0.18	0.50	0.30
MF, FF, CF	0.87	0.88	0.58	0.75

Note that CSFR is decreased slightly by NB treatment and that EPL (as a result of dominant lethal mutations) is increased. It is interesting to observe that there is little effect on male, female or couple factors in this experiment, suggesting that all of the effect on fertility can be explained by the effect on early pregnancy loss (preimplantation loss). Similar human parameters illustrate the difference in human and Sprague Dawley rat reproductive function, with the human much less efficient (see Mattison 1991 for an explanation of the derivation of these parameters). Using this data it would be possible to evaluate the effect of human exposure to NB, assuming that the effect was similar - increased EPL and decreased CSFR. For example if human EPL were increased 3-fold as observed in the rat then CSFR would be decreased from 0.3 to 0.042, and the cumulative percent pregnant after 12 months of unprotected intercourse would decrease from 98.6% to 40.2%. A more detailed hypothetical risk assessment for MMDT is presented next.

Hypothetical Risk Assessment For MMDT

This hypothetical risk assessment for MMDT is laid out in the format suggested by the National Research Council (1983) with four steps; hazard identification, hazard characterization, exposure characterization and risk characterization.

Hazard Identification: This step identifies whether a chemical, physical or biological agent acts through the male to produce developmental toxicity. The hazard identification experiment conducted has demonstrated that at high doses the chemical "PGH" increases the time to the delivery of the first litter of pups in Sprague Dawley rats (Table 5).

Table 5. Effect of "PGH" on Time to Delivery of First Litter in Sprague Dawley Rats

Dose (mg/kg/day)	Time to first litter (days)
0	25
10	27
50	42

In these experiments male rats were treated for 7 weeks before breeding and during the breeding period which lasted an additional 10 weeks. During the breeding experiment males were housed with females until they delivered. After delivery the pups and females were placed in another cage and a new female was placed with the male. After the 10 week breeding period was completed, the males were sacrificed and the reproductive system removed for analysis (Heindel and Chapin 1993).

Hazard Characterization: The minimum data required for this step is characterization of the dose-response relationship between male exposure and developmental toxicity. As described in the hazard identification step "PGH" increases the time to delivery of the first litter in rats (Table 5). Following completion of the 10-week breeding experiment, the male Sprague Dawley rats were sacrificed and their reproductive systems removed for histological analysis. Histological analysis of the testes suggested disruptions of spermatogenesis (Heindel and Chapin 1993), and analysis of the sperm in the caput epididymis revealed an increase in the frequency of morphological alterations (Heindel and Chapin 1993).

Subsequently, more detailed experiments have demonstrated that there is a relationship between dose of "PGH" and sperm damage as measured by morphometric analysis. For example, several shape parameters are combined to form a composite sperm damage parameter. In addition, subsequent experiments have also demonstrated the relationship between initiation of treatment and development of sperm damage, suggesting that the early spermatid stages of spermatogenesis appear to be the most sensitive to the effect of "PGH". Investigations using this technique suggest that damage to the male germ cell is time and dose dependent, with abnormal sperm appearing about three weeks after the initiation of treatment. The post stem cell stages appear to be more sensitive and vulnerable to a broader range of chemicals than the stem cells. In addition the post stem cells vary in vulnerability to chemical attack (Table 6). In addition to variable sensitivity to different chemical agents with cycle of sperm production, the size and type of lesion appears to vary with time of exposure and dose of "PGH".

Table 6. Time and Dose Response Relationships for Sperm Morphological Alterations

Germ Cell Stage	Weeks from start of treatment	Dose (mg/kg/day)		
		0	10	50
		Percent Abnormal Sperm		
Spermatozoa	1	2%	4%	6%
Late spermatids	2	3%	4%	8%
Early Spermatids	3	3%	10%	40%
Diplotene, pachytene	4	4%	12%	45%
Pachytene, leptotene	5	2%	14%	48%
Differentiating gonia	6	3%	14%	50%
Differentiating gonia	7	2%	14%	48%

Following these experiments demonstrating damage to sperm during the diplotene to early spermatid stage of spermatogenesis, a series of experiments were conducted to explore the potential relationship between sperm damage and resorption of the conceptus in Sprague Dawley rats (Table 7). In these experiments, rats were treated for 10 weeks before breeding, at doses which produced 10, 15, 20 and 25% sperm with morphological abnormalities. Treatment then continued for another 10 weeks during which the males were mated with female Sprague Dawley rats as described in the Hazard Identification section. In these experiments the females were checked for vaginal sperm plugs each morning, the day on which a vaginal plug was noted was designated as day 0. The females were removed on day 15, sacrificed and the ovary examined for corpora lutea, and the uterus examined for implantation sites (live pups and resorptions).

Table 7. Relationship between Morphological Abnormalities and Early Pregnancy Loss

	Approximate Dose (mg/kg)	% Abnormal Sperm	% Pregnant	# CL/Preg	Implants/Preg	EPL	% EPL
Control		3%	86%	14.5	13.4	1.1	7.6%
Treated	7	10%	88%	14.2	12.0	2.2	15.5%
	12	15%	91%	14.6	10.2	4.4	30.1%
	18	20%	82%	14.7	8.4	6.3	42.9%
	24	25%	88%	14.9	6.7	8.2	55.0%

In this instance a function describing the relationship between exposure and sperm damage would be prepared for the Male Factor:

Male Factor = MF (sperm damage vs. dose)

This function can be approximated by a linear regression fit to the relationship between dose and percent abnormal sperm as defined in Table 6. In this case the relationship is:

Percent Abnormal Sperm = 0.86*Dose + 3.95

An additional function can be derived describing the relationship between the Male Factor, Female Factor, Couple Factor and Early Pregnancy Loss with this treatment. As indicated by Table 7, treatment appears to impact only on EPL with little effect on the percent of females who are impregnated. In this setting EPL can be described as a function of the percent abnormal sperm:

Early Pregnancy Loss = (MF, FF, CF)

or

EPL = 2.24 * %Abnl Sperm - 2.47

In this example EPL could be expressed either as a function of male exposure or sperm damage. As is apparent the chosen expression is as a function of abnormal sperm in the ejaculate. Note that this is not corrected for the background rate of sperm abnormalities, so another implicit assumption, which is probably incorrect, is that the background abnormality rate and early pregnancy loss rates are related.

Exposure Asseesment

Once the functions defining the relationship between male exposure and sperm damage, and sperm damage and resorption are defined it is necessary to characterize of the amount and duration of exposure for human populations. Given the potential for stage specificity it is also important to define the timing of exposure and reproduction with respect to the male reproductive cycle.

"PGH" is used commercially in a paint product which is typically applied to the exterior of metal buildings. It is estimated that there are approximately 25,000 applicators around the United States that use or are exposed to the product. Because early experiments in both rats and mice suggested that spermatogenesis might be altered, there have been several small epidemiological studies exploring the relationship between exposure and semen characteristics. In one small factory in which lighting fixtures are manufactured and painted with a paint containing "PGH", there were three categories of workers with different but unquantitated exposures to "PGH", engineers, assemblers and painters (Table 8).

Table 8. Relationship between "PGH" exposure and Abnormal Sperm Morphology

Job Classification	No.	"PGH" Exposure	% Abnormal Sperm
Engineer	4	very low	15%
Assembler	12	low	20%
Painter	7	high	35%

The study suggests a relationship between exposure to "PGH" and abnormal sperm morphology with higher exposures (based on a job-exposure matrix) having increased numbers of sperm with abnormal morphology. If the relationship between abnormal morphology and early pregnancy loss observed in the Sprague Dawley rat studies also occurs in humans, the spontaneous abortion rates would be 31%, 42% and 78% among the very low, low and high exposure groups respectively. While the study is small and does not control for confounding factors like smoking and other exposures, it is clearly consistent with the animal studies, and suggests that there is reason for concern about the relationship between exposure to "PGH" and developmental toxicity in the wives and offspring of the painters. For this reason, a risk characterization was conducted. For this risk characterization several questions were explored; what is the population distribution of exposure, are there any uniquely exposed or susceptible groups, can a better understanding of mechanism of action and cross or within species susceptibility be developed?

Risk Characterization

In this step information on dose-response relationships, site and mechanism of action, window of sensitivity during spermatogenesis and early conceptus development, along with the definition of exposure is used to compute risk to the individual or population. The animal data available on "PGH" has been derived from a chronic exposure treatment protocol, and human populations are also exposed in an occupational setting on a chronic basis. These animal experiments and human data allow correlation of the potential risk from MMDT for an occupational exposure setting. Unfortunately, the data are not responsive to questions of greater concern, for example are there effects from short-term exposure and can MMDT occur in a general population, outside of an occupational exposure setting?

There is another set of questions which human data and animal experiments have not addressed, that is, what are the couple factors in determining reproductive success or failure. This question arises because of data suggesting that the oocyte is capable of repairing DNA damage, and that DNA repair varies with different strains of females (Generoso et al 1979a, Generoso 1980, Pederson and Managia 1978). The design of the animal experiments have avoided the question by using male and female Sprague Dawley rats and available human data does not allow the direct characterization of a female factor in the pregnancy outcome. One way to evaluate the influence of female factors on risk of MMDT would be to explore in a rat or mouse model the influence of different female strains on the frequency of preimplantation loss. For example, the experiment illustrating the impact of morphologically abnormal sperm on preimplantation loss frequency might be repeated in several different strains of female rats (Table 9).

Table 9. Influence of Strain of Female Rats on Preimplantation Pregnancy Loss Rate

		Sprague Dawley	Strain B	Strain C	Strain D
% Damaged Sperm		% Preimplatation Pregnancy Loss (EPL)			
Control	3%	8%	2%	8%	5%
Treated	10%	16%	1%	12%	15%
	15%	30%	1%	20%	20%
	20%	43%	2%	30%	25%
	25%	55%	2%	50%	30%

In these experiments the extent of sperm damage is similar within each dose group. However, the impact of the sperm damage, as reflected in EPL, varies substantially with the strain of the female rat used in the protocol. Experiments like these allow characterization of the influence of the female factors in reproductive outcome. In human populations similar data might be available from sperm banking programs by evaluating the pregnancy success of a fixed number of sperm with either artificial insemination from donor semen or *in vitro* fertilization.

Several different approaches are possible for risk assessment given the data on differences across species in EPL. One is to assume that the most appropriate model would be the EPL frequency for male and female rodents from the same strain. Another, more risk averse approach would be to use data from the most sensitive matings across rodent strains. A third approach might be to acknowledge the difference in female factors and use the distribution of outcomes to model what is likely to be a distribution of outcomes in human populations.

Another issue of concern for MMDT risk assessment is the relative sensitivity from rodents to humans (or differential sensitivity across species). The traditional approach assumed a 10-fold variability in sensitivity within species, and a 10-fold variation between species. If that is also the case (or the default assumption) for MMDT, we would shift the animal dose response curve to the left by a factor which may be as small as 1/10 or as large as 1/1000. Note that it is entirely possible that humans may be more resistant to MMDT, by as much as three orders of magnitude, so that for some chemicals, it might be possible to shift the dose-response curve to the right. While a general approach for characterizing risks for MMDT has been summarized, a series of questions remain; how should these risks for early pregnancy loss be extrapolated to human populations, are humans equally, and less or more sensitive to the putative developmental toxicant. Without the direct data to answer these questions the typical approach is to be risk averse; assume the effect occurs in human populations and assume that humans are at least an order of magnitude more sensitive than the rodent model used in hazard identification and characterization.

CONCLUSIONS

A general model has been presented integrating elements necessary for assessment and quantitation of developmental risks following male exposure. Once the relationship between exposure and Male Factors, Female Factors and Couple Specific Factors has been defined, it is possible to explore the relationship to developmental toxicity. The biomarker based model presented is a first step in the development of quantitative approaches for estimation of male mediated developmental risks. Eventually it is hoped that more complete definition of MF, FF, CF could all be combined into one model for reproductive and developmental risk assessment.

REFERENCES

Aafjes JH; Vels JM; Schenck E. Fertility of rats with artificial oligozoospermia. J Reprod Fertil 1980; 58: 345-351.

Amann RP. Use of animal models for detecting specific alterations in reproduction. Fundam Appl Toxicol 1982; 2: 13-26.

Amann RP; Howards SS. Daily spermatozoal production and epididymal spermatozoal reserves of the human male. J Urol 1980; 124: 211-5.

Amann RP; Hammerstedt RH. Validation of a system for computerized measurements of spermatozoal velocity and percentage of motile sperm. Biol Reprod 1980; 23: 647-56.

Ash, P. (1980) The influence of radiation on fertility in man. The Bri J Radiol 53, 271-278.

Baird, D.D., Wilcox, A.J., Weinberg, C.R. (1986) Use of time to pregnancy to study environmental exposures. Amer J Epidemiol 124, 470-480.

Baird, D.D., Wilcox, A.J. (1985) Cigarette smoking associated with conception delay. JAMA 253, 2979-2983.

Barlow SM, and Sullivan FM (1982). Reproductive Hazards of Industrial Chemicals. An Evaluation of Animal and Human Data. Academic Press, New York.

Barnes DG; Dourson M Reference dose (RfD): description and use in health risk assessments. Regul Toxicol Pharmacol 1988:471-486.

Barrett JC, Marshall J. The risk of conception on different days of the menstrual cycle. Pop. Studies 23:455-461, 1969.

Barrett JC Use of a fertility simulation model to refine measurement techniques. Demography 1971a:481-490.

Barrett JC An analysis of coital patterns. J Biosoc Sci 1970:351-357.

Barrett JC Effects of various factors on selection for family planning status and natural fecundability: a simulation study. Demography 1978:87-98.

Barrett, J.C. (1971b) Fecundability and coital frequency. Pop Studies 25, 309-313.

Blazak WF (1989) Significance of cellular endpoints in assessment of male reproductive toxicity. In PK Working (ed) Toxicology of the Male and Female Reproductive Systems Hemisphere Publishing Corp New York pp 157-172.

Brandriff B; Gordon L; Ashworth L; Watchmaker G; and others Chromosomes of human sperm: variability among normal individuals. Hum Genet 1985;70(1):18-24.

Butler WJ Kalasinski LA (1989) Statistical analysis of epidemiological data of pregnancy outcomes. Environ Health Perspect 79, 223 - 227.

CECOS, Schwartz D, Magaux MJ. Female fecundity as a function of age. Results of artificial insemination in 2193 nulliparous women with azospermic husbands. NEJM 306:404-406, 1982.

Chinchilli VM, Clark BC (1989) Trend tests for proportional responses in developmental toxicity risk assessment. Environ Health Perspect 79, 217 - 221.

Clegg ED; Sakai CS; Voytek PE Assessment of reproductive risks. Biol Reprod 1986:5-16.

Creasey DM Foster PMD (1984) The morphological development of glycol ether-induced testicular atrophy in the rat. J. Pathol. 139, 309 - 321.

Crump KS A new method for determining allowable daily intakes. Fundam Appl Toxicol 1984:854-871.

David G; Jouannet P; Martin-Boyce A; Spira A; Schwartz D Sperm counts in fertile and infertile men. Fertil Steril 1979:453-455.

Davis D, B. Goldstein, J. Gibson, R. Henderson, J. Hobbie, P. Landrigan, D.R. Mattison, F. Perera, F. Peter, E. Pfitzer, E. Silbergeld, R. Thomas, D. Wagener, L. Wakefield, and G. Wogan. Biological markers in environmental health research. Environmental Health Perspectives. 74:3-9, 1987.

DeLean, A., Munson, P.J., Rodbard, D. (1978) Simultaneous analysis of families of sigmoidal curves: Application to bioassay, radioligand assay, and physiological dose response curves. Am J Physiol 235, E97 - E102.

Dourson ML; Stara JF Regulatory history and experimental support of uncertainty (safety) factors. Regul Toxicol Pharmacol 1983:224-238.

Dourson ML; Hertzberg RC; Hartung R; Blackburn K Novel methods for the estimation of acceptable daily intake. Toxicol Ind Health 1985:23-33.

Ewing LL and D.R. Mattison. Introduction: biological markers of male reproductive toxicology. Environmental Health Perspectives. 74:11-13, 1987.

Fabro, S. On Predicting environmentally-induced human reproductive hazards: An overview and historical perspective. Fund Appl Tox 1985;5:609-614.

Fabro, S., Schull, G., and Brown, N.A. (1982). The relative teratogenic index and teratogenic potency: proposed components of developmental toxicity risk assessment. Teratogen. Carcinog. Mutagen. 2:61-76.

Faustman EM; Wellington DG; Smith WP; Kimmel CA Characterization of a developmental toxicity dose-response model. Environ Health Perspect 1989:229-241.

Foote RH; Schermerhorn EC; Simkin ME Measurement of semen quality, fertility, and reproductive hormones to assess dibromochloropropane (DBCP) effects in live rabbits. Fundam Appl Toxicol 1986a:628-637.

Foote RH; Berndtson WE; Rounsaville TR Use of quantitative testicular histology to assess the effect of dibromochloropropane (DBCP) on reproduction in rabbits. Fundam Appl Toxicol 1986b:638-647.

Francis EZ; Kimmel CA; Rees DC Workshop on the qualitative and quantitative comparability of human and animal developmental neurotoxicity: summary and implications. Neurotoxicol Teratol 1990:285-292.

Frankos VH (1985). FDA Perspectives on the use of teratology data for human risk assessment. Fund Appl Toxicol 5:615-625.

Freireich EJ; Gehan EA; Rall DP; Schmidt LH; Skipper HE Quantitative comparison of toxicity of anticancer agents in mouse, rat, hamster, dog, monkey, and man. Cancer Chemother Rep 1966:219-244.

Galbraith WM, P. Voytek, M. G. Ryon, in Assessment of Reproductive and Teratogenic Hazards, M. Christian et al., Eds. (Princeton Scientific, Princeton, 1983) Section II pp 1-158.

Gaylor DW (1989) Quantitative risk analysis for quantal reproductive and developmental effects. Environ Health Perspect 79, 243 - 246.

Gaylor DW, D.M. Sheehan, J.F. Young, D.R. Mattison. The threshold dose question in teratogenesis. Teratology 8:389-391, 1988.

Generoso WM, Katoh M; Cain KT; Hughes LA; and others Chromosome malsegregation and embryonic lethality induced by treatment of normally ovulated mouse oocytes with nocodazole. Mutat Res 1989:313-322.

Generoso WM; Huff SW; Cain KT Relative rates at which dominant-lethal mutations and heritable translocations are induced by alkylating chemicals in postmeiotic male germ cells of mice. Genetics 1979a:163-171.

Generoso, W.M. (1980) Repair in fertilized eggs of mice and its role in the production of chromosomal aberrations. Basic Life Sci 15, 411-420.

Generoso, W.M., Cain, K.T., Krishna, M., Huff, S.W. (1979b) Genetic lesions induced by chemicals in spermatozoa and spermatids of mice are repaired in the egg. Proc Natl Acad Sci USA 76, 435-437.

Glickman TS, Gough M. (1990). Readings in Risk. Resources for the Future. 1616 P Street, N.W., Washington, DC 20036.

Guerrero VR, Rojas OI. Spontaneous abortion and aging of human ova and spermatozoa. N. Engl. J. Med. 293:573-575, 1975.

Hamill, P.V.V., Steinberger, E., Levine, R.J., Rodriguez-Rigau, L.J., Lemeshow, S., et al. (1982) The epidemiologic assessment of male reproductive hazard from occupational exposure to TDA and DNT. J Occup Med 24, 985-993.

Hanley TR Jr; Young JT; John JA; Rao KS Ethylene glycol monomethyl ether (EGME) and propylene glycol monomethyl ether (PGME): inhalation fertility and teratogenicity studies in rats, mice and rabbits. Environ Health Perspect 1984 Aug;57:7-12.

Hart WL, Reynolds RC, Krasavage WJ, Ely TS, Bell RH, Raleigh RL. Evaluation of developmental toxicity data: a discussion of some pertinent factors and a proposal. Risk Analysis 8:59-69, 1988.

Heindel JJ, Chapin RE (1993) Female Reproductive Toxicology. Methods in Toxicology, Volume 3, Part B, Academic Press, New York.

Heinrichs, W.L., Juchau, M.R. (1980) Extrahepatic drug metabolism: The gonad. In: Extrahepatic Metabolism of Drugs and Other Foreign Compounds. Gram, T.E. (ed) SP Medical and Scientific Books, New York, pp 313-332.

Hemminki K, Lindbohm M-L, Taskinen H. Transplacental toxicity of environmental chemicals. In: A Milunsky, EA Friedman, L Gluch (eds) Advances in Perinatal Medicine Vol.5, New York, Plenum Publishing Co. 1986, pp. 43-91.

Hemminki K; Vineis P Extrapolation of the evidence on teratogenicity of chemicals between humans and experimental animals: chemicals other than drugs. Teratogenesis Carcinog Mutagen 1985;5(4):251-318.

Horning SJ; Hoppe RT; Kaplan HS; Rosenberg SA Female reproductive potential after treatment for Hodgkin's disease. N Engl J Med 1981:1377-1382.

Janoff S. (1990). The Fifth Branch. Science Advisors as Policymakers. Harvard University Press, Cambridge, MA.

Jelovsek FR, D.R. Mattison, J. Chen. Prediction of risk for human development toxicity: how important are animal studies? Obstetrics and Gynecology 74:624-636, 1989.

Jelovsek FR, D.R. Mattison, J.F. Young. Eliciting principles of hazard identification from experts. Teratology, 42:521-533, 1990.

Katoh M, Cacheiro NL; Cornett CV; Cain KT; and others Fetal anomalies produced subsequent to treatment of zygotes with ethylene oxide or ethyl methanesulfonate are not likely due to the usual genetic causes. Mutat Res 1989:337-344.

Katz DF; Overstreet JW Sperm motility assessment by videomicrography. Fertil Steril 1981 Feb;35(2):188-93.

Katz DF; Diel L; Overstreet JW Differences in the movement of morphologically normal and abnormal human seminal spermatozoa. Biol Reprod 1982;26:566-570.

Katz DF; Overstreet JW; Hanson FW Variations within and amongst normal men of movement characteristics of seminal spermatozoa. J Reprod Fertil 1981;62:221-228.

Kimmel CA; Gaylor DW Issues in qualitative and quantitative risk analysis for developmental toxicology. Risk Anal 1988;8:15-20.

Kimmel CA, Wellington DG, Farland W, Ross P, Manson JM, Chernoff N, Young JF, Selevan SG, Kaplan N, Chen C, Chitlik LD, Siegel-Scott CL, Valaoras G, Wells S (1989) Overview of a workshop on quantitative models for developmental toxicity risk assessment. Environ Health Perspect 79, 209 - 215.

Kimmel CA Perspectives on the concern for and management of prenatal chemical exposure and postnatal effects. Ann N Y Acad Sci 1989;562:1-2.

Kline J, Stein Z, Susser M (1989). Conception to Birth. Epidemiology of Prenatal Development. Oxford University Press, New York, New York.

Koeter HBWM (1983). Relevance of parameters related to fertility and reproduction in toxicity testing. In Mattison DR (ed) Reproductive Toxicology. Alan R. Liss, New York New York pp 81-86.

Lamb JC (1989) Design and use of multigeneration breeding studies for identification of reproductive toxicants. In PK Working (ed) Toxicology of the Male and Female Reproductive Systems Hemisphere Publishing Corp New York pp 131-155.

Leridon H. Human fertility: The basic components. Chicago, IL, Chicago University Press, 1977.

Levine RJ, Symons MJ, Balogh SA, Milby TH, and Wharton MD (1981) A method for monitoring the fertility of workers. 2. Validation of the method among workers exposed to dibromochloropropane. J. Occup. Med. 23, 183-188.

Levine, R.J., Symons, M.J., Balogh, S.A., Arndt, D.M., Kaswandik, N.T., Gentile, J.W. (1980a) A method for monitoring the fertility of workers. 1. Method and Pilot studies. J Occupational Medicine 22, 781-791.

Levine, R.J., Symons, M.J., Balogh, S.A., Milby, T.H., Whorton, M.D. (1980b) A method for monitoring the fertility of workers. 2. Validation of the method among workers exposed to dibromochloropropane. J Occupational Medicine 23, 183-188.

Mattison DR and D.W. Brewer. Computer modeling of human fertility: the impact of reproductive heterogeneity on measures of fertility. Reproductive Toxicology. 2:253-271, 1988.

Mattison, D.R. (1982) The effects of smoking on reproduction from gametogenesis to implantation. Environ Res 28, 410.

Mattison DR, P.K. Working, W.F. Blazak, C.L. Hughes Jr., J.M. Killinger, D.L. Olive, and K.S. Rao. (1990). Criteria For Identifying and Listing Substances Known to Cause Reproductive Toxicity Under California's Proposition 65. Reproductive Toxicology, 4:163-175,1990.

Mattison DR, Jelovsek FR. Pharmacokinetics and expert systems as aids for risk assessment in reproductive toxicology. Environ Health Perspect 1987; 76:107-119.

Mattison DR. Sites of female reproductive vulnerability: implications for testing and risk assessment. Reproductive Toxicology, 7,53-62, 1993.

Mattison D.R. and P.J. Thomford. Mechanisms of action of reproductive toxicants. In: Toxicology of the Male and Female Reproductive Systems. P. Working, Ed. Hemisphere Publishing Corp., New York, pp. 101-129, 1989.

Mattison D.R, D.R. Plowchalk, M.J. Meadows, A.Z. Al-Juburi, J. Gandy, and A. Malek. Reproductive toxicity: the male and female reproductive systems as targets for chemical injury. Medical Clinics of North America, 74:391-411, 1990.

Mattison DR. An overview on biological markers in reproductive and developmental toxicology: concepts, definitions and use in risk assessment. Biomedical and Environmental Sciences, 4:8-34, 1991a.

Mattison DR. Risk Assessment for Developmental Toxicity: Airborne Occupational Exposure to Ethanol and Iodine. Risk: Issues in Health and Safety, 2:227-260, 1991b.

Mattison DR, J. Hanson, D.M. Kochhar, K.S. Rao. Criteria for identifying and listing substances known to cause developmental toxicity under California's Proposition 65. Reproductive Toxicology, 3:3-12, 1989.

Mattison DR, E. Blann and Antoine Malek. Physiological Alterations during Pregnancy: Impact on Toxicokinetics. Fundamental and Applied Toxicology, 16:215-218, 1991.

Mattison DR. Fetal Pharmacokinetic and Physiological Models. In: Developmental Toxicology: Risk Assessment and the Future. R.D. Hood, ed. Van Nostrand Reinhold, New York, New York pp. 137-154, 1990.

Meistrich ML. Calculation of the incidence of infertility in human populations from sperm measures using the two-distribution model. Prog Clin Biol Res 1989a; 302:275-285; discussion 286-290.

Meistrich, M.L. (1989b) Interspecies comparison and quantitative extrapolation of toxicity to the human male reproductive system. In: Toxicology of the Male and Female Reproductive Systems, Working, P.K. (ed), Hemisphere Publishing Company, New York, pp 303-321.

Meistrich, M.L. (1989c) Calculation of the incidence of infertility in human populations from sperm measures using the two-distribution model. In: Sperm Measures and Reproductive Success: Institute for Health Policy Analysis Forum on Science, Health, and Environmental Risk Assessment. Burger, E.J., Tardiff, R.G., Scialli, A.R., Zenick, H. (Eds), Alan R. Liss, New York, pp 275-290.

Meistrich ML; Samuels RC Reduction in sperm levels after testicular irradiation of the mouse: a comparison with man. Radiat Res 1985; 102:138-47.

Meistrich ML Critical components of testicular function and sensitivity to disruption. Biol Reprod 1986; 34:17-28.

Meistrich ML Stage-specific sensitivity of spermatogonia to different chemotherapeutic drugs. Biomed Pharmacother 1984; 38:137-142.

Meistrich, M.L., Brown, C.C. (1983) Estimation of the increased risk of human infertility from alterations in semen characteristics. Fertil Steril 40, pp 220-230.

Meistrich ML. Estimation of human reproductive risk from animal studies: determination of interspecies extrapolation factors for steroid hormone effects on the male. Risk Anal 1988; 8:27-33.

Meistrich ML; Finch M; da Cunha MF; Hacker U; Au WW Damaging effects of fourteen chemotherapeutic drugs on mouse testis cells. Cancer Res 1982; 42:122-31.

Menker, J., Trussell, J., Larsen, U. (1987) Age and infertility. Science 233, 1389-1394.

Milby, T.H., Whorton, D. (1980) Epidemiological assessment of occupationally-related, chemically-induced sperm-count suppression. J Occup Med 22:77-82.

Morrissey RE (1989) Association of sperm, vaginal cytology, and reproductive organ weight data with fertility of swiss (CD-1) mice. In PK Working (ed) Toxicology of the Male and Female Reproductive Systems Hemisphere Publishing Corp New York pp 199-216.

Mosher, W.D., Pratt, W.F. Reproductive impairment among married couples: United Stated Vital and Health Statistics, National Survey of Family Growth Series 23 #11, 1982, (PHS 83-1987).

Mosher, W.D., Pratt, W.F. (1985) Fecundity and infertility in the United States, 1965-1982 NCHS Advance data 104, pp 1-8, publication (PHS) 85-1250.

Mosher WD, W. F. Pratt, Fecundity, Infertility and Reproductive Health in the United States, 1982, (National Center for Health Statistics, Washington) (1987).

National Research Council, (1989) Biologic Markers in Reproductive Toxicology National Academy Press, Washington, DC, 1989.

National Research Council (1984). Toxicity Testing. Strategies to Determine Needs and Priorities. Steering Committee on Identification of Toxic and Potentially Toxic Chemicals for Consideration by the National Toxicology Program. Board on Toxicology and Environmental Health Hazards. Commission on Life Sciences. National Academy Press, Washington DC.

National Research Council (1983). Committee on the Institutional Means for the Assessment of Risks to Public Health. Risk Assessment in the Federal Government: Managing the Process. Commission on Life Sciences, National Research Council. Washington, DC. National Academy Press.

National Research Council (1987) Biomarkers in reproductive and developmental toxicology. Environmental Health Perspectives 74, 1-199.

Nichols AL; Zeckhauser RJ The perils of prudence: how conservative risk assessments distort regulation. Regul Toxicol Pharmacol 1988; 8:61-75.

Office of Technology Assessment (1985). Reproductive Health Hazards in the Workplace. Office of Technology Assessment. Congress of the United States, Washington, D.C.

Overstreet JW; Price MJ; Blazak WF; Lewis EL; Katz DF Simultaneous assessment of human sperm motility and morphology by videomicrography. J Urol 1981; 126:357-60.

Pease W, J. Vandenberg, K. Hooper,(1991) Comparing alternative approaches to establishing regulatory levels for reproductive toxicants: DBCP as a case study. Environ. Health Perspect 91:141-155.

Pederson RA and Manigia F (1978). Ultraviolet light induced unscheduled DNA synthesis by resting and growing mouse oocytes. Mutat Res 49:425-429.

Rao KS; Johnson KA; Henck JW Subchronic dermal toxicity study of trichlorobenzene in the rabbit. Drug Chem Toxicol 1982; 5:249-63.

Rao KS; Schwetz BA Reproductive toxicity of environmental agents. Annu Rev Public Health 1982; 3:1-27.

Rao KS; Burek JD; Murray FJ; John JA; and others Toxicologic and reproductive effects of inhaled 1,2-dibromo- 3-chloropropane in male rabbits. Fundam Appl Toxicol 1982; 2:241-51.

Rao KS; Schwetz BA; Park CN Reproductive toxicity risk assessment of chemicals. Vet Hum Toxicol 1981; 23:167-75.

Rao KS; Cobel-Geard SR; Young JT; Hanley TR Jr; and others Ethylene glycol monomethyl ether II. Reproductive and dominant lethal studies in rats. Fundam Appl Toxicol 1983; 3:80-5.

Rao KS; Burek JD; Murray FJ; John JA; and others Toxicologic and reproductive effects of inhaled 1,2-dibromo-3-chloropropane in rats. Fundam Appl Toxicol 1983; 3:104-10.

Reed NR. (1987) Health Risk Assessment of 1,2,-Dibromo- 3-Chloropropane (DBCP) in California Drinking Water. Department of Environmental Toxicology, University of California, Davis, CA.

Rosenberg, M.J., Wyrobek, A.J., Ratcliffe, J., Gordon, L.A., Watchmaker, G., Fox, S.H., Moore, D.H. II, Whornung, R.W. (1985) Sperm as an indicator of reproductive risk among petroleum refinery workers. Br J Indus Med 42, 123-127.

Schardein JL, Schwetz, BA, Kenel, MF. Species sensitivities and prediction of teratogenic potential. Environ Health Perspect 1985; 61:55-67.

Schardein JL (1993). Chemically Induced Birth Defects. Second Edition, Dekker, New York.

Schwartz D; Laplanche A; Jouannet P; David G Within-subject variability of human semen in regard to sperm count, volume, total number of spermatozoa and length of abstinence. J Reprod Fertil 1979; 57:391-5.

Schwartz, D., Magauz, M.J. (1982) Female fecundity as a function of age. Results of artificial in semination in 2193 nulliparous women with azospermic husbands. N Engl J Med 306, 404-406.

Schwartz D; Mayaux MJ; Martin-Boyce A; Czyglik F; David G Donor insemination: conception rate according to cycle day in a series of 821 cycles with a single insemination. Fertil Steril 1979; 31:226-9.

Schwetz BA; Rao KS; Park CN Insensitivity of tests for reproductive problems. J Environ Pathol Toxicol 1980; 3:81-98.

Selevan SG; Lindbohm ML; Hornung RW; Hemminki K A study of occupational exposure to antineoplastic drugs and fetal loss in nurses. N Engl J Med 1985 Nov 7; 313(19):1173-8.

Selevan SG; Hemminki K; Lindbohm ML Linking data to study reproductive effects of occupational exposures. State Art Rev Occup Med 1986; 1:445-55

Selevan SG, Lemasters GK The dose-response fallacy in human reproductive studies of toxic exposures. J Occup Med 1987; 29:451-4.

Sheehan DM, J.F. Young, W. Slikker, Jr., D.W. Gaylor and D.R. Mattison. Workshop on Risk Assessment in Reproductive and Developmental Toxicology: Addressing the Assumptions and Identifying the Research Needs. Regulatory Toxicology and Pharmacology, 10:110-122, 1989.

Shephard, T.H. (1989) Catalog of Teratologic Agents, 6th ed. Johns Hopkins University Press, Baltimore.

Sparer J; Welch LS; McManus K; Cullen MR Effects of exposure to ethylene glycol ethers on shipyard painters: I. Evaluation of exposure. Am J Ind Med 1988; 14:497-507.

Taskinen H; Lindbohm ML; Hemminki K Spontaneous abortions among women working in the pharmaceutical industry [published erratum appears in Br J Ind Med 1986; 43:432] Br J Ind Med 1986;43:199-205.

USEPA (1986) U.S. Environmental Protection Agency. Guidelines for the health assessment of suspect developmental toxicants. Federal Register; 51:34028-34040.

Welch LS; Schrader SM; Turner TW; Cullen MR Effects of exposure to ethylene glycol ethers on shipyard painters: II. Male reproduction [published erratum appears in Am J Ind Med 1989; 15:239] Am J Ind Med 1988;14:509-26.

Whorton, M.D., Milby, T.H., Stubbs, H.A., Avashia, B.H., Hull, E.Q. (1979) Testicular functions among carbaryl-exposed employees. J Toxicol Environ Health 5, 929-941.

Whorton, M.D., Krauss, R.M., Marshall, S., Milby, T.H. (1977) Infertility in male pesticide workers. Lancet ii, 1259-1261.

Wilcox AJ, Weinberg CR, Wehmann RE, Armstrong EG, Canfield RE, Nisula BC. Measuring early pregnancy loss: laboratory and field methods. Fertility and Sterility 44:366-374, 1985.

Working, P.K. (1989) Toxicology of the Male and Female Reproductive Systems, Hemisphere Publishing Corp., New York.

Wyrobek AJ; Watchmaker G; Gordon L An evaluation of sperm tests as indicators of germ-cell damage in men exposed to chemical or physical agents. Teratogenesis Carcinog Mutagen 1984; 4:83-107.

Wyrobek AJ; Gordon LA; Burkhart JG; Francis MW; and others An evaluation of the mouse sperm morphology test and other sperm tests in nonhuman mammals. A report of the U.S. Environmental Protection Agency Gene-Tox Program. Mutat Res 1983; 115:1-72.

Wyrobek AJ; Watchmaker G; Gordon L An evaluation of sperm tests as indicators of germ-cell damage in men exposed to chemical or physical agents. Prog Clin Biol Res 1984; 160:385-405.

Wyrobek AJ; Gordon LA; Burkhart JG; Francis MW; and others An evaluation of human sperm as indicators of chemically induced alterations of spermatogenic function. A report of the U.S. Environmental Protection Agency Gene-Tox Program. Mutat Res 1983; 115:73-148.

Wyrobek AJ Methods for evaluating the effects of environmental chemicals on human sperm production. Environ Health Perspect 1983; 48:53-9.

Wyrobek, A.J., Gordon, F.L.A., Burkhart, J.G., Francis, M.W., Kapp Jr., R.W., Letz, G., Malling, H.V., Topham, J.S., Whorton, M.D. (1983) An evaluation of human sperm as indicators of chemically induced alterations of spermatogenic function. Mutation Research 115, 73-148.

Wyrobek AJ Identifying agents that damage human spermatogenesis: abnormalities in sperm concentration and morphology. IARC Sci Publ 1984; 387-402.

Wyrobek, A.J., Brodsky, J., Gordon, L., Moore D.H. II., Watchmaker, G., Cohen, E.N. (1981a) Sperm studies in anesthesiologists. Anesthesiology 55, 527-532.

Wyrobek, A.J., Watchmaker, G., Gordon, L., Wong, K., Moore, D.H. II., Whorton, D. (1981b) Sperm shape abnormalities in carbaryl-exposed employees. Environ Health Perspect 40, 255-265.

Zeckhauser RJ; Viscusi WK Risk within reason. Science 1990 4; 248:559-64.

Zirkin BR; Santulli R; Awoniyi CA; Ewing LL Maintenance of advanced spermatogenic cells in the adult rat testis: quantitative relationship to testosterone concentration within the testis. Endocrinology 1989; 124:3043-3049.

PATERNALLY-MEDIATED DEVELOPMENTAL TOXICITY: IMPLICATIONS FOR RISK ASSESSMENT AND SCIENCE POLICY

Harold Zenick[1], Sally Perreault[1], and Jeanne Richards[2]

[1]Health Effects Research Laboratory
Office of Health Research
Office of Research and Development
U.S. Environmental Protection Agency
Research Triangle Park, NC 27711
[2]Office of Pesticide Programs
Office of Prevention, Pesticides and Toxic Substances
U.S. Environmental Protection Agency
Washington, DC 20460

INTRODUCTION

The objectives of this paper are: 1) to provide a brief summary regarding the status of the development of a policy within the U.S. Environmental Protection Agency (EPA) establishing a gender-neutral approach to regulating environmental pollutants; and 2) to discuss the assessment of male-mediated developmental outcomes within the context of the Agency's Guidelines for Developmental Toxicity Risk Assessment (U.S. EPA, 1991). In particular, the definitions and assumptions presented in those guidelines are evaluated as to relevance and applicability to assessing male-mediated developmental outcomes.

STATUS OF GENDER-BASED POLICIES

Background

Historically, gender-restrictive policies have been most prominent in industries where the potential exists for exposure of the pregnant female to workplace chemicals. These policies have sought three objectives: 1) to insure that the fetus is protected from harm resulting from workplace exposures of the mother; 2) to reduce potential liability that might arise from an adverse pregnancy outcome; and 3) to avoid the substantial costs and technical difficulties of achieving the necessary low levels of exposure. Although the courts have previously examined the issue of fetal protection, the Supreme Court ruling in the Johnson Controls case (see Grumet, 1991) applied a stricter standard than employed previously. The Supreme Court found that the Johnson Controls policy constituted sex discrimination in violation of Title VII of the Civil Rights Act of 1964 as amended in 1978 by the Pregnancy Discrimination Act. In reaching this decision, the Court applied the standard of "bona fide occupational qualification" which would allow an employer to exclude a fertile woman from a workplace only if her reproductive potential (i.e., the state of being fertile) prevented her from performing the duties of her job.

The Johnson Controls ruling and previous decisions have involved programs instituted by private employers to reduce reproductive risks. Since EPA is not an "employer" in this context, Title VII does not apply directly to Agency regulations. However, it is possible that this "employer" based ruling could be found to be relevant to EPA regulations if addressed in court.

Evaluation of Gender-Based Policy by the Office of Pesticide Programs

EPA has currently no formal, Agency-wide policy regarding the use of gender-based measures to reduce reproductive risks. However, the Office of Pesticide Programs (OPP) in the Office of Prevention, Pesticides and Toxic Substances has examined extensively this issue over the last decade during the review of various pesticides. In the course of this review, OPP has clarified the legal and scientific arguments regarding the appropriateness of gender-restrictive risk-management approaches to reduce reproductive and/or developmental risks. OPP concluded, in sum, that gender-based restrictions are both legally and scientifically questionable.

From a legal perspective, OPP believed it was important to avoid regulatory actions that would deny equal opportunity for employment or foster sex discrimination in the workplace. OPP also believed it was important to follow previous court rulings, which had generally limited gender-based restrictions instituted by private employers. In addition, OPP found it would be difficult or impossible to implement a gender-based policy that would protect the population of concern without unfairly restricting other employees. For example, a policy that targets women of child-bearing age might be overly broad (i.e., affecting women who are not fertile or do not plan to have children); yet, a more limited policy, such as one that targets pregnant women, would provide inadequate protection during the first few weeks of pregnancy, a time when the fetus is highly susceptible but when many women are unaware that they are pregnant.

Scientifically, OPP found that exposure and hazard data often fail to provide a basis to support a gender-restrictive practice. For example, rather than identifying a mechanism of toxicity that operates only in the female or only in the male reproductive system, the data show marked similarities in the mechanisms controlling reproductive processes across sexes. In fact, data in the literature and information presented at this conference support the hypothesis that male-only exposure can pose developmental risks. The lack of concern about developmental effects transmitted through the male is due at least in part to the fact that testing in this area has focused historically on the pregnant female (e.g., Segment II studies with exposure only during organogenesis).

OPP is now pursuing testing options that provide more data on male reproductive toxicity and will provide better guidance regarding gender-based policies. In the absence of such data, OPP believes the appropriate course is to adopt standards and regulations adequate to protect all exposed individuals, including the most sensitive. Such a policy would encourage equal access to jobs and discourage sex discrimination in the workplace on the basis of potential for chemical exposure. In

1992, the Office of Prevention, Pesticides and Toxic Substances proposed to the Office of the Administrator of EPA that the Agency adopt such an approach as a formal, agency-wide policy.

APPLICATION OF DEVELOPMENTAL TOXICITY RISK ASSESSMENT GUIDELINES

The remainder of this text will examine the appropriateness and applicability of the Agency's Guidelines for Developmental Toxicity Risk Assessment (U.S. EPA, 1991). This document, which provides the basis for assessing developmental risk, is widely used by numerous Federal and State agencies. However, its application to evaluating male-mediated developmental outcomes has not been examined thoroughly.

The Couple as the Unit for Reproductive Risk Analysis

Although the focus of this conference is on paternal-mediated effects, the concept of exposure being limited to a single parent is useful primarily when discussing the occupational or clinical setting. As one shifts to the general environment, the likelihood increases that both partners experience common exposures although dose, duration, and periodicity may differ. Examples where this is the case include ambient air and drinking water contaminants, indoor pollutants, residential and recreational pesticides contact, and potential exposures from neighborhood waste sites. In these instances, to discriminate and assign the male or female contribution to a developmental outcome may be difficult and inappropriate. Since a successful, healthy child is the result of the reproductive competence of both parents, the reproductive risk assessment for environmental pollutants must consider the couple as the unit of analysis. The implication is that studies, especially epidemiologic investigations, that focus solely on single parent contribution (paternal or maternal) to developmental outcomes resulting from environmental exposures may either miss or substantially underestimate the true reproductive risks.

Definition of Developmental Toxicity

The definition of developmental toxicity as provided in the Guidelines is as follows: "Adverse effects on the developing organism that may result from exposure prior to conception (either parent), during prenatal development, or postnatally to the time of sexual maturation." The definition certainly incorporates male-mediated developmental influences. Beyond this statement, the Guidelines also acknowledge the need to ascertain the reproductive competence of each parent as well as the potential for heritable (mutagenic) risk. As such, Agency guidelines for assessing heritable and reproductive risks (U.S. EPA, 1986; U.S. EPA, 1988a,b) should be used in tandem with the developmental guidelines to evaluate, to the fullest extent possible, the developmental risks that may be posed by a given agent.

Assumptions Made in Applying Guidelines for Assessing Developmental Toxicity

For most pollutants, the data set will be incomplete and, as such, create uncertainties associated with assessing risk (e.g., species extrapolation). In the absence of data to the contrary, the Guidelines apply the assumptions presented in Table 1. The applicability or modification of these assumptions as applied to male-mediated developmental toxicity is discussed in the next sections.

Table 1. Assumptions Made in Guidelines for Developmental Toxicity Risk Assessment*

Agents that produce an adverse developmental effect in experimental animal studies will potentially pose a hazard to humans following sufficient exposure during [gametogenesis and/or] development. (Bracket added by authors.)
All four manifestations of developmental toxicity are of concern:
death
structural abnormalities
growth alterations
functional deficits
Types of effects seen in animal studies are not necessarily the same as those that may be produced in humans.
The most appropriate species is used to estimate human risk, when data are available.
In general, a threshold is assumed for the dose-response curve.

*U.S. EPA, 1991

Species Extrapolation. "An agent that produces an adverse developmental effect in experimental animal studies will potentially pose a hazard to humans following **sufficient** exposure during [gametogenesis and/or] development" (bold-face and brackets added by authors).

The key term in this assumption is "sufficient exposure". For traditional developmental toxicity assessment (i.e., exposure during organogenesis), "sufficient" is undefined but potentially could be achieved with only a brief exposure, and timing of the exposure is critical. Likewise, sufficient exposure for male-mediated assessment is a function not only of the dose/duration of exposure but also of the stage(s) of spermatogenesis that are affected, and the timing of conception relative to the appearance of the "affected" sperm in the ejaculate. Two exposure scenarios will be considered, namely acute and continuous, in elaborating on these additional considerations.

Acute Exposure Scenario. The reader can review Dr. Wyrobek's chapter in this book for background on the proliferative and differentiative stages of spermatogenesis through which the male germ cell progresses and the number of days required for these transitions. Suffice it to say that the process in man requires approximately 90 days. For abbreviated exposures (less than 90 days) that impact a specific stage of spermatogenesis, a given amount of time will elapse before those cells appear in the ejaculate. Intercourse leading to conception would have to coincide with the brief presence of the "affected" cells in the epididymides (and thus the ejaculate) to produce an "affected" conceptus. The one exception may be the introduction of a **viable** lesion in the stem cells. In this case, the damage could be incorporated permanently into these mitotically-renewing cells and persist independent of current or subsequent exposures. Although this timing of exposure and assessment can be controlled precisely in laboratory experiments, adverse, male-mediated outcomes following acute exposures would be extremely difficult, if not impossible, to detect in human field studies. Epidemiologists must factor these considerations into the design of field investigations.

An additional factor related to "sufficient exposure" may be viewed as the "numbers game". Simply stated, the number of affected sperm is a function of concentration and duration of exposure (CxT). However, the probability of an affected sperm participating in fertilization is a function of:

$$\frac{\text{no. of affected sperm in epididymis}}{\text{total no. of sperm in epididymis}}$$

Thus CxT must produce a sufficient number of affected sperm to counter the dilution by not only the unaffected sperm in their cohort but also the large number of normal sperm already residing in the epididymis. The greater the dilution, the less the likelihood of producing and/or detecting an adverse, male-mediated event.

Continuous Exposure Scenario. With continuous exposure, CxT may produce a steady state ratio of affected to normal sperm in epididymis after 1-2 cycles of spermatogenesis. In this case, as the ratio increasingly favors the number of affected sperm, the probability of an affected sperm participating in conception increases. Also, with continuous exposure, the timing of conception relative to the appearance of affected sperm in the ejaculate becomes less critical.

It is important to note that, in reality, environmental exposures are neither acute or continuous. Rather, humans may experience some chronic background level of exposure overlaid by periodic spikes in exposure. It is this cumulative experience that may actually define "sufficient exposure" to produce male-mediated developmental toxicity.

Manifestations of Developmental Toxicity. All of the developmental manifestations routinely associated with female exposure have also been observed following male only exposure. A critical issue in risk assessment is the likelihood of occurrence and/or detection of these events in the human population. Most test species produce multiple offspring per litter and as such can concurrently present evidence of a number of developmental outcomes within a given litter. Thus, for the same female [rodent], the investigator can assess pre- and post-implantation loss, live births, birth weight, survival, malformations and so forth associated with a single pregnancy. However, for the human female who primarily has a single offspring per pregnancy, several of these events may represent a temporal sequence wherein the occurrence of one event (e.g., early loss) precludes the observation of a subsequent outcome (e.g., malformation). A more thorough discussion of the relationship of various pregnancy outcomes in humans has been presented by Selevan and LeMasters (1987).

A review of the cyclophosphamide literature indicated that all four of the developmental manifestations (listed in Table 1) have been reported with male-only treatment. The issue of detection of different endpoints has been graphically captured in Figure 1, which represents data from several laboratories on the dose ranges for various developmental outcomes following chronic (>70 days) cyclophosphamide treatment of male rats. These data would suggest that almost total litter loss (pre- and post-implantation loss) occurs at doses lower than those associated with malformations. As a point for additional comparison, the line at the far right of Figure 1 depicts the dose range capable of producing malformations following exposure of the female during organogenesis (Mirkes, 1985). Although the relative dose-response relationship for developmental outcomes may vary for different pollutants, the prospect of competing responses must receive greater consideration in the design of epidemiologic studies of potential male-mediated developmental toxicants.

Another "detection" issue relates to the biological target for male-mediated mutagens. A reasonable assumption is that DNA is the target and that the probability of hitting a given DNA moiety is random. Thus, even under identical exposure scenarios, a myriad of male-mediated developmental outcomes could be produced across individuals. If this hypothesis is true, then human studies (e.g., case-control) that focus on associating a single developmental outcome with male exposures may be doomed to fail. Such studies will most likely not demonstrate an exposure-effect relationship and could underestimate the total developmental risk for that pollutant.

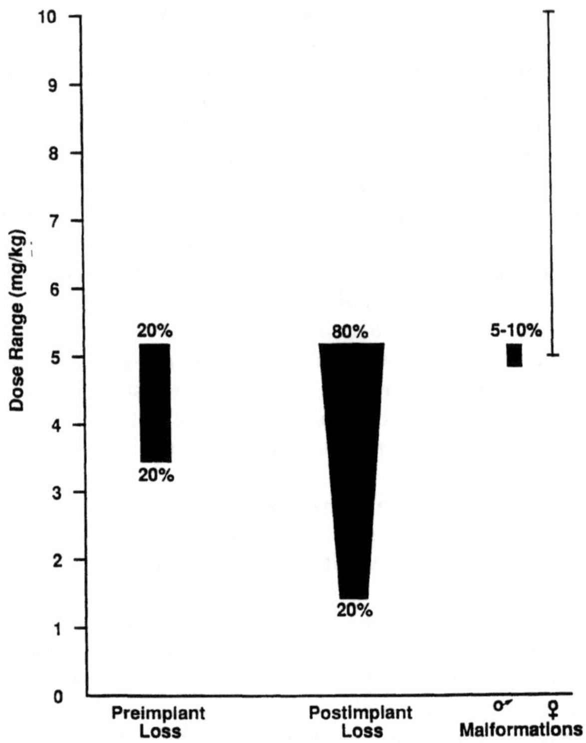

Figure 1. Dose Range for Various Developmental Outcomes Following Cyclophosphamide Treatment. The dose range for various developmental outcomes following subchronic (>70 days) oral cyclophosphamide treatment to male rats. Data summarized across several laboratories for a number of published studies. The single line at the far right is the dose range for malformations following exposure of the female rodent during organogenesis (from Mirkes, 1985).

By focusing on a single developmental event, the sample size required to insure sufficient power to detect an association increases dramatically. Epidemiologists often attempt to obtain adequate sample sizes by collapsing across industries for a given exposure (e.g., workers exposed to hydrocarbon solvents). Since exposure to a given pollutant may vary greatly across populations and occurs in the context of other exposures, this approach is not an acceptable solution. Rather, field studies of male-mediated developmental toxicants may need to "cast a broader net" initially by evaluating a wider spectrum of developmental outcomes.

Different Effects May Be Seen in Animal Versus Human Studies. The developmental guidelines recognize that numerous factors may influence the actual developmental outcomes that result from exposure of the female during organogenesis. These factors include the dose and duration of exposure, critical period of development when exposure occurs (i.e., changing windows of target vulnerability) and species differences in development, metabolism, mechanisms of action and repair. Similar factors would influence the nature of the outcome associated with male exposure. Issues related to critical periods (stages of spermatogenesis), the nature of exposure and the timing of assessment have been discussed in Section III.C.1.

In General, a Threshold is Assumed for the Dose Response. Whereas cancer risk assessment has applied traditionally a "no threshold" assumption (i.e., a single molecule of a toxicant is sufficient to trigger a carcinogenetic sequence), the developmental toxicity guidelines assume that some minimal dose must be exceeded to produce developmental toxicity. Given current understanding about the repair and protective (metabolic) mechanisms that exist in the testes, the threshold assumption would seem applicable to male-mediated developmental toxicity. Moreover, the existence of repair mechanisms in the oocyte would also argue for the existence of a threshold.

In addition to requiring sufficient hits per sperm to override these repair/protective capabilities, there exists a practical threshold, based on the necessity of a sufficient number of "hit" cells to overwhelm the dilution by "normal" cells in the epididymides, as discussed in Section III.C.1. Even if the "affected" cells have a numerical advantage, they must also retain sufficient cellular integrity to survive in the female tract and compete successfully during fertilization.

CONCLUSIONS

With inclusion of the male-specific dimensions described in this paper, the definitions and assumptions presented in the Guidelines for Assessing Developmental Toxicity seem appropriate for evaluating male-mediated developmental toxicity. However, a review of the literature in this area suggests an issue of greater concern from a risk assessment perspective; namely, the lack of significant interaction between the genetic toxicology, reproductive/developmental toxicology, and epidemiology communities in investigating male-mediated developmental outcomes. Much of the published literature reflects limited consideration of the spermatogenic process including genetic and other biologic and toxicologic processes. Yet, such understanding is critical to the design and interpretation of both animal and human studies (Perreault *et al.*, 1991). Increased communication between these communities will be vital to improve understanding and assessment of male-mediated developmental toxicity. Such interactions should directly guide laboratory efforts to provide insights, hypotheses and tools that can be tested more realistically in human populations. Similarly, human studies must integrate laboratory insights regarding reproductive/developmental processes and mechanisms into the design and focus of field studies. The synthesis of these disciplines holds great promise for assessing and potentially anticipating the developmental risks that may be associated with exposure of the male.

ACKNOWLEDGMENT

This document has been reviewed in accordance with U.S. Environmental Protection Agency policy and approved for publication. Mention of trade names or commercial products does not constitute endorsement or recommendation for use.

REFERENCES

Grumet, B.R., 1991, Fertile women may now apply: Fetal protection policies after Johnson controls, *Risk-Issues in Health and Safety* 261:261-276.

Mirkes, P.E., 1985, Cyclophosphamide teratogenesis: A review, *Teratogenesis, Carcinogenesis, and Mutagenesis* 5:75-88.

Perreault, S., 1991, In vitro assessment of gamete integrity, in: "In Vitro Toxicology: Mechanisms and New Technology," A.M. Goldberg, ed., Mary Ann Liebert, Inc., New York, NY, pp. 63-75.

Selevan, S.G., and Lemasters, G.K., 1987, The dose-response fallacy in human reproductive studies of toxic exposures, *J. Occupational Med.* 29:451-454.

U.S. Environmental Protection Agency, 1986, "Guidelines for Mutagenicity Risk Assessment," Federal Register 51(185):34006-34012.

U.S. Environmental Protection Agency, 1988(a), "Proposed Guidelines for Assessing Male Reproductive Risk," Federal Register 53:24850-24869.

U.S. Environmental Protection Agency, 1988(b), "Proposed Guidelines for Assessing Female Reproductive Risk," Federal Register 53:24834-24847.

U.S. Environmental Protection Agency, 1991, "Guidelines for Developmental Toxicity Risk Assessment," Federal Register 56:63798-63826.

PHYSICIAN AND PATIENT EDUCATION

Jan M. Friedman

Department of Medical Genetics
University of British Columbia
226 - 6174 University Boulevard
Vancouver, Canada V6T 1Z3

CURRENT STATUS OF PHYSICIAN AND PATIENT UNDERSTANDING OF MALE-MEDIATED EFFECTS

The group agreed that current understanding of male-mediated reproductive effects among both health professionals and patients is generally poor. Knowledge of possible male-mediated effects among physicians is worse than their knowledge of female reproductive toxicology. The same is true for many other health professionals such as midwives and nurse practitioners, to whom patients turn for this information. Male-mediated reproductive toxicology is not part of most health science curricula and is not usually included in postgraduate medical training, either.

There is a gender bias in both the information that is available and the presentation of available information on reproductive toxicology. This bias reflects the widely-held but erroneous notion that reproduction is an issue that concerns women but not men.

A particular lack of knowledge about male-mediated reproductive toxicity is superimposed on a generally poor understanding of science among the general public. Many people have serious misperceptions about reproductive toxicology. They frequently lack understanding of such basic concepts as the importance of exposure level and the presence of certain irreducible background risks. Moreover, many people cannot accurately interpret statements of probability.

ROLE OF THE MEDIA IN COMMUNICATION REGARDING MALE-MEDIATED EFFECTS

Most knowledge about reproductive toxicity among the general public is obtained from the media. The media have done a good job in publicizing certain

Male-Mediated Developmental Toxicity, Edited by D.R. Mattison
and A.F. Olshan, Plenum Press, New York, 1994

important reproductive hazards, such as that associated with maternal isotretinoin treatment during pregnancy. On the other hand, much of the information the media provide on reproductive toxicology is presented in an alarmist and sensational manner. This problem is most serious on television because the impact of an image of a child with birth defects may be so powerful that it overwhelms any qualifying statements provided in a story.

Most journalists want to present an accurate story, but they may have a poor understanding of the science involved and be working under time constraints that prevent them from learning the background information necessary to understand the story fully. It is important that knowledgeable experts in male reproductive toxicology make themselves available to the press so that when a relevant story comes up, the journalist will turn to them for comment rather than to someone less well informed.

It is best to stick to facts relevant to the issue at hand and not to speculate or digress when discussing a story with a journalist. Most journalists welcome comments from experts and respond well to a telephone call from an expert who believes that the journalist has made a factual error.

When talking with a journalist, it is important to speak clearly, simply and, to the extent possible, in a manner that cannot be misinterpreted. Journalists generally do not let experts review or approve stories prior to publication.

APPROACHES AND GOALS FOR PATIENT, PHYSICIAN, AND PUBLIC EDUCATION

The participants made the following recommendations:

1) Scientists and clinicians who are knowledgeable about male reproductive toxicology should encourage relevant professional societies to which they belong to include information on male-mediated effects in their meetings. Appropriate target groups include obstetricians, family physicians, andrologists, fertility specialists, geneticists, genetic counsellors, nurse practitioners, and public health personnel.

2) Local and regional programs to train health educators and health professionals regarding male and female reproductive toxicology may be particularly effective. The March of Dimes is often receptive to proposals to establish such programs.

3) Efforts should be made to include general principles of both male and female reproductive toxicology in the curricula of medical and other health professional schools and in the postgraduate training of obstetricians, geneticists, endocrinologists, urologists, and family physicians. Physicians and other health professionals should know how to take an adequate history from patients regarding exposures that may produce reproductive toxicity, should be able to recognize high-risk situations, and should know how to refer patients with such exposures for appropriate evaluation and counselling.

4) High school biology and health teachers should be encouraged to include basic information about both male and female reproductive toxicology in their courses. Concepts such as the effects of dosage and background frequencies of adverse reproductive outcomes should be discussed.

5) Clinicians who counsel men and women regarding reproductive hazards should establish professional relationships with occupational medicine specialists and occupational health nurses to help evaluate environmental and occupational exposures that may be of concern with respect to reproduction.

TENTATIVE RECOMMENDATIONS FOR REPRODUCTIVE RISK COUNSELLING OF PATIENTS

This is an area of substantial controversy, and no clear consensus exists among knowledgeable clinicians. However, most participants supported the following recommendations:

a) In general, it is better to provide reproductive risk counselling to a couple together rather than to either the man or woman alone.

b) Counselling related to male reproductive risks should begin with a statement about the limited amount and quality of information that is available on this topic. The couple should understand that any counselling provided has substantial uncertainty.

c) The full spectrum of adverse reproductive outcomes due to environmental or occupational exposures in both the man and woman should be considered. This is especially important in preconceptional counselling or when infertility or recurrent miscarriage is an issue. Such counselling should be provided in the context of the general background risks of various adverse reproductive outcomes and with due consideration of possible adverse reproductive effects of other factors such as family history, parental age, nutrition, and use of tobacco, alcohol or illicit drugs by either partner.

d) Once a couple have conceived, prior paternal exposure to mutagens or other potential reproductive toxins should not usually alter the management of the pregnancy. There was general agreement that ultrasound examination can be offered to such couples for reassurance. There was a lack of agreement on the advisability of making amniocentesis or chorionic villus sampling for chromosomal abnormalities available in such cases. On the one hand, parental anxiety is an accepted indication for prenatal diagnosis in some centers; on the other hand, most abnormalities arising from paternal exposures are unlikely to be detected by amniocentesis or chorionic villus sampling.

e) Men or adolescent boys who have been diagnosed as having a malignancy and in whom radiotherapy or chemotherapy is being planned should be offered the opportunity of storing semen prior to beginning treatment. Similar recommendations are reasonable for males about to undergo cytotoxic therapy for other conditions such as collagen vascular or renal disease.

CHARACTERISTICS OF MALE-MEDIATED TERATOGENESIS

Tetsuji Nagao

Department of Reproductive and Developmental Toxicology,
Hatano Research Institute, Food and Drug Safety Center,
Ochiai 729-5, Hadano, Kanagawa 257
Japan

INTRODUCTION

Although induction of germinal mutations by chemicals is well documented in animals, there is no firm evidence that any agent has induced germinal mutations in men. Direct study of chemically-induced transmitted genetic effects in humans is virtually impossible, so the genetic risk must be estimated from animal experiments. Fortunately, we have many data on spontaneously occurring and chemically-induced congenital malformations in mice (Nagao, 1987, 1988; Nagao and Fujikawa, 1990). Many of these congenital malformations are useful in studying the development of genetic disorders that are similar to those found in humans. However, the genetic basis of such congenital malformations is uncertain or complex. Therefore, more information on *male-mediated teratogenesis* is needed to be established as endpoints of male- mediated effects and to estimate the genetic hazards to men. In this report, I describe the characteristic of spectrum of fetal abnormalities detected as endpoints of male-mediated developmental toxicity.

EXPERIMENTS

In all the experiments ICR mice from Shizuoka Agricultural Cooperative Association for Laboratory Animals (Shizuoka, Japan) were used. All animals received a minimum of 2 weeks acclimatization period before the start of any experimental procedure. They were kept under standard spf conditions in an animal room with controlled dark-light cycle (dark period 7:00 p.m. to 7:00 a.m.), temperature (24 ± 1°C) and relative humidity ($55\pm5\%$) and allowed to feed CA-1 and tap water ad lib. In the experiments on male-mediated F_1 congenital malformations, male mice were injected i.p. with ethylnitrosourea (ENU, Nakarai) and methylnitrosourea (MNU, Nakarai). The treated males were individually caged with two untreated virgin females of the same strain. The mating intervals in the ENU and the MNU experiments were days

Male-Mediated Developmental Toxicity, Edited by D.R. Mattison
and A.F. Olshan, Plenum Press, New York, 1994

1-21 and 64-80 after the last dosing. Copulations during these periods involve, respectively, treated postmeiotic cells and spermatogonial stem cells. In the experiments on preimplantation embryos, the ENU or MNU solution was i.p. injected into female mice at 12:00 p.m. on day 2 of pregnancy. The post-mating interval of 2 days corresponds to 4-cell to morulae stages. In the experiments on organogenic embryos, the ENU or MNU solution was i.p. injected into female mice on day 8 of pregnancy. In all the experiments, presence of vaginal plug defined day 0 of pregnancy and congenital abnormalities were inspected on day 18 of pregnancy. Congenital abnormalities inspected were gross external and skeletal abnormalities. In the case where male-mediated teratogenicity was concerned, dwarfism was included in the category of external abnormalities; but data on dwarf were excluded when the spectrum of external abnormalities in the germ-cell treatment experiments and those in the embryo-treatment experiments were compared. In any of skeletal examination, cervical and lumbar ribs were not scored as skeletal abnormalities because these abnormalities occur spontaneously with an appreciable frequency in fetuses of the ICR strain.

RESULTS AND DISCUSSION

Comparative induction of abnormal fetuses

Figure 1 shows dose vs. frequency relations of the malformed fetuses observed after treatment of spermatogonial stem cells, preimplantation embryos on day 2 of gestation and organogenic embryos on day 8 of gestation with ENU or MNU. Irrespective of the kind of mutagens applied and the developmental stage treated, frequency of external abnormalities showed a dose-dependent increase. The embryos at the organogenic stage are the most susceptible to the teratogenicity of ENU and MNU, and paternal germ cells are the most insusceptible among the developmental stages treated.

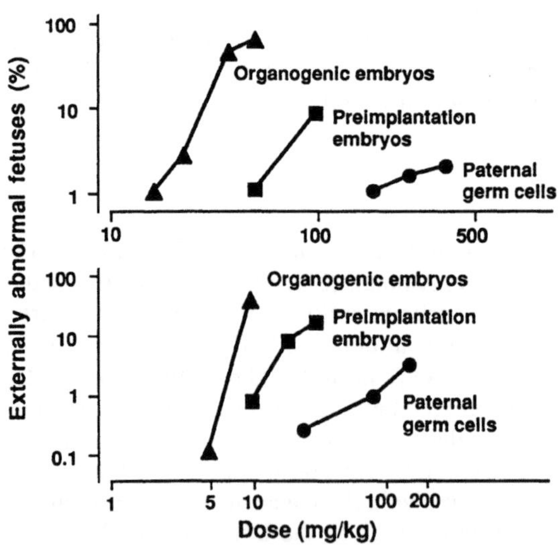

Figure 1. Dose vs. frequency relations of the externally malformed fetuses after treatment of spermatogonial stem cells, preimplantation embryos and organogenic embryos with ENU (upper panel) or MNU (lower panel).

From these dose-response data, I estimate $AD_{0.04}$, the dose level necessary to produce externally abnormal fetuses in F_1 with a frequency of 4% (Table 1). The dose value for either mutagen is highest when paternal germ cells are treated and lowest when embryos at the organogenic stage are treated. Since the reciprocal of the $AD_{0.04}$ value provides a measure of susceptibility in each developmental stage, it is clear from the data given in Table 2 that, treatment of paternal spermatogonial stem cells with ENU or MNU shows one twentieth that of organogenic embryos with respect to the induction of malformations. This means that teratogenic risk associated with acute exposure of male parents to a mutagen is not so high as the risk associated with the exposure of pregnant females. However, if a spermatogonial stem cell carries a stable genetic lesion, then gametes bearing the change will be produced continually throughout the stem cell's life span. In mice, the duration of period at risk for the induction of genetic damage which may result in production of congenital malformations in F_1 is 600 days, whereas the duration of period at risk in preimplantation embryos and organogenic embryos is at longest 4 and 8 days, respectively (Table 2). Taken together, paternal germ cells can be a main target of environmental chemicals with respect to the teratogenic threat.

Table 1. The Dose Level Necessary to Produce Externally Abnormal Fetuses with a Frequency of 4%

Target	ENU (mg/kg)	MNU (mg/kg)
Paternal germ cells	450	132
Preimplantation embryos	73	15
Organogenic embryos	21	7

Table 2. Characteristics of Paternal Germ Cells as Targets to Teratogenic Threat of Environmental Chemicals

	Relative $1/AD_{0.04}$	Duration at Risk (Days)
Paternal germ cells	1	600
Preimplantation embryos	6-9	4
Organogenic embryos	20	8

Spectrum of congenital malformations

Table 3 summarizes the type-distribution of external abnormalities observed after treatment with ENU or MNU at the paternal germ cell, the preimplantation and the organogenic stages. In order to compare with type-distribution of spontaneously induced abnormalities, the data pooled for the concurrent and the historical controls are also shown in this table. The data on paternal germ cells are pooled for all the dose levels tested and all the germ-cell stages tested.

Among external abnormalities induced after treatment of paternal germ cells, cleft palate was the most common, followed by dwarf, exencephaly and abnormal limbs (c.f., polydactyly and syndactyly) in both the ENU- and MNU- treated groups. Spectra of external abnormalities recorded as F_1 defects in the ENU- and MNU- treated groups are indistinguishable from each other, showing no evidence of a mutagen-specificity. More importantly, spectra determined after treatment of paternal germ cells with ENU or MNU do not differ from the spectrum of spontaneously arisen malformations. These results clearly characterize male-mediated teratogenesis, for the spectra of fetal external abnormalities induced after treatment of embryos at the organogenic stage with ENU or MNU showed marked differences from the control spectrum. Thus, I am inclined to hypothesize that a large fraction of external malformations in fetuses from mutagenized paternal germ cells are results of increased yields of spontaneously occurring malformations. In other words, spontaneous fetal malformations may also arise, at least partly, as a genetic disease. Interestingly, spectra of external abnormalities observed in F_1 fetuses derived from mutagenized germ cells and from ENU- and MNU-treated preimplantation embryos are indistinguishable, suggesting that teratogenic damage induced in preimplantation embryos is mutation in nature (Vogel *et al.*, 1989; Nagao *et al.*, 1991).

Table 4 summarizes the type-distribution of external abnormalities detected in fetuses derived from postmeiotic germ cells and spermatogonial stem cells treated with ENU and MNU. Data are pooled for all dose levels tested. In both mutagens, main types of fetal abnormalities were cleft palate, dwarf, exencephaly and abnormal limbs in both treatments of postmeiotic germ cells and spermatogonial stem cells. The spectra in different stages of paternal germ cells are indistinguishable, irrespective of the kind of mutagens used, showing no evidence of a germ-cell stage dependent variation. In summing up all the data on external abnormalities, it seems that the spectrum of external abnormalities in fetuses derived from mutagen-treated paternal germ cells is a mimic of the spectrum of spontaneously arisen malformations and, as far as external abnormalities are concerned, evidence for male-mediated teratogenesis can be obtained only from quantitative comparison of the incidence of abnormal fetuses between the treated and the control groups. Since spontaneous spectrum of fetal abnormalities often shows a strain-dependent variation, the spectrum of induced abnormalities may also depend on the strain of mice used. In the present study with ICR strain, the predominant types were cleft palate and dwarf. The ratio of the average frequency was 2:1 in the ENU experiment (Table 3). In the experiments reported by Kirk and Lyon (1984), who used (C3H/HeHx101/H)F_1 hybrids, the ratio was 1:25.

The situation was quite different when data on skeletal abnormalities are analyzed. Table 5 summarizes the spectrum of skeletal abnormalities in fetuses derived from paternal germ cells, preimplantation embryos, and organogenic embryos treated with ENU and MNU. Data are pooled for all dose levels tested. Among skeletal abnormalities induced after treatment of paternal germ cells, abnormal axial skeleton was the most common, followed by flexion of appendicular skeleton in the ENU-treated group. Fused ribs was the most common, followed by abnormal axial skeleton in the MNU-treated group. The spectra in F_1 fetuses varied drastically depending on the kind of mutagens used, as it was the case for teratogenesis induced at the organogenic stages. The spectrum recorded with MNU is similar to the control spectrum; but this was not the case for the spectrum recorded with ENU. Therefore, I speculate occurrence of skeletal abnormalities of a type that is specific to male-mediated teratogenesis. However, further experiments and analyses are needed to characterize male-mediated teratogenesis in relation to the type-distribution of skeletal abnormalities. In conclusion, in the experimental system used for studying male-mediated teratogenesis, abnormalities that are specific to male-mediated teratogenesis could not be found, and external abnormalities as compared with skeletal abnormalities showed more stable type-distribution. For the purpose of screening of environmental chemicals with an ability to cause male-mediated teratogenesis, the use of external malformations in F_1 fetuses as endpoint and quantitative analysis of the incidence should be recommended.

Table 3. Types and Number of External Abnormalities Detected on Day 18 of Gestation After Treatment of Paternal Germ Cells, Preimplantation Embryos and Organogenic Embryos with ENU and NNU

Treatment	Abnormalities[1]																Total No. of Fetuses Examined
	CP	DW	EC	MP	BH	AL	AH	KT	SB	AA	HG	OC	GE	HC	HT	GS	
Control[2]	43	7	8	2	2	1	1	1	-[3]	-	-	-	-	-	-	-	9054
Paternal germ cells																	
ENU (125-375 mg/kg)	80	44	13	-	-	8	2	-	2	2	1	-	-	-	-	-	7284
MNU (5-125 mg/kg)	57	29	7	-	-	9	1	1	-	-	-	1	-	-	-	1	5230
Preimplantation embryos																	
ENU (50-100 mg/kg)	11	ND[4]	4	-	-	-	-	-	-	-	3	2	3	-	-	-	606
MNU (10-30 mg/kg)	61	ND	24	1	-	9	-	2	1	1	-	6	2	-	-	-	2681
Organogenic embryos on day 8 of gestation																	
ENU (25-75 mg/kg)	2	ND	3	73	-	2	-	1	-	-	-	-	-	2	-	-	266
MNU (5-10 mg/kg)	32	ND	-	39	-	2	-	1	-	-	-	-	-	15	1	-	364
on day 12 of gestation																	
ENU (50-100 mg/kg)	8	ND	-	-	-	-	-	2	-	-	-	-	-	-	-	-	180
MNU (5-20 mg/kg)	9	ND	1	-	-	2	-	-	-	-	-	-	-	-	-	-	471

1) Abbreviations are: CP, cleft palate; DW, dwarf; EC, exencephaly; MP, microphthalmus; BH, brain hernia; AL, abnormal limb; AH, abdominal hernia; KT, kinked tail; SB, spina bifa; AA, anal atresia; HG, hypognathia; OC omphalocele; GE, general edema; HC, hydrocephaly; HT, hematoma; GS, gastroschisis. 2) Pooled data for all the experiments concerned. 3) Not detected.
4) Not determined.

Table 4. Types and Number of External Abnormalities Detected on Day 18 of Gestation after Treatment of Paternal Postmeiotic Cells and Spermatogonial Stem Cells with ENU and MNU

	Abnormalities[1]												
Treatment	CP	DW	EC	AL	SB	AH	KT	AA	HG	OC	GE	GS	Total No. of Fetuses Examined
ENU (125-375 mg/kg)													
Postmeiotic cells	35	22	4	4	2	1	1	-	-	-	-	-	3737
Spermatogonial stem cells	45	22	9	4	-	4	-	2	1	-	-	-	3547
MNU (5-125 mg/kg)													
Postmeiotic cells	35	21	4	6	-	-	-	-	-	-	-	-	3226
Spermatogonial stem cells	22	8	3	3	-	1	1	-	-	1	1	1	2004

1) Abbreviations are: CP, cleft palate; DW, dwarf; EC, exencephaly; AL, abnormal limb; AH, abdominal hernia; KT, kinked tail; AA, anal atresia; HG, hypognathia; OC, omphalocele; GE, general edema; GS, gastroschisis.

Table 5. Types and Number of Skeletal Abnormalities Detected on Day 18 of Gestation After Treatment of Paternal Germ Cells, Preimplantation Embryos and Organogenic Embryos with ENU and MNU

	Abnormalities[1]						
Treatment	AAS	FAS	FR	WR	AMM	CST	Total No. of Fetuses Examined
Paternal germ cells							
ENU (125-375 mg/kg)	16	14	11	5	2	2	6098
MNU (25-125 mg/kg)	4	-	12	1	-	-	3614
Preimplantation embryos							
ENU (50-100 mg/kg)	14	10	4	4	1	-	606
MNU (10-30 mg/kg)	33	11	11	8	4	-	2681
Organogenic embryos on day 8 of gestation							
ENU (25-75 mg/kg)	134	-	19	-	51	-	266
MNU (5-10 mg/kg)	88	-	15	-	50	-	364
on day 12 of gestation							
ENU (50-100 mg/kg)	-	-	-	-	-	-	180
MNU (5-20 mg/kg)	-	53	-	35	4	-	471
Control[2]	1	-	6	1	-	-	10883

1) Abbreviations are: AAS, abnormal axial skeleton; FAS, flexion of appendicular skeleton; FR, fused ribs; WR, wavy ribs; AMM, abnormal maxilla and mandibula; CST, cleft sternum.
2) Pooled data for all the experiments concerned.

REFERENCES

Kirk, K.M., and M.F.Lyon (1984) Induction of congenital malformations in the offspring of male mice treated with X-rays at pre-meiotic and post-meiotic stages, *Mutation Res.*, 125, 75-85.

Nagao, T. (1987) Frequency of congenital defects and dominant lethals in the offspring of male mice treated with methylnitrosourea, *Mutation Res.*, 177, 171-178.

Nagao, T. (1988) Congenital defects in the offspring of male mice treated with ethylnitrosourea, *Mutation Res.*, 202, 25-33.

Nagao, T., and K. Fujikawa (1990) Genotoxic potency on mouse spermatogonial stem cells of triethylenemelamine, mitomycin C, ethylnitrosourea, procarbazine, and propyl methanesulfonate as measured by F_1 congenital defects, *Mutation Res.*, 229, 123-128.

Nagao, T., Y. Morita, Y. Ishizuka, A. Wada and M. Mizutani (1991) Induction of fetal malformations after treatment of mouse embryos with methylnitrosourea at the preimplantation stages, *Teratogen. Carcinogen. Mutagen.*, 11, 1-10.

Vogel, R., I. Granatal and H. Spielmann (1989) Cytogenetic studies on preimplantation mouse embryos exposed to methylnitrosourea *in vivo*, *Reprod. Toxicol.*, 3, 23-26.

ANEUPLOIDY STUDIES IN SPERM: POST-MEIOTIC SELECTION AGAINST ANEUPLOID SPERM

Judith H. Ford, Tie Lan Han, Greg Peters,
Anthony Correll, Maureen Tremaine, and Graham Webb

Genetics Department
The Queen Elizabeth Hospital
Woodville
South Australia, 5011

INTRODUCTION

Observations on the parental transmission of chromosomes in trisomic conceptions have shown that the disomic contribution is usually maternal in origin (Hassold & Takaesu, 1989). It is well known that trisomic conceptions increase with maternal age and it is usually assumed that this reflects an increase in meiotic error with maternal age. No equivalent age-related change in aneuploidy with paternal age is known. If there is no such change, then the proportion of male contributions to trisomy should decrease with increasing maternal age (Sved & Sandler, 1981) unless the rate of trisomic conceptions is the same at all ages and the maternal age effect is post-meiotic, possibly one of decreased early recognition or correction of trisomic conceptions.

Early analyses of parental contributions to trisomies used cytogenetic markers to identify individual chromosomes. However, studies of DNA polymorphisms are now showing that the paternal contribution to autosomal trisomies is generally less than was previously thought (Sherman *et al.*, 1991). There is not yet sufficient data to discover whether there is a paternal age-related change in contribution.

The paternal contribution to sex chromosome aneuploidy is much greater than to autosomal trisomy. For 47,XXY and 47,XXX, it has been shown to be 50% (Jacobs *et al.*, 1988, May *et al.*, 1990) and to 47,XYY is obviously 100%. This significant contribution suggests that there may not be selection against sperm which carry extra sex chromosomes and that sex chromosomes may be the preferred chromosomes to study possible paternal age effects.

To examine the question whether there is selection against cells with sex chromosome aneuploidy, we have considered data from our own laboratory and from the literature. These data estimate the paternal rate of error for sex chromosomes at various stages; the frequency of X-Y univalents at meiosis I, the frequency of sex

Male-Mediated Developmental Toxicity, Edited by D.R. Mattison
and A.F. Olshan, Plenum Press, New York, 1994

chromosome hyperploidy in sperm and the incidence of sex chromosome aneuploidy in spontaneous abortions and livebirths.

FREQUENCY OF MEIOTIC IRREGULARITY IN FERTILE AND INFERTILE MEN

Skakkabaek *et al.* (1973) examined the frequency of meiotic irregularity in testicular biopsies of 18 control men and 74 with fertility problems (Table 1). They found that major pairing failure was restricted to infertile men but that each group had 15% unpaired sex chromosomes.

Table 1. Frequency of Meiotic Irregularity in Testicular Biopsy*

Subjects	Major Pairing Failure	Minor Pairing Failure	X-Y Unpaired	2N
Controls (18)	0	6%	15%	1%
Infertile (74)	4%	4%	15%	1%

*Data from Skakkabaek *et al.* (1973). Major pairing failure is defined as failure to form a chiasma in an F group, or larger, pair of chromosomes. Minor pairing failure is defined as failure of pairing in one or both G group pairs of chromosomes.

In our laboratory, we examined meiotic cells in ejaculated semen. Preparations were made according to the method of Sperling & Kaden, (1971) and typical cells are shown in Figure 1. Because of the relatively low numbers of meiotic forms in semen ejaculates (Peters in preparation) this assessment was restricted to a small sample of men who were oligospermic. We found somewhat different results to Skakkabaek *et al.* (1973) with much higher frequencies of major pairing failure and polyploidy and less X-Y univalent formation. This difference is unlikely to be significant given the small sample of men and the very significant variance between males (not shown).

Table 2. Meiotic Irregularity in Ejaculated Meiotic Cells

Subjects	Major Pairing	Minor Pairing	X-Y Unpaired	2N
Oligospermic men (5)	19%	7.4%	6.4%	19.2%

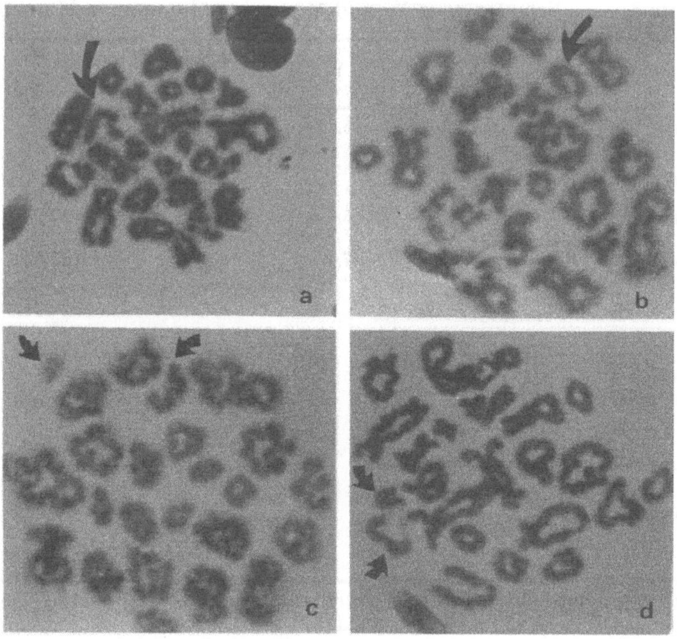

Figure 1. Cells in Diakinesis in ejaculated semen. Arrows in cells (a) and (b) show the X-Y bivalent, each with a single chiasmata in the pseudoautosomal region. In cell (c), the X and Y chromosomes are unpaired and in cell (d) there is incomplete pairing.

SIMULTANEOUS DETECTION OF X- AND Y- BEARING HUMAN SPERM BY DOUBLE FLUORESCENCE IN SITU HYBRIDIZATION

Double fluorescence in situ hybridization was used to detect sex chromosomes in decondensed human sperm nuclei (Han *et al.*, 1992). Biotinylated X chromosome-specific (TRX) and digoxigenin-labelled Y chromosome-specific probes were simultaneously hybridized to sperm preparations from twelve healthy donors. After the hybridization, the probes were detected immunocytochemically using two different and independent affinity systems. The study was first conducted on the washed ejaculated semen of 12 healthy donors. Since these data showed quite a high frequency of diploid cells, some of which might not have been of germ-line origin, the study was repeated on the washed ejaculate and those sperm which were able to "swim-up," from a further 5 healthy donors. The results for the two studies are shown in Table 3. The results of two other unpublished studies are also shown for comparison.

The results of the different studies are rather variable with up to four-fold differences in frequency. Some of these may be due to technical factors or scoring, but it is also likely that they reflect between individual differences in X and Y aneuploidy. Where numbers of aneuploid cells are so low, it is difficult to test for between-individual differences. However, the variation between subjects in our study was quite marked.

Table 3. Frequency of X- and Y-bearing Gametes in Human Sperm*

Specimen	haploid-X-	Y-	haploid-XX	YY	XY
whole semen (12 donors: 12,636 cells)	47.4%	46.8%	0.28%	0.21%	0.21%
whole semen	46.7%	46.8%	0.22%	0.20%	0.08%
motile sperm (5 donors: 5,000 cells each test)	48.8%	47.4%	0.16%	0.22%	0.12%
Wyrobek (pers comm 10,000 total cells)	?	?	0.06%	0.06%	0.09%
Hulten (pers comm 3 donors: 6,000 cells each)	?	?	0.11%	0.13%	0.21%
Mean frequency			0.17%	0.16%	0.14%

*The frequencies of the different cell types, where known, are shown in the Table. The mean frequency of the hyperploid types for all studies is shown in the last row.

FREQUENCY IN TRISOMY AFTER CONCEPTION

Jacobs (1990) analyzed the frequency of X and Y aneuploidy in spontaneous pregnancy loss and in livebirths. The findings of her study are shown in Table 4.

Table 4. Frequency of X and Y Aneuploidy After Conception

Sex-chromosome Aneuploidy	Spontaneous Abortion	Livebirths	Total
XXX	0.03%	0.05%	0.08%
XXX	0.02%	0.05%	0.07%
XYY	0.00%	0.05%	0.05%
Total	0.05%	0.15%	0.20%

Since the male contribution to XXX and XXY are found to be approximately 50%, the male contribution to these becomes 0.04% and 0.035% respectively and for XYY, it is 0.05%.

Table 5. Reduction in Male Transmission of Aneuploidy with Sperm Maturation and Fertilization

Stage	Percent X + Y Univalents or Disomy
TESTES: MEIOSIS	15%
SEMEN:MEIOSIS	*6.40%
SPERM	0.47%
MALE CONTRIBUTION TO TOTAL CONCEPTIONS	0.13%

*This group of cells is anomalous in that they will not contribute to the sperm population, as post-meiotic development is extremely unlikely. This figure should be taken as equivalent to the 15% figure found in the testes.

The data presented in Tables 4 and 5 demonstrate that there is a reduction in sex chromosome aneuploidy from sperm to conception. This reduction may occur during passage through the female genital tract or at fertilization. Data from hamster fertilization could have been added to attempt to define this further but it was felt that because the system is both artificial and uses only a relatively small number of sperm per individual, it would not add any useful information.

The frequency of cells showing failure of sex chromosome pairing at meiosis I is considerably higher than the frequency of XY aneuploidy detected in sperm (XX and YY will be generated by errors at meiosis II). It is likely that a reasonable proportion of unpaired sex chromosomes still segregate normally but we are unaware of data on this point. However, in the absence of selection, it would be necessary for 99.9% of unpaired univalents to segregate normally to achieve the levels of aneuploidy detected in sperm. This seems to be unlikely and we suggest that there is significant selection against paternally-derived sex chromosome aneuploidy between male meiotic division and conception.

SUMMARY

The data presented here compare the frequencies of failure of sex chromosome pairing with aneuploidy in sperm and aneuploidy after conception. There is considerable reduction in the observed aneuploidy which appears to reflect considerable selection during spermiogenesis or fertilization. The study is limited by the biases which inevitably occur in the assessment of very low frequency events. Nevertheless, the considerable numbers of cells scored would indicate that the averaged observed frequencies are close to the real frequencies. The study provides some insight into the differences between male and female contributions to aneuploidy in humans.

REFERENCES

Han, T.H., Ford, J.H., Webb, G.C., Flaherty, S.P., Correll, A., and Matthews, C.D.(1992) Simultaneous detection of X- and Y- bearing human sperm by double fluorescence in situ hybridization. *Mol Reprod Dev* 34:308-313.

Hassold T.J. and Takaesu, N., (1989) Analysis of non-disjunction in human trisomic spontaneous abortions. in:"Molecular and Cytogenetic studies of Non-disjunction,". T.J. Hassold, Epstein C.J. eds., Alan R Liss, New York, pp: 115-134.

Jacobs, P.A. (1990) The role of chromosome abnormalities in reproductive failure. *Reprod Nutr Dev* (1990) suppl 1, 63s-74s.

Jacobs, P.A., Hassold, T.J., Whittington, E., Butler, G, Collyer, S., Keston, M., and Lee, M. (1988) Klinefelter's syndrome: an analysis of the origin of the additional chromosome using molecular probes. *Ann Hum Genet*, 52: 93-109

May, K.K., Jacobs, P.A., Lee, M., Ratcliffe, S., Robinson, A., Neilsen, J., and Hassold, T.J. (1990) The parental origin of the extra X chromosome in 47, XXX females. *Am J Hum Genet* 46: 754-761.

Sherman, S.L., Takaesu, N., Freeman, S.B., Grantham, M., Phillips C., Blackston R.D., Jacobs P.A., Cockwell, A.E., Freeman, V., Uchida, I., Mikkelsen, M., Kurnit, D.M., Buraczynska, M., Keats, B.J.B., and Hassold T.J. (1991) Trisomy 21: Association between reduced recombination and nondisjunction. *Am J. Hum Genet* 49: 608-620.

Skakkebaek, N.D., Bryant, J.I and Philip, J. (1973) Studies on meiotic chromosomes in infertile men and controls with normal karyotypes. *J. Reprod Fertil* 35: 23-36

Sperling, K. and Kaden, R. (1971) Meiotic studies of the ejaculated seminal fluid of humans with normal sperm count and oligospermia. *Nature* 232: 481.

Sved, J. and Sandler, L. (1981) Relation of maternal age effect in Down syndrome to nondisjunction. in: "Trisomy 21 (Down Syndrome)", F de la Cruz, P.S. Gerald eds.,University Park Press, Baltimore, pp. 95-98.

ASSOCIATION OF PATERNAL AND MATERNAL EXPOSURE WITH LOW BIRTH WEIGHT AND PRETERM BIRTHS AMONG WOMEN TEXTILE WORKERS*

Xiping Xu[1], Min Ding[2], Baolue Li[3], and David C. Christiani[1]

[1]Department of Environmental Health
Harvard School of Public Health
Boston, Massachusetts, 02115, USA
[2]Suixi Nursing School, Suixi, Anhui, China
[3]Peking Union Medical College, Beijing, China

INTRODUCTION

Data from 845 women textile workers and their husbands in Suixi, China were analyzed to investigate the association of paternal and maternal exposure with birth weight and preterm birth. The paternal agents examined in this study were occupational exposure to dusts, gases or fumes, and cigarette smoking. The maternal agents were occupational exposure to dusts, gases or fumes, stress, carrying and lifting of heavy loads, working in a squat position, and rotating shift work during the pregnancy, indoor coal combustion for heating or cooking. Information on reproductive history, paternal and maternal exposures and other covariates including age at pregnancy, time and duration of leave from job since pregnancy, mill location was obtained by trained nurses using a standardized questionnaire. Generalized estimation equation method with logit link was used in the risk assessment. Rotating shift work was marginally associated with low birth weight (< 2,900 g) (OR=1.49, 95% CI:0.96-2.31), and significantly with preterm birth (<37 weeks) (OR=1.99, 95% CI:1.14-3.48), and the union of the two outcomes (OR=2.01, 95% CI: 1.33 to 3.02). Working in a squatting position and use of a coal stove for heating were significant for the union of low birth weight and preterm birth. No significant association was found between paternal exposures with birth weight or preterm births.

Low birth-weight and preterm births have been extensively evaluated (Institute of Medicine, 1985). However, only a few studies have considered maternal occupation, and even less studies examined occupational exposures of both parents simultaneously. Low birth weight was found to be associated with maternal employment in the food and drink manufacturing industry, in the metal and electrical manufacturing industries,

* Supported in part by grants OH02421 from NIOSH and ES0002 from NIEHS.

and employment as cleaners or maids after adjusting for several confounding factors including smoking, age, education and ethnicity (McDonald, *et al.*, 1987). Preterm delivery was increased among women in psychiatric nursing, food and beverage service, and metal and electrical manufacturing (McDonald, *et al.*, 1988). A recent report on female physicians suggests that long hours of physical work were not associated with any adverse reproductive outcomes including low birth weight, preterm delivery and intrauterine growth retardation (Klebanoff, *et al.*, 1980). A prospective population-based study found no association between maternal physical activity and birth weight within the range of activities performed by pregnant women studied (Rabkin, *et al.*, 1990). However, a significant association between the prolonged standing on the job and preterm birth was demonstrated from a sample of 1,206 women in New Haven, Connecticut (Teitelman, *et al.*, 1990).

Evidence for an influence of paternal exposure on pregnancy outcome is scanty and indirect, with no known causative associations between paternal occupational exposure and adverse reproductive effects other than infertility (Office of Technology Assessment, 1985). There are several posited mechanisms by which paternal exposure could adversely affect fetal development, including genetic alterations to sperm, exposure of the mother (and indirectly, of the fetus) through seminal fluid, and exposure of the mother through contamination of the home environment (Knishkowy and Baker, 1986). Paternal exposure in laboratory animals has been reported to cause growth retardation (Joffe and Soyka, 1982) and increased fetal death and malformations (Trasler, *et al.*, 1985). In humans, dibromochloropropane (DBCP) (Goldsmith, *et al.*, 1984) and lead (Lancranjan, *et al.*, 1975) may be considered as unequivocal causes of male infertility. Evidence for paternal effects on later pregnancy outcomes is even more limited. Stillbirth was over twice as common among the offspring of male workers in a copper smelter as compared to the offspring of office workers (Beckman and Nordstrom), but only 21 stillbirths were available for study. Work in textile machining was associated with stillbirth on the basis of four exposed cases from Collaborative Perinatal Project (Hartz and Jacques, 1978). A twofold increased risk of preterm delivery was found with paternal employment in the glass, clay, and stone; textile; and mining industries (Savitz, *et al.*, 1989). Paternal employment in the art and textile industries was associated with delivery of intrauterine growth retarded infants (Savitz, *et al.*, 1989). Paternal exposure has also been examined with respect to infant mortality (Savitz, *et al.*, 1984), with inconclusive results.

Data collected from 1,035 married women workers and their husbands in three textile mills in Suixi, China provide an opportunity to investigate the association of parental exposure at work and at home with low birth weight and prematurity. The data also offered unique opportunity to examine rotating shift work and pregnancy outcomes, as more than half of the women workers in this sample rotated in four shifts, that is, each cycle consisted of two days of morning shift: 6:00am-2:00pm, two days of daytime shift: 2:00pm-10:00pm, and two days of night shift: 10:00pm-6:00am, and two days of rest.

METHODS

We surveyed 1,035 married women workers and their husbands in three textile mills (Huaibei First Textile Mill, Huaibei Second Textile Mill, and Suzhou Textile Mill) in Anhui, China in 1992. Huaibei First Textile Mill and Huaibei Second Textile Mill were built in 1978, and Suzhou Textile Mill in 1984. All three mills were comparable with respect to facilities, equipments, manufacturing process, and products. Table 1 presents the sample selection frame.

A standardized questionnaire was administrated by trained nurses to obtain information on maternal reproductive history, cigarette smoking, alcohol consumption, occupational exposures, indoor air pollution, and demographic characteristics. The major occupational agents of interest included exposures to dusts, gases/fumes, physical activity and position on the job (e.g. working in a squat position, carrying or lifting heavy loads), rotating shift work, and stress at work. If a woman reported a history of pregnancy, detailed information on pregnancy outcomes, occupational exposures during pregnancy, and the time and duration of leave from the job since pregnancy was collected.

Paternal information on cigarette smoking and occupational exposures was obtained by a self-administrated questionnaire. Each interviewed woman was asked to bring home a questionnaire to her husband, and then to bring back the completed questionnaire to the textile mill.

Table 1. Sample Selection Frame

Total Participants	Total Sample		Excluded (Alcohol drinker)
1,035	**1,028**	Non-smoking Non-drinking	7
First order live birth	Second order live birth	Third order live birth	No live birth or incomplete information on reproductive history
886	49	2	142
	Education Level		
Middle/high school*	Illiterate	Primary school	College
887*	11	31	6

*Observations used in the analysis.

In this study, low birth weight was defined as birth weight equal or less than 2,900 grams, which is the 20th percentile of the total births. We chose not to use the conventional definition ($\leq$ 2,500 grams) in order to include more potentially compromised births, and to increase statistical power. Preterm birth was defined as gestational age less than 37 weeks.

Logistic regression was used to assess the association of maternal and paternal exposures with low birth weight and preterm birth. The variances of the estimates from logistic regression were corrected by robust method (Liang and Zeger, 1986) because the outcomes of consecutive births given by the same woman may be correlated. Besides maternal and paternal occupational exposures, mill location, maternal age at pregnancy, time and duration leaving the job since pregnancy, parity, use of coal stove for heating at home, and paternal cigarette smoking was also included in the regression model.

RESULTS

There were a total of 845 women included in the analyses. They were all middle or high school graduates, never-smokers, non-alcohol drinkers, and with maximum two live births. The means and standard deviations of birth weight and gestational age in this sample were 3,274 (± 490) grams, and 38.8 (± 2.3) weeks, respectively.

The distribution of paternal and maternal exposures and crude rates of low birth weight and preterm births are shown in Table 2. Among the maternal factors examined in the study, manufacture workers, work in a squat position, and rotating shift work appear to be associated with increased prevalence of low birth weight and preterm births, while stress, and carrying and lifting heavy loads associated only with low birth weight. Paternal smoking was related to higher rates of low birth weight and

Table 2. Prevalence of Low Birthweight (<= 2,900 grams), Preterm Birth (< 37 weeks), and Union of Low Birthweight and Preterm Birth by Maternal and Paternal Exposure

			Low birthweight		Preterm Birth		LBW or preterm	
Variables	Status	No. live births	No.	%	No.	%	No.	%
Total		887	177	20.0	165	18.6	292	32.9
			Maternal occupational agents					
Dust/gas/fume level	High	280	54	19.3	52	18.6	94	33.6
	Low	607	123	20.3	113	18.6	198	32.6
Manufacture workers	Yes	771	158	20.5	148	19.2	259	33.6
	No	116	19	16.4	17	14.7	33	28.5
Stress	Severe	382	84	22.0	72	18.9	132	34.6
	Mild	505	93	18.4	93	18.4	160	31.7
Carrying heavy load	Yes	168	40	23.8	29	17.3	62	36.9
	No	719	137	19.1	136	18.9	230	32.0
Lifting heavy load	Yes	122	29	23.8	22	18.0	45	36.9
	No	765	148	19.4	143	18.7	247	32.3
Squat position	Yes	103	26	25.2	25	24.3	43	41.8
	No	784	151	19.3	140	17.9	249	31.8
Rotating shift work	Yes	635	138	21.7	127	20.0	225	35.4
	No	252	39	15.5	38	15.1	67	26.6
			Paternal exposure					
Occupational exposure to dust/gas/fume	Yes	144	25	17.4	23	16.0	44	30.6
	No	587	115	19.6	31	19.9	190	32.4
Cigarette smoking	Yes	535	102	19.1	104	19.4	173	32.3
	No	196	38	19.4	30	15.3	61	31.1
Unknown		156	37	23.7	37	20.0	58	37.2
			Indoor air pollution					
Use of coal stoves for heating	Yes	313	61	19.5	76	24.3	122	39.0
	No	574	116	20.2	89	15.5	170	29.6

preterm births than non-paternal-smoking. Noticeably, women using coal stove for heating had much higher rates of preterm births than those using other heating systems. No association was found between parental dust and gas/fume exposures and adverse pregnancy outcomes.

Figure 1 presents odds ratios and their 95 percent confidence intervals of low birth weight and preterm birth for each agent, adjusting for other occupational agents, age at pregnancy, duration of leave from job since pregnancy, parity, and mill location. Shift work was significantly associated with preterm birth (OR=1.99, 95% CI:1.14-3.48), and marginally significant for low birth weight (OR=1.49, 95% CI:0.96-2.31). Squat position and use of coal stove for heating was marginally significant for preterm births. Parental exposures to dust and gases/fumes and paternal cigarette smoking were insignificant for both low birth weight and preterm birth. To obtain quantitative estimate of effect for each agent, the similar analyses using birth weight and gestational age as continuous variables were also performed. Shift work was associated with 0.44 week deduction in gestation (P=0.03), and 80 gram deduction in birth weight (P=0.06).

Since low birth weight and preterm birth were correlated outcomes (P=0.0002), further analysis was performed by using union of low birth weight and preterm birth, that is, an indicator variable for low birth weight or preterm birth (Table 3). Shift work was again significantly associated with union of low birth weight and preterm birth (OR=1.80, 95% CI: 1.22 to 2.67). The union of low birth weight and preterm birth was also significantly associated with squat position (OR=1.54, 95% CI: 1.00 to 2.38), and use of coal stove for heating (OR=1.40, 95% CI: 1.03 to 1.91). Finally, we repeated all the above analyses by limiting to first order live births to eliminate the influence of parity, and got similar results.

Table 3. Relative Odds of Low Birthweight (<=2,900 grams)/ Preterm Birth (< 37 weeks) for Selected Maternal and Paternal Exposures

		LBW or Preterm		
Variables	Status	OR	95% CI	p value
	Maternal occupational agents			
Manufacture workers	Yes vs No	0.89	0.53-1.47	0.64
Dust/gas/fume level	High vs Low	1.00	0.71-1.40	0.99
Stress	Severe vs Mild	1.04	0.76-1.42	0.80
Carrying heavy load	Yes vs No	1.14	0.71-1.85	0.58
Lifting heavy load	Yes vs No	1.20	0.70-2.06	0.51
Squat position	Yes vs No	1.54	1.00-2.38	0.055
Rotating shift work	Yes vs No	1.80	1.22-2.67	<0.01
	Paternal exposure			
Occupational exposure to dust/gas/fume	Yes vs No	0.90	0.60-1.36	0.63
Cigarette smoking	Yes vs No	1.03	0.71-1.50	0.87
Unknown	Unknown vs No	1.21	0.75-1.96	0.43
	Indoor Air pollution			
Coal stove for heating	Yes vs No	1.40	1.03-1.91	0.03

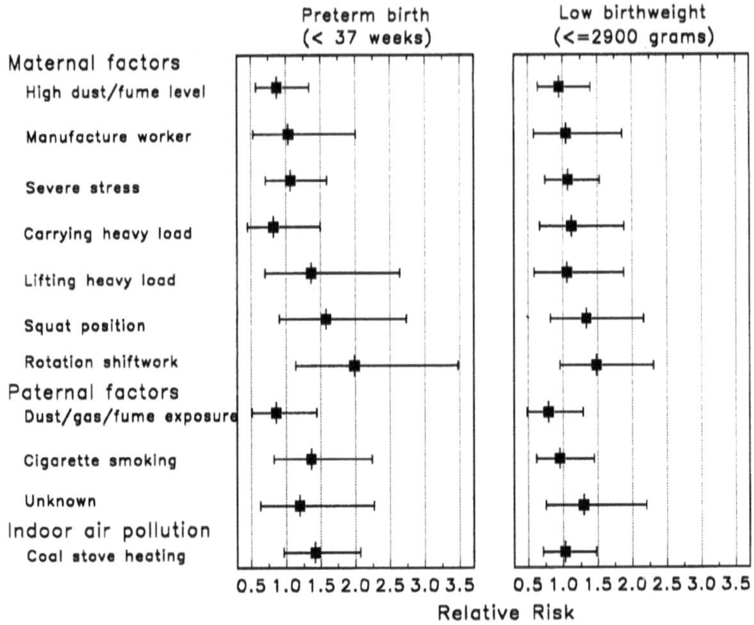

Figure 1. Relative odds (± 95% confidence limits) of preterm births (<37 weeks) and low birthweight (<=2900 g) with maternal and paternal factors and use of coal stoves for heating.

DISCUSSION

There have been a number of reports on respiratory symptoms and lung function among textile workers (Christiani, *et al.*, 1986a; Christiani, *et al.*, 1986b. This is one of few studies on pregnancy outcomes among a large number of textile women workers, with special attention to parental exposure to specific occupational agents. We found that the adjusted rates of preterm birth or low birth weight were higher among women whose jobs required rotating shift, squatting position, or with indoor coal combustion for heating as compared to those without these factors. On the other hand, paternal cigarette smoking and parental occupational exposures to dust and gas/fume were not significantly associated with low birth weight or preterm births.

Our study design has several important features. All the women workers in this study were non-smokers, non-alcohol drinkers, with similar educational level, and with maximum two live births, which minimize the confounding of these factors to the analyses. All the participants were relatively young and currently employed in the textile mills, and therefore the healthy worker effects should be minimal. The women workers were selected independent of their spouse's occupations, and more than half of their husbands did smoke, which provides an opportunity to examine the effects of paternal smoking and occupational exposures on pregnancy outcomes.

There are potential sources of misclassification in this study. Since the survey was conducted retrospectively, there may be selection bias. If some women with history of exposure to occupational hazards left or changed jobs before pregnancy, or some women with history of adverse pregnancy outcomes left or switched jobs, the association of low birth weight and preterm births presented in this report may have been under-estimated. However, such selection bias is likely to be independent of their spouse's occupations and smoking status.

Secondly, since maternal and paternal occupational information was self-reported, the misclassification of exposure is possible. The net effect of such misclassification has biased the results towards null. However, such misclassification is unlikely to occur for the variables such as rotation shift work, work in a squat position, and use of coal stoves for heating, because these are relatively objective measurements and the participants had no knowledge about the potential adverse effects of these factors on pregnancy outcomes.

The circadian system is a pervasive biologic phenomenon which influences diverse physiologic functions among all species of animals. The circadian system and the female reproductive process interface at many levels, but the exact relationship varies greatly from species to species (Turek, *et al.*, 1984). Many medical and psychosocial problems associated with rotating shift work have been reported, with gastrointestinal, sleep and mood disturbances the most common (Moore-Ede, *et al.*, 1983; Weitzman and Pollack, 1979; Hak and Kampan, 1980; Czeisler, *et al.*, 1983) Data from a National Center for Health Statistics survey indicate that women working rotating shifts report higher rates of use of sleeping pill, tranquilizer, and alcohol as well as more job and emotional stress, compared with women not working variable shifts (Gorden, *et al.*, 1986). Considerable nursing literature is concerned with adaptation to rotating shifts in terms of health and sleep, though few studies actually document objectively the adverse health parameters (Cohn, 1981; Hoskins, 1981; Eaves, 1980). To our knowledge, this is the first study to examine and demonstrate the association of rotating shift work with low birth weight and preterm births based on a homogeneous occupational sample.

An association between standing jobs and the rate of preterm births has been reported from several studies in French (Turek, *et al.*, 1984; Moore-Ede, *et al.*, 1983; Weitzman and Pollack, 1979; Hak and Kampan, 1980; Czeisler, *et al.*, 1983; Gorden, *et al.*, 1986; Cohn, 1981; Hoskins, 1981; Eaves, 1980; Mamelle, *et al.*, 1984; Saurel-Cubizolles, *et al.*, 1985) and United States (Teitelman, *et al.*, 1990). The result for low birth weight was inconclusive (Teitelman, *et al.*, 1990) The relations between preterm birth/low birth weight and working in a squat position have not been reported. Our study demonstrated a significant association between union of low birth weight and preterm births and working in a squat position. In the separate analyses for these two outcomes, the association with preterm births (OR=1.58, 95% CI: 0.91 to 2.74, p=0.10) appears greater than with low birth weight (OR=1.34, 95%CI: 0.82 to 2.17, p=0.24), although neither reached a statistically significant level.

Indoor use of a coal stove for heating has been demonstrated to increase chronic respiratory symptoms (Xu, in press) and cause accelerated decline of pulmonary function (Xu, 1991). This is the first study to examine and demonstrate the association of indoor coal combustion with low birth weight and preterm births. Further study is needed to confirm our findings.

REFERENCES

Beckman, L., Nordstrom S., 1982. Occupational and environmental risks in and around a smelter in northern Sweden. Fetal mortality among wives of smelter workers. *Hereditas* 97:1-7.

Christiani, D.C., *et al.*, 1986a. Respiratory Disease in Cotton Textile Workers in the People's Republic of China, I. Respiratory Symptoms. *Scand J Work Environ Health* 12:40-45.

Christiani, D.C., *et al.*, 1986b. Respiratory Disease in Cotton Textile Workers in the People's Republic of China, II. Pulmonary Function. *Scand J Work Environ Health* 12:45-50.

Cohn, M.R., 1981. A case against inflexible routine shift rotation. *Nursing Leadership* 4:14-8.

Czeisler, C.A., Moore-Ede, M.G., Coleman, R.M., 1983. Resetting circadian clocks in man: applications to sleep disorders, medicine and occupational health in sleep/wake disorders, natural history, epidemiology and long term evolution. Raven Press: New York. 243-60.

Eaves, D., 1980. Time for a change...circadian rhythm disturbances experienced by nurses on night duty. *Nursing Mirror* 150:22-4.

Goldsmith, J.F., Potashnik, G., Israeli R., 1984. Reproductive outcomes in families of DBCP-exposed men. *Arch Environ Health* 39:85.

Gorden, M.P,, Cleary P.D., Parker C.E., Czeisler C.A., 1986. The prevalence and health impact of shiftwork. *Am J Public Health* 76:1225-8.

Hak, A., Kampan, R., 1980. Working irregular hours: complaints and state of fitness of railway personnel. *Chronobiologia* 7:398.

Hartz, S.C., Jacques, P.F., 1978. Occupational health surveillance of reproductive outcomes: a case-control study. Final report for NIOSH contract no. 210-78-0057. Cincinnati, National Institute for Occupational Safety and Health.

Hoskins, C.N., 1981. Chronobiology and health. *Nursing Outlook* 29:572-6.

Institute of Medicine, Committee to Study the Prevention of Low Birth Weight. Preventing low birth weight. Washington, DC, National Academy Press.

Joffe, J.M., Soyka, L.F., 1982. Paternal drug exposure, effects on reproduction and progeny. *Semin Perinatal* 6:116-24.

Klebanoff, M.K., Shiono, P.H., Rhoads, G.G., 1980. Outcomes of pregnancy in a national sample of resident physicians. *New Engl J Med* 323:1040-1045.

Knishkowy, B., Baker, E.L., 1986. Transmission of occupational disease to family members. *AM J Ind Med* 9:545-50.

Lancranjan, I., Popescu, H.I., Gavanscu, O., Klepsch, I., Serbanesu, M., 1975. Reproductive ability of workmen occupationally exposed to lead. *Arch Environ Health* 30:396.

Liang, K-Y, Zeger, S.L., 1986. Longitudinal data analysis using generalized linear models. *Biometrika* 73:13-22.

Mamelle, N., Laumon, B., Lazar, P., 1984. Prematurity and occupational activity during pregnancy. *Am J Epidemiol* 119:309-22.

McDonald, A.D., McDonald, J.C., Armstrong, B., *et al.*, 1987. Occupation and pregnancy outcome. *Br J Ind Med* 44:521-526.

McDonald, A.D., *et al.*, 1988. Prematurity and work in pregnancy. *Br J Ind Med* 45:56-62.

Moore-Ede M.G., Czeisler G.A., Richardson G.S., 1983. Circadian timekeeping in health and Disease Part 2. Clinical implications of circadian rhythmicity. *New Engl J Med* 309:530-6.

Office of Technology Assessment, 1985. Reproductive health hazards in the workplace. Washington, DC: US GPO.

Rabkin C.S., Anderson H.R., Bland J.M., Brooke O.G., Chamaberlian G., Peacock J.L., Maternal activity and birth weight: A prospective population-based study. *Am J Epidemiol* 1990;131:522-31.

Saurel-Cubizolles, M.J., Kaminsmk M., Lladp-Arkhipoff J., *et al.,* 1985. Pregnancy and its outcomes among hospitals personnel according to occupation and working conditions. *J Epidemiol Community Health* 39:129-34.

Savitz D.A., Whelan E.A., Kleckner R.C., 1989. Effect of parents' occupational exposures on risk of stillbirth, preterm delivery, and small-for-gestational-age infants. *Am J Epidemiol* 1989;129:1201-18.

Savitz D.A., Harley B., Krekel S., *et al.*, 1984. Survey of reproductive hazards among Oil, Chemical, and Atomic workers exposed to halogenated hydrocarbons. *Am J Ind Med* 6:253-64.

Teitelman A.M., Welch L.S., Hellenbrand K.G., Bracken M.B., 1990. Effects of maternal work activity on preterm birth and low birth weight. *Am J Epidemiol* 131:104-13.

Trasler J.M., Hales B.F., Robaire B., 1985. Paternal cyclophosphamide treatment of rats causes fetal loss and malformations without affecting male fertility (letter). *Nature* 316:144-6.

Turek F.W., Swann J., Earnest D.J., 1984. Role of the circadian system in reproductive phenomena. *Recent Prog Horm Res* 40:143-83.

Weitzman, E.D., Pollack, C.P., 1979. Disorders of the circadian sleep-wake cycle. *Medical Times* 1979;107:83-94.

Xu, X., (in press). Effects of indoor and outdoor air pollution on respiratory symptoms in adults: A multivariate Analysis. *Am Rev Res Dis.*

Xu, X., Dockery D.W., Wang L., 1991. Effects of Air pollution on Adult Pulmonary Function. *Arch Environ Health* 46:198-206.

GENOTOXIC CONSEQUENCES OF TESTICULAR LOCALIZATION OF INDIUM-114m

Katharine P. Hoyes[1], N.Colin Jackson[2],
Harold Jackson[2], Harbans L. Sharma[2],
Jolyon H. Hendry[1], and Ian D. Morris[3]

[1]Department of Experimental Radiation Oncology
Paterson Institute for Cancer Research
Manchester
M20 9BX United Kingdom
Departments of [2]Medical Biophysics, and
[3]Physiological Sciences
University of Manchester
Manchester
M13 9PT United Kingdom

INTRODUCTION

The incidence of childhood leukaemia in the offspring of workers at the Sellafield nuclear reprocessing plant has been associated with preconceptional paternal radiation exposure (Gardner *et al.*, 1990). One possible causative factor for this effect may be that in addition to external radiation, workers may also be subject to internal radionuclide contamination. Local testicular exposure to radionuclides offers a route for increased potential for transmission of genetic damage via sperm, irradiated during their developmental stages, which could be a contributory factor to childhood leukaemogenesis.

In the literature, the main emphasis of investigations into the genetic effects of occupational radiation exposure of human males has been directed towards the hazards posed by radionuclides emitting alpha particles. Radionuclides producing beta and gamma radiation have been considered to present a similar hazard to externally administered X-rays and to produce no unique genetic effects. Over recent years, however, the radiobiological effects of low energy Auger emitting radionuclides have attracted attention. In particular, studies by Rao and colleagues (1988, 1991) have demonstrated that damage caused to the spermatogenic epithelium by Auger emitters may be comparable to that produced by alpha emitting radionuclides. In our laboratories the potential of cell targeted In-114m (half life = 50 days) as a

Male-Mediated Developmental Toxicity, Edited by D.R. Mattison
and A.F. Olshan, Plenum Press, New York, 1994

radiotherapeutic adjunct for treatment of acute lymphocytic leukaemia has been investigated in a rat T-cell model (Jackson *et al.*, 1984). These studies involved extensive examination of the distribution and clearance of the radionuclide after administration of labelled lymphoid cells (Jackson *et al.*, 1989). It was observed that a small proportion of the radioactive dose (approximately 0.3%) became localised within the testes by 24h after administration and remained associated with the organ for up to 250 days (Jackson *et al.*, 1991). Autoradiographic studies revealed a portion of the testicular radionuclide was associated with spermatids and their derived spermatozoa. This association with the sperm appeared to be maintained during epididymal transport and maturation.

These observations provide experimental evidence that the indium radionuclide can pass across the blood testis barrier and gain access to the spermatogenic epithelium and bind to mutagenically sensitive germ cells. The indium ion is known to bind strongly to the iron transport protein transferrin and it is possible that In-114m utilises the physiological iron pathway to gain entry to the seminiferous epithelium. The aims of our current work, therefore, are to study to what extent indium and other Auger emitting radionuclides can utilise the iron transferrin pathway to gain entry to the spermatogenic epithelium and to investigate the consequences of such radioactive exposure in terms of spermatogenic toxicity and the frequency of mutagenic events.

METHODS

Animals used in these experiments were adult male Sprague Dawley rats weighing between 150g and 200g at commencement. For tissue distribution studies In-114m, Fe-55 or Cd-109 were administered by intraperitoneal injection of radionuclide incubated with rat plasma. At the time points shown animals were killed and tissues were dissected, weighed and assayed for radioactivity. In a parallel experiment In-114m was administered as a preformed In-114m transferrin complex dissolved in saline.

To investigate the biological consequences of testicular localisation of the Auger emitter In-114m rats were injected i.p. with In-114m-plasma (14.8MBq/kg body weight) or plasma alone as a control. At 21 day intervals animals were killed, testes and epididymides were dissected, weighed and sperm reserves were counted in accordance with previously described methods (Rao *et al.*, 1988).

For the mating trials a total of 30 male Sprague Dawley rats of proven fertility were randomly distributed into 3 experimental groups (n=10). Animals in Group 1 were dosed with 14.8MBq/kg In-114m, those in Group 2 were given 3.7MBq/kg In-114m and the animals in Group 3 served as controls. Starting 3 days post-treatment each male was housed with 2 virgin females for a total of 5 successive periods of 19 days, each followed by a resting period of 2 days.

Immediately after birth the size and weight of each litter were recorded. The females were killed and the number of implantation sites and corpus luteum were counted. The percentage of dominant lethal mutations was calculated from the number of pre- and post-implantation losses (Partington and Bateman, 1964).

RESULTS AND DISCUSSION

The uptake of In-114m (administered both as In-114m-plasma and In-114m-transferrin), Fe-55 and Cd-109 into the testes, epididymides, liver and spleen are summarised in Table 1. Approximately 0.13%/ g of tissue of the injected dose of indium was localised within the testes by 24 hours and remained constant. Testicular and epididymal uptake of In-114m was markedly similar to that of the other transferrin binding radionuclides (Fe-55, Cd-109), and also remained constant when the dose was administered as a preformed indium-transferrin complex. As previously reported (Jackson *et al.*, 1991), autoradiography showed that tubules were not uniformly labelled in histologically comparable stages of the spermatogenic cycle; some tubules exhibited a considerable degree of radionuclide concentration within the developing germ cells, whilst others were unlabelled.

Iron, indium and cadmium all bind strongly to transferrin. Plasma transferrin does not have direct access to the seminiferous epithelium due to the blood-testis barrier. The physiological delivery of iron to developing spermatogenic cells involves the receptor mediated endocytosis of plasma transferrin leading to the delivery of iron to the Sertoli cell. The iron is then bound by a second immunologically distinct transferrin, synthesised by the Sertoli cell, and is so delivered to the germ cells (Morales *et al.*, 1987). It seems reasonable to expect that transferrin binding radionuclides may utilise the physiological iron pathway into the seminiferous epithelium. A homeostatic regulating mechanism for iron (Skinner and Griswold 1982) would explain the observation of a relatively constant proportion of a systemic dose of a transferrin binding radionuclide localizing in the testis with maintenance via turnover from the liver.

Table 1. Tissue Distribution of In-114m, Cd-109 and Fe-59 After i.p Administration

Time	Radionuclide	Testis	Epididymis	Liver	Spleen
24hrs	In-114m-plasma	0.13±0.01	0.15±0.03	0.69±0.07	0.72±0.16
	Cd-109-plasma	0.09±0.04	-	1.03±0.26	0.33±0.07
	In-114m-Tf	0.16±0.02	0.19±0.06	0.64±0.07	0.85±0.21
7 day	In-114m-plasma	0.09±0.03	0.06±0.03	0.40±0.16	0.89±0.07
	Cd-109-plasma	0.07±0.01	0.04±0.01	2.99±0.49	0.18±0.02
	Fe-55-plasma	0.05±0.02	0.03±0.01	0.77±0.09	0.44±0.07
	In-114m-Tf	0.15±0.05	0.08±0.01	0.44±0.07	0.83±0.03
63 days	In-114m-plasma	0.16±0.06	0.07±0.03	0.21±0.09	0.75±0.23
126 days	In-114m-plasma	0.17±0.06	0.11±0.04	0.19±0.02	0.77±0.06

Data show the % of injected radioactivity/g of tissue (mean ± s.e.m.) for 5 animals

Administration of a single dose of 14.8MBq/Kg In-114m-plasma resulted in decreased testicular weight and sperm head count. Maximal reduction was observed at about 84 days after treatment. Thereafter there was recovery towards control values but neither parameter had returned to normal at 105 days. This large decrease in testicular weight and sperm content was compatible with a loss of mature germ cells due to a maturation depletion effect after damage to spermatogonia. Similar changes in epididymal sperm reserves were also observed with the nadir occurring at 105 days (Table 2). When a substantially lower dose of In-114m-plasma (3.7MBq/kg) was administered no changes in testicular or epididymal weight or sperm content were observed.

Table 2. Changes in Testicular Weight (g) and Testicular and Epididymal Sperm Content (x10^8) After a Single Injection of 14.8MBq/Kg In-114m

Days after treatment	21	42	63	84	105
Testicular weight(g)					
Treated	1.68±0.02	1.55±0.06	1.47±0.08[a]	1.29±0.07[b]	1.53±0.1[a]
Control	1.69±0.01	1.79±0.03	1.82±0.01	1.86±0.02	1.79±0.03
Testicular sperm content(10^8)					
Treated	1.01±0.06	0.85±0.05	0.88±0.01[a]	0.76±0.04[b]	0.89±0.07
Control	0.87±0.07	0.97±0.03	1.04±0.04	1.16±0.04	1.04±0.04
Epididymal sperm content (10^8)					
Treated	3.80±0.12	3.80±0.51	4.43±0.37	2.50±0.55[a]	1.32±0.2[b]
Control	4.23±0.26	3.90±0.21	4.48±0.49	3.76±0.27	4.06±0.08

[a] $P<0.05$, [b] $P<0.01$, in comparison with respective control value (Students t test).

To investigate further the mutagenic consequences of such radioactive exposure, mating trials were instigated to study the reproductive capacity of In-114m-treated male rats. Treatment of male rats with In-114m resulted in a reduction in litter size relative to controls. This was most marked in the fourth and fifth mating trials where litter size dropped to 65% of control values in the 14.8MBq/kg group and to 87% in 3.7MBq/kg group. Fertility remained at over 95% in the control and 3.7MBq/kg groups throughout the trials. In the 14.8MBq/kg group fertility dropped to 80% in the fourth and fifth mating trials. This was consistent with the decrease in testicular and epididymal weight and sperm content previously observed at this dose (Table 2). At birth none of the F1 generation, of any experimental group, presented any obvious gross external abnormalities. Body weights were also normal.

A significant increase in the frequency of dominant lethal mutations was observed in females mated with the treated males. Furthermore, there was a significant increase in mutation rate at the 3.7MBq/kg dose at which no effect on testicular weight or sperm reserves was noted. This was most marked in the fifth mating trial (87 days after treatment) where the mutation rate was 25.29 ± 3.87% in the 14.8MBq/kg group and 22.79 ± 2.9% in the 3.7MBq/kg group compared with a control value of 6.94 ± 2.1% ($P<0.001$, n=10 males per group). A significant increase ($P<0.01$) in the incidence of dominant lethal mutations was apparent, however, from the second trial onwards in the irradiated animals in both treatment groups. This suggests that In114m may induce lethal mutations in all classes of developing germ cells although this is most marked in cells that have been exposed to the radionuclide at the Type A spermatogonia stage and throughout their subsequent development.

CONCLUSIONS

These data suggest that the ability of indium-114m and certain other radionuclides to transcend the blood-testis barrier and gain access to developing germ cells, can be related to their ability to utilise the physiological serum and testicular transferrin pathway. The consequences of such concentration of radionuclide within radiosensitive germ cells are time-dependent cytotoxic and genotoxic damage. The possibility therefore exists that transferrin binding radionuclides emitted by nuclear plants, such as plutonium (Lehmann *et al.*, 1983), may be deposited in human testes by similar mechanisms resulting in mutagenic changes to emergent sperm.

ACKNOWLEDGEMENTS

We thank Mark Cox and Mrs A.M. Smith for technical assistance. This work is funded by the United Kingdom Coordinating Committee for Cancer Research.

REFERENCES

Gardner, M.J., Snee, M., Hall, A., Powell, C., Downs, S., and Terrell, T.D., 1990, Results of a case-control study of leukaemia and lymphoma amoung young people near Sellafield nuclear plant in West Cumbria, *Brit. Med. J.* 300: 423-429.

Jackson, C., Jackson, H., Bock, M., Morris, I.D., and Sharma, H.L.,1991, Metal radionuclides and the testis, *Int. J. Radiat. Biol.* 60: 851-858.

Jackson, H., Jackson, N.C., Bock, M., Perera, A., Ramsden, C., and Sharma, H.L., 1989, Experimental investigation into the clearance, targeting and stability of In-114m labelled lymphoid cells, in Radiolabelled Cellular Blood Elements, H. Sinzinger and M.L. Thakur, ed, Wiley-Liss, Inc., New York.

Jackson, H., Jackson, N.C., Bock, M., and Lendon, M., Testicular invasion and relapse and meningeal involvement in a rat T-cell leukaemia, 1984, *Brit. J. Cancer.* 50:617-624.

Lehmann, M., Culig, H., and Taylor, D.M., 1983, Identification of transferrin as the principal plutonium-binding protein in blood serum and liver cytosol of rats: immunological and chromatographic studies. *Int. J. Radiat. Biol.* 44:65-74.

Morales, C., Sylvester, S.R., and Griswold, M.D., 1987, Transport of iron and transferrin synthesis by the seminiferous epithelium of the rat in vivo, *Biol. Reprod.* 37:995-1005. Partington,M., and Bateman A.J., 1964, Dominant lethal mutations induced in male mice by methyl methanesulphonate, *Heredity* 19:191-200.

Rao, D.V., Narra, V.R., Howell, R.W., Lanka, V.K., and Sastry K.S.R., 1991, Induction of sperm head abnormalities by incorporated radionuclides: dependence on subcellular distribution, types of radiation, dose rate and presence of radioprotectors, *Radiat. Res.* 125: 89-97.

Rao, D.V., Sastry, K.S.R., Grimmond, H.E., Howell, R.N., Govelity, G.F., Lanka, V.K., and Mylavarapu, V.B., 1988, Cytotoxicity of some indium radiopharmaceuticals in the mouse testes. *J. Nucl. Med.* 29:375-384.

Skinner, M.K., and Griswold, M.D., 1982, Secretion of testicular transferrin by cultured Sertoli cells is regulated by hormones and retinoids, *Biol. Reprod.* 27:211-21.

MALE-MEDIATED DEVELOPMENTAL AND REPRODUCTIVE TOXICITY OF SYMM-TRIAZINE PESTICIDES*

Margaret V. Vartanian[1], Rita M. Khetchumova[2],
Aida S. Makaryan[2]

7007 Hadlow Drive,
[1]Springfield, VA 22152, USA
[2]Research Institut of Environmental Hygiene and Preventive Toxicology, Yerevan, Armenia (former name: All Union Institut of Hygiene and Toxicology of Pesticides and Polymers)

INTRODUCTION

The Metazin and Cotofor pesticides belong to the symmetrical triazine group. This group of triazines has been shown to be moderately toxic when introduced by intravenous, intraperitoneal routes or by ingestion (Geovorkyan, *et al.*, 1977a; Georvorkyan and Khechumova, 1977; Makaryan, 1980). Mutagenic and reproductive effects have been shown in experiments (Gevorkyan, *et al.*, 1977b; Vartanian and Konobeeva, 1986; Vartanian and Gevorkyan, 1978; Makarian and Konobeeva, 1979; Pavlova, 1986).

There is little evidence, from studies on humans, pointing to hereditary effects of these chemicals resulting from short term exposure. Consequences of primary damages of the DNA usually do not lead to hereditary mutations (Venitt and Parry, 1984). A cell which has sustained DNA damage may respond in several ways: DNA repair mechanism; elimination and prevention of DNA replication or transcription; or death of the cell. The residue of pesticides collected from long term exposure, have the ability to accumulate and circulate in the biospheric chain of soil-water-plant -animal-human. For example, high doses of DDT residue present, after more than two decades, in chicken eggs and animal and human milk is a warning to us to avoid mistakes in the application of new pesticides.

Damaged DNA and repair mechanisms do not only lead to somatic mutations and cancer. The mutant forms of insects which have developed resistance to

---

* The materials presented in this article are part of a study conducted at The Research Institut of Environmental Hygiene and Preventive Toxicology at The Department of Health of Armenia in 1976-77.

pesticides, warn us against germ cell hereditary mutations. The fact that the male germ cells go through several hundred premeiotic mitoses and are at a high risk of mutations (Thompson, *et al.*, 1991) emphasizes the importance of studies on male mediated genotoxicity of chemicals, particularly of pesticides.

Human exposure to pesticides in low, chronic doses comes mainly from agricultural products and water. For that reason we performed long-term exposure experiments of 3 and 9 month duration on male and female rats to establish permissible remnant quantities in food and water (Gevorkyan, *et al.*, 1977a, Gevorkyan and Khechumova, 1977; Makaryan, 1980; Gevorkyan, *et al.*, 1977b; Vartanian and Konobeeva, 1986; Vartanian and Gevorkyan, 1978; Makarian and Konobeeva, 1979; Pavlova, 1986; Venitt and Parry, 1984). However, for the purposes of this article, we will discuss only the data collected from experiments on male rats.

MATERIALS AND METHODS

Three month old, male albino rats from St. Petersburg animal nurseries, after going through acclimatization, were randomly divided into 14 groups. Each group consisted of ten animals. The animals were exposed for 3 and 9 months, 5 days per week, to either Metazin (M) or Cotofor (C) water solutions. Metazin is a synonym of Sulfadimethyldiazine; $C_{12}H_{14}N_4O_2S$; mw: 278.36; crystals; which is odorless and soluble in acetone, water, ether; slightly soluble in alcohol. Cotofor is one of the synonyms of Dipropetryn; $C_{11}H_{21}N_5S$; mw: 255.39; powder herbicide; soluble in water and organic solvents (Budavari, 1989).

The experiment included three doses of the tested compounds and the control group. For Metazin LD_{50} 1/10,000 (0.17 mg/kg/day), 1/1,000 (1. mg/kg/day) and 1/100 (17.0 mg/kg/day) were used. For Cotofor the doses were 0.47, 4.78 and 47.8 mg/kg/day, respectively. The water diluted agents were administered orally, using a syringe. The animals in the control group were administered with clear water.

On the final day of exposure, two unexposed female rats, in proestrus or estrus cycles, were placed in the cage of each male. Starting the following day, the females were checked (through vaginal smears) for presence of sperm in the vagina. If no copulation had occurred, a new pair of females was introduced until copulation, followed by pregnancy, were noticed. The date on which sperm were detected was considered as the first day of gestation period. Females were sacrificed on gestational day 20. Control animals were sacrificed parallel to the experimental ones. The parameters utilized to characterize dominant lethality data included:

- Number of implantations (I)
- Number of early (ED) and late deaths (LD)
- Number of live fetuses per pregnancy (LF)
- Corpora lutea per pregnancy (CL)

The following formulas (Sanotskii, 1970; Venitt and Parry, 1984) were used in the study (Assume: m = mean number, c = control group, t = test group):

$$\textit{Pre-implantation losses are calculated by:}\quad Pre\text{-}IL = 1 - \frac{mI}{mCL}$$

$$\textit{Post-implantation losses:}\quad Post\text{-}IL = \frac{mLD}{mI} \times 100$$

Percentage of Mutagenic Index: $MI\% = \frac{mED}{mI} \times 100$

The percentage of frequency of dominant lethality: $FL\% = \left(1 - \frac{mLFt}{mLFc}\right) \times 100$

Skeletal examinations were performed by the Dawson (Hurley, 1965) method (Alcian Red S skeletal stain), and visceral examinations were done by the Wilson method (Wilson and Warkany, 1965). The Chi-square and Student's t-test methods were used for the statistical analysis.

RESULTS

Effects of Metazin and Cotofor on Reproductive Organs (Germ Cells and Testis) of Male Rats

Administration in three doses (1/10,000; 1/1,000; 1/100 LD_{50}) of Metazin and Cotofor (3 and 9 months exposure time) revealed that both of these compounds are significantly toxic in the highest studied doses about 1/100 LD_{50} ($p<0.05$).

Data presented in Table 1 shows toxic effects on sperm quality and function. It is principally expressed through oligospermia.

At 1/100 LD_{50} of Cotofor, concentration of sperm was at 0.95 ± 0.20 million versus 5.3 ± 0.91 million for the control. In addition, there was reduced sperm mobility at 1/100 LD_{50} for Metazin - 96.2 ± 16.4 minutes and for Cotofor - 104.40 ± 16.90 minutes vs. 171.00 ± 19.40 for the control. The number of viable forms of sperm in the ejaculate were reduced. In this category, the control group animals had 91.0 ± 1.50 percent sperm viability, while the animals in the Metazin and Cotofor groups (1/100 LD_{50} at 3 months exposure) show 83.12 ± 0.80 and 55.10 ± 9.90 percent, respectively. Abnormality of sperm head or tail were also observed in the (3 months at 1/100 and 9 months at 1/1000 LD_{50}) Cotofor groups (Table 1).

At the highest dose of Cotofor the test showed only significant reduction in the development of testis in terms of weight and volume. At 1/100 LD_{50}, 3 month exposure the average of testis was: 1.53 ± 0.23 grams and 0.86 ± 0.09 cm^3 versus 2.37 ± 0.70 gr and 1.37 ± 0.005 cm^3 in the control group.

The evaluation of function and quality of seminiferous tubules in testes show dose related statistically significant changes. Namely, reduction in the index of spermatogenesis and an increase in the number of atrophic seminiferous tubules (Figure 1.). The results of studies of exfoliation of germinal epithelia and number of giant cells in tubules show non-significant differences compared to the control group.

These studies suggest that the quality and function of sperm and seminiferous tubules during spermatogenesis are sensitive to chemical exposure and could be good indicators of genotoxic changes.

Table 1. Sperm Quality and Function in Rats Exposed to Metazin and Cotofor*

Expos. Time (mos.) Dose (mg/kg)	Movement duration (minute)	Concen-tration (million)	Osmotic resist. (% NaCL)	Normal forms (%)	Viability (%)	Immovable forms (%)
Control 3 months -	171.0±19.4	5.9±0.91	3.64±0.15	98.7±1.8	91.1±1.5	52.3±4.9
Metazin 3 months 1.7 mg/kg	140.0±10.7	4.3±0.86	3.8±0.06	95.3±1.2	94.2±1.0	62.8±5.7
Metazin 3 months 17.0 mg/kg	96.2±16.4*	5.1±1.40	3.65±0.29	94.7±0.7	83.1±0.8*	50.0±5.9
Cotofor 3 months 0.48 mg/kg	182.0±10.2	4.4±0.59	3.84±0.20	97.1±2.3	95.9±1.0	63.5±5.9
Cotofor 3 months 4.76 mg/kg	129.7±11.8*	5.4±0.59	3.40±0.12	93.8±2.1	90.8±2.2	40.2±4.9
Cotofor 3 months 47.6 mg/kg	104.4±16.9*	0.9±0.20*	2.92±0.45*	85.8±3.8*	55.1±9.9*	58.9±8.4*
Control 9 months -	167.0±8.2	7.0±0.87	5.40±0.04	99.6±0.1	100±0.0	-
Cotofor 9 months 0.48 mg/kg	152.0±6.7	5.6±0.72	5.40±0.08	98.2±0.08	99.9±0.04	-
Cotofor 9 months 4.76 mg/kg	128.5±6.1*	6.9±0.9	4.8±0.08*	95.2±0.7*	97.6±0.31*	-

* $P < 0.05$

Male-Mediated Effects of Metazin and Cotofor on The Outcome of Pregnancy in Unexposed Females

The number of females that were exposed to each male for copulation was recorded. Thus, the effect of the chemicals on libido was also measured. Our studies showed slight, but not statistically significant, changes in the sexual activity of males. In addition, we could not find significant changes in the ratio of female to male fetuses per litter, or in the number of dead fetuses, and in the number of corpora lutea per pregnant female. For this reason we have omitted these data.

However, the exposure of male rats to Metazin and Cotofor has a detrimental effect on the number of pre-implantation embryos (from zygote to late blastocyst) (Moore, 1989). There was a dose-related statistically significant increase in the rate

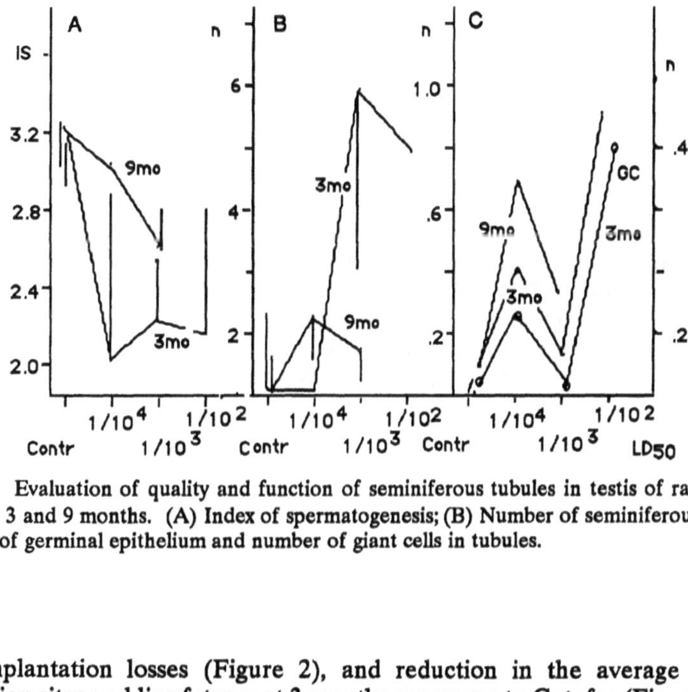

Figure 1. Evaluation of quality and function of seminiferous tubules in testis of rats exposed to Metazin for 3 and 9 months. (A) Index of spermatogenesis; (B) Number of seminiferous tubules; (C) Exfoliation of germinal epithelium and number of giant cells in tubules.

of pre-implantation losses (Figure 2), and reduction in the average number of implantation sites and live fetuses at 3 months exposures to Cotofor (Figure 3). In the rate of post-implantation losses, however, there were some changes which were not significant.

The calculations, for 3 month exposure, suggest that one of the main causes of the increase in pre-implantation losses is the increase of dominant lethal mutations (Figure 4). The percentage of dominant lethal mutations from the results of the 9 month exposure study, were not significant.

In summary, relatively long-term (3 month) oral intake (doses 1/1,000 and 1/100 LD_{50}) of Metazin and Cotofor by male rats has toxic effects on the reproductive outcome, and shows increases in dominant lethal mutations.

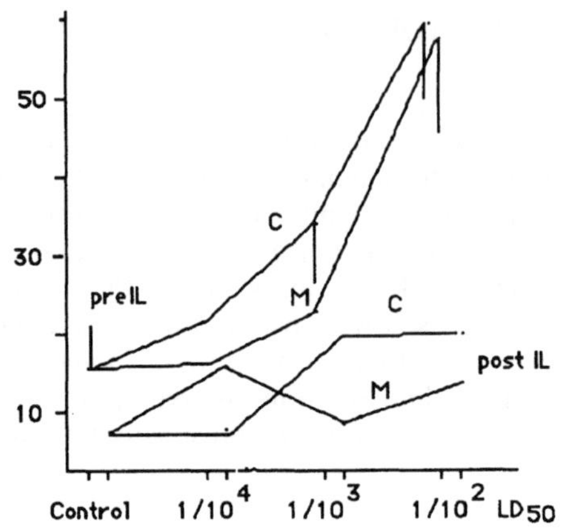

Figure 2. Effect of Metazin and Cotofor on embryonic and fetal loss based on 3 month exposure of male rats.

Male-Mediated Toxicity on The Fetal Development

The effects of Metazin and Cotofor on fetal weight and craniocaudal size, and on the development of fetal ossification were also studied based on three month exposure of male rats. Significant reductions were observed in the mean body weight and size of the fetuses (Figure 4). Interestingly, the dramatic changes were observed at the lower doses, rather than the higher ones. Similar correlation was observed in the development of fetal ossification at 20th day of pregnancy at 9 month exposure (Table 2). Increased incidence of reduced (average) number of fetal vertebrae, ribs, sternebrae and metacarpus-metatarsals was observed in cases of both pesticides.

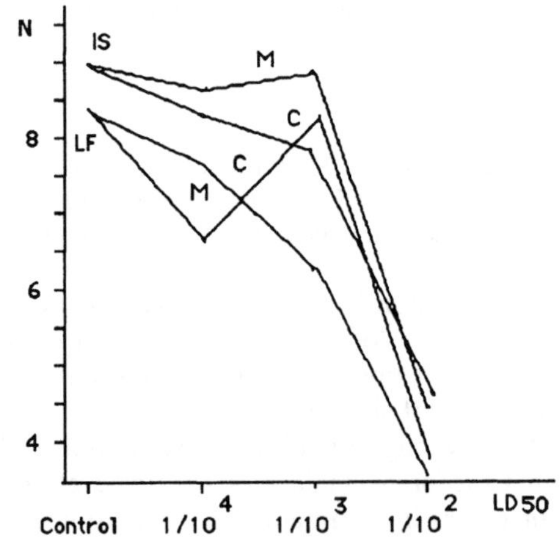

Figure 3. Effect of Metazin and Cotofor on the average number of implantation sites and live fetuses per pregnant female based on 3 month exposure to male rats.

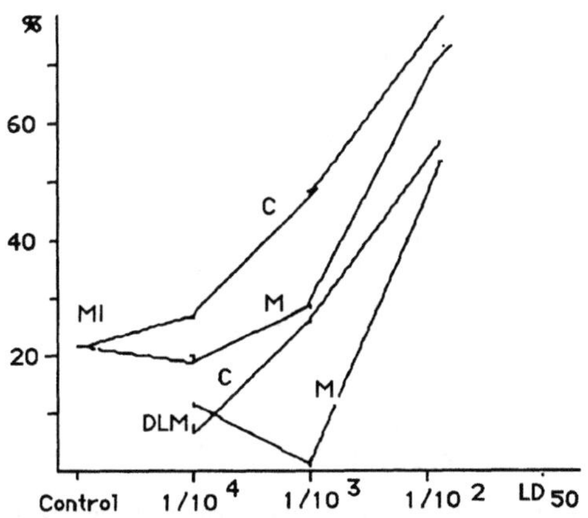

Figure 4. Effect of Metazin and Cotofor frequency (%) of dominant lethality and mutagenic index based on 3 month exposure of male rats.

Table 2. Fetal Ossification at 20th Day of Pregnancy

Group/No. of Animals	Time+ Dose (mg/kg)	Vertebrae	Ribs	Sternebrae	Tail	Metacarpus	Metatarsal
		Average number of bones per fetus					
Control n=42	3	19.5±0.13	12.2±0.04	5.9±0.04	1.8±0.13	3.5±0.04	4.0±0
Metazin n=33	3 1.7	19.9±0.38	11.6±0.11*	5.2±0.11*	1.4±0.22	3.3±0.05*	4.0±0
Metazin n=33	3 17.0	19.2±0.16	12.0±0.60	5.6±0.08*	1.6±0.08*	3.2±0.08*	4.0±0
Cotofor n=42	3 0.48	19.9±0.27	12.6±0.11*	5.9±0.08*	1.9±0.11	3.5±0.05*	4.0±0
Cotofor n=31	3 4.78	19.2±0.13	12.8±0.08	5.7±0.04*	1.2±0.08	3.2±0.08*	3.9±0.04*
Cotofor n=11	3 47.8	18.9±0.22*	12.4±0.22	5.4±0.22*	2.2±0.22	3.4±0.11	4.0±0
Control n=51	9 -	19.9±0.10	12.9±0.03	5.8±0.07	1.6±0.07	3.4±0.03	4.0±0
Cotofor n=38	9 0.48	18.3±0.19	12.6±0.12	6.0±0	1.7±0.12	3.5±0.06	4.0±0
Cotofor n=30	9 4.78	17.4±0.18*	12.1±0.13*	5.9±0.06	1.5±0.13	3.3±0.06	4.0±0

*P <0.05; + Exposure time in months.

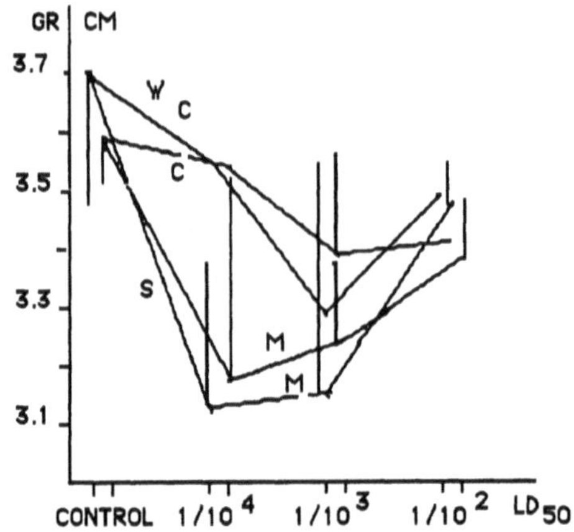

Figure 5. Effect of Metazin and Cotofor on fetal weight and size (craniocaudal) based on 3 month exposure of male rats.

Evaluation of external and internal organs for malformations by Wilson's method (Wilson and Warkany, 1965) did not present notable abnormalities. Hence, the data is not presented.

To find out the effect of the paternal genome on the formation and development of placenta, we measured the weight and diameter of placenta. No significant changes were noted in those observations either (Figure 6).

DISCUSSION

The mechanism of paternal exposure which causes embryonic loss and developmental delay of fetuses is ambiguous. One of several hypothesis, concerning this phenomenon is genomic imprinting. Specificity of the effect of paternal and maternal genome can be the cause in the differences of gametes' input in the embryo (Thompson, *et al.*, 1991). Since there are many more stem cell divisions in males than in females, there are more opportunities for new paternal mutations. According to Thompson and Thompson (1991), approximately one in ten sperm carries a new deleterious mutation. Fortunately, most of these mutations are recessive or lethal and thus are not phenotypically apparent in subsequent conceptions. Nevertheless, some genetic disorders are transmitted more often paternally than through the mother.

Our studies revealed male mediated reproductive toxicity and fetal developmental delay. We do not know the reason why the effects of these pesticides on fetal body weight and ossification is more pronounced in the three month studies, than in the nine month studies. If the post-implantation losses were higher in the nine-month experiment, we could hypothesize that the larger size of the fetuses was due to greater available space and less competition for nourishment in the uterus for the remaining fetuses. However, the post-implantation losses did not differ significantly in the experiments. Nevertheless, the test data of the three month experiments is compelling enough to suggest the existence of male-mediated developmental toxicity in this experiment.

The studies also indicated that the sperm quality and function and observation of the germ cells in their developmental stages are valuable and sensitive venues for

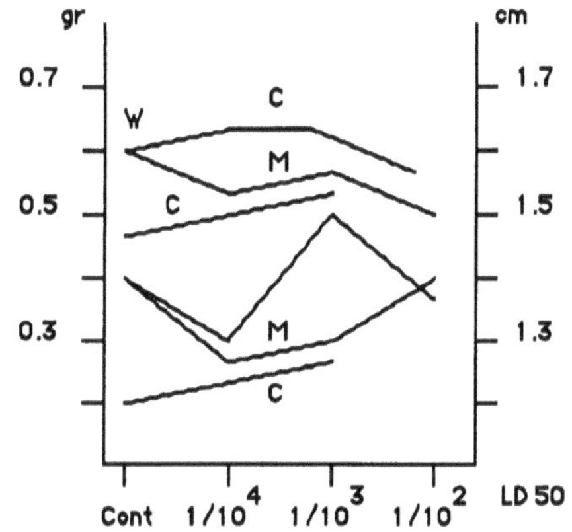

Figure 6. Effect of Metazin and Cotofor on the weight (gr) and diameter of placenta (cm).

determination of male mediated reproductive toxicity. The primordial germ cells in the male go through many hundreds of divisions throughout its adult life. In contrast, oogenesis is suspended in prophase I (dictyotene) at birth. The whole mitotic division process (only about 30 mitoses to develop an ova from the primordial germ cells) is completed by the third month of neonatal life of the human female (Thompson, *et al.*, 1991). Therefore the DNA of male germ cells are exposed to endogenous and exogenous factors (genotoxicants) much more, throughout the male life span. DNA damage and reduced response of DNA repair mechanisms can make the male germ cells life-long targets to environmental factors which can often produce changes in quality and function of sperm and testicles, developmental delays of fetuses and pre-implantation losses, as suggested by this study.

REFERENCES

Budavari, S., editor, (1989) The Merck Index, An Encyclopedia of Chemicals, Drugs and Biological Substances 11th edition, Merck and Co, Rahway, N.J.,USA.

Gevorkyan, S.G., Makaryan, A.S., Khechumova, R.M., Vermishyan, Z.M., Aghajanyan, S.M., (1977a) Several aspects of toxicity of herbicides from symm-triazine group. In the materials of symposium "*Okhrana truda i zdorovie selskogo naselenia*," 16-17 April, Baku, 85-87.

Gevorkyan, S.G., and Khechumova, R.M., (1977) Cotofor. Results of studies of toxicity and hygiene of new pesticide, *Materiali 17 ogo plenuma*, Moscow.

Gevorkyan, S.G., Vartanian, M.V., Mosyan, I.A., Movsesyan, M.A., and Stepanyan, G.S., (1977b) Experimental data of gonadotropic action of symmtriaszine pesticides, *Actualnie Voprosi Gigieni Primenenia Pestisidov v Razlichnikh Klimto-Geograficheskikh Zonakh*; 155.

Hurly, L.S., (1975) "Demonstration of A. Alizarin Staining of Bone" (Revised), *Supplement to Teratology Workshop Manual, Berkley*, California January 25-30, pp 121-122.

Makaryan, A.S., (1980) Establishment of hygiene standards for Metsin in the food and water, *Hyginia i Sanitaria*, # 8 , Moscow, p 74-75.

Makarian, A.S., and Konobeeva, G.I., (1979) Studies of mutagenic herbicide Metazin *Biologicheskii journal Armenii*, v. 32, #10, p 1039.

Moore, K.L., (1989) "Before We Are Born: Basic Embryology and Birth Defects," W.B. Saunders Co., Philadelphia, London.

Pavlova, A.V., ed., (1986) Directory of pesticides (Hygiene implementation and toxicology), 3rd edition, Ourajai, Kiev.

Sanotskii, I.V., (1970) editor, "Methods of Determination Toxicity and Danger of the Chemicals," Moscow.

Thompson, M.V., McInnes, R.R., and Willard, H.F., (1991) "Thompson and Thompson: Genetics in Medicine," W.B.Saunders Co. Philadelphia, London.

Vartanian, M.V., and Konobeeva, G.I., (1977) The study of mutagenic properties of symm-triazine pesticides. Man and Technology of the USSR, Kiev 12.

Vartanian, M.V., and Gevorkyan, S.G., (1978) Mutagenic activity of symm-triazinepesticides; *15th International Congress of Genetics*. Moscow (Abstract).

Venitt, S., and Parry, J.M., edited, (1984) Mutagenicity testing a practical approach, IRL Press, Oxford, Washington DC.

Wilson, J.L., and Warkany, Z., (1965) "Teratology principles and techniques," University of Chicago Press, Chicago, Illinois.

NATIONAL TRANSPLANTATION PREGNANCY REGISTRY: OUTCOMES OF PREGNANCIES FATHERED BY MALE TRANSPLANT RECIPIENTS

Karl M. Ahlswede[1], Beth Anne Ahlswede[1], Bruce E. Jarrell[2], Michael J. Moritz[1], and Vincent T. Armenti[1]

[1]Department of Surgery, Thomas Jefferson University,
605 College Building, 1025 Walnut Street
Philadelphia, PA 19107
[2]Department of Surgery, College of Medicine
University of Arizona
Tucson, AZ 85724*

INTRODUCTION

Previous reports of male renal transplant recipients who have fathered pregnancies total 71 pregnancies. All report greater than 80% live births and none reported complications or malformations in the offspring. Only eleven pregnancies were fathered by renal transplant recipients maintained on cyclosporine A (Sandimmune).

The National Transplantation Pregnancy Registry has been established to study pregnancy outcome in transplant recipients. The registry population includes female transplant recipients who have become pregnant and male transplant recipients who have fathered a pregnancy. Data from the male fathered pregnancies will be reported herein.

METHODS

Data were derived from recipient questionnaires and discharge summaries. In the first 15 months, the registry accumulated data from 204 male recipients who fathered 288 post-transplant pregnancies resulting in 290 evaluable outcomes (2 pairs of twins). All recipients were maintained on one or more of the following immunosuppressants: cyclosporine A, azathioprine, and prednisone. All but 14 of the 204 recipients were taking at least 2 immunosuppressant drugs.

---

* Supported in part by a grant from Sandoz Pharmaceuticals.

Male-Mediated Developmental Toxicity, Edited by D.R. Mattison
and A.F. Olshan, Plenum Press, New York, 1994

RESULTS

The mean age at conception was 32±6.2 years with a mean transplant to conception interval of 3.7±3.6 years. Stratification of the study population by organ transplanted is shown in Table I.

Table 1. Male Transplant Recipient Population by Organ Transplanted

Organ Transplanted	Number of Recipients
Kidney	163
Heart	30
Liver	8
Pancreas-Kidney	3
Total	204

Pregnancy outcomes appear similar to the general population with respect to live birth, miscarriage, stillbirth and therapeutic abortion (Table 2). The incidences of premature, low and very low birthweight infants are also similar to the general population (Table 3).

Table 2. Outcomes of Pregnancies Fathered by Male Transplant Recipients

Live births	273	94%
Miscarriage	14	5%
Therapeutic abortion	1	0.3%
Stillbirths	2	0.7%
Total	290	100%

Table 3. Live Births in Pregnancies Fathered by Male Transplant Recipients

Premature (less than 37 weeks gestation)	28	10%
Low birthweight (less than 2500 grams)	11	4%
Very low birthweight (less than 1500 grams)	0	0

Malformations were noted in nine (3.3%) live births (Table 4), while complications were noted in nineteen (7%) different live births (Table 5).

Table 4. Malformations in Live Births Fathered by Male Transplant Recipients (n=9)

Ventricular septal defect (n=2)
Other (n=1)
Ventricular septal defect, Patent ductus arteriosus
Small, dysfunctional kidneys
Absence of left radius and bilateral first metacarpals
Bilateral polycystic kidney disease, mild pulmonic stenosis
Down's syndrome
Potter's syndrome
incomplete aortic arch

Table 5. Complications in Live Births Fathered by Male Transplant Recipients (n=19)

Mild jaundice (n=5)
Respiratory distress (n=4)
Pyloric stenosis (n=2)
Other (n=1)
Hematuria
Seizures
Pneumomediastinum, mild respiratory distress
Heart murmur, jaundice
Psoriasis
Herpes, shoulder dystocia
Newborn pneumonia
Newborn fever of unknown origin

Neonatal deaths were seen in two live births. One in the infant with Potter's syndrome and the other in the smaller twin of a pair born prematurely with respiratory distress at birth.

Mean gestational ages and birthweights of infants of fathers maintained on cyclosporine A regimens (39 ± 1.9 weeks, 3370 ± 574 grams) did not significantly differ from infants of fathers on non-cyclosporine A regimens (39 ± 1.9 weeks, 3444 ± 543 grams).

DISCUSSION

Pregnancies fathered by male transplant recipients have outcomes, complication rates and malformations which appear similar to those of the general population. Evaluation of the offspring of our male recipients has not to date identified any specific problems related to immunosuppressive regimen. Our data are in agreement with previously published reports, but document a larger number of subjects taking a wider array of immunosuppressive regimens.

Fertility of male transplant recipients both pre and post-transplant is an issue which deserves further analysis. Sexual dysfunction has been reported in male renal transplant recipients, but the incidence of reproductive ability is unknown. Problems that originate with faulty spermatogenesis, maturation and transport of the sperm, or pre and post-implantation losses of the gamete may be suspected if male transplant recipients were found to have a high rate of infertility. Further studies are needed to determine fertility rates in this population.

Surveillance of the offspring of male transplant recipients through the transplant registry will provide a means to identify potential male-mediated effects related to the original disease process, the function of the transplanted organ, chronic pharmaceutical immunosuppression and other associated conditions.

REFERENCES

Cockburn, Krupp, P., Monka, C., 1989, Present experience of Sandimmune in pregnancy, *Transplant Proc.* 21(4):3730-3732.

Golby, M., 1970, Fertility after renal transplantation, *Transplantation*. 10(3):201-207.

Penn, I., Makowski, E.L., 1981, Parenthood in kidney and liver transplant recipients *Transplant Proc.* 13(1):36-39.

Schover, L.R., Novick, A.C., Steinmuller, D.R., Goormastic, M., 1990, *J of Sex and Marital Ther.* 16(1):3-13.

OCCUPATIONS OF FATHERS BEFORE CONCEPTION AND THE RISK OF TESTICULAR CANCER IN THEIR SONS

Julia A. Knight, Loraine D. Marrett, and Hannah K. Weir

Department of Preventive Medicine and Biostatistics
McMurrich Building, 3rd floor
University of Toronto
Toronto, Ontario, Canada M5S 1A8
and
Division of Epidemiology and Statistics
Ontario Cancer Treatment and Research Foundation
7 Overlea Boulevard
Toronto, Ontario, Canada M4H 1A8

INTRODUCTION

Testicular cancer is a disease which occurs largely among young adult, white males. The incidence rate in Ontario in 1988 was 4 per 100,000 men per year. Among men aged 15-29 the rate was 7 and among men aged 30-44 the rate was 8 per 100,000. Testicular cancer has attracted recent attention because the incidence has been increasing in high incidence populations in Europe, North America, and Australia, yet the etiology remains largely unknown. Nearly all of these cancers arise in the germ cells. Germ cell tumours of the testis can be subdivided into distinct histologic subgroups. About half of the tumours are seminomas while the other half are often grouped together as nonseminomas and includes teratomas and choriocarcinomas (a small proportion of tumours have mixed seminoma and nonseminoma histology). These two groups have distinct age distributions with nonseminomas occurring more commonly at younger ages than seminomas which peak during the 30's. This suggests that they may have different etiologies or, if they have the same etiology, that the type of tumour depends on the age at which it develops.

Many studies have investigated possible associations between paternal occupation and childhood cancer. Although testicular cancer is not a childhood cancer, it does occur at much younger ages than most adult cancers. It is therefore possible that testicular cancer may be initiated in childhood or earlier, but require the occurrence of puberty to be expressed. In particular, events which occur in the germ

Male-Mediated Developmental Toxicity, Edited by D.R. Mattison
and A.F. Olshan, Plenum Press, New York, 1994

cell of the father may result in malignancy in the germ cells of the son. This might be true for some or all histologies.

METHODS

A case-control study of testicular cancer was carried out in Ontario. Cases were defined as all those aged 16 to 59 who were diagnosed in Ontario with a malignant germ cell tumour of the testis from 1987 to 1989. Nearly all of the cases had a confirming pathological report. Cases were identified through the Ontario Cancer Registry and permission to invite their participation in the study was obtained from their physicians. Controls were identified through provincial enumeration composite records maintained by the Ontario Ministry of Revenue and were randomly selected to frequency match the expected age distribution of cases in 5-year age strata. The control to case ratio was 2:1.

A mail questionnaire sent to all subjects included questions on demographic variables as well as possible risk factors. Subjects were also asked for permission to contact their mothers. If permission was granted the mother was sent a questionnaire which requested information on exposures before and during pregnancy and during childhood as well as the mother's reproductive history. Included in this was a request for the job title and industry or business of the father for up to two jobs held in the year before the pregnancy with the subject. Data were collected predominantly by telephone interviews with mothers, although a few mothers filled in their responses and mailed the completed questionnaires to the study office.

Job titles and industries was entered alphanumerically into the computer and alphabetized lists were generated for coding. Jobs were assigned 4-digit codes using the 1980 Canadian Standard Occupational Classification and industries were given 2-digit codes using the 1980 Canadian Standard Industrial Classification. All coding was done without knowledge of case-control status. Codes were then amalgamated into 54 job groups and 49 industry groups. The job groups roughly conform to the three-digit codes.

Each job and industry group was analyzed separately. Fathers were considered "exposed" to a given job or industry group if they had worked in that group during the year before pregnancy; all others were "unexposed" to that job or industry. This was done for all cases combined as well as for seminomas and nonseminomas separately. In the histologic subgroup analysis the control to case ratio was close to 3:1. Analysis was done using unconditional logistic regression except where there were no case fathers or no control fathers, when Fisher's exact test was used to obtain p-values. Jobs and industries whose contribution to the univariate model had $p \leq .15$ were then adjusted for the education level and/or the age of the subject. Age was incorporated as a continuous variable and education was defined as any post-secondary versus no post-secondary education. Jobs, industries, age, and education were combined in multivariate models to see which were the most important. The models were reduced by backwards stepwise elimination until all factors had $p \leq .10$.

RESULTS

There were 620 cases of primary testicular germ cell cancer and 1532 controls identified. Responses were received for 495 cases and 974 controls, representing 80% of cases and 64% of controls. There were 343 cases (55%) and 524 controls (34%)

who also had information provided by their mothers. The most common reason for no maternal information was that the mother had died. Of the 343 cases with mothers' data, 189 (55%) were seminomas and 154 (45%) were nonseminomas. Those with mixed histology were included in the nonseminoma group.

Table 1 shows the distributions of the fathers in the three groups, seminomas, nonseminomas, and controls, according to the age and education level of the subject. A higher proportion of seminomas had some post-secondary education compared with controls. This difference is not seen when nonseminomas are compared with controls. However, there are differences in age distributions which might account for this. Nonseminomas tended to be younger than controls and seminomas to be older; therefore the proportion who have completed their education would be expected to be the highest in seminomas and the lowest in nonseminomas.

Table 1. The Numbers of Fathers in Each Group According to the Age and Education Level of Their Sons

	Seminomas		Nonseminomas		Controls	
Age						
<20	1	(1%)	11	(7%)	51	(10%)
20-24	13	(7%)	55	(36%)	105	(20%)
25-29	55	(29%)	41	(27%)	123	(24%)
30-34	55	(29%)	20	(13%)	108	(21%)
35-39	37	(20%)	16	(10%)	72	(14%)
40-44	16	(9%)	10	(7%)	36	(7%)
45-49	8	(4%)	1	(1%)	15	(3%)
50+	4	(2%)	0	(0%)	14	(3%)
Total	189		154		524	
Education						
Elementary	3	(2%)	1	(1%)	15	(3%)
High School	54	(29%)	66	(43%)	232	(44%)
Community College	81	(43%)	52	(34%)	170	(32%)
University	51	(27%)	33	22%	107	(20%)

Jobs and industries for which the unadjusted p was $\leq .15$ when all cases were combined and in which least one case father and one control father worked are shown in Table 2 with odds ratios (OR) and 95% confidence intervals (CI) included. The most significant occupations were social workers, food and beverage preparers and servers, and stationary engineers, but no control fathers and four case fathers were social workers. Among textile, fur, and leather fabricators there were no case fathers and six control fathers while three case fathers and no control fathers were lawyers. The most significant industries were food products, metal products, and food and beverage services.

Seminoma

The results for seminomas alone are shown in Table 3. Food and beverage preparers and servers, metalworkers, and material handlers represented the three job groups with unadjusted OR significantly ($p \leq .05$) different from one. There were two case fathers and no control fathers who were social workers. Food products and food and beverage services were the most significant industries for seminomas. Tobacco products had two case fathers and no control fathers. When adjusted for the age and education of the subject, OR estimates for the jobs remained essentially unchanged while the OR for the industries increased somewhat.

Table 2. Jobs and Industries of Fathers of all Cases for which Unadjusted $p \leq .15$ and in which at least One Case Father and One Control Father Worked

	Cases n	Controls n	Odds Ratio	95% Confidence Interval
Jobs				
Engineers	21	19	1.74	0.92-3.29
Food & Beverage Preparers & Servers	6	1	9.36	1.12-78.06
Food & Beverage Processors	12	10	1.87	0.80-4.38
Wood Processors	5	1	7.77	0.90-66.83
Metalworkers	8	5	2.49	0.81-7.68
Rail Transport Workers	1	7	0.22	0.03-1.77
Material Handlers	10	7	2.23	0.84-5.91
Stationary Engineers	9	4	3.52	1.08-11.52
Industries				
Food Products	20	13	2.45	1.20-4.99
Metal Products	10	3	5.24	1.43-19.17
Financial Industry	5	18	0.42	0.15-1.13
Business Services	6	18	0.50	0.20-1.28
Food & Beverage Services	12	5	3.78	1.32-10.83

All of the jobs and industries shown in Table 3 were combined in a model along with the age and education of the subject. The results of the model which includes all factors having $p<.10$ are shown in Table 4. The age and education of the subject remain predictive of the occurrence of seminoma, along with some of the occupations and industries. Metalworkers and material handlers as well as those who worked in the food products industry or in food and beverage services had significantly increased risk of having a son with seminoma. There was also some evidence of increased risk for fathers who were engineers or who worked in the metal products industry, although OR were not significantly greater than one.

Table 3. Jobs and Industries of Fathers of Seminoma Cases for which Unadjusted p≤.15 and in which at least One Case and One Control Father Worked

	Cases n	Controls n	Odds Ratio	95% Confidence Interval
Jobs				
Engineers	12	19	1.82	0.87-3.82
Food & Beverage Preparers & Servers	4	1	11.40	1.27-102.7
Metalworkers	6	5	3.43	1.04-11.38
Material Handlers	8	7	3.29	1.18-9.21
Industries				
Food Products	12	13	2.69	1.21-6.02
Metal Products	4	3	3.79	0.84-17.11
Food & Beverage Services	8	5	4.64	1.50-14.36

Table 4. A Multivariate Model for Seminoma Including the jobs and Industries of Fathers

	Odds Ratio	95% Confidence Interval
Age	1.04	1.02-1.07
Post-secondary education	2.20	1.51-3.22
Jobs		
Engineers	2.09	0.96-4.59
Metalworkers	4.18	1.20-14.58
Material Handlers	3.55	1.18-10.65
Industries		
Food Products	3.86	1.66-8.94
Metal Products	4.21	0.80-22.10
Food & Beverage Services	5.86	1.81-18.96

Nonseminoma

Univariate results for nonseminomas can be seen in Table 5. Jobs with the most significant OR included social workers and lawyers of which there were two and three case fathers respectively and no control fathers, wood processors, and stationary engineers. For three industries the p-value was ≤.05, namely metal products, communications, and education. After adjustment for age most OR changed little. The OR for teachers decreased the most, but still exceeded 3.00, while the OR for food and beverage preparers and servers increased.

Results of the multivariate modelling retaining factors with p≤.10 are shown in Table 6. The OR for age was significantly less than one while the OR for wood processors, stationary engineers, and four industries (primary metal, metal products, communications, and education) were significantly high. There was also a suggestion of increased risk for fathers who were food and beverage preparers and servers or who worked in the food or beverage products industries.

Table 5. Jobs and Industries of Fathers of Nonseminoma Cases for which Unadjusted p≤.15 and in which there was at least One Case Father and One Control Father

		Cases n	Controls n	Odds Ratio	95% Confidence Interval
Jobs					
	School Teachers	4	4	3.47	0.86-14.04
	Salesmen	8	48	0.54	0.25-1.18
	Food & Beverage Preparers & Servers	2	1	6.89	0.62-76.47
	Farmers and Farm Workers	8	48	0.54	0.25-1.18
	Wood Processors	3	1	10.40	1.07-100.7
	Stationary Engineers	6	4	5.28	1.47-18.95
Industries					
	Agriculture	8	48	0.56	0.26-1.20
	Food Products	8	13	2.15	0.87-5.28
	Beverage Products	3	3	3.44	0.69-17.22
	Wood Products	6	8	2.61	0.89-7.63
	Paper Products	6	9	2.31	0.81-6.61
	Primary Metal	12	24	1.76	0.86-3.60
	Metal Products	6	3	7.02	1.73-28.42
	Communications	7	9	2.72	0.99-7.42
	Wholesale & Retail	11	62	0.57	0.29-1.11
	Education	7	9	2.72	0.99-7.42
	Food & Beverage Services	4	5	2.76	0.73-10.41

Table 6. A Multivariate Model of Nonseminoma Including the Jobs and Industries of Fathers

		Odds Ratio	95% Confidence Interval
	Age	0.96	0.94-0.99
Jobs			
	Food & Beverage Preparers & Servers	10.09	0.90-112.7
	Wood Processors	13.44	1.35-133.4
	Stationary Engineers	6.63	1.81-24.24
Industries			
	Food Products	2.50	0.99-6.30
	Beverage Products	4.30	0.85-21.81
	Primary Metal	2.18	1.05-4.52
	Metal Products	8.86	2.16-36.39
	Communications	3.21	1.16-8.91
	Education	3.08	1.11-8.56

DISCUSSION

Methodologic Issues

One issue that needs to be discussed is the small proportion of mothers from whom information was obtained. There are two potential problems. One is that the mothers from whom information was available were not representative of all case and control mothers. For this analysis in particular this would require that wives of husbands in some occupations were more or less likely to have been interviewed than wives of husbands in other occupations. This would result in a nonrepresentative occupational distribution. It is possible that this could have occurred, but it is not obvious how this might have affected the results. Several steps were required to obtain information from the mothers: the subject had to be located and agree to participate, the mother had to be alive, the son had to give permission to contact the mother, and the mother had to agree to participate. The last step was not a problem since over 90% of mothers contacted agreed to participate. However, a number of factors may operate at each step to eliminate some mothers and this can affect the occupational distribution of the father in different ways. The other major problem which is related is that there could be systematic differences between case and control respondents which creates a bias. This is a concern here since there were higher proportion of mothers who provided information for the cases than for the controls which may mean that the control mothers were less representative than the cases. This could lead to spurious associations or could obscure associations. For these reasons as well as the possibilities of chance associations, these results should be considered to be preliminary and the hypotheses generated should be subjected to further study.

Confounding must always be considered, although there are no known factors which are likely to confound the results. The only well-established risk factor for testicular cancer is cryptorchidism. However, only a small proportion of cases have a history of this. In addition, if paternal occupation were associated with this it would be more likely to be part of the same mechanistic pathway rather than being affected by confounding.

Since there were 54 jobs and 49 industries examined, one would expect a number of significant results to occur by chance alone. However, it is possible that some of these results are "real". This analysis is intended to generate hypotheses. Current knowledge and examination of the data can be used to try to evaluate results suggesting which ones might represent real phenomena, but only further studies can provide evidence confirming or rejecting hypotheses. Additionally, laboratory studies may indicate biological plausibility. It may also be that workplace exposures of fathers constitute a risk factor for testicular cancer, but not a major one.

When considering these results it should be kept in mind that job and industry titles give only a rough indication of exposure. Within each job or industry group individuals will experience heterogeneous exposures since a variety of jobs may be included in one job group or one industry group. Job and industry classifications are two different methods of grouping the same people according to the work they do and for what purpose. Therefore, using job and/or industry titles as a surrogate for exposure results in misclassification which tends to obscure associations. Both jobs and industries were included in models because the different codes tend to describe different groups of people even though there may be considerable overlap between some job and industry groups. Those with the same job code will be doing similar work, but the industry may or may not vary, while those in the same industry will work in a variety of jobs. The goal was to find which groups had the strongest associations with testicular cancer in order to determine exposures which might be relevant. When there was overlap one group would take precedence. There are three possible reasons for this. An industry might dominate because it includes a number of jobs with similar exposures and, therefore, the larger number of subjects results in higher significance compared to individual jobs. Alternatively, a job group might be more significant because the individuals included are more homogeneous than the industry groups and, therefore, the "noise" is reduced. Finally, it may just depend on who is being grouped together. An example of this is telecommunications installers and repairers who are grouped with electricians in the job classification and grouped with others who work in telecommunications and the post office in the industrial classification. The number of electricians overwhelms the number of telecommunications workers in the former. In general, jobs and industries associated with food and metals tended to be important and these will be discussed separately.

Food and Beverages

There were two subgroups in the food products industry in which fathers of subjects with seminoma predominated. Three of the case fathers worked in bakeries and none of the control fathers despite the fact that there are about four times as many control as case fathers. Bakers would be exposed to heat and agents used to keep the work area clean and free of pests. The second subgroup was the dairy industry, where five case fathers and only three control fathers worked; only about one case father would be expected on the basis of the frequency of control fathers. However, most of the men in this group drove delivery trucks. If there is risk associated with driving a truck it should have appeared in the occupational group "truck drivers", although it is possible that truck drivers in the dairy industry are exposed to materials used to keep the trucks clean and which may differ from exposures of other truck drivers. The jobs included in the food and beverage services industry were cook, waiter, dishwasher, restaurant manager, hotel manager, and deliverer. Although no particular category stands out, when all those who definitely worked in restaurants or equivalent are combined, there is a large excess of case fathers. These people would all have some exposure to food and agents used to keep the premises clean and free of pests.

Among fathers of nonseminomas there were only three who had jobs involving preparing and serving food, but the two case fathers were both cooks while the control father was a bartender. The association with food related to jobs and industries seems to be common to both histological subgroups, although the actual jobs the fathers held varied between the two groups. Both the food products and the beverage products industries had elevated OR for nonseminomas. As with seminoma, there were more case fathers than expected who worked for a dairy, although not to the same extent. Two of the three case fathers in the dairy industry were actual dairy workers while the third case father and the three control fathers were involved in the transportation. Unlike seminoma, there were slightly more case fathers than expected who worked in meat packing. There were only a small number of fathers in beverage products. However, two of the case fathers were brewery workers compared to only one control father who worked as a brewery salesman. It is not clear whether these subgroups are important or whether there is some exposure related to food handling in general that is important for both seminoma and nonseminoma.

Metals

Among the metalworkers, there was an excess of welders among fathers of seminomas. It is not known what type of welding they did, but welders do have a variety of exposures which should be considered. There were four seminoma case fathers who worked in the metal products industry: a packer, a machinist in the tool and die industry, a tool and die maker, and a tool and die supervisor. The three control fathers worked as a plant manager, a tool designer, and an accountant. There seems to be decreased risk with higher socioeconomic status (SES) jobs, but metal compounds might constitute a relevant exposure which would be more common in the lower SES jobs.

The primary metal industry as well as the metal products industry produced significantly increased OR for nonseminoma. Both industries include people working in a wide variety of jobs. Among those in the metal products industry both case and control fathers tended to have higher SES jobs which probably have fewer exposures. The only excess of case fathers seems to be among foundry workers. Heat may be important here, although there would also be potential for exposure to metal compounds. Like food, work in metal-related jobs appears to increase risk of both seminoma and nonseminoma in sons.

Other Exposures

Occupations in engineering included a wide variety of jobs. Possibly noteworthy is the fact that there were two chemical engineers among fathers of cases and only one related job, namely a draftsman in the chemical industry, among the fathers of controls. Material handlers included crane operators, hoisting engineers, fork lift operators, and packers; no category and no particular industry seems overrepresented among case fathers.

There were a number of other jobs and industries in the nonseminoma model. There were only four fathers classified as wood processors and three of them were case fathers. There are a variety of exposures which can result from wood treatment, including heat, fungicides, and tar and other petroleum-based products. Pentachlorophenol and metal compounds have been commonly used to treat wood. Another job group with an elevated OR is stationary engineers. Although this group includes other types of jobs, the large OR for this group is clearly due to the excess of stationary engineers among case fathers: there were four case fathers who were stationary engineers and one who was a steam plant attendant, when less than one would have been expected based on the number of control fathers in this group. One exposure which might be relevant for this group is heat. The communications industry includes those who work in telecommunications, radio and television broadcasting, and the post office. The excess of case fathers was clearly associated with telecommunications. Also, two of the four control fathers were office managers while the case fathers included two installers, a lineman, a construction foreman, and an engineer. Electromagnetic fields is one possible relevant exposure. The other industry group which may be important for nonseminoma was education. Most, but not all, of the fathers in this group were teachers. It is not clear why this group should be at increased risk.

CONCLUSIONS

In summary, it appears that there are many similarities between the results for seminoma and for nonseminoma. This could indicate either that there are not major etiologic differences between the two groups or that the cancer initiation process is similar for the two and that later events are responsible for the histologic differences. There are, however, a few differences between the results for the two groups. Some of the recurrent themes are heat and metals as well as exposures associated with food which may include disinfectants and pesticides. In future studies it may be more useful to pursue these types of specific exposures rather than concentrating on job and industry titles, although confirmation of these results is warranted.

ACKNOWLEDGMENTS

The authors would like to acknowledge the financial support of the Ontario Cancer Treatment and Research Foundation and the Health Care Systems Research Program of the Ontario Ministry of Health. In addition, the authors would like to note that the opinions contained in this paper are those of the authors, and no official endorsement by the Ministry is intended or should be inferred.

REFERENCES

Boyle, P., Kaye, S.B., and Robertson, A.G., Changes in testicular cancer in Scotland, *Eur. J. Cancer Clin. Oncol.* 23:827 (1987).

Brown, L.M., Pottern, L.M., Hoover, R.N.,. Devesa, S.S., Aselton, P., and Flannery, J.T., Testicular cancer in the United States: trends in incidence and mortality, Int. J. Epidemiol. 15:164 (1986).

Gudlaugsson, E., Magnússon, B., and Hallgrímsson, J., Tumours in Iceland. 13. Malignant tumours of the testis, *APMIS* 98:173 (1990).

Heimdal, K., Fosså, S.D., and Johansen, A., Inreasing incidence and changing stage distribution of testicular cancer in Norway 1970-1987, *Br. J. Cancer* 62:277 (1990).

National Cancer Institute of Canada. "Canadian Cancer Statistics 1990," Toronto (1990).

Nethersell, A.B.W., Drake, L.K., and Sikora, K., The increasing incidence of testicular cancer in East Anglia, *Br. J. Cancer* 50:377 (1984).

Østerlind, A., Diverging trends in incidence and mortality of testicular cancer in Denmark, 1943-1982, *Br. J. Cancer* 53:501 (1986).

Pike, M.C., Chilvers, C.E.D., and Bobrow, L.G., Classification of testicular cancer in incidence and mortality statistics, *Br. J. Cancer* 56:83 (1987).

Roush, G.G., Holford, T.R., Schymura, M.J., and White, C, "Cancer Risk and Incidence Trends: The Connecticut Perspective," Hemisphere, Washington (1987).

Spitz, M.R., Sider, J.G., Pollack, E.S., Lynch, H.K., and Newell, G.R., Incidence and descriptive features of testicular cancer among United States whites, blacks, and Hispanics, 1973-1982, *Cancer* 58:1785 (1986).

Stone, J.M., Cruickshank, D.J, Sandeman, T.F., and Matthews, J.P., Trebling of the incidence of testicular cancer in Victoria, Australia (1950-1985), *Cancer* 68:211 (1991).

Swerdlow, A.J., and Skeet, R.G., Occupational associations of testicular cancer in southeast England, *Br. J. Ind. Med.* 45:225 (1988).

TWO-DIMENSIONAL ELECTROPHORESIS OF PROTEINS: DETECTION AND CHARACTERIZATION OF MALE-MEDIATED GENOTOXICITY

Carol S. Giometti

Biological and Medical Research Division
Argonne National Laboratory
Argonne, IL 60439

INTRODUCTION

Damage to structural or regulatory genes in germ cell DNA is expected to produce abnormalities in the abundance or structure of the proteins whose synthesis is controlled by those genes in subsequent progeny. Detection of such protein abnormalities could be useful for monitoring mutation frequencies in the offspring of exposed populations, for the molecular characterization of DNA damage caused by different environmental or occupational exposures (e.g., ionizing radiation, toxic chemicals), and for the correlation of such damage with deleterious effects in subsequent generations. The biochemical specific locus test, in which the activities of approximately 30 different proteins can be monitored by one-dimensional electrophoresis, has shown that heritable protein effects can be detected in mice sired by chemically exposed males (Johnson and Lewis, 1981).

Two-dimensional gel electrophoresis (2DE) of proteins coupled with computer-assisted data analysis provides a tool to enlarge the number of proteins that can be examined per sample to at least several hundred (Celis and Bravo, 1984). The capability to resolve so many proteins from each sample suggests that effects on hundreds of structural and regulatory genes can be monitored. This is in contrast to the one-dimensional electrophoresis system which, due to the limited number of proteins analyzed, allows monitoring of only those genes related to the functional integrity of the 30-40 proteins included in the detection system. 2DE would seem, therefore, to offer the opportunity to monitor a much larger sector of the genome for mutation events than other available techniques, although the exact number and identity of a majority of the genes and protein gene products are unknown.

MUTATION DETECTION USING 2DE OF PROTEINS

The Protein Mapping Group at Argonne National Laboratory is using 2DE to characterize the types of mutations that can be detected as alterations in protein expression, thus evaluating the applications of this approach to both the detection and

Male-Mediated Developmental Toxicity, Edited by D.R. Mattison
and A.F. Olshan, Plenum Press, New York, 1994

characterization of genotoxic events. The studies involve male-mediated heritable mutations in mice, induced by exposure of sires to chemicals (*N*-ethyl-*N*-nitrosourea; ENU) or ionizing radiation (gamma rays or fission-spectrum neutrons). By using inbred mice, results from the 2DE studies can be readily compared with those obtained using the seven specific locus test (Ehling, 1978; Russell *et al.*, 1979) or the biochemical specific locus test (Johnson and Lewis, 1981). In addition, protein expression in mice carrying mutations detected by other methods can be analyzed to characterize the detection limits of 2DE and to further define the molecular nature of these mutations.

The general strategy for detection of mutations as alterations in 2DE patterns of mouse liver proteins has been described in detail elsewhere (Giometti and Taylor, 1991) and is shown schematically in Figure 1.

Briefly, male mice (C57BL/6JANL; control or exposed to mutagen) are bred with untreated female mice (BALB/cJANL). At an appropriate age, partial hepatectomies are done on the F_1 offspring. The liver is homogenized directly in a solution (pH 9.5) containing 9 M urea, 1% dithiothreitol, 2% ampholytes (LKB, pH 9-11), and 4% Nonidet P40 (a nonionic detergent) in a ratio of 1 ml per 125 mg of wet tissue weight. An aliquot of the supernatant recovered after centrifugation of the homogenate (435,000 × *g* in a Beckman TL100 ultracentrifuge) is then used for 2DE. The first-dimension isoelectric focusing (IEF) and second-dimension sodium dodecyl sulfate polyacrylamide gel electrophoresis (DALT) separations are done essentially as described by Anderson and Anderson (1978a, 1978b). The proteins are detected in the gels by using Coomassie Blue R250 staining (Giometti *et al.*, 1987). The protein patterns are digitized by scanning the destained gels using an Eikonix 1412 flat-bed scanner interfaced with a VAXstation 3200. Computer software developed by Dr. John Taylor at Argonne National Laboratory (Anderson *et al.*, 1981; Taylor *et al.*, 1981) is used to generate spot lists from the image data. The individual spot lists are

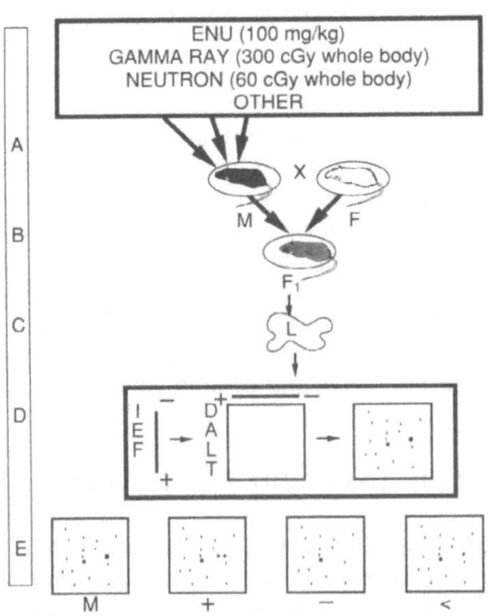

Figure 1. Schematic of experimental design for mouse mutagenesis studies done by using two-dimensional gel electrophoresis.

then matched to a common master spot list (M in Fig. 1E) so that each protein spot is assigned an identification number. Once matched, the spot lists can be searched for new proteins ("+" in Fig. 1E), missing proteins ("-" in Fig. 1E), and proteins altered in abundance ("<" in Fig. 1E). These searches are done using the GR42 software for interactive data analysis of 2DE patterns, also developed by Dr. Taylor at Argonne (Giometti and Taylor, 1991).

Using this general approach, a database containing 2DE patterns from approximately 1500 CBF_1 mice has been constructed. Figure 2 shows a typical pattern of mouse liver proteins displayed using a spot list. Searches of these patterns have revealed five isoelectric point variants among the F_1 offspring of ENU-treated sires, three quantitative variants among the F_1 offspring of neutron-exposed sires, and no protein variants of any type among the F_1 offspring of sires exposed to gamma rays or of untreated sires (Giometti *et al.*, 1987; Giometti *et al.*, 1988; Giometti and Taylor, 1991; Giometti *et al.*, 1992b; Taylor and Giometti, 1992). On a per mouse basis, the yield of ENU-induced mutation events detected by 2DE (1 mutation per 65 offspring screened; Giometti and Taylor, 1991) is comparable or slightly higher than that found using the seven specific locus test (1 mutation per 217 offspring screened; Russell *et al.*, 1979) or the biochemical specific locus test (1 mutation per 119 offspring screened; Johnson and Lewis, 1981). On a per locus basis, however, assuming that the 2DE pattern shown in Fig. 2 represents several hundred distinct genetic loci (a conservative estimate), the number of mutations detected by 2DE is small relative to these other two tests.

In addition to analysis of liver proteins in the F_1 offspring sired by mice with known exposures, 2DE studies have been done on the liver proteins of mice known to carry heritable mutations originally induced by either chemical or radiation exposure of sires. Mixed results have been obtained in these studies. Searches of liver protein patterns from mice known to be heterozygous for eight different recessive lethal mutations induced by exposure of sires to either triethylene melamine, X-rays, or gamma rays failed to reveal any protein abnormalities (Giometti *et al.*, 1990). Similarly, no differences were detected in the fetal livers of mice heterozygous for the c^{3H} deletion around the albino locus, originally detected as a specific locus mutation

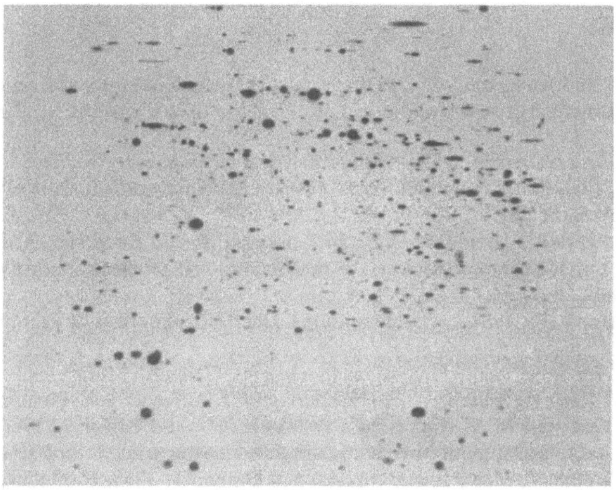

Figure 2. Computer display of two-dimensional electrophoresis pattern of CBF_1 mouse liver proteins.

in an offspring sired by a male exposed to X-rays. Numerous protein differences relative to wildtype littermates were observed, however, in the fetal and newborn livers of mice homozygous for the c^{3H} deletion (Giometti *et al.*, 1992a). These data suggest that mutation events involving regulatory regions of DNA produce multiple protein effects that are visible in the homozygous but not the heterozygous carriers.

PROSPECTS FOR 2DE IN ANALYSIS OF MALE-MEDIATED GENOTOXICITY

Unlike mutation detection methods which depend on the altered function of the products of well defined loci, 2DE allows analysis of hundreds of unidentified protein gene products based simply on their quantity within a given tissue. The relatively conservative number of ENU-induced mutations detected in the 2DE study of mouse liver proteins, compared to seven specific locus and biochemical specific locus test results, indicate that 2DE provides a significantly different perspective of genetic effects than that covered by these other methods. This unique perspective is further demonstrated when tissue from mice of a specific mutant phenotype is examined at the level of protein expression, revealing an entire subpopulation of proteins altered in homozygous carriers. This latter finding suggests that analysis of the protein expression in affected individuals using 2DE can provide detailed information relevant to the molecular lesions that cause phenotypic alterations. Such knowledge of the molecular lesions is an important part of defining the mechanisms of genotoxicity relevant to transmission of genetic damage to offspring. The mutation studies done thus far using 2DE would suggest that although mutation events can be readily detected as quantitative changes in protein expression, the technique is equally useful and perhaps more valuable as a tool for the characterization of the mechanics of mutation induction and transmission.

ACKNOWLEDGMENTS

This work was supported by the United States Department of Energy, Office of Health and Environmental Research, under contract number W-31-109-ENG-38.

REFERENCES

Anderson, N.G., and Anderson, N.L., 1978a, Analytical techniques for cell fractions. XXI. Two-dimensional analysis of serum and tissue proteins: Multiple isoelectric focusing, *Analyt. Biochem.* 85:331.

Anderson, N.L. and Anderson, N.G., 1978b, Analytical techniques for cell fractions. XXI. Two-dimensional analysis of serum and tissue proteins: Multiple gradient slab-gel electrophoresis, *Analyt. Biochem.* 85:341.

Anderson, N.L., Taylor, J., Scandora, A.E., Jr., Coulter, B.P, and Anderson, N.G., 1981, The TYCHO system for computer analysis of two-dimensional gel electrophoresis patterns, *Clin. Chem.* 27:1807.

Celis, J.E. and Bravo, R., 1984, "Two-Dimensional Gel Electrophoresis of Proteins," Academic Press, New York.

Ehling, U.H., 1978, Specific-locus mutations in mice, *in*: "Chemical Mutagens: Principles and Methods for Their Detection," A. Hollaender and F.J. deSerres, eds., Plenum Press, New York.

Giometti, C.S., Gemmell, M.A., Nance, S.L., Tollaksen, S.L., and Taylor, J., 1987, Detection of heritable mutations as quantitative changes in protein expression, *J. Biol. Chem.* 262:12764.

Giometti, C.S., Gemmell, M.A., Tollaksen, S.L., and Grahn, D., 1988, Heritable protein variants induced by exposure to ethylnitrosourea: Heritability, subcellular location, and tissue distribution, *Mut. Res.* 202:9.

Giometti, C.S., Gemmell, M.A., Taylor, J., Tollaksen, S.L., Angeletti, R., and Gluecksohn-Waelsch, S., 1992a, Evidence for regulatory genes on mouse chromosome 7 that affect the quantitative expression of proteins in the fetal and newborn liver, *Proc. Natl. Acad. Sci. USA* 89:2448.
Giometti, C.S., Taylor, J., and Tollaksen, S.L., 1992b, Mouse liver protein database: A catalog of proteins detected by two-dimensional gel electrophoresis, *Electrophoresis* (in press).
Giometti, C.S., and Taylor, J., 1991, The application of two-dimensional electrophoresis to mutation studies, *in*: "Advances in Electrophoresis," M.J. Dunn and B.J. Radola, eds., VCH Verlagsgesellschaft, New York.
Giometti, C.S., Tollaksen, S.L., Gemmell, M.A., Taylor, J., Hawes, N., and Roderick, T., 1990, The analysis of recessive lethal mutations in mice by using two-dimensional gel electrophoresis of liver proteins, *Mut. Res.* 242:47.
Johnson, F.M. and Lewis, S.E., 1981, Electrophoretically detected germinal mutations induced in the mouse by ethylnitrosourea, *Proc. Natl. Acad. Sci. USA* 78:3138.
Russell, W.L., Kelly, E.M., and Hunsicker, P.R., 1979, Specific locus test shows ethylnitrosourea to be the most potent mutagen in the mouse, *Proc. Natl. Acad. Sci. USA* 76:5818.
Taylor, J., Anderson, N.L., and Anderson, N.G., 1981, A computerized system for matching and stretching two-dimensional gel patterns represented by parameter lists, *in*: "Electrophoresis '81," R. Allen and P. Arnaud, eds., Walter de Gruyter and Co., New York.
Taylor, J., and Giometti, C.S., 1992, Use of principal components analysis for mutation detection with two-dimensional electrophoresis protein separations, *Electrophoresis* 13:162.

WORKSHOP REPORT ON MECHANISMS*

Robert L. Brent

Department of Pediatrics
Jefferson Medical College
Alfred I. DuPont Institute
1600 Rockland Road, Box 269
Wilmington, DE 19899

There are multiple and diverse mechanisms that may be involved in male mediated reproductive toxicity because of the large number of diseases that are included in the category of reproductive failure. The span of reproductive problems include:

- Spontaneous abortions
- Congenital malformations
- Genetic diseases in the offspring
- Malignancy in the offspring
- Prematurity
- Intrauterine growth retardation (IUGR)
- Stillbirth
- Neurobehavioral effects in the offspring
- Infertility

To complicate matters further, each of the above-mentioned reproductive problems can be caused by more than one mechanism. The known etiologies for reproductive problems include both genetic and epigenetic mechanisms. Furthermore, the environmental factors include a host of external chemical, drug, infectious and physical agents as well as intrinsic disease processes that may affect male and female germ cells during their development before fertilization and the embryo during gestation.

What are the possible mechanisms that could be responsible for the spectrum of environmentally induced male mediated reproductive toxicity? They include:

Direct effect on spermatogenesis:
Mutations (M)

---

* Workshop participants included: E. Carney, J. Ford, B. Hales, V. Huff, I. Morris, T. Nomura, B. Robaire, J. Rogers, and G. Szcech.

Male-Mediated Developmental Toxicity, Edited by D.R. Mattison and A.F. Olshan, Plenum Press, New York, 1994

Karyotype abnormalities (K)
Cytotoxicity (C)
Imprinting pattern alterations (I)
Epigenetic factors (E)
Sperm cells or ejaculate as vectors
Ejaculate containing toxic substances (Ej)
Sperm containing toxic substances (Sp)

Mutations can be induced by chemical or physical agents that produce reactive oxygen or hydroxyl radicals, toxic epoxides that reach the nucleus, electrochemical reactions from ionization in tissues and many other chemical reactions that could damage the genome, directly, or indirectly. Mutations in the paternal genome can be causally related to congenital malformations, genetic disease, cancer, stillbirth, prematurity, IUGR and sterility in the offspring. The same reproductive problems can be causally related to karyotype abnormalities in the offspring. But mutations and karyotype abnormalities are not the only mechanisms for producing male mediated reproductive problems. A more complete list of mechanisms include:

Mutations (M): Base deletion, base substitution, frameshift, inversion, translocations and large deletions. While these changes are considered gene mutations they are not necessarily translated into clinically recognized disease.

Some mutations may result in loss of function because of:

a) Gene loss; b) Mis-sence.

Some mutations may result in a gain of function because of:

a) Mis-sence; b) Regulatory effect in a gain of function

Karyotype abnormalities (K): Trisomy, monosomy, triploidy, tetraploidy, deletion, inversion, translocation etc.

Cytotoxicity (C): Many mutagens and other toxic chemicals exert their effects simply by killing proliferating cells. The cell killing effect can have many serious developmental consequences.

Imprinting (I): The concept of imprinting relates to the fact that the same gene can have different manifestations depending on whether the gene is contributed by the male or female germ cell. A corollary to the imprinting concept is that the maternal and paternal genome exhibit different patterns of DNA methylation which may influence the male genome's susceptibility to mutation by environmental agents.

Epigenetic factors (E): It is conceivable that the alterations of molecules adjacent to DNA, such as histones, or molecules interacting with DNA (e.g. transcription factors) can affect genetic regulation and gene function. Proving the existence of such phenomena is difficult, although theoretically possible.

Ejaculate containing toxic substances (Ej): It is reasonable to assume that chemicals and drugs can be present and may even be concentrated in the semen, seminal vesicles and prostate. While it may seem incredulous that one ejaculation at the time of fertilization could result in a spectrum of reproductive failures, it is not inconceivable that it might account for embryonic loss. Furthermore, since the deposition of ejaculate containing toxic compounds could occur on many occasions during gestation, multiple exposures could occur via this route during one pregnancy. It is likely that only chemicals and drugs that are potent reproductive toxicants and that can be stored and/or concentrated in seminal fluid would have the potential for affecting the developing human embryo or fetus.

Sperm containing toxic substances (Sp): Toxic substances can affect the sperm by killing the sperm or interfering with the sperm's motility or ability to fertilize the ovum. It has been suggested that toxic substances could be contained in the sperm and that these substances could be delivered to the ovum and the developing zygote, resulting in developmental defects in the embryo. This hypothesis has such an untenable basis that it can be rejected as a plausible line of investigation. The basis of this rejection is the tremendous dilution of the toxic substance at the time of fertilization because the egg has a volume 4100 times greater than the sperm. Secondly, congenital malformations first become readily inducible after the second week of embryonic development. Therefore, the toxic substance would have to remain in the embryo at least two weeks before it could be teratogenic. It is inconceivable that reproductive toxicants contained in the sperm could affect the developing embryo and be responsible for inducing congenital malformations.

Whether there is evidence in humans or animals for a reproductive toxic effect or one can postulate a reasonable mechanism to explain an effect, there is no way to categorically dismiss any of the suggested reproductive or developmental toxic effects as being possible. In fact, many of the male mediated reproductive effects have multiple mechanisms that would explain such effects. Questions have been raised with regard to the actual existence of any of these effects in humans, because of the inability to definitively document some of these effects in humans.

The reproductive or developmental toxic effects that have been discussed in this conference are listed in Table 1. The plausible mechanisms head the columns and are designated with a plus (1-3) or minus depending on the evidence or plausibility. It can be seen that we do not have information about the plausibility nor do we have sufficient data pertaining to many of the mechanisms and reproductive problems.

Table 1. Probable Mechanisms of Reproductive Failure Endpoints

Reproduction Effect	M	K	C	I	E	Ej	Sp
Spontaneous Abortion	+	++	++	?	?	+/-	-
Congenital Malformation	+	+	-	?	?	+/-	-
Genetic disease	+	++	-	?	?	-	-
Malignancy F_1	+	+	-	?	?	-	-
Prematurity	+	+	+/-	-	-	+/-	-
IURG*	+	++	-	?	?	-	-
Stillbirth	+	+	-	?	?	+/-	-
Neurobehavioral effects in F_1	-	-	-	-	-	-	-
Sterility	+	+	+++	?	?	+	-

*IURG = Intrauterine growth retardation.

Animal and human studies dealing with male mediated developmental toxicity have not provided consistent results. Thus investigators are uncertain about the magnitude of this problem. Basic research concerned with the mechanisms of reproductive toxicity, the prevention of toxicity and the augmentation of repair would provide important information toward the understanding of this problem.

BIOMARKERS AND HEALTH ENDPOINTS OF DEVELOPMENTAL TOXICOLOGY OF PATERNAL ORIGIN: SUMMARY OF WORKING GROUP DISCUSSIONS*

A. J. Wyrobek[1], D. Anderson[2], S. Lewis[3], T. Nagao[4], S. Perreault[5], B. Robaire[6], and S. Schrader[7]

[1]Biology and Biotechnology Research Program
Lawrence Livermore National Laboratory, L-452, PO Box 808,
7000 East Avenue, Livermore CA 94550
[2]BIBRA Toxicology International, Woodmansterne Road,
Carshalton Surrey SM5 4DS Great Britain
[3]Research Triangle Institute, PO Box 12194,
Research Triangle Park, NC 27709
[4]Hatano Research Institute, Food and Drug Safety Center,
729-5 Ochaiai, Hadano, Kanagawa 257, Japan
[5]U.S. Environmental Protection Agency,
Research Triangle Park, NC 27711
[6]McGill University, 3655 Drummond Street,
Montreal Quebec, Canada H3G 1Y6
[7]National Institute of Occupational Safety and Health,
4676 Columbia Parkway, Cincinnati, OH 45226

SUMMARY

The working group on biomarkers and endpoints discussed the status of biological markers for assessing male reproductive toxicology, germinal mutagenesis of paternal origin, and male-mediated developmental toxicity. Because of the complexity of reproduction, health endpoints that are related to paternally mediated effects on development involve measurements in the father, his offspring, and to some extent, the mother. Health endpoints specific for the father include changes in his hormonal system, his sexuality, specific parts of his reproductive tract including accessory organs, and his semen. Abnormal reproductive outcomes include infertility due to undetected early embryo death, fetal death resulting in miscarriage, malformations evident at birth, and a variety of functional and developmental deficits that may not become evident until later in life. However, the association between

* This report was prepared after substantial discussions at the working group meetings with important contributions made by the attendees. This is the chair's summary, and should not be interpreted as a consensus document.

Male-Mediated Developmental Toxicity, Edited by D.R. Mattison and A.F. Olshan, Plenum Press, New York, 1994

damage to the male reproductive system, specifically semen, and paternally mediated abnormal reproductive outcome is not well understood. Efficient biomarkers of physiological and genetic changes are needed (a) to measure germinal exposure of the male parent, (b) to investigate the progression from early biological effects in the male to changes in his semen, and (c) to identify biological effects in the male that will predict abnormal reproductive outcomes. In addition, new approaches are needed to characterize and categorize the molecular and genetic defects underlying abnormal reproductive outcomes (i.e., spontaneous abortion, birth defects, childhood cancers), to characterize the types of defects transmitted via the germ cells of the father, and to develop appropriate biomarkers of both exposure and effect.

INTRODUCTION

In studies of both human beings and animals, there is compelling evidence that abnormal reproductive outcomes can be of paternal origin. Three major lines of evidence indicate that exposure of the male parent to toxicants before fertilization may have detrimental effects on the viability, morphology, genetics, and/or health of the resulting embryo and offspring. First, investigations in rodents (primarily mice) during the last four decades strongly indicate that genetic and chromosomal abnormalities can be induced in the germ cells of male mice exposed to germinal mutagens and that these abnormalities can be transmitted to the offspring, thus resulting in detectable gene mutations and chromosomal abnormalities (see Russell, this volume). Second, exposure of male rodents to germinal mutagens (e.g., cyclophosphamide, X-rays) before fertilization resulted in loss of embryonic viability and in morphological defects, behavioral changes, or cancer in their offspring (see this volume). Third, an increasing number of epidemiological reports show that paternal exposure (which can be assessed by exposure to certain agents such as tobacco; by job category such as painter or welder; or by industry such as the aircraft industry) is associated with various abnormal reproductive outcomes such as spontaneous abortions, malformations, and cancer (see this volume).

Recent developments in molecular cytogenetics have provided clear evidence that certain cases of human spontaneous abortions, sex chromosomal aneuploidies, and gene mutations are paternal in origin. Such defects are generally thought to be caused by *de novo* errors that occur in the germ cells of the male before mating. These may be caused by spontaneous errors, exposure to endogenous agents, or exposure to exogenous agents. However, overall we have only a limited understanding of the molecular details of paternal contribution to abnormal reproductive outcomes in humans and this is a growing research area.

Endpoints in male reproductive and developmental toxicology fall into two broad classes: (1) those directly involving the male reproductive system and male germ cells and (2) those involving biological and genetic processes of fertilization and development. This chapter is a summary of the meetings held by the working group on biomarkers and endpoints of male-mediated developmental toxicity. To provide a context for the discussion on male effects on development, the working group began by discussing male effects on fertility, an area that has a considerable research history in man. This report is a summary of selected aspects of the group discussions and is not intended to be a consensus document.

FERTILIZATION DEFECTS OF PATERNAL ORIGIN

Historically, human male reproductive toxicity has dealt primarily with concerns for male fertility and factors that might diminish it. This research dealt with effects on the major biological process and systems of male reproduction, including the male hormonal axis, male sexuality, the reproductive tract (divided into the testicular, epididymal, and other extra-testicular components), and semen. For example, more than 100 chemical agents and chemical mixtures have been evaluated for their effects on semen quality (see Wyrobek, this volume), and more than half of these have reported detrimental effects on sperm concentration, motion, or morphology.

Biomarkers are important in assessing exposure of the male to toxicants and its effects on the male reproductive system. Biomarkers applicable to male reproductive toxicology have been reviewed elsewhere (e.g., NRC 1989). The working group concluded that we have only a superficial understanding of the targets and mechanisms of male reproductive toxicity or of how changes may affect a man's ability to fertilize an egg. The working group discussed several mechanisms and targets of toxicity, with special reference to semen analysis. Semen analysis is the most common approach to assessing male reproductive toxicity but, by itself, it cannot readily distinguish among abnormalities that are caused by (1) effects of toxicant exposure, (2) effects from one's inherited genetic background, or (3) random errors that occur in spermatogenesis and other parts of the male reproductive system.

Effects measured in sperm (e.g., changes in sperm count, motion, morphology, biochemistry) may result directly or indirectly from exposure to a toxicant. There are several possibilities. Indirect effects on semen may be mediated via toxicity to any of a large number of possible cell and organ targets, for example: (1) the pituitary or other brain components of the male hormonal axis; (2) Sertoli cells, Leydig cells, or other somatic cells of the testis; (3) epididymis; or (4) glands of the male reproductive tract (e.g., seminal vesicle). Semen effects may also result directly from effects of a toxicant on germ cells, and the initial lesion may have occurred in the testis, the epididymis, or other parts of the efferent ducts. In addition, the initial lesion may have occurred very early in the life of the father, even as early as when he was developing *in utero*. Studies have shown that several chemical agents affect the developing embryo *in utero* (e.g., exposures to DES and alcohol).

The working group identified a critical need to identify and characterize the relationships between specific semen changes and alterations in male fecundity.

DEVELOPMENTAL DEFECTS OF PATERNAL ORIGIN

Genetically, the role of the father goes beyond fertilization and, therefore, the father may be responsible for a substantial fraction of abnormal development. In its broadest sense, abnormal development includes any morphological and functional defect or disease in any organ system that arose as a genetic defect in the germ cells of one of the parents or that occurred during development *in utero*. Onset of the defect or disease could be any time from first cleavage to an adult in advanced years of life.

In environmental toxicology, a disease or defect is generally considered to be the end of a process that began with an exogenous exposure, which was followed by a number of early and intermediate biological effects, and resulted finally in the late effects (i.e., the disease or abnormality). The probability of proceeding along this process is modified at an undetermined number of steps (expected to be large) by host factors and genetic susceptibilities. However, so little is known about the etiology of spontaneous abortions, birth defects, and childhood cancers that the working group was concerned about the limited perspective provided by a model that focuses on exogenous exposures. Figures 1 and 2 present a broader conceptual framework for male mediated effects on development that includes four categories of the possible causes of abnormal reproductive outcomes: (1) genetic origins, (2) endogenous factors, (3) exogenous factors, and (4) random errors in differentiation. Figure 1 deals with processes that occur in the father before fertilization of his offspring and Figure 2 with processes that may occur after fertilization and during development. Key features of exposure-disease models of toxicology are preserved: exposure, host factors, and susceptibilities remain critical issues in this framework.

Determining the genesis of individual abnormal reproductive outcomes may be a more complex problem than even carcinogenesis because each abnormal outcome involves genetic and physiological interactions among two to three individuals (father, mother, and child) and numerous molecular mechanisms may be involved. There are several potential target organs of the primary toxicity in the male and numerous potential target organs of effects in the offspring. Adding to the complexity, the organs of exposure, early biological effect, and late biological effect are not necessarily the same and do not necessarily reside in the same person. For example, a toxicant may act on developing male germ cells in the testis, on other parts of the male reproductive tract (Figure 1), or directly on the developing offspring with exposure mediated via the pregnant mother (Figure 2). Testicular germ cells may be the primary target of a toxicant, while an organ system of the offspring shows the resulting developmental effect, such as cancer (Anderson *et al.*, this volume). Also, evidence from dominant lethality studies in mice and from the human-sperm/hamster-egg cytogenetic technique shows that the genome of the egg can process damage induced in male germ cells before fertilization and that different genotypes can process this damage to varying extents (see Wyrobek, this volume).

The working group enumerated potential pathways of male-mediated effects on the developing embryo. First, a distinction was made between effects on the embryo via exposure of the pregnant female to agents of paternal origin (Figure 2) versus effects mediated via the fertilizing semen sample (Figure 1). Second, semen is a complex fluid made up of germ cells, somatic cells, and secretions from somatic glands. Therefore, effects transmitted via semen may involve alterations to germ cells, the somatic cells of the testis, the post-testicular reproductive tract, or the glands producing most of the seminal fluid. One of the greatest challenges in using semen analysis to assess heritable effects is to determine the predictive value of specific semen measurements for abnormal reproductive outcomes. The predictive value of an individual semen biomarker is expected to vary depending on the specific health effect. For example, sperm concentration changes are expected to be highly predictive of the ability to fertilize an egg but not predictive of malformations in the embryo.

Rather than identifying semen biomarkers for each potential morphological or health defect during development or after birth, the working group suggested that more emphasis may be warranted on identification of sperm carrying specific genetic and cytogenetic defects (such as aneuploidy, structural abnormalities in chromosomes, genetic deletions, imprinting changes). Mechanism-based biomarkers of development are also needed. To accomplish this goal, researchers will need an improved understanding of intermediates (metabolites, mutation, repair, early effects) in the processes that lead to abnormal reproductive outcomes and of the host and susceptibility factors that affect the processes.

A promising approach in investigating paternal effects may be to identify the specific genetic defects associated with abnormal reproductive outcomes, then to determine the parental origin of the chromosomes involved and whether the defect occurred before or after fertilization. For example, paternal and maternal origins of numerical abnormalities in autosomes and sex chromosomes have been investigated

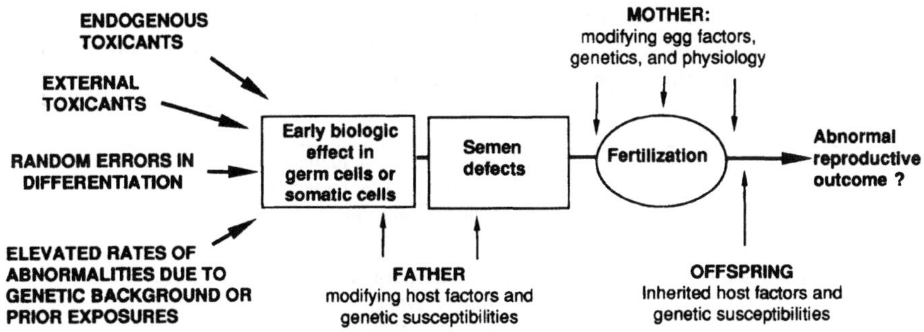

Figure 1. Schematic of possible pathways leading to abnormal development of paternal origin: pre-fertilization effects in the male.

in cases of spontaneous abortion. Maternal contributions predominate for autosomal aneuploidies, whereas paternal contributions are substantial for sex chromosomal aneuploidy. Recent studies show that even in normal healthy men, a fraction of their sperm carries sex chromosomal abnormalities (see Wyrobek this volume for details).

BIOMARKERS FOR ASSESSING DEVELOPMENTAL DEFECTS OF PATERNAL ORIGIN

Biomarkers in toxicology are measurements of adverse effects that may occur in any step in the pathway that begins with exposure and/or the earliest biological changes and progresses to the occurrence of the abnormal reproductive outcomes (NRC, 1989). The working group listed examples of existing and promising biomarkers for assessing male-mediated effects on development and categorized them by their application: biomarkers of exposure, early effects in male, and effects measured in embryos or offspring. Table 1 lists biomarkers of exposure and early effects in the male parent. The endpoints measured in offspring were grouped as genetic or phenomenological measurements in animals (Table 2, 3) and human beings (Tables 4, 5).

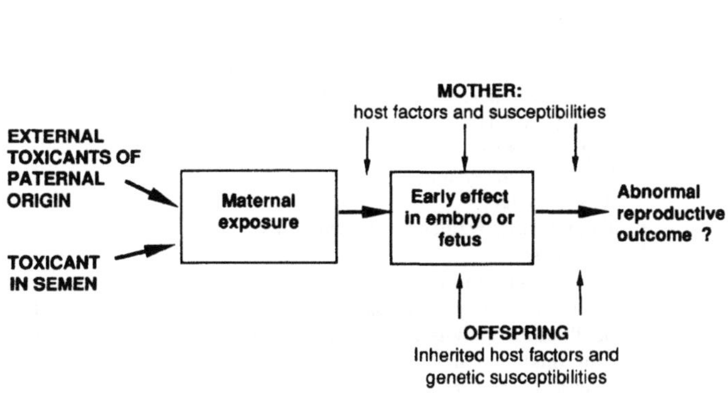

Figure 2. Schematic of possible pathways leading to abnormal development of paternal origin: post-fertilization events.

Table 1. **Human and Animal Biomarkers of Biologic Effects in Testicular Tissue or Semen which are Candidate Biomarkers of Male-Mediated Abnormal Reproductive Outcomes**

Tissue source	Biomarker
Testicular tissue	Abnormal meiotic metaphase I [a] Abnormal meiotic metaphase II [a] Spermatid micronuclei Unscheduled DNA synthesis [a]
Genome of male germ cells	Abnormal sperm karyotype - first cleavage cytogenetics [a] - hamster cytogenetic technique FISH for aneuploidy Spermatid micronuclei Abnormal chromatin stability Abnormal DNA structure by alkaline and neutral elutriation Unscheduled DNA synthesis [a] PCR for chromosomal alterations or gene mutations DNA and protamine adducts [a,b]
Other sperm components	Changes in membrane, cytoplasmic, or tail components [c] Adducts to non-nuclear proteins [a,b,c] Abnormal sperm morphology [c]

[a] endpoints specifically relevant in animal studies
[b] possible candidates for exposure biomarkers
[c] not specific for genetic effect but may provide an indication of altered spermatogenesis

Table 2. **Examples of Approaches used in Animals for Identifying Genetic Abnormalities in Developing Embryos and Offspring after Paternal Pre-fertilization Exposures to Germinal Mutagens**

Methods applicable during development

- cytogenetics (first cleavage division and beyond)
- dominant lethality
  - pre-implantation
  - peri-implantation
- changes in gene expression (RNA or protein detection)

Methods applicable at birth or later

- visible specific locus mutations
- dominant visibles
- protein electrophoresis (1 and 2 dimensional)
- protein activity heritable translocations
- transgenic models for mutagenesis
  - "Big Blue" system
  - "Muta-mouse" system
  - "Phi X" system
- proliferative disadvantage using chimeras
- DNA analyses:
  - effects of imprinting on gene expression
  - large lesion analysis for insertions, deletions, or rearrangements
  - small lesion analysis using denaturing gradient gel electrophoresis
  - conformational analysis using SSCP
  - analysis for transposable elements

Table 3. Examples of Approaches Using Animals for Detecting Phenotypic Abnormalities in Embryos and Offspring after Paternal Prefertilization Exposure to Germinal Mutagens

Methods applicable during development	
	malformations (external, skeletal, visceral)
	growth markers
Methods applicable at birth or later	
	birth weight, birth measures
	birth to weaning survival
	neurological battery
	puberty markers
	growth markers
	malformations
	reproductive markers
	cancer

Table 4. Examples of Approaches for Detecting Genetic Defects in Human Beings during Development or after Birth

Cytogenetics	
	at amniocentesis,
	from chorionic villi samples
DNA alterations	
	insertions, deletions, rearrangements
	denaturing gradient gel electrophoresis
	SSCP
	protein electrophoresis
	oncogene and other gene amplifications

Table 5. Examples of Approaches for Detecting Phenotypic Abnormalities in Humans Applied during Development or After Birth

Methods applicable during development	
	gross malformations (external, skeletal)
	growth markers
	clinically recognized abortions
	hormonally detected early fetal loss
	alpha fetal protein
Methods applicable after birth	
	birth weight, birth measures
	morphological defects
	neurological and behavioral defects
	placental investigations
	neurological tests (various)
	growth markers
	childhood cancer

BIOMARKERS OF EXPOSURE

A first question in animal and human studies of paternal effects is usually whether there is or was a paternal exposure to a suspected toxicant. Related questions are whether germ cells or somatic cells were targets, whether DNA or proteins in germ cells were adducted, and other questions related to the type of adduct formed and its repair. Distinctions are usually made among external exposure, internal dose, gonadal dose, and semen chemical concentration. Most of our current knowledge in this area is derived from animal studies. However, there have been very few applications of exposure biomarkers utilizing germ cells in human studies. The working group noted that using semen concentration of toxicants does not answer the question of whether there was a significant testicular dose (because most of the semen is not of testicular origin) and whether there will be a significant dose at the site of fertilization (because little of the semen will arrive at this site). Human studies of adducts in semen using P-32 post-labeling have yielded uncertain results, and further study is warranted.

Sega and coworkers (see Russell, this volume, for references) challenged the importance of DNA adduction in germinal mutagenesis. They observed in male mice that the level of dominant lethality in the offspring of male mice treated post-meiotically with certain germinal mutagens was strongly associated temporally with the levels of adducts in sperm protamine, not sperm DNA. Protamine is the major protein in the mammalian sperm nucleus, and it replaces histones during the post-meiotic steps of spermatogenesis. The importance of protamine adduction was confirmed for four germinal mutagens: MMS, EMS, ETO, and acrylamide. More research will be required to understand the relative roles of protamines and DNA adduction in germinal mutagenesis of post-meiotic cells.

The working group summarized that more research was needed to develop efficient biomarkers of exposure for the earliest biological targets of the male reproductive system and to understand what happens to these earliest biological changes as the cells proceed along their differentiation pathway.

BIOMARKERS OF SUSCEPTIBILITY

Susceptibilities can be viewed as variations in biological responses as cells progress along the exposure - disease pathway, such as the one shown in Figure 1. Variations can also be seen among individuals receiving the same external exposure and there may be variation in the incidence of abnormal outcomes given the same early biological effects. Factors expected to affect susceptibility include anti-oxidant levels, diet, repair competency, metabolism, and immune function status. There are few direct biomarkers of susceptibility, and more typically it is evaluated by comparing differences in response in individuals receiving the same exposures, especially in animal model systems. There is considerable research interest in the development of biomarkers for specific genes for metabolism and repair that may be responsible for differences in susceptibility.

BIOMARKERS OF EARLY BIOLOGICAL EFFECTS IN MALES, ESPECIALLY THOSE MEASURABLE IN SEMEN

The working group listed biomarkers that can be applied to the male directly (Table 1) and that might be related to the risk of producing an abnormal reproductive outcome of paternal origin. Most of the biomarkers discussed have unknown and untested relationships with abnormal reproductive outcomes. Most have not been fully evaluated for their reproducibility, specificity, sensitivity, and predictive value. Some are likely to remain impractical, such as those requiring testicular tissue. Promising biomarkers are those measuring genetic damage in the sperm, using endpoints that have direct correlates at birth, such as the detection of Y-Y of X-Y sperm. Recent studies using fluorescence in situ hybridization (see Wyrobek, this volume) have made considerable progress toward replacing the more tedious human-sperm/hamster-egg system for the detection of aneuploid sperm. See NRC (1989) for a more detailed description of the semen biomarkers and the biomarkers of genetic damage.

BIOMARKERS CORRESPONDING TO EARLY EFFECTS IN THE EMBRYO AND TO LATER HEALTH ENDPOINTS

Compiling a set of biomarkers for all conceivable abnormal health outcomes that might be of paternal origin proved to be a difficult task for the working group because of the complexity of the outcomes that were possible. The group prepared four tables of examples of biomarkers and methods for assessing genetic and phenotypic abnormalities in the germ cells and offspring of animals (Tables 2, 3) or human beings (Tables 4, 5). Detailed reviews of these procedures were beyond the scope of this group but are available elsewhere (NRC 1989).

RESEARCH NEEDS

The discussions of the working group raised the following research needs and suggestions. This list was not intended to be comprehensive.

General research needs in male-mediated developmental toxicology

Biomarkers research

Develop better detection methods for DNA damage, especially for small lesions in sperm and early embryos;

Develop efficient germ cell and embryo markers of sex-specific imprinting;

Develop non-invasive methods to detect subtle morphological defects in embryos to avoid sacrificing the test animals (e.g., MRI);

Evaluate promising biomarkers for their reproducibility, reliability, sensitivity, specificity, and predictive value (*vis-a-vis* specific and relevant health effects).

Mechanism research

Investigate the molecular, cytogenetic, and genetic components of childhood cancer, birth defects, spontaneous abortions, and other abnormal reproductive outcomes;

Evaluate the relative importance of protamine adduction versus DNA adduction in germinal mutagenesis;

Determine whether developmental abnormalities can be caused by abnormalities in paternal imprinting.

Exposure selection in model systems

The exposure regimen and route used in animal models should mimic that experienced by exposed humans. Specifically, chronic and repeated exposures may be more relevant for some human exposures than an acute exposure regimen using the intraperitoneal route. In addition, the effects of exposures *in utero*, as juveniles, and during adolescence should be included if exposures at comparable times occur in humans.

Additional animal species

Individuals and species can differ markedly in their response to chemical exposures. Species that have responses similar to those of human beings would be excellent candidates for studies of the mechanisms of toxicity. A good example of species differences in chemical response of the male reproductive system is dibromochloropropane, a nematocide. Exposed rats, hamsters, and humans appear to experience similar effects on their testis, hormone axis, and sperm quality from exposure to dibromochloropropane. In contrast, male mice show essentially no reproductive response, suggesting that mice may not be a suitable animal model for studying the reproductive or developmental toxicology of this agent. Although the mouse remains the primary animal model for genetic research and germinal mutagenesis, it should not be considered suitable for mutagenesis studies of all compounds. This example emphasizes the need for developing biomarkers of mutagenesis for the rat and other species.

Research focused on priority compounds

The literature on male reproductive and developmental toxicity contains studies of numerous agents; however, few results have been replicated in second laboratories, and only a few biomarkers have been used in any one study (Table 1). This approach is credited with identifying essentially all known reproductive toxicants, developmental toxicants, and germinal mutagens. However, because few agents (with the notable exception of ionizing radiation) have received attention in depth, little is known of interspecies differences in response; the understanding of the molecular mechanisms of toxicity for any compound is extremely limited; and little is known about the relationship between defects in the reproductive system of the father and abnormal reproductive outcomes.

The working group suggested that a concerted effort be considered to select a few priority compounds and to investigate in-depth the mechanisms of their reproductive and developmental toxicity using expert laboratories. The priority compounds could be selected on the basis of their potentially differing mechanisms of toxicity, differing reproductive and developmental effects, and relevance to human exposures.

Several strategies for selecting candidate agents for the study of male-mediated effects were discussed by the working group. Several resources were identified. First, studies of germinal mutagenicity of chemicals in mice have identified numerous agents that induce dominant lethality, germinal mutations and chromosomal abnormalities in male germ cells (for references see papers by Russell, Radcliff, Favor, and others, this volume). Second, candidate agents may be found among those identified as critical pollutants in the Great Lakes Basin ecosystem (Toxic Chemicals in the Great Lakes and Associated Effects, Government of Canada Report). It identified 11 critical pollutants in the lakes, including chlorinated organic chemicals (polychlorinated biphenyls, mirex, hexachlorobenzene, dieldrin, DDT, dioxin, 2,3,7,8-TCDF, toxaphene), polynuclear aromatic hydrocarbons (benzo[*a*]pyrene), and toxic metals (mercury and alkylated lead). Third, the European Communities (CEC) Aneuploidy Programme (Adler 1993) identified 10 compound as candidate germinal aneugens (colchicine, econazole, diazepam, thiabendazole, cadmium chloride, thimerosal, pyrimethamine, hydroquinone, vinblastine, and chloral hydrate). Other agents discussed by the group were cyclophosphamide, lead, and dibromochloropropane. In the selection of two or three candidate compound, it will be important to evaluate completely the evidence for male-mediated effects for each agent and to identify an adequate supply of pure chemical. By limiting the number of compounds, each participating laboratory may devote more of its resources for in-depth investigations of the molecular aspects of male toxicity and the parental origin of developmental defects.

Biomarkers for screening and for studies of mechanisms

For screening chemicals in animal studies, series of phenotype-based biomarkers related to health endpoints have been commonly used: fetal loss, malformation, changes in function, or cancer in the offspring. However, investigations of the mechanism of toxicity of selected priority compounds are expected to include many specialized biomarkers of potential underlying chemical, molecular, biochemical, cellular, and physiological mechanisms. It is hoped that common mechanisms for developmental defects will be discovered and that these discoveries will herald new mechanism-based biomarkers. This would greatly improve the amount and quality of information obtained in screening.

Since such studies could require significant effort in animal costs alone, it is suggested that as many biomarkers as possible be combined in each study. Collaborative efforts between laboratories with different expertise would be advantageous. The study of as many biomarkers as possible within each laboratory would not only make the best use of experimental material but would permit direct correlation between developmental, reproductive and genetic findings.

Role of old versus new methods

The desire to develop and use new technologies in studying male-mediated effects should not result automatically in discarding more classical methods that may still be cost-effective and efficient in detecting such effects. It is important to consider the general usefulness and historical reference value when determining which tests to use and how each test should be used in the risk assessment process or in research. One factor to consider is the cost of each test so that the financial impact of the process is minimized. New technologies focused solely on DNA damage may provide insufficient information for risk assessment for the organism. Many of the older tests provide important information of relevance for the organism as a whole, of the proteins produced, and of their function. The working group suggested that multiple endpoints, including whole animal systems, are essential and should be used in a balanced way in the risk assessment process.

ACKNOWLEDGMENTS

Work performed by the Lawrence Livermore National Laboratory under the auspices of the U.S. Department of Energy under contract W-7405-Eng-48, with partial support from the National Institute of Environmental Health Sciences (Y01-ES-10203-00) and State of California Tobacco Related Disease Research Program (3RT-0223). I thank Carolin Middleton for editing and formatting.

REFERENCES

Adler, I-D. (1993) Synopsis of the *in vivo* results obtained with the 10 known or suspected aneugens tested in the CEC collaborative study. *Mutation Research* 287:131-137

Committee on Life Sciences, National Research Council. Biologic Markers in Reproductive Toxicology. National Academy Press, Washington D.C; 1989:37-146.

EPIDEMIOLOGIC APPROACHES

David A. Savitz

Department of Epidemiology
Campus Box #7400
School of Public Health
University of North Carolina
Chapel Hill, NC 27599-7400

INTRODUCTION

The general goal of the breakout group discussion was to identify epidemiologic strategies that would advance understanding of male-mediated developmental toxicity. In particular, we sought to identify methodological approaches specific to this phenomenon, perhaps differing in design from studies of maternal influences on development or paternal influences on fertility.

A critical question at present is whether the hypothesized phenomenon actually occurs in humans. In spite of some epidemiologic research on this issue, there is not yet a single unequivocal link between a male exposure and a developmental outcome. The limited quality and quantity of research done to date certainly does not establish the absence of such effects in human populations, but rather it is inadequate to draw a conclusion. A single persuasively established association could provide a model in the search for other links and would respond to those who argue that such effects are unlikely to occur.

A convincing demonstration of the phenomenon might be attained in several different ways. A unique cause-effect linkage between a specific environmental agent and a unique or at least highly unusual endpoint could be discovered, analogous to the link between vinyl chloride and angiosarcoma or asbestos and mesothelioma. In those instances, the disease virtually does not occur in the absence of the exposure.

Another avenue would be through descriptive studies of paternal characteristics such as age or lifestyle factors, in which a replicated observation of an independent association of a paternal attribute with a reproductive outcome might be identified. Such observations from descriptive epidemiology often precede an understanding of the specific agent or process that is responsible.

Finally, certainty might be achieved by identifying a process through epidemiologic studies that is closely parallel to an established toxicologic phenomenon with respect to exposure, outcome, and time course. Where laboratory investigations offer areas of certainty, the credibility of epidemiologic studies indicating the same phenomenon would be enhanced.

None of these suggested approaches are "safe bets" in that the most effective

Male-Mediated Developmental Toxicity, Edited by D.R. Mattison
and A.F. Olshan, Plenum Press, New York, 1994

research strategies remain unclear. In fact, with the current state of knowledge, it is preferable to pursue a diverse array of strategies rather than arbitrarily select one or even just a few to which all the resources should be devoted. It is important for research to be of sufficient quality to be capable of both encouraging pursuit of positive leads but also capable of discouraging further work in the face of negative evidence. Finally, the need for interchange of epidemiologists with toxicologists and geneticists is critical in many of the specific avenues noted below. Laboratory-based hypotheses need to be evaluated epidemiologically and, conversely, epidemiologic suggestions need to be examined and refined in the laboratory.

STRATEGIES TO SELECT EXPOSURES TO STUDY

Much of the epidemiologic research on male-mediated developmental toxicity has been concentrated on a relatively small group of occupational exposures. A broader evaluation of male exposures is needed.

Descriptive Epidemiology of Paternal Factors and Reproductive Outcomes

In the tradition of descriptive epidemiology, in which rates of events are tabulated for groups defined on the basis of location or time, much more data are needed that simply describe paternal attributes and reproductive outcomes. General characteristics of the father such as age, ethnicity, diet, smoking, etc. are needed in relation to such outcomes as miscarriage, preterm delivery, birth defects, and indices of child development. The patterns identified through evaluating such arrays of data have yielded useful etiologic hypotheses virtually whenever they have been generated, and there is every reason to believe that such information would advance the study of male-mediated developmental toxicity as well.

Pursue Epidemiologic Leads

In spite of the limitations in all epidemiologic studies of paternal factors to date, there are some avenues that warrant confirmation or refutation. Agents that have been most suggestive of having adverse reproductive effects mediated by paternal exposures are heavy metals, particularly lead and mercury, with associations found for miscarriage, birth defects, and cancer. Paternal exposure to ionizing radiation has been related to both birth defects and cancer as well. Solvents and paints, a broader class of agents, has been linked to neural tube defects and brain cancer. Specific associations have been reported for other agents such as chloroprene, vinyl chloride, and some pesticides, but not as extensively as the others noted above.

Study Biologically Plausible Agents

Both biological theory and empirical observations in the laboratory point to some agents that warrant epidemiologic evaluation. Specific agents that have been insufficiently studied by epidemiologists given the extent of experimental or theoretical support are heat, chemotherapeutic agents, ethylene oxide, and acrylamide. In addition, mutagens as a class of agents are all of concern.

Infectious Agents and Paternal Illness

In spite of their known influence on male infertility and ability to alter DNA, infectious agents have not been examined with regard to males and reproductive health. A number of chronic diseases have diverse systemic effects and the potential for influences on reproduction is similarly unexplored. Factors associated with testicular cancer such as cryptorchidism, for example, might be evaluated for their potential influence on reproductive toxicity as well.

METHODOLOGICAL ISSUES REGARDING EXPOSURE

In addition to decisions about which specific agents to study, epidemiologists face a number of methodological challenges that apply to any agent. Also, suggestions for some general approaches to improving exposure data collection are needed to facilitate future epidemiologic studies.

Identify Appropriate Time Windows

The etiologically relevant time window for male-mediated developmental effects has not been clearly defined. When exposures are not appropriately restricted to the relevant time window, measures of association will be diluted. Most research has assumed that the window encompassing the later stages of spermatogenesis is critical, typically including the 3 to 6 months prior to conception. However, the question of whether stem cell changes can produce long-delayed effects is uncertain. Such changes could occur at any time from gestation, when the germ cells are initially being formed, to the period several months before conception, when the final stages of spermatogenesis are being completed. Also, for specific stages in the sperm cycle, narrower time windows of a particular week or even day before conception may be critical. Laboratory research is critical in the identification of periods of vulnerability.

Incorporate Paternal Factors into Large Data Bases

To expand the realm of inquiry regarding paternal exposures, a minimal set of such factors should be incorporated into some routine data collection as well as conducting focused studies. Obstetrical data bases might add paternal age, smoking, alcohol use, occupation, and chronic illnesses to facilitate wide-ranging evaluations of paternal factors and pregnancy outcome. National surveys of reproductive health should incorporate paternal information when feasible as well. Birth certificates should include at least minimal paternal information, such as age and occupation. The Congressional mandate for funding of studies of birth defects should include paternal factors in the realm of inquiry. Hypothesized associations could be examined through such efforts and new avenues would be suggested.

Identify Highly Exposed Populations

Much of the epidemiologic research has been concentrated on populations of public concern but whose exposure was not notably high or was poorly documented. Informative studies should be pursued that evaluate risks to heavily exposed men in which even negative results would be valuable in exonerating agents or exposure circumstances. Behavioral exposures (diet, smoking) and therapeutic drugs are typically more intense exposures than occupational agents and may well be more amenable to accurate assessment. Within the category of environmental contaminants in the workplace or community, developing countries often encounter markedly higher exposures than those presently occurring in North America and Western Europe.

Add Reproductive Outcomes to Cohort Studies of Men

Studies undertaken to evaluate the health consequences of exposures to men often include large populations with well-documented exposures. Although many studies of chronic disease are focused on older men who are past their reproductive years, some younger populations may be available, e.g., men exposed to polybrominated biphenyls through environmental contamination.

Evaluate Temporal Stability of Paternal Exposures

For logistical reasons, it is extremely difficult to identify and study exposures in men around the time of conception. If the pregnancy is unplanned or if the period required to conceive is prolonged, then the time around conception is only known in retrospect. A practical methodological question is whether exposures to men, based on biological markers, remain sufficiently stable from conception to the onset of prenatal care or the time of delivery to justify use of later measures as a surrogate for earlier exposures. In contrast to women who experience profound behavioral and physiological changes over the course of pregnancy, male exposures may well be rather stable, though this requires empirical evaluation.

Interview Males in Reproductive Outcome Studies

Although women are needed to report reproductive outcomes and can report a minimal set of information about paternal exposures, more detailed information on lifestyle or occupational hazards requires interviews directly with the fathers. The logistics of identifying, locating, and recruiting the father to participate are acknowledged, but it is essential for the validity of many types of exposure data.

Consider Home Contamination and Transmission Through Semen

The biological pathway of greatest interest to many investigators is a sperm-mediated effect on fetal and child development. However, another recognized pathway is for non-volatile workplace contaminants to produce maternal and thus fetal exposure, demonstrated in the past for lead and asbestos. What is not known in any detail is how widespread or important the phenomenon actually is.

Laboratory studies and limited human evidence also indicate that some agents are concentrated in semen and can thus result in exposure to the woman and fetus, but the scope of the phenomenon is unclear. Since these pathways are known to operate in humans (unlike the one based on genetic alterations in the sperm) and would be amenable to public health intervention if found to be important, more thorough evaluation is needed. Related to this is a need to carefully distinguish between exposures shared by the couple and those that are unique to the male.

STRATEGIES TO SELECT ENDPOINTS TO STUDY

Reproductive and developmental processes comprise a wide range of events. Known maternal risk factors tend to be specific, with even agents as broadly toxic as tobacco smoke only affecting a subset of developmental outcomes. It also would seem likely that if paternal exposure to a particular agent affects reproduction, it would influence some but not all outcomes. Therefore, a strategy is needed for selecting the events most deserving of attention. Past research has concentrated on miscarriage, birth defects, and childhood cancer, but a broader array of outcomes should be considered.

Expand Scope of Reproductive and Developmental Health Outcomes

If paternal factors influence reproductive and child health, there is no reason to believe that such influence is limited to classic "diseases" such as cancer or birth defects. Instead, a wide variety of outcomes should be considered, even including those of limited clinical significance, for example, minor variations in birth weight. Demonstration of an influence of paternal exposures on any aspect of reproduction, even if not of direct public health consequence, would be of value.

The male genetic influence on placental development is well known, but the influence of male exposures on placental weight has not been explored. Other aspects of placental size and function could also be considered. A paternal role in twinning or in influencing the sex ratio is also worthy of evaluation. Paternal genetic factors have also been implicated in the etiology of some pregnancy complications, most notably preeclampsia.

The broad realm of neurodevelopment, including cognitive and intellectual development, is generally a sensitive process and some animal studies suggest a paternal influence. A broader look at size at birth, including height and head circumference as well as weight, and child growth would be useful. Susceptibility to infection, chronic diseases of children such as diabetes, and other health conditions should also be considered.

Group Endpoints for Studying Male-Mediated Effects

Among the reproductive endpoints, decisions must be made about grouping both within a broad category like birth defects or across categories, such as nervous system disorders which might include selected birth defects and childhood cancer. Most organizational schemes are based on organ system or functional impact, yet the processes postulated to link paternal factors to reproductive endpoints may warrant entirely different approaches to aggregation. Given the rarity of many individual reproductive and developmental outcomes, scientifically based approaches to grouping are needed.

Study Outcomes Predicted Based on Imprinting

Recognition that the male genome is the unique contributor to certain reproductive outcomes encourages closer examination of male exposures in relation to those outcomes. Although rare and thus not easily studied, events like Prader-Willi Syndrome and certain forms of neurofibromatosis are known to be paternally derived and thus worthy of the effort to establish collaborative research to accrue sufficient numbers of cases for study. Some efforts to develop such registries have been undertaken.

Follow Leads from Dominant Lethal Assay and Multi-Generation Animal Studies

Although not a clear model for human miscarriage, the extensive body of data derived from the dominant lethal assay should be used for whatever suggestions it can provide. The most direct implication would concern peri-implantation fetal loss but any other human events analogous to the dominant lethal effect should be considered. Laboratory studies of the consequences of exposures across generations also provide suggestions for epidemiologic studies. Specifically, such studies point to cancer in children and adults as potential consequences of exposure. Childhood cancer has and will continue to receive a great deal of attention, but adult cancer has received much less attention.

Encourage Toxicologists to Examine a Broader Array of Endpoints

Male exposures should be considered as potentially capable of influencing virtually all reproductive and developmental endpoints, and laboratory studies would be most beneficial to epidemiologic study design if they included a comprehensive array of such outcomes. By extending the array of endpoints to include both the traditional semen parameters and developmental outcomes, there would be the opportunity to address the relationships among exposure, semen quality, and development. Even if mechanisms are not fully understood, evaluation of many outcomes would inevitably encourage or discourage study of specific outcomes in humans.

ADDITIONAL RECOMMENDATIONS

In addition to the avenues outlined above some rather specific suggestions were made that would at least have the potential to advance the field of research:

1) Encourage clinicians to be aware of a potential male influence and perhaps increase the likelihood of making a fortuitous observation regarding an association. Historically, many discoveries have been made by astute clinicians, and such help would be welcome here as well.

2) Clarify the relationship of semen quality and function to reproductive endpoints, particularly in order to define markers of reproductive outcome that can be directly discerned through study of semen. If a sperm attribute could be identified which was clearly related to reproductive risk, such markers could be used in lieu of clinical outcomes in prospective studies of exposed populations. Such studies could be much smaller and shorter in duration than those required to examine traditional reproductive outcomes.

3) Conduct large, wide-ranging epidemiologic studies of potentially important paternal exposures and reproductive outcomes that are capable of identifying modest but credible associations.

4) Exploit infertility clinic data to evaluate the relation of semen characteristics to outcomes as well as study biological markers of exposure in semen samples.

5) Seek favorable study settings based on the cooperativeness of the population, planning and documentation of pregnancy, or unique record-keeping.

6) Improve exchange of information among investigators working in the area of paternal influences on reproduction, both across disciplines and within epidemiology.

The last of these points, better communication among researchers in this field, warrants more detailed comment. Transmission of information from toxicologists and geneticists to interested epidemiologists might be enhanced through development of tutorial documents or short courses. Within the community of involved epidemiologists, many persons who participated in the discussion knew one another, but we had never gathered together before to address this particular theme. The conference and discussion were thought to be of real benefit to epidemiologists and there was enthusiasm for some type of continued communication. Such contact would provide an opportunity to share wisdom and insights, facilitate coordination among studies, avoid duplication, increase the probability of having comparable study results, and provide a potential mechanism for proposing and conducting collaborative studies. A number of suggestions were offered for achieving this goal, including creation of an interest group within an existing professional society.

ACKNOWLEDGMENTS

Although this document should not be interpreted as a consensus of all participants, the ideas it contains were developed as a group that included Drs. Jonathan Buckley, Brenda Eskenazi, Paul Garbe, Robert Miller, Leslie O'Leary, Ninfa Redmond, Eve Roman, David Savitz, Sherry Selevan, Lowell Sever, Gary Shaw, Anne Sweeney, Gina Terracciano, Margaret Vartanian, Allen Wilcox.

LABORATORY RESEARCH METHODS IN MALE-MEDIATED DEVELOPMENTAL TOXICITY

M.D. Shelby[1], L.B. Russell[2], R.P. Woychik[2], J.W. Allen[3], L.M. Wiley[4], J.B. Favor[5]

[1]National Institute of Environmental Health Sciences
P.O. Box 12233
Research Triangle Park, NC 27709
[2]Oak Ridge National Laboratory
P.O. Box 2009
Oak Ridge, TN 37831-8077
[3]U.S. Environmental Protection Agency
Genetic Toxicology Division MD-68
Research Triangle Park, NC 27711
[4]University of California
Old Davis Road
Davis, CA 95616
[5]GSF-Institut für Säugetiernenetik, Neuherberg
D-85764 Oberschleissheim, Germany

INTRODUCTION

It would be useful to have a clear definition of Male-Mediated Developmental Toxicity (MMDT) in order to make suggestions or recommendations for its investigation in the laboratory, but this not available. For the purposes of this document, MMDT will be considered to include all factors contributed by the male parent to the conceptus that can result in other-than-normal development.

Mammalian spermatozoa are motile cells that function to deliver a haploid genome to the oocyte. With this critical reproductive function, spermatozoa consist of little more than a haploid nucleus and a system of motility. Because of the highly specialized nature of sperm and their relative simplicity with regard to cellular components, limited attention has been given to the possibility that the male could contribute to abnormal development other than through a genetic mechanism.

The role of the male's genetic contribution to the development of the conceptus has been studied for decades and the fact that abnormal development can result from mutations, both new and pre-existing, carried in the male gamete is clearly established

Male-Mediated Developmental Toxicity, Edited by D.R. Mattison
and A.F. Olshan, Plenum Press, New York, 1994

in laboratory animals as well as in humans. However, in recent years, reports have been published that associate exposure of the male parent with adverse developmental outcomes that are not fully consistent with genetic damage induced in male germ cells, i.e., effects associated with exposure to chemicals not shown to be germ cell mutagens or in the absence of demonstrable genetic damage. These reports include laboratory animal experiments and human epidemiology studies and deal with developmental abnormalities that include prenatal effects such as embryonic death and fetal malformations and postnatal effects including neurobehavioral anomalies and cancer.

Because all the effects reported in these studies can result from transmitted genetic damage, the primary question raised by these reports is whether or not exposure of the male parent can result in developmental abnormalities that do not result from alterations of the genetic material of the male gamete. Toxicants in the seminal fluid or bound to the sperm have been discussed as possible nongenetic mechanisms of such effects, although no clear evidence to this effect are available to date. Other possible nongenetic targets for toxicants that affect development include the plasma membrane and mitochondria of the sperm. Effects on methylation patterns or other imprinting systems, or changes in the protein component of chromatin have been mentioned as possible means by which paternal exposure could influence development of the offspring through mechanisms that involve the DNA but not necessarily a change in the base sequence. This chapter will focus on experimental methods for discriminating between genetic and nongenetic mechanisms of abnormal development resulting from paternal exposures.

DETERMINATION OF TARGETS AND MECHANISMS

There are three general ways in which paternal exposure to a toxicant might influence the development of his offspring: (1) via effects on the germ-cell DNA, (2) via effects on non-DNA components of the sperm, and (3) via toxicants transmitted through the father's semen to the mother and thereby into the environment of the conceptus.

The genetic contribution to abnormal development through damaged DNA encompasses numerous types of DNA lesions that are induced in germ-cell precursors and in germ cells at various stages of their development. It also encompasses changes in the numerical complement of chromosomes, e.g., the loss or addition of a specific chromosome that could occur as a result of nondisjunction during meiotic divisions of the father's germ cells. Finally, it can include non-permanent changes in DNA, such as changes in methylation patterns that might lead to altered expression of genes in the conceptus.

Nongenetic components of the sperm that might possibly affect development if altered by a chemical exposure include the sperm plasma membrane, the sperm tail, or mitochondria that enter the zygote cytoplasm; any such effects incompatible with sperm penetration of the oocyte would, obviously, be inconsequential. It is not known whether non-DNA components of the sperm can affect the development of the conceptus, but it seems likely that such effects, if they exist, would be of short duration and would cease by the time of implantation or before.

Little is known of the amounts of toxicants that can be transmitted by the father into the environment of the conceptus but it presently seems unlikely that they would be present in high enough concentrations at any one critical developmental period, or through long-sustained periods, to pose a significant risk for abnormal development.

Two classes of experimental methods are summarized in this report: those that distinguish genetic from non-genetic effects and those that address the relation between the sperm's DNA contribution and the development of the conceptus.

Distinguishing Genetic From Nongenetic Effects

Genetic alterations induced in male germ cells by exposure to environmental agents must be transmitted to the conceptus to produce "male-mediated developmental toxicity". Therefore, genetic damage that will lead to death of the germ cell either at the stage of exposure or in a subsequent spermatogenic or spermiogenic stage will not be considered in the context of this report.

Genetic lesions transmitted to the conceptus must be dominant in order to have a phenotypic effect on the immediate offspring of the exposed male. Effects can include death in early embryonic stages (before, during, or shortly after implantation), morphological or physiological abnormalities detectable *in utero*, at birth, or postnatally (some of which may lead to death), functional defects (e.g., reduced reproductive capacity), or neoplasms of various types developing at various ages.

Transmission of a phenotype from the immediate offspring of the exposed male to the next generation can provide unequivocal proof of the genetic origin of the phenotype. Such proof is not always obtainable, but the failure to obtain it does not necessarily argue against the effect being of genetic origin. As is apparent from the list in the preceding paragraph, many dominant phenotypes are not compatible with the offspring's survival to reproductive age and/or the offspring's fertility. Without these prerequisites, proof of transmission of a paternally induced genetic change cannot be obtained. Even where a transmission test is feasible, a negative result does not necessarily disprove the genetic origin of the phenotype, since many dominant mutations are known to have incomplete penetrance and/or variable expressivity.

While it is not always possible to provide direct proof of the genetic origin of a particular phenotype, some types of evidence lend strong support to such a conclusion. For example, when epidemiologic studies suggest a causal link between exposure to an environmental agent and a male-mediated effect, laboratory evidence that the agent induces transmissible genetic damage in mammalian test systems such as the morphological specific locus test points to a strong case for the male-mediated effect being mutationally based. A direct method for determining if the induced, male-mediated effect is associated with the father's DNA is through nuclear transfer. Following fertilization, it is possible to reciprocally transfer male pronuclei between zygotes derived from exposed and control males. DNA-based effects would be observed only in offspring arising from nuclei of exposed males while effects mediated through non-nuclear components of the sperm would be observed in offspring arising from the zygotes initially fertilized by sperm of exposed males, even though the pronuclei contributed by those sperm were replaced by pronuclei from control males.

Epidemiological evidence itself may provide clues as to genetic versus non-genetic origin of the cases under consideration. Genetic lesions induced in paternal DNA are generally thought to be randomly distributed throughout the genome although minor differences in the frequencies of mutations induced may vary among gene loci. Thus, an assortment of phenotypes is expected to result, e.g., a variety of developmental anomalies. The finding of a preponderance of a single type of abnormality could conceivably indicate a non-genetic mechanism of MMDT acting via the dysfunction of an active component of the sperm contributed to and required for a specific aspect of the normal development of the conceptus or via a toxicant transferred by the father to the intrauterine environment of the conceptus at a critical period in development.

There are targets other than DNA structure/sequence that toxicants can affect to induce heritable changes. The complex processes of chromosome distribution, and of imprinting, in the germ-line may afford diverse opportunities for damage to result in altered gene dosage and expression that control embryonic growth.

The high frequencies of numerical chromosome aberrations in human gametes that lead to embryo loss and aneuploidy syndromes are believed to be primarily caused by chromosomal malsegregation during meiotic divisions. In experimental organisms, such errors can result from gene and chromosomal mutations as well as from damage to the spindle apparatus. A number of compounds are known to specifically interfere with microtubule assembly, or impair the structure/function of centrioles, cytoplasmic membranes, or kinetochore proteins. The unique behavior of meiotic chromosomes to synapse and recombine, with associated synaptonemal complex formation, is also viewed as an important pathway by which induced alterations to non-DNA targets may predispose chromosomal nondisjunction and/or loss.

Imprinting in germ cells refers to epigenetic modifications of DNA, which program the differential expression of maternal and paternal genomes during development. Although the mechanisms for such modifications are unclear, patterns of DNA methylation may effect a heritable imprint serving to regulate gene activity in the embryo. Just as DNA-damaging agents may induce classical forms of heritable mutation due to changes in base sequences, they may also induce epimutations (e.g., loss of methylation to reactivate transcription) transmissible to offspring as defects in scheduled regulation of gene expression. It has been suggested that such epigenetic defects may be operative in both teratogenesis and carcinogenesis.

Non-Nuclear Components

Methods that are proposed for the study of male-mediated developmental toxicity must account for effects mediated by interactions of toxicants with non-nuclear components of the sperm that can persist for a time following fertilization. These components include the plasma membrane, mitochondria, and the tailpiece of the sperm. The tendency to focus on nuclear DNA effects is, in part, due to the fact that our knowledge of genetics and our ability to manipulate the genome and genome expression are highly advanced. However, many environmentally-encountered developmental toxicants are non-mutagens. If it is confirmed that paternal exposure to such toxicants can lead to developmental abnormalities, it becomes necessary to consider that their cellular targets may be extra-nuclear components of the sperm.

One of these components is the plasma membrane. The plasma membrane of the zygote appears to be a mosaic of oocyte and sperm plasma membranes. Sperm-specific cell surface antigens persist up to blastocyst formation and exist in sufficient quantity to mediate complement-mediated cell lysis. This fact supports the hypothesis that alterations of sperm-derived embryonic cell surface components might be able to affect plasma membrane function in subtle ways to impair embryo viability prior to implantation. Understanding this contribution of sperm in relation to the failure of the embryo is especially important in humans where up to 50% of conceptions fail during preimplantation development.

Two other non-nuclear components of the sperm that are transferred to the zygote are the tailpiece and mitochondria. Both appear to break down in the first few cell divisions of the conceptus and are not considered to be of any functional significance. However, a better understanding of the fate of their breakdown products and of their potential to act as carriers of toxicants into the zygote would help resolve questions regarding non-nuclear contributions of the male to developmental toxicity.

Relation Between Sperm DNA and Development of the Conceptus

Three important questions regarding paternal exposures and developmental abnormalities are: (1) Is the agent capable of inducing DNA alterations in sperm or its precursor cells? (2) If so, what is the nature of these alterations, i.e., chromosome rearrangements, chromosome misassortment, multi-locus deletions, intragenic alterations including single-base-pair-changes, methylation or demethylation of specific genes? (3) What dominant phenotypes are likely to be associated with specific types of DNA alterations?

Whether or not an agent is capable of inducing DNA alterations in male germ cells can be determined by a number of genetic methods such as the morphological or biochemical specific-locus tests, the heritable translocation test, transgenic-shuttle-vector systems, and dominant-lethal assays. In addition, tests carried out directly in male germ cells such as detection of DNA adducts, measurement of unscheduled DNA synthesis, or detection of DNA strand breakage by alkaline elution, provide good evidence of induced DNA alterations. Standard tests for changes in methylation patterns or other possible imprinting effects remain to be developed.

The specific nature of DNA alterations cannot be revealed by all tests in the above list. For example, the dominant-lethal test is not suitable for molecular analyses of induced genetic changes, and the shuttle-vector systems are currently incapable of detecting multi-locus deletions, gross chromosome rearrangements, or numerical chromosome anomalies. However, some of the tests are designed to detect specific endpoints, e.g., aneuploidy or heritable translocations. The specific-locus tests can detect lesions of various types and methods are now available to characterize these mutations at the molecular level.

While direct studies of dominant phenotypes provide relevant health endpoints, little is known at present about the associations between specific types of abnormalities and the types of DNA alterations that produce them. An increased knowledge of such associations could lead to testing shortcuts, since information obtained in experiments discussed above could then be used to make predictions of phenotypic outcomes.

Molecular Experiments to Determine the Basis of Male-Mediated Developmental Effects

Early mammalian development is regulated by the precise orchestration of a complex series of molecular events, both genetic and nongenetic. Any change in the profile of the individual molecules within the early embryo could potentially disrupt the normal sequence of events and lead to detectable developmental abnormalities in the later-stage embryo. Such changes could result from mutations induced in paternal germ cells or from effects on imprinting of paternally-contributed genes. Therefore, in attempts to evaluate the molecular basis of any male mediated toxicological effects on development, efforts should be directed at the generation of methods to monitor changes in the normal population of molecules present within the embryo at several stages of early development.

Methodologies for the efficient production of complete cDNA libraries have been developed and are being utilized to characterize several stages of early mouse development, including the 2-cell, 4-cell, and 8-cell embryos, as well as the blastocyst. Every library is comprised of hundreds of thousands of DNA clones representing the mRNA molecules that are present at a particular stage of the embryo. These libraries are being compared at the molecular level by subtracting each with another in a pair-wise manner to identify those clones that are present within one library and not another, e.g., subtracting cDNA common to both the 2-cell library and the zygote library. Overall, one of the most noteworthy advantages of this subtraction hybridization approach stems from the fact that very little starting material is necessary to conduct the experiments.

Utilizing a cDNA library subtraction strategy, we recommend that cDNA libraries, prepared from embryos whose fathers were treated with an environmental agent, be subtracted with normal libraries derived from embryos from untreated fathers. In this way, those mRNAs that are uniquely produced in or are missing from the treated embryo as a result of paternal exposure could be identified and characterized at the molecular level. Additional experiments utilizing the reverse-transcriptase PCR (RT-PCR) methodology could be used to further characterize the normal pattern of expression of these genes, and ultimately, targeted mutagenesis procedures could be used to evaluate the roles these individual genes play in normal mammalian development.

COLLABORATION BETWEEN EPIDEMIOLOGISTS AND LABORATORY RESEARCHERS

Perhaps the most easily achievable and productive activity that can be recommended is closer communication and collaboration between researchers conducting laboratory experiments and those conducting epidemiological studies.

Laboratory scientists working with animal model systems can identify environmental agents that affect development in offspring of exposed fathers, can characterize these effects with respect to such factors as dose response and dose-rate effects and germ-cell stages affected, can conduct mechanistic studies to determine the precise nature of the interaction between the toxicant and the cells affected, and may be able to define, at the molecular level, the nature of the induced change leading to abnormal development. Such information provides important clues as to exposed populations of humans that are most likely to be at an elevated risk of producing offspring with developmental abnormalities and, thus, should be primary candidates for epidemiological studies.

Conversely, epidemiological studies can identify populations that exhibit increased risks of developmental abnormalities and may associate these effects with occupations, chemical exposures, personal habits or other risk factors. Epidemiologists can, thus, provide laboratory researchers with clues as to environmental agents that have a high probability of being human reproductive toxicants and are most deserving of laboratory testing and research.

By integrating the efforts of laboratory researchers and epidemiologists, resources can be more effectively directed toward identifying and characterizing those exposures that are most likely to pose a risk to human development.

CONCLUSIONS

1. The concept of MMDT requires clearer definition. 2. It needs to be determined if paternal exposures can result in developmental toxicity through mechanisms other than genetic ones. 3. If so, a) research should be conducted to understand targets and mechanisms, b) animal models should be developed to permit identification of such toxicants, and c) the extent of this phenomenon as a public health issue should be determined. 4. The study of genetic causes of MMDT should be enhanced to identify agents capable of inducing DNA alterations in paternal germ cells, determine the nature of such alterations, and develop the capacity to predict developmental outcomes resulting from various classes of DNA alterations.

PHYSICIAN AND PATIENT EDUCATION

J. M. Friedman

Department of Medical Genetics
University of British Columbia
226 - 6174 University Boulevard
Vancouver, Canada V6T 1Z3

INTRODUCTION

It is appropriate that the critical scientific reviews of various aspects of male-mediated developmental toxicity that have been presented at this meeting are being followed by a session on risk assessment and policy issues because these are the contexts in which this science is applied in our society. Similarly, it is appropriate to close the session on risk assessment and policy issues with a discussion of physician and patient education because this is the context in which most people who encounter male-mediated developmental toxicity in a personal way do so.

It is important to point out that "Physician and Patient Education" really encompasses public education as well. People spend only a tiny fraction of their lives as patients, and they bring to that experience the knowledge and understanding gained in the rest of their lives.

When one speaks of knowledge and understanding of developmental toxicity among the general public, one is struck that there seems to be precious little of either, even among people who are otherwise well-educated and well-informed. In fact, much of what the public seems to "know" about developmental toxicity is incorrect. As Will Rogers said, "It's not what we don't know that hurts. It's what we know that ain't so." To illustrate this point, I shall consider ten commonly-held misconceptions* in developmental toxicity.

* The pun is intentional, but Dr. Victor McKusick deserves the credit (or blame) (McKusick, 1971).

Male-Mediated Developmental Toxicity, Edited by D.R. Mattison
and A.F. Olshan, Plenum Press, New York, 1994

MISCONCEPTIONS IN DEVELOPMENTAL TOXICOLOGY

The background risk of birth defects is zero. Many people do not know that every pregnancy for every couple is associated with a risk of about 5% for producing a baby with a serious congenital anomaly or mental retardation that is apparent by a year of age (Heinonen *et al.*, 1977; Baird *et al.*, 1988). This risk pertains even to pregnancies for young healthy couples with no family history of congenital anomalies in which there have been no exposures to any teratogens or mutagens.

All birth defects have an identifiable cause. The cause of most congenital anomalies is unknown. Purely genetic factors account for about 25% of all birth defects, teratogenic factors for about 10%, and a combination of poorly characterized genetic and non-genetic factors (so-called "multifactorial etiology") for another third (Beckman and Brent, 1984). The cause of the remaining birth defects is completely unknown. In fact, some may result from stochastic "noise" in the complex process of embryonic development and not have a definable cause as such (Kurnit *et al.*, 1987).

Some chemicals are reproductive toxins; others are not. The occurrence of reproductive toxicity depends not just upon the chemical nature of an agent but also upon its dose, route of exposure, frequency of exposure, and timing of exposure with respect to gametogenesis and/or embryonic development. Factors related to metabolism, genetic background, and other concurrent exposures often are relevant as well. It is the exposure in this broader context, not just the chemical, that may pose a risk of reproductive toxicity.

A toxin is a carcinogen is a mutagen is a teratogen. Substances may produce many different kinds of toxicity, and there is no necessary relationship of any one kind to another. Nevertheless, many people believe that if an agent is "toxic" in any sense, it has the capacity for reproductive toxicity as well.

Reproductive toxicity is reproductive toxicity. Some reproductive toxins produce infertility, some miscarriage, and some birth defects. Although there are agents that may cause any of these adverse effects, there are many that produce just one or two of them. Particular confusion may result when rare events (e.g., a specific malformation) and common ones (e.g., miscarriage) are considered together within a small group of women, producing an apparent "cluster" of "adverse reproductive outcomes".

An agent that affects a man's sperm will affect his offspring. The relationship between sperm morphology or number and the genetic content of the sperm is unclear. While the physical characteristics of sperm in an ejaculate may be related to the likelihood of fertilization, this does not mean that the spermatozoon that actually succeeds in fertilizing the ovum is itself abnormal.

Animal studies have no relevance to humans. This idea results at least in part from repeated sensational accounts in the media of findings in animal studies that do not square with the experience in humans. While most scientists understand that there are usually substantial differences between exposure conditions encountered by humans and those imposed upon animals in reproductive toxicology experiments and that the ability to extrapolate the results of animal experiments to humans is limited, many members of the public lack such understanding.

Birth defects are the mother's (or the doctor's) fault. Most congenital anomalies are neither predictable nor preventable with current technology. They are no one's fault. This is not surprising, given that so many congenital anomalies are of unknown etiology.

It is possible to avoid exposure to chemicals. Many people think that chemicals are limited to manufactured products and fail to realize that not only our entire world but we ourselves are composed entirely of chemicals. Even manufactured chemicals are ubiquitous. One cannot avoid exposure to chemicals, although it is certainly prudent to avoid toxic exposures as much as possible.

There is nothing that we can do to prevent birth defects. Given the frequency, variety, and largely unknown etiology of congenital anomalies, prevention of birth defects may seem impossible. Nevertheless, some things can be done, and should. For example, congenital rubella, Rh hemolytic disease, and isotretinoin embryopathy are largely preventable teratogenic disorders. Although male-mediated developmental toxins may not substantially increase the risk of congenital anomalies among the children of an individual man who has been exposed to them, such agents may nevertheless produce an important effect at the population level. It is, therefore, necessary to identify and control exposures to male-mediated developmental toxins appropriately.

Recognizing that many public misperceptions regarding developmental toxicology exist is an important first step, but it only increases the need for knowledgable scientists to disseminate the information in this field effectively. How can this best be done?

EDUCATING PHYSICIANS AND PATIENTS ABOUT DEVELOPMENTAL TOXICOLOGY

In order to educate physicians and the public about developmental toxicology, we must provide them with information in a simple, consistent, and accurate fashion. We must also concentrate on the issues that concern them, which may or may not be the same aspects of developmental toxicology that concern us as investigators.

The key question that most people have is: "Will this exposure harm my children?" Thus, in interpreting our knowledge of developmental toxicology to the public we should concentrate on absolute rather than relative risks. In epidemiological studies, this means explaining results in terms of their effect on the overall background risk of birth defects. In animal studies, this means trying to provide the results in terms of their likely implications for individual human couples.

The messages we provide the public regarding reproductive toxicology must be presented in a clear and simple fashion. In general, the scientific sophistication of the public is poor, and the subtleties and qualifications that characterize scientific discussions tend only to confuse most people. The information we provide must be accurate and, as far as possible, unequivocally factual. Scientists expect their "facts" to be subject to continual revision; the public do not. Speculation is essential to the progress of science but is frequently counterproductive in public education.

THE BOTTOM LINE

To this point, what I have said seems unlikely to evoke substantial controversy. I shall now put forth what I believe to be the messages regarding male-mediated developmental toxicology that are most important for us to transmit to physicians and their patients. In doing so, I expect to elicit discussion which I hope will lead to the development of a consensus that can be widely supported within the scientific community. I suggest that we concentrate our public and physician education regarding male-mediated developmental toxicity on the following three points:

1) A man's exposure to toxic agents is unlikely to affect the risk of birth defects in his children substantially.

2) Decisions to limit exposures to toxic agents should usually be based on the risk to the man himself.

3) When in doubt, find out!

The data reviewed at this meeting clearly indicate that exposure of male experimental animals to some chemical and physical agents may adversely affect their reproductive capacity and produce congenital anomalies among their offspring. Even though the direct evidence in humans is not compelling, there can be no doubt that germinal genic or chromosomal mutations result from certain paternal exposures in our species, too. However, given the relatively low frequency of congenital anomalies due to new constitutional mutations in the general population compared to the total frequency of birth defects, it seems unlikely that a substantial increase in the overall rate of congenital anomalies would be produced by exposure of a man to a germ cell mutagen (Narod *et al.*, 1988; Rüdiger, 1991). Mechanisms other than mutation may account for adverse effects of some paternal exposures on the development of the offspring, but there is no reason to suspect that any or all of these mechanisms produce congenital anomalies among the offspring of men exposed to such agents more frequently than new mutations do.

Chemicals and physical agents may cause many kinds of toxicity in exposed adults. It is, of course, desirable to avoid or limit such exposures appropriately. Since exposures of a male are unlikely to increase his risk of having a child with birth defects substantially, decisions regarding limitation of toxic exposures should usually be based on the risk to the man himself rather than on the risk to his children.

Finally, people and their physicians need to know that developmental toxicology is a science that has something to offer both in terms of public policy and individual risk assessment. Current authoritative information on the potential developmental toxicity of various exposures is available in databases such as REPROTOX (Reproductive Toxicology Center, Columbia Hospital for Women, Washington, DC), REPRORISK (Micromedex, Inc., Denver, CO) and HSDB (Toxicology Information Program, U.S. National Library of Medicine, Bethesda, MD). Expert consultation can also be obtained when necessary.

REFERENCES

Baird, P.A., Anderson, T.W., Newcombe, H.B., and Lowry, R.B., 1988, Genetic disorders in children and young adults: A population study, *Am. J. Hum. Genet.* 42:677-693.

Beckman, D.A. and Brent, R.L., 1984, Mechanisms of teratogenesis. *Ann. Rev. Pharmacol. Toxicol.* 24:483-500.

Heinonen, O.P., Slone, D., and Shapiro, S., 1977, "Birth Defects and Drugs in Pregnancy," PSG Publishing Company, Littleton, Massachusetts.

Kurnit, D.M., Layton, W.M., and Matthysse, S., 1987, Genetics, chance, and morphogenesis, *Am. J. Hum. Genet.* 41:979-995.

McKusick, V.A., 1971, Fourteen genetic misconceptions. *Ann. Intern. Med.* 75:642-3.

Narod, S.A., Douglas, G.R., Nestmann, E.R., and Blakey, D.H., 1988, Human mutagens: Evidence from paternal exposure, *Environ. Molec. Mutagen.* 11:401-415.

Rüdiger, H.W., 1991, Clinical, genetic and regulatory consequences of exposure to mutagens, *Ann. Génét.* 34:173-178.

RISK ASSESSMENT AND RISK MANAGEMENT

Paul B. Selby

Biology Division
Oak Ridge National Laboratory
Oak Ridge, TN 37831-8077

INTRODUCTION

The members of the Breakout Group on Risk Assessment and Risk Management were Donald R. Mattison, Marvin L. Meistrich, William L. Russell, Harold Zenick, and myself. Many issues related to risk assessment and risk management were discussed. Although the group attempted to consider the broad range of issues represented at this conference, considerably more emphasis was given to genetic effects because quantitative risk estimation is much further advanced for genetic effects than it is for the other endpoints discussed.

Quantitative genetic risk estimation has also been developed more extensively for ionizing radiation than it has been for other physical agents or for chemicals. For that reason the group's discussion was influenced considerably by the experiences and reports of the United Nations Scientific Committee on Atomic Radiation (i.e., UNSCEAR) and of the Committees on the Biological Effects of Ionizing Radiations (i.e., the BEIR Committees) of the United States National Research Council.

This conference was especially valuable because it focused attention on the importance of effects found in progeny following exposure of parents, particularly fathers. A current, much more common, focus of attention regarding health effects of environmental chemicals and physical agents is upon the individuals who are actually exposed, especially regarding whether they are likely to develop cancer. That preoccupation with somatic effects is reflected in research spending. In recent years, however, the press has shown heightened interest in genetic diseases, partly as the result of the revolution occurring in molecular genetics. That increased attention, as well as the many remaining uncertainties regarding genetic risk estimation, will probably, over time, restore what used to be considerable public concern over genetic effects—that is, over health effects in children, and later descendants, of exposed individuals. We feel that much too little research is currently being funded to understand such effects.

Male-Mediated Developmental Toxicity, Edited by D.R. Mattison and A.F. Olshan, Plenum Press, New York, 1994

The meeting of the Breakout Group on Risk Assessment and Risk Management is summarized in the following recommendations and discussion.

DOMINANT LETHAL MUTATIONS

We suggest that more work be done to understand the mechanism of induction of dominant lethal mutations by chemicals, both concerning preimplantation and postimplantation loss. It would undoubtedly be important to identify physical agents or chemicals that induce miscarriages or that decrease, by more than a slight extent, the chances of having children. However, in this regard, it is of interest that the risk estimation committees for radiation mentioned earlier do not now include dominant lethal mutations in the genetic risk estimates, and they have not included them for many years. Indeed, the charge to these committees has typically been to estimate the risks of genetic disease among the live-born offspring. As recently as 1986, UNSCEAR discussed the implications for genetic risk estimation of data on dominant mutations causing death in mice at any time between conception and three weeks of age. However, only those deaths that occurred after birth were considered relevant to genetic risk estimation. The probable reason for this emphasis is that much of the dominant lethality is expected to occur so early in embryogeny that women would not even be aware that they were pregnant at the time of the conceptus's death. Furthermore, the risk estimates are generally made for small levels of exposure, such as 0.01 Gray. For such exposures, the frequency of induced miscarriages, for example, would be extremely small in comparison to the high normal incidence of miscarriages of about 15% of recognized pregnancies.

The dominant-lethal test is undoubtedly a valuable assay for identifying certain types of mutagens, but more needs to be known, particularly for chemicals, about how induction of dominant lethality relates to human health hazards.

THE NEED FOR QUANTITATIVE GENETIC RISK ESTIMATION

The primary concern in genetic risk assessment is to develop methods for estimating *quantitatively* the risk of having deleterious mutations in offspring following exposure of their parents to physical agents or to chemicals. This concern is especially warranted for known or suspected mutagens for which there is widespread exposure, or for which there is less exposure but mutagenicity is suspected to be high. Short-term tests for mutagenicity, epidemiological studies, and other toxicity studies can be useful for identifying chemicals or physical agents for which quantitative genetic risk estimation should be developed. However, short-term assays and the other studies sometimes suggest high risks even when, after more is known, the genetic risks prove to be small or nonexistent. One reason for this is that short-term tests for mutagenicity have been found to be limited in their ability to predict which chemicals induce *germinal* mutations in mammalian germ cells. Furthermore, epidemiological studies are often fraught with serious problems regarding the exposure levels of the subjects and possible confounding factors. If a significant correlation is found between genetic diseases and the exposure of fathers to some substance, one of the most important questions that arises is whether the statistically significant finding is biologically plausible. In many cases, too few data are available from the truly relevant test systems to evaluate the findings carefully in this regard.

Controlled experimentation in the laboratory is necessary to resolve many of the above questions. For this reason, it seems especially important that a solid base of laboratory evidence on germ-cell mutation induction be collected for several chemicals that were shown to be effective mutagens in other assays. Such information is needed to elucidate the mechanisms of induction—for example, the many factors that influence the mutational response, such as differences in germ-cell stages within and between the sexes, dose-response, the effect of protracting the exposure, the effect of fractionating the exposure, and so on.

NEAR-TERM SOLUTIONS TO QUANTITATIVE GENETIC RISK ESTIMATION

Various advances in human and experimental genetics, to be discussed later, should eventually permit much more precise quantitative genetic risk estimation. However, such a level of precision is probably unnecessary for making reasonable risk assessment and risk management decisions. The experience that has been gained in making quantitative genetic risk estimates for radiation—the subject of committee deliberations for more than thirty years—suggests ways to deal with chemicals or other physical agents using presently available methods. In this regard, it seems useful to consider the two methods of quantitative genetic risk estimation that were used in UNSCEAR's most recent report in 1988.

One method, called the indirect method, is often referred to as the doubling-dose method. It has been used in various forms since the 1950s, and it is based on the assumption that induced mutations are similar to spontaneous mutations. This has long been known not to be true, since a given agent—radiation or chemical—induces qualitatively different kinds of mutations in different germ-cell stages. In the indirect method of genetic risk estimation, an estimate is made, for each pattern of inheritance, of the amount of induced genetic damage expected at genetic equilibrium if there is a constant exposure in all generations until that time. (Genetic equilibrium would occur when a steady state is reached between newly induced mutations and the selection of such mutations out of the population.) Then, based on assumptions about the persistence in the population, from generation to generation, of different types of mutations, an estimate of the genetic damage in the first generation can be derived from the equilibrium estimate. A major limitation of this method is that, as now applied, it is used to estimate risk for only those genetic diseases that have simple patterns of inheritance. Such diseases constitute only about 2% of the genetic diseases discussed in the 1988 UNSCEAR report. Primarily because of its present application to only a small fraction of serious genetic diseases, because of its roundabout method of calculation, and because of the considerable uncertainty of its underlying assumption that induced mutations are similar to spontaneous mutations, we prefer the other method of quantitative genetic risk estimation, which is called the direct method.

The direct method is based directly on the amount of damage seen in first-generation mouse progeny following exposure of their fathers to radiation or to a chemical. As it is currently applied, the method is based on frequencies of induced dominant mutations causing either malformations in the skeleton or cataracts. The induced frequency of mutations causing serious effects is then extrapolated to the entire genome by using a multiplication factor that expands the frequency of damage in the body system studied to an estimate of the total damage in all body systems. For the skeleton, the multiplication factor that has been used to estimate total damage is 10.

NEW DATA FOR USE IN THE DIRECT METHOD

A long-term experiment presently being conducted at Oak Ridge National Laboratory should provide many improvements in the data available for quantitative genetic risk estimation for radiation by the direct method. It will also provide information on the relationship between different types of molecular damage and the extent of dominant damage. Four different treatments are being tested, one with X rays at high dose rate, one with gamma rays at low dose rate, and two with chemicals, namely ethylnitrosourea (ENU) and chlorambucil (CHL). Each chemical treatment is applied to the germ-cell stage known to be most sensitive to that treatment. ENU is known to induce primarily base pair changes in spermatogonial stem cells, CHL induces large lesions in postspermatogonial stages, and radiation induces both types of molecular damage, as well as others, in spermatogonia. One goal of this work is to determine whether there is a strong correlation between specific-locus mutation frequencies and frequencies of dominant phenotypic damage that has clinical relevance. The endpoints being used in these experiments include skeletal malformations, cataracts, other eye abnormalities, stunted growth, dominant visibles, and survival to 11 weeks of age. Early results already provide strong reasons to believe that the various assumptions that had been used in earlier applications of the direct method did not lead to any large underestimation of genetic risk. Many of the assumptions used in the earlier application of the skeletal data in the direct method will no longer be needed once these new data are available. Since the results to date suggest no important change from the earlier risk estimates, they greatly increase confidence in those estimates.

While the above-mentioned experiments cover a rather broad range of types of phenotypic damage, there are still many types of damage for which there are no data available. Some potentially useful types of damage have never been tested, and for other types of damage, such as lung tumors and malformed fetuses, there is, for various reasons, much more uncertainty about how to extrapolate the available results to genetic diseases in humans.

To improve our understanding of the reliability of extrapolating from mice to humans, efforts should be made to collect data from a broader array of biological systems. We could better understand the degree of uncertainty in extrapolating from mice to humans by determining, for example, how well we can extrapolate the extent of induced skeletal damage between different mouse strains, or from mice to other small mammals such as hamsters or rats. Valuable studies of this type, on cataract mutations in hamsters, are in progress at the Gesellschaft für Strahlen- und Umweltforschung in Germany.

Experiments must also be conducted to determine the extent of induced phenotypic damage detectable in first-generation progeny of exposed female mammals.

HUMAN DATA ON GENETIC RISK

Results of the massive epidemiological cohort study of Dr. James V. Neel and coworkers on the children of the atomic bomb survivors in Hiroshima and Nagasaki greatly increase confidence in estimates of genetic risk based on the mouse model, described in the previous section. In over four decades of intensive study, no evidence has been found of induced genetic effects in those populations in Japan. Detailed analyses of children of exposed and unexposed survivors of the atomic bombings at Hiroshima and Nagasaki have revealed no statistically significant differences for the following indicators of possible effects: stillbirth, major congenital defect, death among live-born children up to an average age expectancy of 26 years, sex ratio, physical development at various ages, cancer with onset prior to the age of 20 years, presence of certain chromosomal abnormalities, and occurrence of mutations affecting the electrophoretic mobility or physiological activity of a series of 30 polypeptides. In regard to this finding of no evidence of any effect, two conclusions from the direct method of genetic risk estimation based on the mouse model are worthy of note. If the risk estimate based on mutation studies in mice is applied to the doses and sample sizes in the human populations, it is predicted (a) that some children would have radiation-induced genetic diseases and (b) that so few of these diseases would be expected that the finding of a statistically significant increase over the control would be most unlikely. The human and mouse data are thus not inconsistent. Although the negative human data cannot rule out the rather small level of risk suggested by the mouse model, they at least provide no reason to think that the risk is higher than that predicted using that model.

With this valuable comparison in mind, we recommend that an attempt be made to identify chemicals of special importance, in terms of human exposures, for which it might be worthwhile to collect extensive information regarding genetic risk both in the laboratory and in epidemiological studies (perhaps using cohort studies as in Japan). If the phenotypic damage predicted from the mouse model were found to be consistent with the findings in humans, we could be much more confident of predictions from similar controlled experiments on other chemicals. Possible chemicals that might be considered were suggested by the group. These were methanol (because of its possibly greatly expanded use as a fuel), ethylene oxide, and ethanol. An epidemiological study that could prove useful in this kind of comparison is the one discussed by John Mulvihill (this volume) on the children of patients who survived after receiving therapy for cancer.

DETECTION OF GENETIC EFFECTS IN EPIDEMIOLOGICAL STUDIES

Another recommendation is that geneticists work with epidemiologists to suggest ways in which they might improve selection of endpoints in their studies. For certain agents, laboratory studies might suggest the inclusion or exclusion of particular types of disorders. For example, if a chemical were known to induce trisomy in animal experiments, it would become especially relevant to look for Downs syndrome or Trisomy 18 in an epidemiological study.

In epidemiological studies, an important distinction must be recognized between the effects expected in exposed individuals and those expected in the offspring of exposed individuals. In the former case, it would not be surprising if only a narrowly restricted kind of effect were induced. For example, this might be expected if a cell type in a particular tissue had a marked sensitivity to the exposing agent. In the latter case, unless there is a situation such as that mentioned for trisomies above, the presumed occurrence of dominant mutations anywhere in the genome indicates that it would be highly unlikely that increases in genetic diseases would be restricted to one or a few of the thousands of possible genetic diseases.

One approach sometimes considered when searching for induction of mutations in human populations is to look for what are termed sentinel phenotypes among the progeny. This is a group of about 40 diseases, each of which is caused by a single highly penetrant mutation. These diseases are individually so rare that even their combined total is still so small that extremely large samples would be needed to provide any chance of finding an increase, unless the mutation frequency were extremely high, indicating a serious risk from the exposure. It would, however, seem impractical to limit a genetic epidemiological study to sentinel phenotypes. A more practical approach would be to look for an effect on a broad array of congenital malformations and other known genetic diseases. Unless a broad grouping of congenital malformations and other genetic diseases showed an increase, a mutational explanation would seem extremely unlikely.

POSSIBLE FUTURE IMPROVEMENTS IN QUANTITATIVE GENETIC RISK ESTIMATION

The rapid advances in human and experimental genetics, including the Human Genome Project and attempts to learn about the structure and functions of numerous genes, should eventually provide methods for determining, (a) the specific changes induced in the germ-cell DNA of clinically relevant genes by particular treatments, and (b) which of these changes would be transmitted to the offspring and cause phenotypic changes of clinical importance. Such information would undoubtedly greatly increase the precision of quantitative genetic risk estimates. It is also expected that much will be learned about the relationships between genes and phenotypes in both humans and experimental mammals. Such information should, for example, provide a firmer basis for extrapolating phenotypic damage detected in first-generation progeny of exposed experimental mammals to humans. Such advances, together with more precise methods for measuring human exposure levels, may well lead to impressive improvements in the precision of quantitative estimates of genetic risk.

It is important to realize, however, that a rather precise method of genetic risk estimation, such as that described above, will not be possible for several decades. Indeed, presently we are only just beginning to learn how to relate particular kinds of DNA damage to serious health effects in the progeny. Some types of molecular damage may be much more likely than others to cause serious abnormalities in offspring, and quantitative estimates of genetic risk can only be extremely crude until such relationships are understood.

The difficulty of this problem is illustrated by the following example which shows that quantitative estimates of genetic risk are vastly more complex than just simply counting all mutations. Based on the results of W. L. Russell's experiment in which male mice were injected with four 100 mg/kg fractions of ENU, it is known that each first-generation offspring must carry a large number of newly induced mutations throughout the genome. (Indeed, on the average, one out of every 94 offspring had a specific-locus mutation at one or another of the seven genetic loci that Russell studied, and there are probably several tens of thousands of other genes in the haploid mammalian genome.) Nonetheless, even though every conceptus must have carried many mutations, Russell found that the number of conceptuses that failed to survive until 3 weeks after birth was only 11% higher than in the untreated control. Furthermore, studies of various types of dominant mutations by others, including myself, using similar or identical ENU regimes, suggest that probably fewer than 10% of the live-born offspring exhibit effects of clinical relevance that can be detected by careful examination of the offspring, including detailed examinations of their entire skeletons and slit-lamp examinations of their eyes. Furthermore, the high mutation frequency found by Russell was for visible mutations. Many additional DNA lesions would undoubtedly have consequences far too subtle to result in detectable phenotypes. The above illustration makes it apparent that much more must be learned about the relationship of DNA damage to phenotypic damage before precise methods of genetic risk estimation will become possible.

A further illustration of the difficulty of this problem is illustrated by reference to the Human Genome Project. If that project is successful, the complete sequence of human DNA will be known in an estimated ten years. (Of course, everyone except identical twins will have a different sequence and could respond differently to the same mutation.) Even if the structural sequence for each person could be easily determined, the functions will be known for only a very small fraction of the tens of thousands of genes thought to be present in a single haploid genome. A further high degree of complexity will result from the interactions of certain genes in producing phenotypes.

Obviously, a rather precise method of genetic risk estimation, such as that described earlier in this section, cannot become practical until genetic knowledge, and experimental techniques, become vastly more advanced.

Fortunately, as noted earlier, such precision in quantitative genetic risk estimation is unnecessary for reaching many important decisions in risk assessment and risk management. Some of the suggestions made in this paper could lead to reasonable estimates of genetic risk for many mutagens in the foreseeable future.

CHOICE OF TERMINOLOGY

To avoid confusion, it would seem worthwhile to return to the use of the term germ-cell mutagenesis, instead of male-mediated developmental toxicity, when the intention is to refer to genetic changes. Everything discussed so far in this paper relates to mutagenesis. A major component, and the best understood part, of male-mediated developmental toxicity is undoubtedly mutagenesis. Perhaps a new term, such as "non-mutational male-mediated developmental toxicity," should be coined to encompass the non-mutagenic phenomena described at this conference.

CONCLUSION

The recommendations in this paper apply only to risk assessment for germ-cell mutagenesis. It is hoped that some of the ideas advanced might stimulate progress in dealing with risks from non-mutational male-mediated developmental toxicity.

During discussion of the report of the Breakout Group on Multidisciplinary Approaches, a request was made for references that would help those trying to learn about new areas. In response to the request, several selected reports and reviews are listed below. These sources provide more detailed information on material presented in this report.

ACKNOWLEDGMENTS

Research jointly sponsored by the Office of Health and Environmental Research, U.S. Department of Energy under contract DE-AC05-84OR21400 with Martin Marietta Energy Systems, Inc., and by the National Institute of Environmental Health Sciences under IAG No. 222Y01-ES-10067.

SELECTED READINGS

BEIR Report, 1980, "The Effects on Populations of Exposure to Low Levels of Ionizing Radiation: 1980," National Academy of Sciences, National Research Council, Washington, D.C.

BEIR Report, 1990, "Health Effects of Exposure to Low Levels of Ionizing Radiation—BEIR V," National Academy of Sciences, National Research Council, Washington, D.C.

Meistrich, M.L., 1992, A method for quantitative assessment of reproductive risks to the human male, *Fundamental and Applied Toxicology* 18:479.

Selby, P.B. Radiation genetics, in: "The Mouse in Biomedical Research, Vol. 1 History, Genetics, and Wild Mice," H.L. Foster et al., eds., Academic Press, New York, (1981).

Selby, P.B. Experimental induction of dominant mutations in mammals by ionizing radiations and chemicals, in: "Issues and Reviews in Teratology, Vol. 5," H. Kalter, ed., Plenum Press, New York (1990).

United Nations, 1986, "Genetic and Somatic Effects of Ionizing Radiation. United Nations Scientific Committee on the Effects of Atomic Radiation, 1986 Report to the General Assembly, with annexes. United Nations sales publication E.86.IX.9," United Nations, New York.
United Nations, 1988. "Sources, Effects and Risks of Ionizing Radiation. United Nations Scientific Committee on the Effects of Atomic Radiation, 1988 report to the General Assembly, with annexes. United Nations sales publication E.88.IX.7," United Nations, New York.

MULTIDISCIPLINARY APPROACHES: WORKSHOP REPORT

Jennifer M. Ratcliffe*

Department of Epidemiology
The School of Public Health
McGavern-Greenberg Hall, CB# 7400
University of North Carolina
Chapel Hill, NC 27599-7400

INTRODUCTION

It may seem almost axiomatic to state that a multidisciplinary approach to the study of male-mediated developmental toxicity (or other research topics) is necessary, or useful, or desirable, much as we routinely state that "further research is needed" in the conclusions of scientific papers. On a practical level, however, it is arguable that little consideration has been given to questions such as: Why and in what situations are multidisciplinary approaches needed or advantageous? What are the major obstacles to putting such approaches into effect? What recommendations can be made to improve the implementation of multidisciplinary approaches? In an attempt to focus on these questions, the group drew extensively on individual experiences in a number of different disciplines, as well as on the collective experience of participants at this multidisciplinary conference.

WHY MULTIDISCIPLINARY APPROACHES ARE NEEDED

Broadly speaking, the need for multidisciplinary approaches to the study and control of male-mediated reproductive toxicants can be said to rest on two major factors:

a) the need to delineate (i) the interrelationships between the endpoints of mutagenicity, carcinogenicity and reproductive toxicity in different species in vivo or different test systems in vitro, and (ii) the interrelationships between specific reproductive endpoints (such as between male germ cell-mediated embryotoxicity and spermatotoxicity). Such information is critical to identifying suspect male reproductive toxicants, extrapolating across species, understanding underlying mechanisms and evaluating the reproductive toxicity of given agents;

---

* Significant contributions to this workshop were made by A. Olshan, L. Anderson, K.S. Kasprzak, G. Bunin, M. Kharrazi, Jianping Qin and Xiping Xu, among others.

b) the need to involve a number of different disciplines in (i) the design and conduct of studies of male reproductive toxicants, (ii) the dissemination and interpretation of information on reproductive toxicity, and (iii) the setting of research priorities, policies and regulations regarding the public health impact of male-mediated reproductive toxicity.

Situations in which these needs arise can be identified in three major areas. In the first, the area of **implementing individual research projects**, it has increasingly become the case that collaboration between researchers in several different disciplines is required to carry out a given project. A typical example would be a cytogenetic and reproductive study of men occupationally exposed to an industrial chemical. The initial **choice of agent to study** would (should) rely extensively on e.g. any available toxicologic evidence regarding its mutagenicity, spermatotoxicity, and effects on reproductive outcomes, combined with data on available biomarkers of exposure and industrial hygiene data on the estimated extent and level of human exposure. The actual **design and conduct** of the study would rely on the combined disciplines of, for example, cytogenetics, andrology, epidemiology and biostatistics, analytical chemistry and industrial hygiene. In this area, then, multidisciplinary approaches are clearly already being used, and here the main obstacles to the efficiency with which they are used may be a lack of familiarity with the terminology, premises, etc. of other disciplines; this would also affect the intelligent **interpretation** of the results of a given multidisciplinary study.

The second area concerns the **integration, summarization, evaluation and dissemination** of data from different disciplines on reproductive hazards for a) researchers in other fields and b) individuals and groups who may be involved in the clinical treatment, education, or counseling of people with reproductive health problems and/or exposure to suspect reproductive toxins. Such people include physicians, pharmacologists, nutritionists, health educators, factory health and safety personnel, genetic counselors and health educators, most of whom have training in disciplines different from each others' and from those of the people actively engaged in research. The process of integrating and evaluating such data may be done by individuals to produce books, guidelines, summaries etc., by groups of technical experts in order to develop data bases that are readily accessible (e.g. the Reprotox, reproductive toxicity data base from the Columbia Hospital for Women's Medical Center in Washington, or the Toxicology Information Response Center at Oak Ridge, Tennessee), or by agencies such as the International Agency for Research on Cancer (IARC) which reviews the mutagenicity, genotoxicity and carcinogenicity of chemical agents in its yearly monographs.

It is in the third area, in **research program planning and the development of public health policies and regulations**, that the need for multidisciplinary approaches may arguably be most critical. Program directions and policy decisions, including regulations, may need to be decided by bodies as small as an individual laboratory or department, whose members need to determine what research projects to submit for funding and to what funding sources, through to research institutes, government agencies, funding bodies and regulatory agencies, which need to determine which agents to study or regulate, which programs to fund and so on. In all cases, knowledge of the status of research in several disciplines is often required to make intelligent program directives, policies or regulations and to ensure the efficient use of scarce resources. For example, the National Institute of Environmental Health Sciences recently constituted several multidisciplinary committees to decide on issues such as high priority areas for future research, and on specific research projects to include in the institute's new research program on lead. It is also clear, for example, that knowledge of any available epidemiologic data on male-mediated reproductive effects would usefully inform the process of determining agency priorities for animal studies on reproduction by bodies such as the National Toxicology Program, and vice versa. Similarly, the decision by a regulatory agency such as the Occupational Safety and Health Administration to, e.g., limit exposure to glycol ethers based on their reproductive effects requires the combined multidisciplinary knowledge of dose-response relationships, interspecies extrapolation, degree and routes of human exposure, control technology and so on.

OBSTACLES TO THE IMPLEMENTATION OF MULTIDISCIPLINARY APPROACHES

Overall, the main areas in which constraints on the development of multidisciplinary approaches were identified are the following: **perception** of the need for such approaches; **communication** between individuals and groups in different disciplines; **dissemination of information** across disciplines; and **comprehension and integration** of such information. At least at the level of the individual, a number of factors were identified as contributing to limitations in these areas. Perhaps the most obvious is the exponential growth in scientific information that has occurred over the past few decades combined with time (and/or financial) constraints on individual researchers' abilities and motivation to keep abreast with information in fields other than their own. Secondly, the increase in the degree of specialization in different disciplines makes the communication and comprehension of information, even in closely related fields, more difficult to achieve. A third factor is the ability (and willingness) of individuals or groups in a given discipline to explain their techniques, findings, etc. to researchers in other fields or to laypeople, but, perhaps more importantly, also to delineate what the implications of research findings in one field have for other disciplines and groups, including policy makers, regulators, physicians, health educators etc.

One of the experiences gained from this multidisciplinary conference that was expressed by a number of participants was, firstly, that we tend not to attend presentations in fields other than our own unless we are required to, and secondly, that although data from other disciplines could clearly illuminate findings in other fields and influence the direction of subsequent research, such data can sometimes be semi-incomprehensible to the non-specialist.

RECOMMENDATIONS FOR FUTURE MULTIDISCIPLINARY APPROACHES

There are a number of ways in which multidisciplinary approaches could be instituted or improved upon in the areas delineated above. They include the following:

a) the recognition by high level administrators that, in many cases, risks can be accurately evaluated and/or mechanisms adequately understood only by an integrated multidisciplinary effort; this recognition would lead to the decision to commit resources earmarked specifically for carrying out this effort, including the appointment of people with multidisciplinary backgrounds and interests to organize, coordinate, integrate and report on this effort;

b) the appointment of people with multidisciplinary backgrounds to key decision-making positions where research priorities, project funding, and standards are to be determined, to ensure the coordination of research programs involving different disciplines and the efficient use of resources;

c) the setting up of multidisciplinary committees, panels and working groups to carry out the functions outlined in a) and b) (e.g. IARC Monograph working groups to evaluate chemical genotoxicity and carcinogenicity);

d) the organization of multidisciplinary conferences and meetings;

e) attendance and/or presentations by specialists in different fields at professional society meetings and conferences (e.g. of the American Society of Human Genetics, The Environmental Mutagenicity Society, the Society for Epidemiologic Research); dissemination of calendars of events to interested parties;

f) attendance at training sessions specifically designed to teach basic principles of a given field to those in interrelated fields; the institution of policies at e.g. government agencies to sponsor such events and/or require attendance by specific personnel;

g) the production and dissemination of on-line or hard copy data on reproductive toxicity of given agents, such as via the Reprotox data base mentioned above, the Genetic Activity Profile, the Toxicology, Occupational Medicine and Environmental Series (TOMES) data base, and sources such as the Environmental Mutagenicity Information Center;

h) the formation of multidisciplinary clubs or networks through which relevant information can be disseminated (e.g. the Perinatal Carcinogenesis Network run by Dr. L. Anderson of the National Cancer Institute, which annually mails out details of risk assessments, brief summaries of recent clinical, laboratory and epidemiologic studies, bibliographies, c.v's and research interests of participants, contents pages of relevant journals, news of meetings etc.);

i) the publication of multidisciplinary reviews in journals and books.

In summary, multidisciplinary approaches should help ensure that research efforts in any given field contribute more effectively to the overall goal of establishing the toxicity and implementing the control of hazards to male reproduction. They should also ensure the more efficient use of human and material resources. Ultimately, however, the success of these approaches rests on the perceived need for them and the willingness, on the part of individuals through to agencies, to meet that need by implementing steps such as those outlined above.

AUTHOR INDEX

SUBJECT INDEX

MIX
Papier aus verantwortungsvollen Quellen
Paper from responsible sources
FSC® C105338

If you have any concerns about our products,
you can contact us on
ProductSafety@springernature.com

In case Publisher is established outside the EU,
the EU authorized representative is:
Springer Nature Customer Service Center GmbH
Europaplatz 3, 69115 Heidelberg, Germany

Printed by Libri Plureos GmbH
in Hamburg, Germany